STRONG AND ELECTROWEAK MATTER 2004

Proceedings of the SEWM2004 Meeting

STRONG AND ELECTROWEAK MATTER 2004

Proceedings of the SEWM2004 Meeting

Helsinki, Finland 16 – 19 June 2004

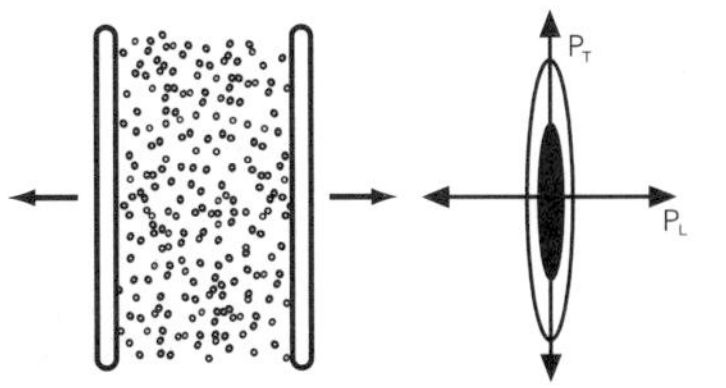

Editors

Kari J. Eskola
Kimmo Kainulainen
University of Jyväskylä, Finland

Keijo Kajantie
University of Helsinki, Finland

Kari Rummukainen
CERN, Switzerland & University of Oulu, Finland

World Scientific

NEW JERSEY • LONDON • SINGAPORE • BEIJING • SHANGHAI • HONG KONG • TAIPEI • CHENNAI

Published by

World Scientific Publishing Co. Pte. Ltd.

5 Toh Tuck Link, Singapore 596224

USA office: 27 Warren Street, Suite 401-402, Hackensack, NJ 07601

UK office: 57 Shelton Street, Covent Garden, London WC2H 9HE

British Library Cataloguing-in-Publication Data
A catalogue record for this book is available from the British Library.

STRONG AND ELECTROWEAK MATTER 2004
Proceedings of the SEWM2004 Meeting

ISBN 981-256-135-8

Printed in Singapore by World Scientific Printers (S) Pte Ltd

PREFACE

The conference "Strong and Electroweak Matter 2004", SEWM04, was organised on 16-19 June 2004 by the Theoretical Physics Division, Department of Physical Sciences, University of Helsinki, Finland. This was the sixth in a series of conferences; the previous ones were organised in Sintra (Portugal, 1994), Eger (Hungary, 1997), Copenhagen (Denmark, 1998), Marseille (France, 2000) and Heidelberg (Germany, 2002).

The physics covered by this series of theoretical conferences is that of spatially and temporally extended systems at finite temperature or density, governed by the laws of the standard model of particle physics, including both its QCD and electroweak sectors. Connections to beyond-the-standard model physics also arise naturally. Empirical manifestations of these systems are ultrarelativistic heavy ion collisions, cosmology (e.g., generation of net baryon number, cosmological phase transitions) and compact astronomical objects (e.g., quark stars). Since 2000 the relativistic heavy ion collider RHIC at Brookhaven National Laboratory has been producing new, accurate and exciting data, which has lead to a shift of emphasis towards the QCD sector in this series of conferences. Nevertheless, one of the main ideas behind this conference series is that there is lots to gain by joint discussions of strong and electroweak interactions. Both are local gauge theories, one in the strong coupling and the other in weak coupling limit, but general principles of manybody physics, field theory, thermodynamics and kinetic theory apply and unify.

The meeting was organised by the Local Organising Committee consisting of Kari Rummukainen (chair, CERN and University of Oulu), Kari Enqvist (Helsinki), Kari J. Eskola (Jyväskylä), Kimmo Kainulainen (Jyväskylä), K. Kajantie (Helsinki) and Mischa Sallé (Helsinki). The members of the Advisory Committee were P. Arnold, W. Buchmüller, Z. Fodor, F. Karsch, C. Korthals-Altes, A, Patkós, A. Rebhan, K. Rummukainen, M. Schmidt, M. Shaposhnikov and J. Smit. The practical help given by Antti Gynther, Ari Hietanen, Tuomas Lappi, Mikko Vepsäläinen and Aleksi Vuorinen was invaluable.

At the meeting 15 invited longer and 35 contributed shorter talks were delivered and 27 posters were presented. Almost all of these are represented in these proceedings.

The home page of the meeting is http://www.physics.helsinki.fi/~tfo_sewm/. It contains, among other things, all the talks in presentation format, also lots of photos from the meeting. Particular thanks go to Mischa Sallé, who alone was responsible for making this home page beautifully functional and totally stable with different browsers and, specifically, for putting all the talks and photos therein.

The meeting was sponsored by the Magnus Ehrnrooth Foundation, the University of Helsinki, the Helsinki Institute of Physics and various research projects at the Academy of Finland.

The Editors
Helsinki, July 2004

CONTENTS

Contributed Talks

Posters

RHIC EXPERIMENTAL SUMMARY: THE MESSAGE FROM pp, d + Au AND Au + Au COLLISIONS

M. CALDERÓN DE LA BARCA SÁNCHEZ
Bldg. 510A, STAR Group
Brookhaven National Laboratory,
PO Box 5000, Upton, NY 11973, USA
E-mail: mc@bnl.gov *

We review here the data from the RHIC experimental programme in view of the first available summary of Au+Au, d+Au and pp collisions taken since the start-up RHIC collision. We will focus here on several highlights of the experimental data, including the striking disappearance of angular correlations at high transverse momentum ($p_\perp$). The view from the experiments is optimistic, yet reserved, regarding the question of the formation of a Quark-Gluon Plasma at RHIC. We discuss here some of the questions that would need to be addressed in the future, specifically addressed to theorists, to help shed light on the possible paths to answer the question to the satisfaction of the community.

1. Introduction

The Relativistic Heavy Ion Collider (RHIC) was planned as a facility which would have as one of its primary goals the creation and study of a Quark-Gluon Plasma (QGP) in the laboratory. It is also thought that this stage of matter played an important role during at the early stages of the Big Bang. RHIC would allow us to perform measurements of at least some of the detailed properties of the QGP if formed, and to test these measurements against predictions of Quantum Chromodynamics (QCD). At this stage, RHIC has gone through 3 data taking runs in which the 4 RHIC experiments, BRAHMS, PHENIX, PHOBOS and STAR, have harvested a wealth of data, which the community is currently digesting. The hope is that a coherent picture of what is happening in the collisions will emerge. As this is meant to be an experimental review of RHIC as a whole, we

*Currently at Indiana University, Physics Department, Swain Hall West, Bloomington, IN 47405

will present results learned from all 4 collaborations; a slight bias towards presentation of STAR results might be gleaned in this writing simply for the reason that the writer, being from STAR, is more familiar with, and has easier access to, the STAR results. All collaborations have contributed significantly to the present state of our knowledge, and it would be impossible to do justice to even a fraction of the results. With these caveats in mind, we aim to address here, some of the results that have surfaced in the past 3 years. We will also present a particular view, not necessarily held by everyone in the community, but certainly voiced by a majority of the experimentalists, regarding the question whether the observed behaviour permits a compelling discovery claim for the QGP.

In addressing this question, it has been critical to define more clearly what we mean by the QGP, since the theoretical expectations about its properties have evolved significantly since the first case for RHIC was made. It is perhaps a good point in time for the community to reassess this question. For our purposes here, in the past year of discussions within STAR, it seems that a reasonable ansatz for this definition could be phrased as follows: we take the QGP to be a locally thermally equilibrated state of matter in which quarks and gluons are deconfined from hadrons, so that color degrees of freedom become manifest over *nuclear*, as opposed to nucleonic, volumes. We note that this relaxes or modifies some of the expectations that might have been built up regarding the QGP.

With this frame of mind, we now turn to discussion some of the experimental observables, and the evidence that can be gleaned from them.

2. Selected Experimental Highlights

The observed hadron spectra and correlations at RHIC reveal roughly three transverse momentum ranges with distinct behaviour: a soft range ($p_\perp \lesssim 1.5$ GeV/c) containing the vast majority of produced hadrons, representing most of the remnants of the bulk collision matter; a hard-scattering range ($p_\perp \gtrsim 6$ GeV/c), providing partonic probes of the early collision matter; and an intermediate range ($1.5 \lesssim p_\perp \lesssim 6$ GeV/c) where hard processes coexist with softer ones. The behaviour in each of these ranges is quite different than would be expected from an incoherent sum of independent nucleon-nucleon collisions; for the hard sector in particular, this is one of the most important new observations. We summarize below the major findings within the soft and the hard sector. This is only in the interest of space, as the intermediate sector has yielded also significant insights, chiefly the apparent constituent quark scaling of the elliptic flow parameter, and the scaling of the yields in groups of baryons and mesons: a very strong hint at the important role that constituent quarks play in the evolution of the system.

2.1. *Bulk Properties*

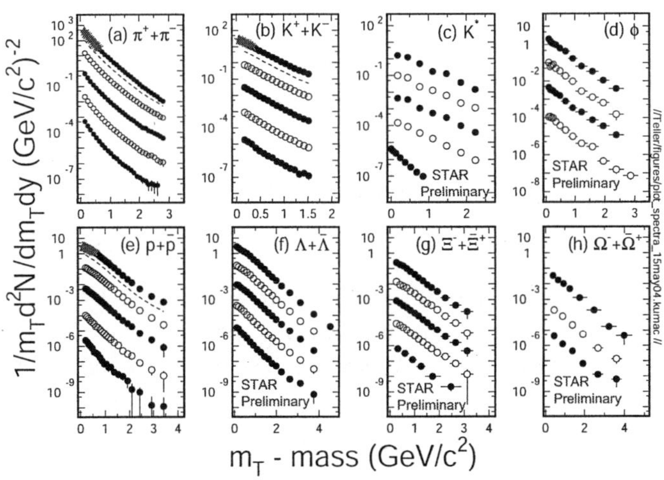

Figure 1. Mid-rapidity spectra for hadrons from $\sqrt{s_{\mathrm{NN}}} = 200$ GeVAu + Au collisions [3,4].

All experiments have been very successful in measuring a wealth of particle spectra. Figure 1 shows some of the $m_\perp$ distributions for various hadron species. These data are fit with a simple blast wave approximation (a quick way of approximating a hydrodinamic fit), where one of the parameters is a common radial flow velocity.

Results for these parameters are shown for different centrality bins and different hadron species in Fig. 2. As the collisions become more central, the bulk of the system, dominated by the yields of π, K, p, appears to grow cooler at kinetic freeze-out and to develop stronger collective flow. These results may indicate a more rapid expansion after chemical freeze-out with increasing collision centrality. The spectra for multi-strange particles ϕ and

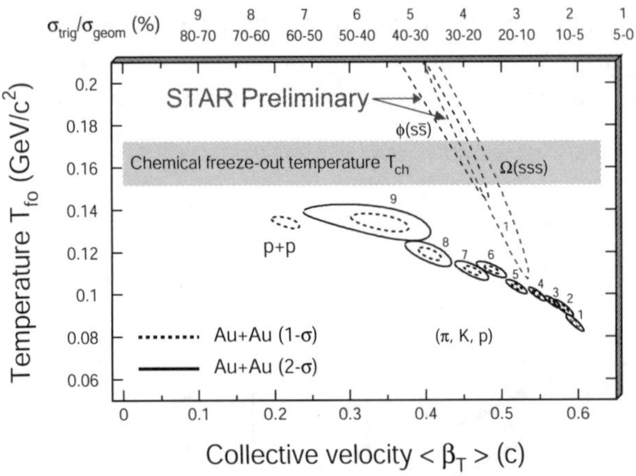

Figure 2. The χ^2 contours, extracted from thermal+radial flow fits for the most abundant hadrons, and for multi-strange baryons [5, 6].

Ω appear, within the large uncertainties, to reflect a higher temperature, suggesting diminished hadronic interactions with the expanding bulk matter after chemical freeze-out, as predicted for hadrons containing no valence u or d quarks.

By looking at the systematic dependence of the elliptic flow strength on the mass of the hadron, a similar trend including a common collective flow velocity is also clearly observed. Figure shows the low-$p_\perp$ v_2 parameter for various hadrons, together with hydrodynamic fits to the data (which were done together with the $m_\perp$ spectra).

In addition, the HBT identical particle correlations, and the recent non-identical particle correlations [9], also support the picture of a common velocity field. One of the conclusions we can draw from these measurements is that the matter produced exhibits strong collective flow. Most of the hadrons at low $p_\perp$ reflect a communal transverse velocity field resulting from conditions early in the collision, when the matter was clearly expanding rapidly under high, azimuthally anisotropic, pressure gradients and there were frequent interactions among the constituents. All of these features fit naturally, at least qualitatively, within a hydrodynamic description of the system evolution, and are consistent with calculations including a phase transition. However, there is at least one significant result which cannot be reconciled so far with the current hydrodynamic calculations.

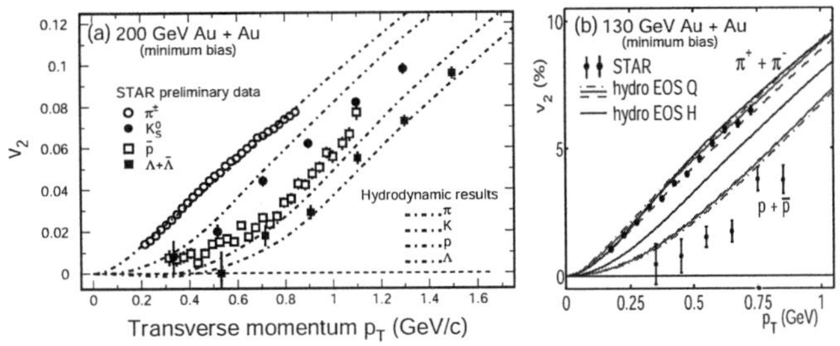

Figure 3. The $p_\perp$ dependence of the elliptic flow parameter $\check{2}$ for charged $\pi^+ + \pi^-$, K_s^0, $\bar{p}$ and Λ [7]. Hydrodynamic calculations [8] assuming early thermalization, ideal fluid expansion, an equation of state including a phase transition, and a sharp kinetic freeze-out at 120 MeV, are shown.

The extracted lengths of homogeneity from HBT correlations, specifically the ratio $R_{\text{out}}/R_{\text{side}}$, which is commonly related to the emission duration, cannot be accommodated so far in the calculations, as shown in Figure 4. The success of the hydrodynamical description in describing many of the

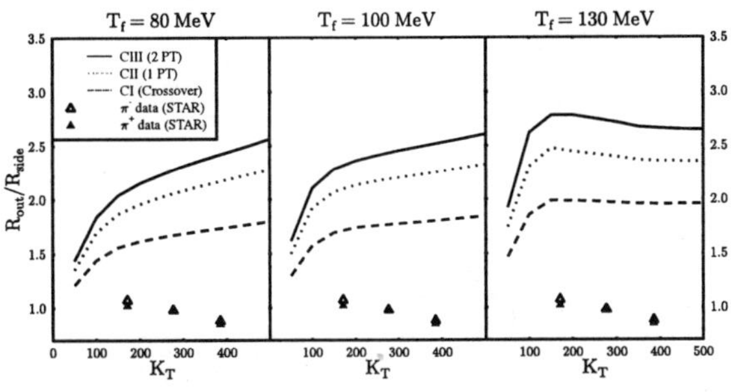

Figure 4. STAR measurements [10] of $R_{\text{out}}/R_{\text{side}}$ from pion HBT correlations, plotted *vs.* the pion pair transverse momentum $k_\perp$. The experimental results are identical in the 3 frames, but are compared to hydrodynamics calculations [11] for a variety of parameter values.

features of the data is quite striking. Simply the fact that the hydrodynamical limit of zero mean free path seems to work at RHIC, which is in stark contrast to a description based on hadronic degrees of freedom, has been taken by some as evidence for reaching the sought after partonic phase.

While these results are certainly encouraging, we believe that the present evidence is indirect, and in addition the failure to describe consistently the time evolution inferred from the HBT measurements might point to a possible flaw. More work is needed in this direction to arrive at a consistent picture.

2.2. *Hard Probes*

The dominant characteristic of the hard regime is the strong suppression of hadron yields in central Au + Au collisions, in comparison to expectations from *pp* or peripheral Au + Au collisions, scaled by the number of contributing binary (nucleon-nucleon) collisions, as illustrated in Fig. 5 (Refs. [12–15]). Such suppression sets in already in the intermediate sec-

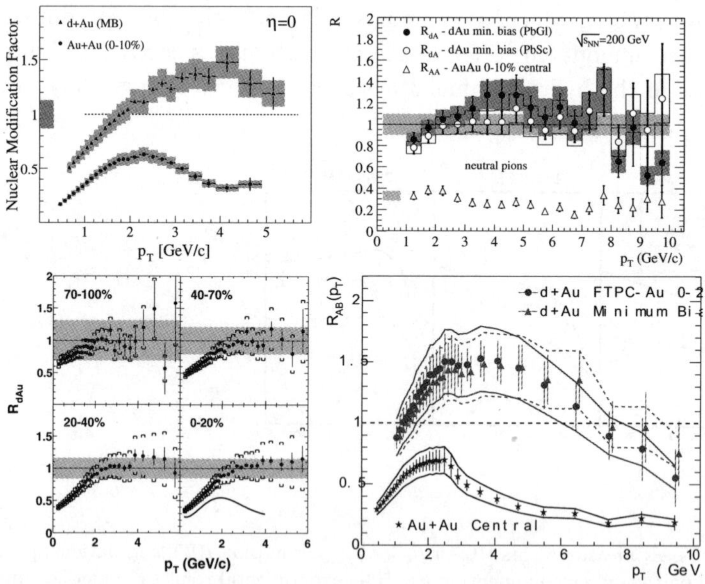

Figure 5. Binary scaled ratio $R_{AB}(p_{\perp})$ of charged hadron and π^0 inclusive yields from 200 GeV Au + Au and d + Au relative to that of *pp* collisions, from BRAHMS (upper left), PHENIX (upper right), PHOBOS (lower left, d + Au only) and STAR.

tor, but saturates and remains constant as a function of $p_{\perp}$ throughout the hard region explored to date. One of the most important lessons gleaned from the experimental data is that such suppression was not seen in d + Au collisions at RHIC, indicating that it is a final-state effect associated with the collision matter produced in Au + Au. It is consistent with effects of parton energy loss in traversing dense matter, predicted before the data were available [16, 17]. This has been observed by all 4 collaborations.

In addition, azimuthal correlations of moderate- and high-$p_\perp$ hadrons exhibit clear jet-like peaks on the near side. However, the anticipated away-side peak associated with dijet production is suppressed by progressively larger factors as the Au + Au centrality is increased, and for a given centrality, as the amount of (azimuthally anisotropic) matter traversed is increased. Again, no such suppression is observed in d + Au collisions (see Fig. 6, Ref. [15, 18]).

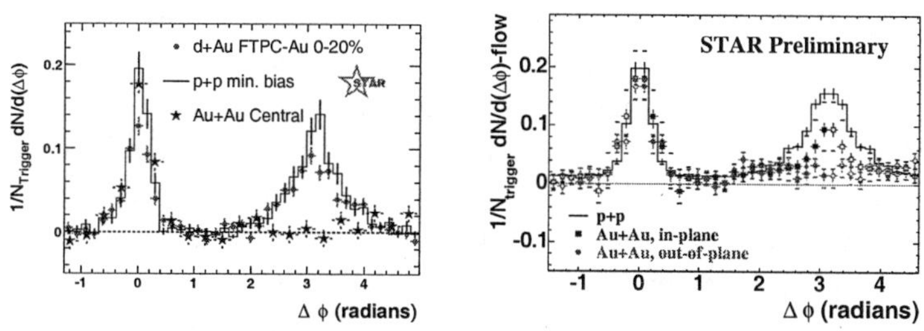

Figure 6. Dihadron azimuthal correlations at high-$p_\perp$. Left panel shows correlations for pp, central d + Au, and central Au + Au collisions (background subtracted). Right panel shows the same measurement for Au + Au separated for different orientations of the trigger hadron relative to the reaction plane.

The suppression of hadron yields and back-to-back correlations firmly establish that jets are quenched by very strong interactions with the matter produced in central Au + Au collisions. The jet-like near-side correlations survive presumably because one observes preferentially hard fragments of partons scattered outward from the surface region of the collision zone. Effects of interaction with the bulk matter are nonetheless still seen on the near side, primarily by the broadened distribution in pseudorapidity of softer correlated fragments [19].

Many features of the observed suppression of high-$p_\perp$ hadrons, including the centrality dependence and the $p_\perp$ independence, can be described efficiently by perturbative QCD calculations incorporating parton energy loss in a thin, dense medium. To reproduce the magnitude of the observed suppression, despite the rapid expansion of the collision matter the partons traverse, these treatments need to assume that the initial gluon density when the collective expansion begins is more than an order of magnitude greater than the that characteristic of cold, confined nuclear matter [16]. These results indicate, with fairly modest reliance on theory, that RHIC collisions produce highly dissipative and dense matter that behaves collectively. The magnitude of the density inferred from parton energy loss treatments, together with the hints of constituent quark collective flow, argue against the effectiveness of a purely hadronic treatment of this matter.

The following overall picture of RHIC collisions is emerging: Interactions of very short mean free path within the gluon density saturation regime lead to a rapidly thermalized partonic system at energy densities and temperatures above T_c. This thermalized matter expands collectively and cools as an ideal fluid, until the phase transition (or crossover) back to hadronic matter begins, leading to a significant pause in the build-up of v_2. During and after the phase transition, constituent quarks appear to emerge as the effective degrees of freedom in describing hadron formation at medium $p_\perp$ out of this initially partonic matter. Initially hard-scattered partons traversing this matter lose substantial energy to the medium via gluon radiation, and thereby approach equilibration with the bulk matter. Thus, some evidence of degraded jets survives, depending on the amount of matter traversed.

3. Conclusions and Open Questions

The consistency noted above of many RHIC results with a QGP-based theoretical framework is and important and highly non-trivial statement! Indeed, it is the basis of some publicized claims [20–22], that the QGP has been "nailed down" at RHIC. In our judgement, there is still cause at this juncture for healthy skepticism about these claims on several grounds.

Currently, the theoretical claims of QGP discovery *e.g.* Ref. [22], is mainly a collection of different treatments applied in succession, where each invokes a separate model for its interpretation: (i) statistical model fits to measured hadron yields to infer possible chemical equilibrium across the u, d and s sectors; (ii) hydrodynamics calculations of elliptic flow to suggest early thermalization and soft EOS; (iii) quark recombination models to highlight the role of thermalized constituent quarks in intermediate-sector v_2 scaling; (iv) parton energy loss models to infer an initial gluon density from high-$p_\perp$ hadron suppression observations; (v) gluon saturation model fits to observed hadron multiplicities and yields at large rapidity, to suggest how high-density QCD may predetermine the achieved initial gluon densities. Each piece has its own assumptions, difficulties, parameters and uncertainties; and they fit together somewhat uneasily. Until they are assimilated into a self-consistent whole, it may be difficult to assess theoretical uncertainties quantitatively or to make quantitative predictions whose comparison with data have the potential to prove the theory wrong.

The indirect evidence for a phase transition and for attainment of local thermal equilibrium in the produced matter are intertwined in the hydrodynamics account for RHIC hadron spectra an elliptic flow results. The uniqueness of the solution involving early thermalization and an EOS with a soft mixed phase is not yet demonstrated. Nor is its robustness against changes in the freeze-out treatments or the introduction of viscosity, as might be needed, for example, to reduce discrepancies from HBT measurements.

The indirect evidence for deconfinement rests primarily on the large initial gluon densities inferred from parton energy loss fits to the observed hadron suppression at high $p_\perp$. It has not yet been demonstrated that the extracted values *demand* deconfinement in the matter traversed. The agreement with initial gluon densities suggested by Color Glass Condensate approaches is encouraging, but is still at a basically qualitative level.

Our assessment at the present stage is that in the absence of a direct signal of deconfinement revealed by experiment alone, a QGP discovery claim must rest in comparison with a promising, but still not yet mature, theoretical framework. We certainly recognize that the results we have seen are striking, but we judge that a QGP discovery claim based on RHIC measurements to date would be premature. We do not propose that a comprehensive theoretical understanding of all observed phenomena must be attained before a discovery claim is warranted, but only that at least some of the serious open questions posed above be successfully answered.

The answer to some of the questions above, as well as others with which the community is currently wrestling, both on the theory side on the experimental side, outline a several-year long programme of research. This is not to imply that the answer to all of the questions must precede a discovery announcement for the Quark-Gluon Plasma. With additional clarification of some of the open issues, the preponderance of evidence may well point inescapably to the conclusion that RHIC collisions have produced thermalized, deconfined quark-gluon matter. At that point, the programme of research will be recast as a mapping of the properties of the QGP. While we do not believe we are yet at that point, we are convinced that RHIC's first three years have already led to enormous progress in addressing this complex, but profoundly important set of questions regarding the nature of strongly interacting matter and the QCD vacuum. We hope that this assessment communicates clearly that the progress is real and extensive, that we have travelled a long way in understanding the the properties of matter at extreme energy densities, and that the remaining questions are well defined.

References

1. J.W. Harris and B. Müller, *Annu. Rev. Nucl. Part. Sci.* **46**, 71 (1996).
2. S.S. Adler, *et al.*, (PHENIX Collaboration), *Phys. Rev.* **C69**, 034909 (2004); nucl-ex/0307022
3. E. Yamamoto, *et al.*, (STAR Collaboration), *Nucl. Phys.* **A715**, 466c (2003).
4. K. Schweda, *et al.*, (STAR Collaboration), *J. Phys.* **G30**, S693 (2004); nucl-ex/0403032.
5. O. Barannikova (STAR Collaboration), nucl-ex/0403014.
6. C. Adler, *et al.*, (STAR Collaboration), *Phys. Rev. Lett.* **89**, 092301 (2002); nucl-ex/0206008; J. Adams, *et al.*, (STAR Collaboration), nucl-ex/0306029; nucl-ex/0406003.
7. J. Adams, *et al.*, (STAR Collaboration), *Phys. Rev. Lett.* **92**, 052302 (2004); nucl-ex/0306007.
8. P. Huovinen, *et al.*, *Phys. Lett.* **B503**, 58 (2001).
9. J. Adams, *et al.*, (STAR Collaboration), *Phys. Rev. Lett.* **91**, 262302 (2003).
10. C. Adler, *et al.*, (STAR Collaboration), *Phys. Rev. Lett.* **87**, 082301 (2001).
11. S. Soff, nucl-th/0204040; D Zshiesche, *et al.*, *Phys. Rev.* **C65**, 014902 (2002); S. Soff, S. Bass, and A. Dumitru, *Phys. Rev. Lett.* **86**, 3981 (2001)
12. I. Arsene, *et al.*, (BRAHMS Collaboration), *Phys. Rev. Lett.* **91**, 072305 (2003).
13. S.S. Adler, *et al.*, (PHENIX Collaboration), *Phys. Rev. Lett.* **91**, 072303 (2003).
14. B.B. Back, *et al.*, (PHOBOS Collaboration), *Phys. Rev. Lett.* **91**, 072302 (2003).
15. J. Adams, *et al.*, (STAR Collaboration), *Phys. Rev. Lett.* **91**, 072304 (2003).
16. X. N. Wang, nucl-th/0305010.
17. I. Vitev and M. Gyulassy, *Phys. Rev. Lett.* **89**, 252301 (2002)
18. J. Adams, *et al.*, (STAR Collaboration), nucl-ex/0407007.
19. F. Wang (STAR Collaboration), nucl-ex/0404010.
20. M. Gyulassy, nucl-th/0403032.
21. M. Gyulassy and L. McLerran, nucl-th/0405013.
22. RIKEN-Brookhaven Research Center Scientific Articles, Vol. 9: *New Discoveries at RHIC: the current case for the Strongly Interactive QGP* (May, 2004).

HYDRODYNAMIC ASPECTS OF RELATIVISTIC HEAVY ION COLLISIONS AT RHIC

PETER F. KOLB
Physik Department
Technische Universität München
D-85747 München, Germany
E-mail: peter.kolb@ph.tum.de

The current status of the application of hydrodynamics to ultrarelativistic heavy ion collisions is reviewed. We elaborate on the arguments for strong transverse flow and rapid thermalization and discuss future applications and trends in hydrodynamics.

1. Overview

Among the many predictions[1] made for observables of collisions at the Relativistic Heavy Ion Collider (RHIC), hydrodynamic calculations regarding the expected flow characteristics[2,3] turned out to be surprisingly accurate already a few months after running the new facility.[4] In fact, those predictions were so convincing that they were awarded a bottle of wine in the RHIC prediction competition.[5] That RHIC creates systems which – *for the dominant part of their lifetimes* – expand according to hydrodynamic principles is of utmost importance: The dynamical evolution is governed by the nuclear equation of state at extreme energies, whose thermodynamic features are thus accessible through collisions in the laboratory.

In the following, I will review indications of the strong transverse expansion that can be deduced from the observable final state of central and non-central Au+Au collisions. Thereby, I will pay particular attention to signals of collectivity at mid-rapidity, i.e. in the plane perpendicular to the beam-axis in the center of mass of the reaction (at vanishing pseudorapidity η). Although interesting phenomena are to be expected at $\eta \neq 0$, to which I will return later, the midrapidity region is of particular interest for the observation of collectivity: Before the reaction, there is no transverse component of motion. Observing collectivity in the transverse plane at midrapidity must thus be a result of the dynamical evolution of the fireball *after* the initial impact. In contrast, at rapidities $\eta \neq 0$, it is difficult to disentangle collectivity created by the fireball region and the longitudinal dynamics of the nuclei that was present already before the reaction.

To address the questions of the how and when those collective features in the transverse plane are established, I will turn to dynamical models and specifically introduce the formalism of hydrodynamics to study the temporal evolution of the fireball matter in this framework. This analysis makes the success of the simple formulas of blast wave parameterizations understandable and validates such a simplistic approach as a tool to rapidly characterize and summarize the main thermodynamic information of the final state.

I will then highlight the most important comparisons of experimental data and hydrodynamic calculations and their interpretation, before I give a detailed account on recent and future developments that can be expected in the field. Almost every single observable of heavy ion physics is influenced by the bulk matter, as the bulk provides the background of the evolution through which even the rare probes have to propagate. Now, with hydrodynamics, we have a tool to calculate the dynamics of this evolution.

2. Indications of strong transverse flow

If the microscopic constituents of a flowing medium share a common average flow velocity $v_\perp$, heavier particles in this medium gain larger momenta than the lighter constituents. In the case of the late stages of heavy ion collisions, the flowing medium is the hadronic soup consisting of hadrons of widely varying masses. From experimental evidence we find that the hadronic abundancies reflect a composition of a hadronic gas at a temperature $T_{\text{chem}} \sim 175\,\text{MeV}$.[6] Refinements of the simple blast wave model[7,8] assume the knowledge of a certain flow profile at the break up stage of the reaction as well as the shape of this surface of decoupling from which particles supposedly do not interact any further. In this way one achieves a handy formalism,[9] that can easily be used to fit to the vast amount of data currently available from RHIC experiments, in particular to single particles spectra and eventually their anisotropies.[10] Although the temperature parameter of these fits differs widely from 175 MeV [9] to 100 MeV,[10] the common finding of these model calculation is that a vast amount of transverse flow (with an average velocity greater than $c/2$) is achieved on a rather short time scale of about 10 fm/c. While the stunningly good description hints that there possibly is some physical truth underlying these parameterizations, these simple descriptions of the final state do not offer any insight *how* and *when* such strong transverse flow develops during the course of the evolution. For this, we have to turn to a fully dynamical description of the transverse expansion stage of the reaction.

3. Hydrodynamic formalism and phenomenology

There are two main philosophies to model the dynamical evolution of a system. If the system is small and the scattering processes can be treated individually, one can adapt a microscopic viewpoint, treating all the collisions of the constituents individually.[11] In a very dense, strongly interacting medium, however, this approach becomes quickly impractical, particularly if one also were to consider low-momentum transfer processes consistently. Alternatively, in this limit of high density and strong rescattering, one can give up the particles' individual personality, and characterize the system in terms of density fields and continuity equations. Ideally, if scattering is strong enough to allow for local thermodynamic equilibrium, the equations of energy momentum conservation can be formulated by the thermodynamic equivalent, the conservation of the thermodynamic energy-momentum tensor, $\partial_\mu \left[(e+p)u^\mu u^\nu - p\, g^{\mu\nu}\right] = 0$. The energy density e and the pressure p are related by the thermodynamic equation of state of nuclear matter. u^μ is the collective four velocity, whose time evolution we can study within this formalism. Once supplied with appropriate initial conditions, which are often taken from geometric considerations[12] or more fundamental models such as the saturation model[12] or the color glass condensate,[13] the system's evolution is fully determined. At this point, the computer takes over to solve for the time evolution of the thermodynamic fields.

Fig. 1 shows the temporal evolution of the entropy density on a double logarithmic scale for three different positions in the fireball of a central collision, 0, 3 and 5 fm away from the fireball center.[14] The tangents drawn to the curve stress the occurring transition from the initial one-dimensional, longitudinal expansion to a fully 3-dimensional expansion at late stages. The right plot of Fig. 1 shows the radial flow profile at different times throughout the evolution. Striking here is the fast transition to a linear profile, which persists throughout time. Still, as matter is continually moved to larger radii, the mean transverse velocity of the medium increases and transverse flow thus continues to increase. The main part of the fireball matter is know to freeze-out over a rather short time frame as the rescattering rate drops sharply as a function of temperature, and therefore even more so as function of time. This fact, together with the rapid development of a flat flow profile is at the heart of the applicability of the simple and the improved blast wave parameterizations.

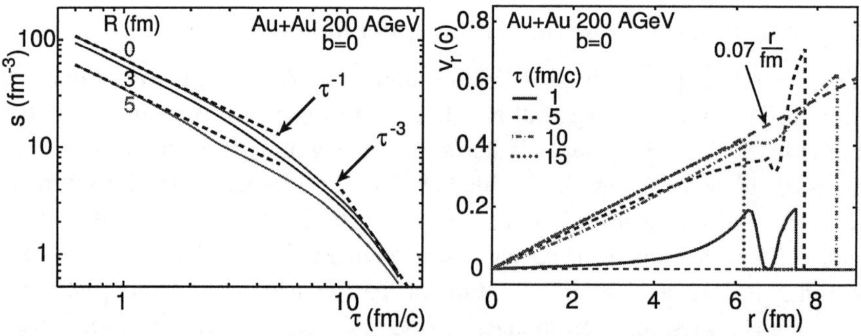

Figure 1. Left panel: Time evolution of the entropy density at three different points in the fireball (0, 3, and 5 fm from the center). Dashed lines indicate the expectations for pure one-dimensional and three-dimensional dilution, respectively.[14] Right panel: The radial velocity profile at different times during the evolution.[14] Shown is also the linear profile $v(r) = 0.07\ r/\mathrm{fm}$.

In contrast to the overall radial flow which increases throughout the lifetime of the system, anisotropies in the transverse flow profile are generated during the earliest instances of the reaction. Such anisotropies arise from the eccentricity in coordinate space of non-central collisions. Larger pressure gradients in the short direction (the direction of the impact parameter) lead to a larger transport of matter in this direction, thereby reducing the eccentricity and undermining the source of its own origin.[15] In the course of this process more particles are transported into the direction of the impact parameter and the mean transverse momentum of the particles is greater, both of which manifests itself experimentally in anisotropies v_n of the transverse momentum spectra $\frac{dN}{p_T dp_T dy d\varphi} = \frac{dN}{2\pi p_T dp_T dy}\left(1 + \sum_n v_n \cos(n\varphi)\right)$. It has been shown in many studies, both microscopically[16] as well as macroscopically[2,14] that those momentum anisotropies are generated during the first 5 fm/c of the reaction, where most of the matter is still at temperatures exceeding the critical temperature of nuclear matter. Anisotropies in the transverse momentum spectra of hadrons thus originate from the partonic stage of the fireball and signals their collective behavior.[2]

4. Application to experimental observations

As mentioned before, the transverse momentum distribution at moderate p_T are the natural choice of observables to study collective dynamical effects in the medium. We use the transverse momentum spectra of identified particles to determine the parameters of the calculation for RHIC at 200 GeV center of mass energy per nucleon. Those spectra are well reproduced applying an equilibration time of $\tau_{\text{equ}} = 0.6\,\text{fm}/c$ and a central fireball temperature of 360 MeV.[17] Although the particle ratios reflect chemical equilibrium at a temperature of about 175 MeV,[6] the slopes in particular of heavier baryons require further transverse accelertion down to temperatures of about 100 MeV.[18] To keep the particle ratios fixed to their value at 175 MeV, chemical potentials seem to be dynamically generated throughout the evolution.[19,20,21]

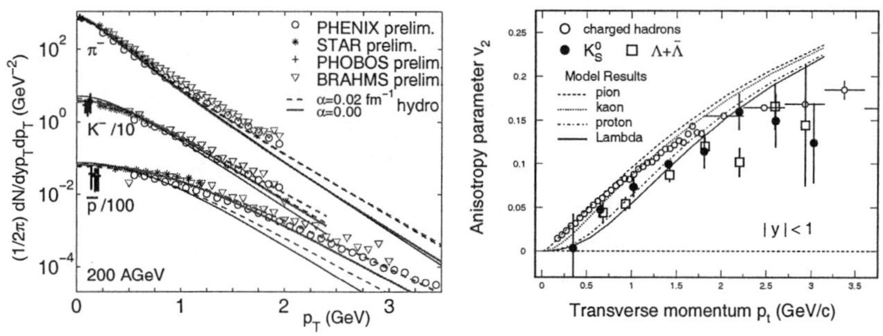

Figure 2. Left panel: Transverse momentum spectra of Pions, Kaons and Antiprotons, from all four RHIC experiments combined[22] in comparison to results from hydrodynamics at different freeze-out temperatures (and different initilizations, see[17]). Right panel: Momentum anisotropies of Pions, Kaons and Lambdas. Results from the STAR experiment[23] compared to results from hydrodynamics.

Although the good reproduction of the spectra of central collisions is reassuring, it has been achieved to a great deal by adjusting the parameters of the calculation. Important *predictions* come into play in non-central collisions. Changing centrality, which introduces a breaking of the azimuthal symmetry inherent in central collisions, renders essentially all observables sensitive on the azimuthal angle of observation. This literally opens up an entirely new dimension of observable space. However, for fully dynamical simulations no additional parameters are introduced when stepping out to explore observables in this new dimension. The modified initialization for non-central collisions is stricktly determined by the geometry of the underlying initialization scenario. The only appearing *parameter* – the impact parameter b – is estimated experimentally by giving the range of centrality for certain collisions.

We already mentioned the great significance of momentum anisotropies as they provide a signal from the first few fm/c of the collision. Microscopic studies[11] in comparison with experimental data show that the fireball constituents must undergo an incredible amount of rescattering among themselves. Hydrodynamic calculations, which exploit the limit of negligible mean free pathlengths, provide an impressive account of a large collection of momentum anisotropies.[18] They give a good overall description of momentum anisotropies up to transverse momenta of 2 GeV (Pions) and even 4 GeV for Baryons. The centrality dependence of the hydrodynamic elliptic anisotropy compares well with experimental data as long as the impact parameter does not get too large. Finally, the mass characteristics, a flatter onset of momentum anisotropies at small transverse momenta,[25] is confirmed by experiment in stunningly good agreement. A recent compilation of momentum anisotropies of heavy particles is reprinted in Fig. 2 (right panel).[23] Still more recently the momentum anisotropies of Cascades and Omegas have become available.[24] Again, those anisotropies have been stunningly large, as predicted 3 years ago by hydrodynamic calculations.[25] As the multistrange resonances are believe to not interact strongly in a hadron gas, this is the clearest signal that strangeness anisotropy is generated in the partonic stage of the reaction. It appears that all quark flavors share the same flow anisotropy on the partonic level. This fact could be most prominently (dis-)proved by investigating momentum anisotropies of higher order and their ratios.[26] That the momentum anisotropies are as large as observed is a strong indication that the thermalization of the medium is achieved very rapidly.[27] The fact that hydrodynamics overestimates anisotropies at large transverse momenta and in peripheral collisions is well understood, considering that high p_T particles escape the fireball too rapidly to follow the collective motion of the bulk, and that small systems do not provide enough rescattering in the limited volume. Here, it is also important to remark that while hydrodynamics does not seem to be fully applicable at the lower beam energies of the SPS in peripheral collisions, there might still be a fair chance for its quantitative application in *central* collisions. This could be in particular true for the early, plasma part of the reaction, whereas the later hadronic stage might drop out of local thermal equilibrium. Only detailed quantitative comparisons of hydrodynamic calculations and the most recent results[29] from SPS experiments can answer this question and address important topics close to thermal equilibrium and just above the transition temperature.

5. Current trends and future requirements

Most of the hydrodynamic calculations that are applied to experimental data at present include some great simplifications in order to keep the numerical efforts to a reasonable degree. One of these is the strong bias to investigations of observables at midrapidity. Although the collectivity observed at midrapidity is clearly produced during the evolution of the produced fireball and therefore contains clean probes of the dynamics, a broader view of the full collision volume is highly desirable, in particular in the light of recent data extending the analysis of momentum anisotropies to large rapidities. Fully three-dimensional calculations have been successfully performed in recent years.[30] However not too much is gained by applying the *ideal* hydrodynamic calculations at large forward rapidities as the system is thought reach thermal equilibrium only at later times[31] or not at all[32] in regions far away from the center of mass of the reaction.

Nevertheless, there is a lot to be gained from investigations of hydrodynamic calculations at non-zero rapidity. Due to the (net)-baryon content of the initial nuclei, one can expect that the (net) baryon-number density increases with increasing rapidity. As one leaves the 'clean' midrapidity region with its antibaryon to baryon ratio of about 0.75, one gets closer to the initial nuclei region where the antibaryon content effectively vanishes. Thus, at larger rapidity, a hydrodynamical system evolves according to the equation of state at larger baryon chemical potential than at mid-rapidity. This offers the exciting possibility of tracing signals of the expected tricritical point at finite rapidity. The effect of the tricritical point on the hydrodynamic evolution of RHIC collisions is currently explored by Nonaka.[33] It was found that such a point acts as an attractor for the thermodynamic paths of a constant fraction of entropy and baryon-density. Clearly, due to the large fluctuations that appear around the critical point, the ideal hydrodynamic treatment will become invalid, but it will be interesting to explore how the system drops out of equilibrium, and which quantitative experimental measurements will be predicted.

The less efficient energy density production and deposition at large rapidity eventually leads to a breakdown of the assumption of full thermalization and an ideal expansion. Viscous effects will become more and more important in smaller and more dilute systems. It is of fundamental interest to get estimates of the viscosity of the QCD plasma state. Viscous hydrodynamics can deliver valuable information in this direction. The realization of fully dynamical hydrodynamical calculations is however difficult. The viscous terms have to be solved dynamically, increasing the number of differential equations by up to 5 (depending on the symmetries one exploits explicitly). Although difficult to solve,[34] first results on the dynamical evolution of central collisions have recently been presented.[35] The message of these dynamical studies is that viscous matter sticks together for a longer period, just to blow up quicker in the late stages. The left panel of Fig. 3 shows plots of the energy density as a function of the transverse coordinate for a viscous calculation compared to the ideal Euler calculation.[35] Although the densities change slower at first, velocities build up faster due to a restructuring of the pressure components, which is also observed in the case of only transverse thermalization and longitudinal free streaming.[36] Finally, in the late stage of the reaction, the left over matter dissipates more rapidly, due to the larger velocities that were achieved. The net result is that the system stays longer in the hot phase of its reaction and disperses faster through the late stages. The late stages, for which viscosity become increasingly important, thus have a shorter lifetime and freeze-out occurs more rapidly. Comforting is the observation that viscosity does not lead to self-enhancing features!

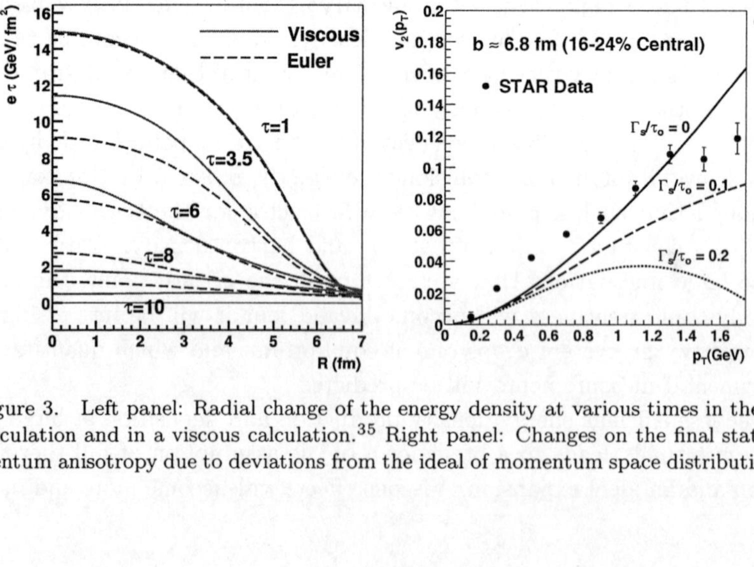

Figure 3. Left panel: Radial change of the energy density at various times in the ideal calculation and in a viscous calculation.[35] Right panel: Changes on the final state momentum anisotropy due to deviations from the ideal of momentum space distributions.[37]

In this context it is also interesting to cast doubt on the assumption of having particles distributed in momentum space according to the simple ideal distribution laws of Fermions and Bosons. It has been shown[37] that a large deviation from these distributions smears out the directed collective motion observed in the momentum anisotropies and is thus not supported by the data (see right panel of Fig. 3.)

6. Summary

We have learned *a lot* about the collective phenomena exhibited in RHIC data of the first few years. The data clearly show many characteristics of rapid thermalization and an extended stage of hydrodynamic expansion of the fireball into the surrounding vacuum. On a linear timescale, the hydrodynamic stage is the most significant stage of the reaction, thereby influencing many other relevant observables. Hydrodynamics is a clean (meaning well defined) tool, with only a small number of parameters and easily reproducible by anybody having access to a computer. The two most important facts learned from hydrodynamics and the experimental data so far is the need for rapid thermalization (in a time of less than 1 fm/c) where matter obtains a large transverse push due to a hard equation of state ($p \sim e/3$). Many quantitative questions are still unanswered: What is the frame of error around the equilibration time, what different kinds of equations of state and phase transitions would still account for the data, how does one feed back on the other? What is the role of viscosity in the expansion? Why does viscosity of high temperature QCD matter seem to be so low as to allow for an ideal hydrodynamic description of the data?

Acknowledgements

I would like to thank the organizers for inviting me to present this overview. I deeply acknowledge the contributions of my collaborators and friends without whom I would not have received this honor.

References

1. K. Eskola, *Nucl. Phys.* **A698** 78 (2002).
2. P.F. Kolb, J. Sollfrank, and U. Heinz, *Phys. Lett.* **B459** 667 (1999); *Phys. Rev.* **C62** 054909 (2000).
3. D. Teaney, J. Lauret, and E.V. Shuryak, *Phys. Rev. Lett.* **86** 4783 (2001).
4. K.H. Ackermann et al. (STAR Collaboration), *Phys. Rev. Lett.* **86** 402 (2001).
5. See *http://www-hpc.lbl.gov/prediction* and *http://int.phys.washington.edu/-PROGRAMS/int_rhic02.html*

6. P. Braun-Munzinger, D. Magestro, K. Redlich, and J. Stachel, *Phys. Lett.* **B518** 41 (2001).
7. P.J. Siemens and J.O. Rasmussen, *Phys. Rev. Lett.* **42** 880 (1979).
8. E. Schnedermann, J. Sollfrank, and U. Heinz, *Phys. Rev.* **C48** 2462 (1993).
9. W. Broniowski and W. Florkowski, *Phys. Rev. Lett.* **87** 272302 (2001).
10. F. Retiere and M. Lisa, nucl-th/0312024.
11. D. Molnar and M. Gyulassy, *Phys. Rev.* **C62** 054907 (2000).
12. P.F. Kolb, U. Heinz, P. Huovinen, K. Eskola, and K. Tuominen, *Nucl. Phys.* **A696** 197 (2001).
13. T. Hirano and Y. Nara, nucl-th/0404039.
14. P.F. Kolb, arXiv:nucl-th/0304036.
15. H. Sorge, *Phys. Rev. Lett.* **78** 2309 (1997).
16. B. Zhang, M. Gyulassy, and C.M. Ko, *Phys. Lett.* **B455** 45 (1999).
17. P.F. Kolb and R. Rapp, *Phys. Rev.* **C67** 044903 (2003).
18. P.F. Kolb and U. Heinz, in *Quark Gluon Plasma*, edited by R.C.Hwa and X.N. Wang (World Scientific, Singapore, 2004) Vol. 3, p. 634, nucl-th/0305084.
19. R. Rapp, *Phys. Rev.* **C66** 017901 (2002).
20. D. Teaney, arXiv:nucl-th/0204023.
21. T. Hirano and K. Tsuda, *Phys. Rev.* **C66** 054905 (2002).
22. T. Chujo et al. (PHENIX Collaboration), *Nucl. Phys.* **A715** 151c (2003); B. Wosiek et al. (PHOBOS Collaboration), *Nucl. Phys.* **A715** 510c (2003); D. Ouerdane et al. (BRAHMS Collaboration), *Nucl. Phys.* **A715** 478c (2003); O. Barannikova et al. (STAR Collaboration), *Nucl. Phys.* **A715** 458c (2003).
23. C. Adler et al. (STAR Collaboration), *Phys. Rev. Lett.* **89** 132301 (2002).
24. J. Castillo et al. (STAR Collaboration), nucl-ex/0403027
25. P. Huovinen, P.F. Kolb, U. Heinz, P.V. Ruuskanen, and S.A. Voloshin, *Phys. Lett.* **B503** 58 (2001).
26. P.F. Kolb, *Phys. Rev.* **C68** 031902(R) (2003); P.F. Kolb, L.-W.Chen, V. Greco, and C.M. Ko, *Phys. Rev.* **C69** 051901(R) (2004).
27. U.W. Heinz and P.F. Kolb, hep-ph/0204061, and *Nucl. Phys.* **A702** 269 (2002).
28. P.F. Kolb, P. Huovinen, U.Heinz, and H. Heiselberg, *Phys. Lett.* **B500** 232 (2001).
29. C. Alt et al. (NA49 Collaboration), *Phys. Rev.* **C68** 034903 (2003).
30. K. Morita, S. Muroya, C. Nonaka, and T. Hirano, *Phys. Rev.* **C66** 054904 (2002).
31. P.F. Kolb and U. Heinz, nucl-th/0403044.
32. T. Hirano, *Phys. Rev.* **C65** 011901(R) (2001).
33. C. Nonaka at RIKEN-BNL workshop *Collective Flow and QGP properties*, November 17-19, 2003. Talks available at *http://tonic.physics.sunysb.edu/flow03*
34. A. Muronga, *Phys. Rev.* **C69** 034903 (2004).
35. D. Teaney, nucl-th/0403053.
36. U. Heinz and S. Wong, *Phys. Rev.* **C66** 014907 (2002).
37. D. Teaney, *Phys. Rev.* **C68** 034913 (2003).

PHOTON EMISSION IN A HOT QCD PLASMA

P. AURENCHE

Laboratoire d'Annecy-le-Vieux de Physique Théorique, LAPTH,[*]
BP. 110, F-74941, Annecy-le-Vieux Cedex, France
E-mail: aurenche@lapp.in2p3.fr

Various mechanisms of thermal photon production are reviewed and their implications for heavy ion collisions are briefly sketched.

1. Introduction

It is often mentioned that photons, or more generally electromagnetic probes, can be used to signal the formation of quark-gluon plasma (QGP) in ultra-relativistic heavy ion collisions. They could also be used to measure the temperature of this plasma at the time of its formation. Data relevant for this problem will soon cover a wide energy range from a nucleon-nucleon center-of-mass energy about 20 GeV (Pb-Pb collisions at CERN) to 200 Gev (Au-Au collisions at RHIC, Brookhaven) and 5.5 TeV (Pb-Pb collisions at LHC after 2007).

Many production mechanisms of electromagnetic probes exist in heavy ion collisions. On a very qualitative level, one distinguishes the emission of photons in the initial stage of the collision (primary production): it is similar to the case of proton-proton collisions and goes via QCD Compton scattering, quark-antiquark annihilation and bremsstrahlung emission, involving partons in the incoming hadrons. It is calculated in perturbative QCD to the next-to-leading order (NLO). The energy spectrum is power behaved and it is expected to dominate the high momentum region. Hadrons such as π^0 and η are also produced by perturbative mechanisms and they decay into photons which contribute an important background to the photon signal at large momentum. In the heavy ion collision many partons are produced which eventually thermalise into a quark-gluon plasma. This hot QGP bubble expands and cools until a hadronic phase is formed. Photons

[*]UMR5108 du CNRS associée à l'Université de Savoie

are produced in the QGP phase as well as in the hot hadronic phase with a rate which is expected to be exponentially decreasing with the energy of the photon (secondary production). This thermal contribution is expected to be important mainly in the "intermediate" energy region region up to 10 GeV, say. Thermally produced π^0's and η's again contribute an important background.

In the following we review recent progress in the calculation of photon rates in the QGP phase and present phenomenological applications for RHIC and LHC heavy ion collisions.

2. Thermal production of hard photons

We assume in this section that the plasma is in chemical and thermal equilibrium at temperature T. The rate of production of a real photon of momentum $P = (E, \mathbf{p})$, per unit time and volume, is

$$\frac{E\, dN}{dtd\mathbf{x}d\mathbf{p}} = -\frac{1}{(2\pi)^3}\, n_{_B}(E)\, \mathrm{Im}\, \Pi_\mu{}^\mu(E + i\epsilon, \mathbf{p}) \ , \tag{1}$$

where $\Pi_\mu{}^\mu(E+i\epsilon, \mathbf{p})$ is the retarded photon polarisation tensor. The Bose-Einstein factor $n_{_B}(E)$ provides an exponential damping when $E \gg T$. A similar formula holds for lepton pair production.

The two-point correlation function is calculated in the framework of the hard thermal loop (HTL) resummed theory of Braaten and Pisarski[1]. In this approach one distinguishes two scales: the "hard" scale, typically of order T or larger (the energy of quarks and gluons in the plasma) and the "soft" scale of order gT where g, the strong coupling, is assumed to be small. Collective effects in the plasma modify the physics at scale gT, *i.e.* over long distances of $O(1/gT)$. Propagators and vertices are modified by the summation of higher order diagrams. This is easily illustrated with the example of the fermion propagator, $S(Q)$, which in the "bare" theory is simply $1/p$ (we neglect spin complications and make only a dimensional analysis)[a]. The thermal contribution to the one loop correction is found to be $\Sigma(Q) \sim g^2T^2/q$ which is of $O(gT)$ when $q \sim gT$. The resummed propagator $^*S(Q) = 1/(q - \Sigma(Q))$ is then deeply modified, compared to the bare propagator, for soft momenta of $O(gT)$ whereas thermal corrections appear essentially as higher order effects for hard momenta. Thermal resummation affects the propagator in two ways: 1) in the space-like region ($Q^2 < 0$), the propagator acquires an imaginary part, due to Landau

[a] A low case letter denotes the modulus of the 3-momentum.

damping (a new feature compared to the $T = 0$ theory) characterised by a thermal mass scale; 2) in the time-like region a thermal mass, of asymptotic value $m_{\rm q}^2 = g^2 T^2 C_F/4$, is generated. In both cases thermal effects screen potential soft or collinear singularities. Likewise, the gluon propagator is modified: Debye screening mass $m_{\rm D}^2 = g^2 T^2 (N_c + N_f/2)/3$ in the space-like region and quasi-particle masses in the time-like region.[b] One-loop corrections also modify the vertices when the external momenta are soft.[1] One can construct an effective Lagrangian[2] in terms of effective propagators and vertices and calculate observables in perturbation theory.

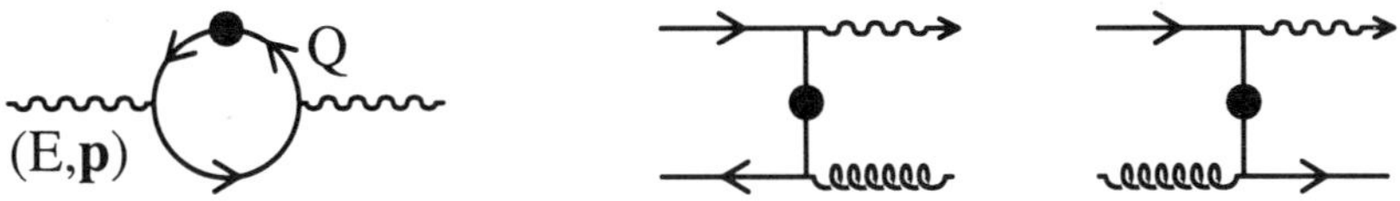

Figure 1. Left diagram: one-loop contribution; right diagrams: annihilation and Compton scattering processes obtained when cutting the one-loop diagram.

In the one-loop approximation, the production rate of hard photons is given by the diagram shown in Fig. 1 where the symbol • means that effective propagators are used. For hard photon momentum it is indeed enough to consider bare vertices and only one effective fermion propagator, since one propagator is necessarily hard and needs no resummation. The result is (for one quark flavor)[3]

$$\mathrm{Im}\,\Pi_\mu{}^\mu(E,\mathbf{p}) = \frac{5}{9}\,\alpha\, g^2\, T^2 \left(\ln(\frac{ET}{m_{\rm q}^2}) + constant\right). \qquad (2)$$

One notices the logaritmic growth of the 2-point function with the energy of the photon while the thermal quark mass acts as a soft cut-off which screens the singularity in the forward scattering of the processes in Fig. 1.

At this point, the HTL picture of photon production, dominated by Compton and annihilation scattering, turned out to be at variance with the result of a semi-classical calculation where it was found that the production was dominated by the bremsstrahlung radiation of photons from quarks.[4] The way out of this paradox is to consider the two-loop diagrams in the effective theory (Fig. 2). A dimensional analysis reveals that the dominant contribution comes from the kinematical region where the quark

[b] N_c is the number of colors, N_f the number of flavors and C_F the Casimir of the fundamental representation.

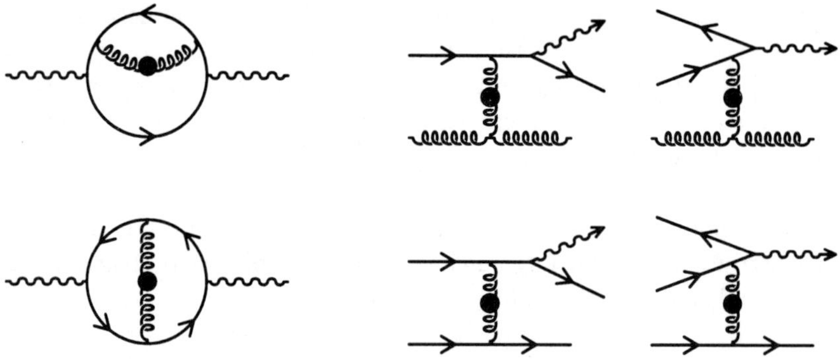

Figure 2. Left diagrams: two-loop contributions; right diagrams: bremsstrahlung and off-shell annihilation processes obtained when cutting through the two-loop diagrams.

momenta in the loop are hard but the gluon is soft and space-like. The physical processes exhibited by cutting through the loop diagrams of Fig. 2 correspond to the bremsstrahlung emission from a quark or an anti-quark scattering in the plasma (as in the semi-classical approximation) and by a new process, the off-shell annihilation of a quark-antiquark pair where one of the (anti-)quark scatters in the plasma. For two flavors, the result of the calculation is very simple[5]:

$$\mathrm{Im}\,\Pi_\mu{}^\mu(E,\mathbf{p}) = \frac{5}{9}\,\alpha\, g^2\,\frac{8}{3}\left[\frac{T^3}{E}+\frac{ET}{\pi^2}\right], \tag{3}$$

where the first term arises from bremsstrahlung and the second one from off-shell annihilation. In the general case the numerical coefficient is replaced by a function of the ratio of the Debye mass over the thermal quark mass.[5] The two-loop HTL result is not suppressed, as expected, by powers of the coupling g compared to the one-loop result. This is due to an enhancement factor of $O(T^2/m_\mathrm{q}^2) \sim 1/g^2$ associated to the collinear emission of the photon.

One may wonder about the usefulness of the loop expansion in the HTL effective theory if the two-loop rate is of the same order as the one-loop rate. It turns out that higher loop diagrams contribute to leading order.[6] The physical reason is the following. Consider a virtual quark of momentum R decaying into a photon of momentum P and a quark of momentum Q. The photon formation time (*i.e.* the life-time of the virtual state) is $\tau_\mathrm{F} \equiv r/R^2 \equiv qr/(p(\mathbf{q}_T^2+m_\mathrm{q}^2))$ with the transverse momentum of the quark measured with respect to the photon momentum. This formation time is long, of $O(T/m_\mathrm{q}^2) \sim 1/g^2T$, and is comparable to the mean free path of the quark in the plasma given by the inverse of the quark damping rate γ. Quark rescattering in the medium should thus be taken into account. This is done by resumming the self-energy diagrams on the quark propagator going beyond the hard thermal loop approximation and including also the imaginary part (related to γ) generated by rescattering. In our case the time-like quark propagator, which in the HTL theory was $1/(q_T^2 + m_q^2)$ becomes approximately $1/(q_T^2 + m_q^2 + i\gamma qr/p)$: this shows that both thermal mass and damping rate act as competing cut-offs. Of course, to respect gauge invariance, one should resum vertex corrections in the same approximation as self-energy corrections[7], which motivates the fact that ladder diagrams such as the graph of Fig. 3 should be summed. Arnold, Moore

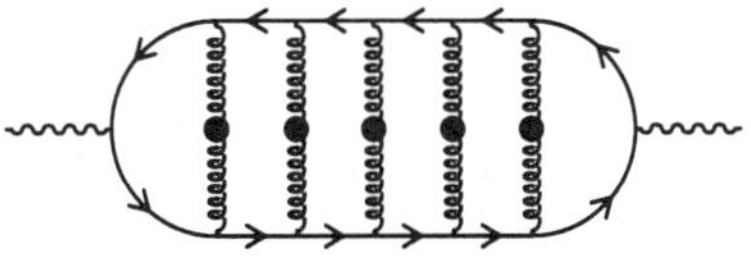

Figure 3. An example of a ladder diagram.

and Yaffe[7] showed that such diagrams indeed contribute to leading order and they derived a very elegant integral equation to perform the resummation. The imaginary part of the photon polarisation tensor takes the form:

$$\mathrm{Im}\,\Pi_\mu{}^\mu(P) \approx \alpha N_c \int_{-\infty}^{+\infty} dq\,[n_{_F}(r) - n_{_F}(q)]\,\frac{q^2+r^2}{q^2r^2}\,\mathrm{Re}\int \frac{d^2\mathbf{q}_T}{(2\pi)^2}\,\mathbf{q}_T\cdot\mathbf{f}(\mathbf{q}_T)\,,$$

with $r = p + q$ and $n_{_F}(q) = 1/(\exp(q/T) + 1)$, the Fermi-Dirac statistical weight. The dimensionless resummed quark-quark-photon vertex, $\mathbf{f}(\mathbf{q}_T)$, is shown by the shaded vertex in Fig. 4 while $\mathbf{q}_T$ is proportional to the bare vertex. The value of $\mathbf{f}(\mathbf{q}_T)$ is obtained by solving the following equation[7]:

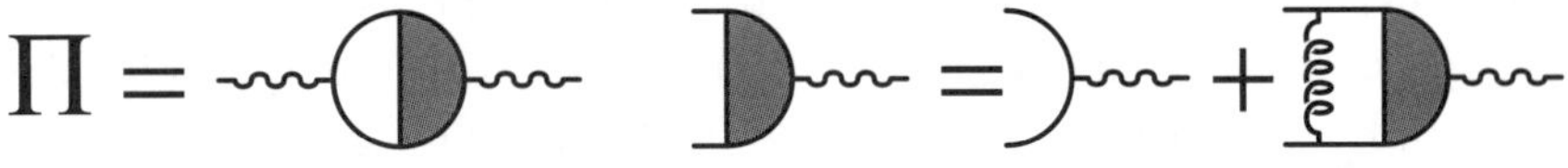

Figure 4. The integral equation resumming the ladder diagrams.

$$\frac{i}{2\tau_{_F}}\mathbf{f}(\mathbf{q}_T) = 2\mathbf{q}_T + g^2 C_{_F} T \int \frac{d^2\mathbf{l}_T}{(2\pi)^2}\, C(\mathbf{l}_T)\,[\mathbf{f}(\mathbf{q}_T + \mathbf{l}_T) - \mathbf{f}(\mathbf{q}_T)]\,, \qquad (4)$$

where $\tau_{_F}$ is the formation time defined above. The collision kernel is $C(\mathbf{l}_T) = m_D^2/\mathbf{l}_T^2(\mathbf{l}_T^2 + m_D^2)$ and it includes the exchange of both longitudinal and transverse gluons.[6] Solving for this integral equation one can obtain a fit, in the large E/T range

$$\mathrm{Im}\,\Pi_\mu{}^\mu(P) \approx \alpha\, g^2\, T^2 \sqrt{\frac{E}{T}}\,, \tag{5}$$

to be compared to the linear in E behaviour of Eq. (3). This is an illustration of the Landau-Pomeranchuck-Migdal (LPM) suppression of hard photons due to multiple scattering of the quark in the plasma.

Rates for small mass lepton pairs produced at large momentum have also been calculated at one-loop[8], two-loop[9] and including the LPM effect.[10]

3. Phenomenology at RHIC and LHC

To make predictions for photon production in heavy ion collisions, the rates calculated above have to be included in a hydrodynamical code which describes the expansion and the cooling of the plasma from chosen initial state conditions. In the following we use the code of Ruuskanen et al.[11] The rate for thermal photon production is shown by the thick dash-dotted line in Fig. 5 for Au-Au central collisions at RHIC and in Fig. 6 for Pb-Pb central collisons at LHC. Further details on the model can be found in Ref. 13. Also shown on the figures are the rates of production of decay photons from thermally produced resonances (mainly π^0 and η) (thick dashed lines) which dominate in the low p_T range. We also display the rates of primary photons produced in perturbative QCD mechanisms (solid lines) contributing in the large p_T range as well as the decay photons from perturbatively produced resonances. In these estimates account has been taken of energy loss effects on the jet fragmentation in the medium using the model of Ref. 14. It is instructive to look at the ratio of "all" produced photons over decay photons and to compare it with the prediction for the same ratio assuming no thermal production. This plot quantifies the excess of direct photons produced by the hot hadronic medium formed in a heavy ion collision. At LHC, thermal effects are important but the ratio of directly produced photon remains only about 10%.

A channel which should suffer less background than direct photon is that of small mass lepton pair at large momentum: indeed choosing a pair mass above the π^0 mass should eliminate a large background. The thermal production rates have been calculated and included in the hydrodynamical code.[13] The result is shown in Fig. 8.

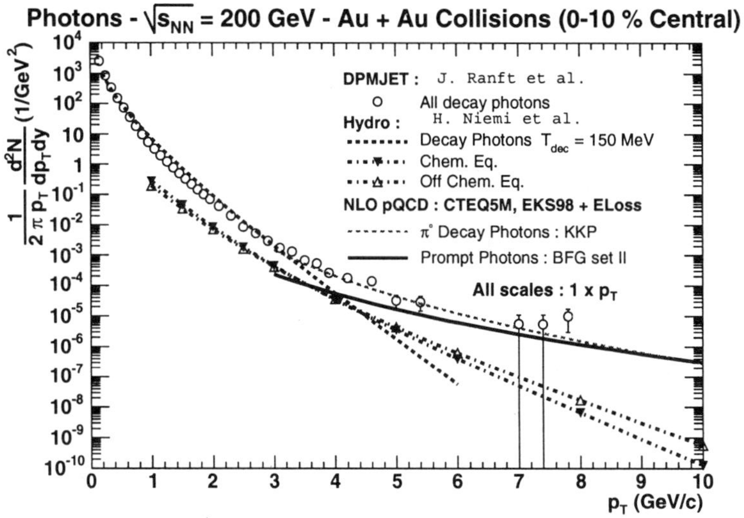

Figure 5. Various contributions to the production of photons in Au-Au collisions at RHIC. The open dots are the predictions of decay photons from the non-thermal DPMJET model[12] for heavy ion collisions, in good agreement with the hydrodynamic + perturbative QCD model.

4. Further developments

In the discussion above we have considered two independent production processes: emission of the photon in the initial stages of the collision and production in the quark-gluon plasma and hot hadronic phase. One can envisage "mixed" production mechanisms where a photon is emitted in the interaction of a hard quark or gluon (calculable in perturbative QCD) with a quark or the gluon of the plasma. Since the rate of hard partons is power behaved while thermal interactions are exponentially damped, the resulting rate of photon production is expected to be power behaved and to dominate over purely thermal processes at large momentum. Such examples of "mixed" interactions have been considered in Ref. 15 who made simplified estimates of the Compton ($gq \to \gamma q$) and annihilation ($q\bar{q} \to \gamma g$) processes when one of the initial q or g is hard and the other parton is thermalised. The result is shown in Fig. 9 by the dashed line to be compared with the dotted line which is the estimate of the purely thermal process under the same hydrodynamical conditions (no radial expansion). Recently B. Zakharov estimated the enhanced bremsstrahlung emission of a photon by a hard quark traversing the plasma.[16] He found that, due to finite size

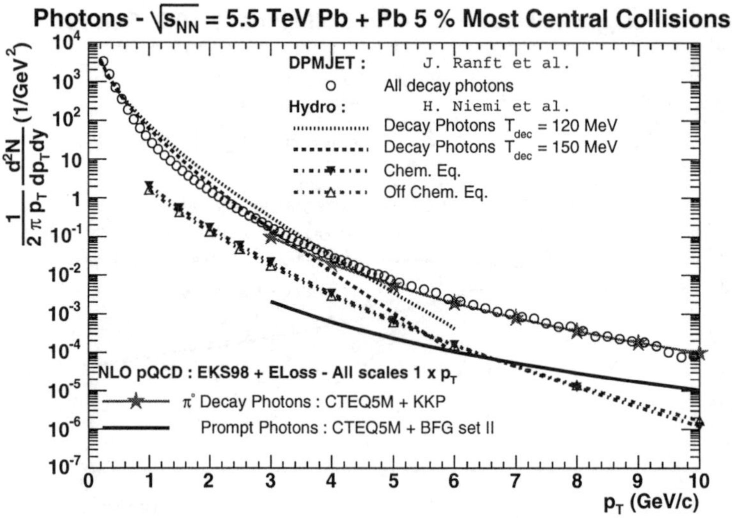

Figure 6. Same as the above figure for Pb-Pb collisions at LHC.

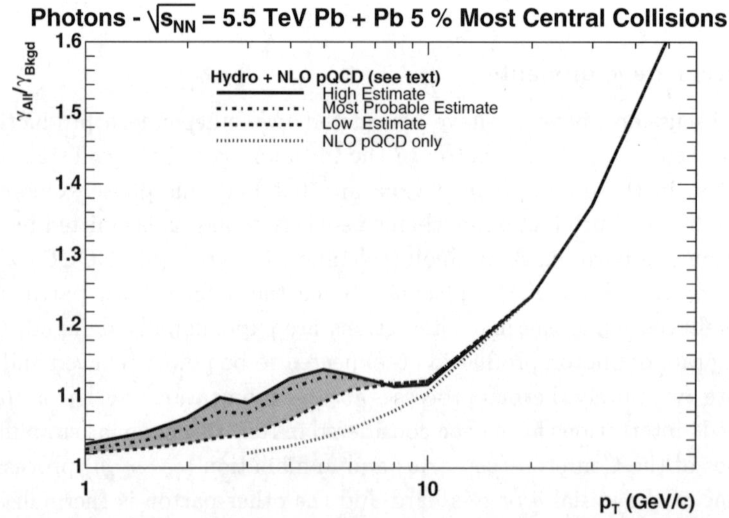

Figure 7. Ratio of all photons over decay photons at LHC with thermal production (thick lines) and without thermal production.

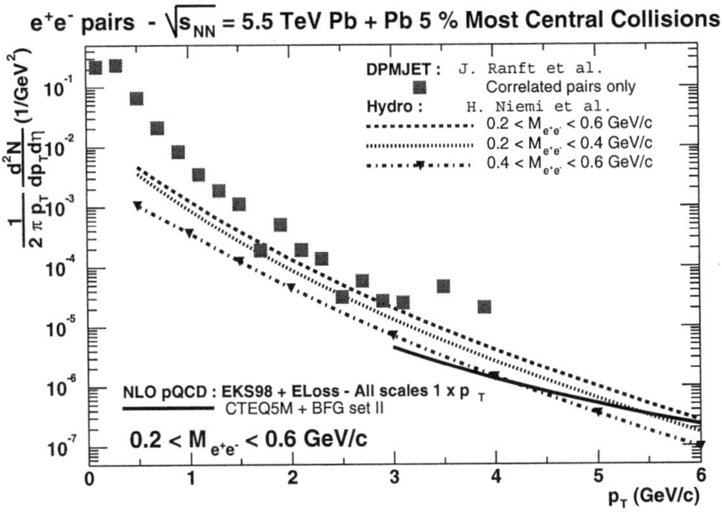

Figure 8. Thermally produced lepton pairs (dashed line) compared to perturbatively produced pairs (solid line) and background from hadronic decays (squares).

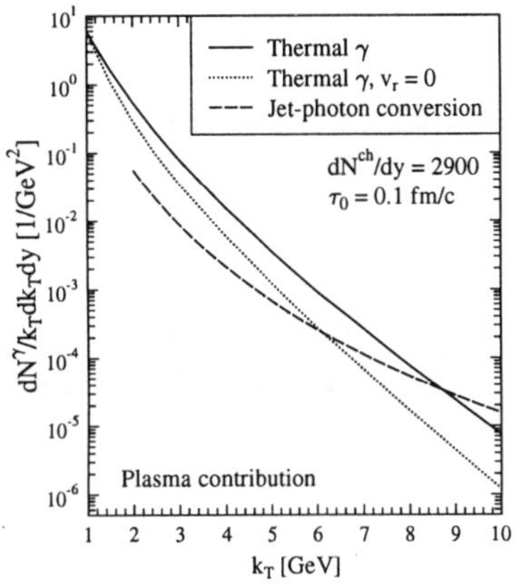

Figure 9. Photon production from "mixed" processes: rate from the Compton and annihilation processes between a hard parton with a parton in the plasma (dashed line); thermal photon production under the same hypothesis of no transverse expansion (dotted line). For comparison the thermal rate with radial expansion, used in the previous section, is also shown (solid line).

effects, the usual $1/x$ spectrum characteristic of bremsstrahlung of a quark in vacuum is replaced by an approximate $1/(1-x)$ spectrum for $x < 1$. This is illustrated in Fig. 10 which compares the fragmentation function of a quark into a photon in vacuum with the same in the medium taking into account only one rescattering in the plasma. The change of shape of the spectrum will lead to an enhanced production of photon by bremsstrahlung compared to production in the vacuum. Multiple scattering effects will somewhat reduce the size of the predicted enhancement.[16] In the figures of the previous section we already included medium effects on bremsstrahlung production of photons and this lead to a suppression of photon emission compared to vacuum which seems to contradict the present result. In fact, there is no contradiction if we consider that in the latter case the photon was emitted by the quark outside the plasma while the former case one estimates the emission of the photon in the plasma.

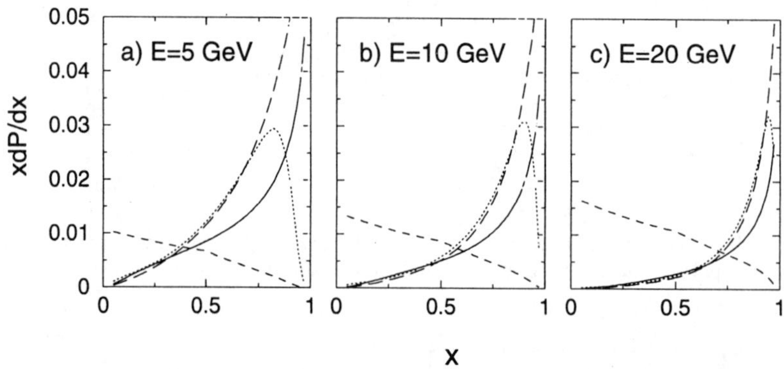

Figure 10. Comparison of the fragmentation function of a quark into a photon in vacuum (short dashed line) and in the medium (dotted line) with one scattering of the hard quark in the plasma. The other curves are for an infinite medium.

It would be important to have a complete and consistent estimate of the mixed production rates, using a realistic hydrodynamical model: in any case, they will increase the thermal production rates calculated in the previous section.

All previous results are obtained assuming local thermal equilibrium and the rates are calculated per unit time. In a series of papers Boyanovsky et al.[17] proposed a model to explicitely account for the finite life-time of the plasma. They assume that photon emission starts at an initial time t_i and ends at time t_f: under this assumption, usually forbidden processes of type $q \to q\gamma$ contribute since energy conservation does not hold. As a consequence a transient term of type $\alpha \ln t$, where $t = t_f - t_i$, appears besides the usual $\alpha\alpha_s t$ term, and it is claimed to dominate the photon production rate at LHC at large enough momenta. Several objections have been raised concerning the validity of this picture.[18,19,20] In particular the hypothesis of a sharp switching on and off of the interactions is probably not adequate for the description of the physics of the problem.

Acknowledgments

I thank F. Gelis and V. Ruuskanen for collaboration and many discussions. I also thank F. Gelis for a careful reading of the manuscript.

References

1. E. Braaten, R.D. Pisarski, *Nucl. Phys.* **B337**, 569 (1990); **B339**, 310 (1990); J. Frenkel, J.C. Taylor, *Nucl. Phys.* **B334**, 199 (1990); **B374**, 156 (1992).
2. J.C. Taylor, S.M.H. Wong, *Nuc. Phys.* **B346**, 115 (1990); E. Braaten, R.D. Pisarski, *Phys. Rev.* **D45**, 1827 (1992).
3. J.I. Kapusta, P. Lichard, D. Seibert, *Phys. Rev.* **D44** (1991) 2774; R. Baier, H. Nakkagawa, A. Niegawa, K. Redlich, *Z. Phys.* **C53**, (1992) 433.
4. J. Cleymans, V.V. Goloviznin, K. Redlich, *Phys. Rev.* **D47**, 989 (1993); V.V. Goloviznin, K. Redlich, *Phys. Lett.* **B 319**, 520 (1993).
5. P. Aurenche, F. Gelis, R. Kobes, H. Zaraket, *Phys. Rev.* **D58**, 085003 (1998); P. Aurenche, F. Gelis, H. Zaraket, *JHEP* **0205**, 043 (2002).
6. P. Aurenche, F. Gelis, H. Zaraket, *Phys. Rev.* **D61**, 116001 (2000); **D62**, 096012 (2000).
7. P. Arnold, G.D. Moore, L.G. Yaffe, *JHEP* **0111**, 057 (2001); *JHEP* **0112**, 009 (2001); *JHEP* **0206**, 030 (2002).
8. T. Altherr, P.V. Ruuskanen, *Nucl. Phys.* **B380**, 377 (1992); M.H. Thoma, C.T. Traxler, *Phys. Rev.* **D56**, 198 (1997).
9. P. Aurenche, F. Gelis, H. Zaraket, *JHEP* **0207**, 063 (2002).
10. P. Aurenche, F. Gelis, G.D. Moore, H. Zaraket, *JHEP* **0212**, 006 (2002).
11. K.J. Eskola, K. Kajantie, P.V. Ruuskanen,*Eur. Phys.J.* **C1**, 627 (1998); K.J. Eskola, P.V. Ruuskanen, S.S. Räsänen, K. Tuominen, *Nucl. Phys.* **A696**, 715 (2001); K.J. Eskola, H. Niemi, P.V. Ruuskanen, S.S. Räsänen, *Nucl. Phys.* **A715**, 561 (2003); *Phys. Lett.* **B566**, 187 (2003).
12. R. Engel and J. Ranft, *Phys. Rev.* **D54**, 4244 (1996); S. Roesler, R. Engel and J. Ranft, hep-ph/0012252; J. Ranft, hep-ph/9911213, hep-ph/9911232.
13. F. Arleo et al., hep-ph/0311131;
14. J. Jalilian-Marian, K. Orginos and I. Sarcevic, *Phys. Rev.* **C63**, 041901 (2001); *ibid.* **A700**, 523 (2002); S. Jeon, J. Jalilian-Marian and I. Sarcevic, *Phys. Lett.* **B562**, 45 (2003).
15. R. J. Fries, B. Muller and D. K. Srivastava, *Phys. Rev. Lett.* **90**, 132301 (2003).
16. B.G. Zakharov, hep-ph/0405101.
17. D. Boyanovsky, H. J. de Vega, *Phys. Rev.* **D68**, 065018 (2003); hep-ph/0311156.
18. I. Dadic, F. Gelis, G.D. Moore, J. Serreau in Ref. 13, Sec. 9.
19. J. Serreau, *JHEP* **0405**, 078 (2004).
20. E. Fraga, F. Gelis, D. Schiff, hep-ph/0312222.

IN SEARCH OF THE SATURATION SCALE: INTRINSIC FEATURES OF THE CGC

HERIBERT WEIGERT

Institut für theoretische Physik, Universität Regensburg,
93040 Regensburg, Germany
E-mail: heribert.weigert@physik.uni-r.de

This is a brief review of the physics ideas behind the Color Glass Condensate. It describes a dense gluonic medium created in collider experiments at high energy. The main characteristic feature is the emergence of a density driven scale, the saturation scale Q_s. This scale will grow with energy and drives the energy dependence of affected phenomena at HERA, RHIC and LHC.

1. Introduction

Modern collider experiments probe matter in the extremely ultrarelativistic region. Although the motivations range from the search for the Higgs and physics beyond the standard model to the creation of a quark gluon plasma as exemplified by the LHC and RHIC, they all share the property that collision energies are gigantic: from 200GeV/nucleon at RHIC via 2TeV/nucleon in the LHC heavy ion program to 7TeV in *pp* collisions. Besides providing the means to test the originally posed questions this leads to the creation of a new, gluon dominated medium in the QCD sector. This underlying mechanism is simply phase space enhanced soft gluon radiation, the consequences are new. Due to Lorentz contraction, one is faced with a system in which the dominant degrees of freedom are frozen on the typical time scales of interaction. Contrary to photons, gluons carry charge and thus the gluonic medium develops a characteristic scale, beyond which saturation effects dominate the QCD degrees of freedom and the gluon fields grow to be as large as $1/g$. This compact list of features has earned it the name Color Glass Condensate (CGC) and I will not defend the name but instead try to explain how these features emerge.

2. DIS at high energies: the idea of a saturation scale

DIS of leptons off protons or nuclei at high energies is characterized by two kinematical invariants. The large spacelike momentum transfer $Q^2 := -q^2$ sets the transverse resolution and x sets the energy or rapidity separation of projectile and target as shown in Fig.1. Best under control is the Bjorken

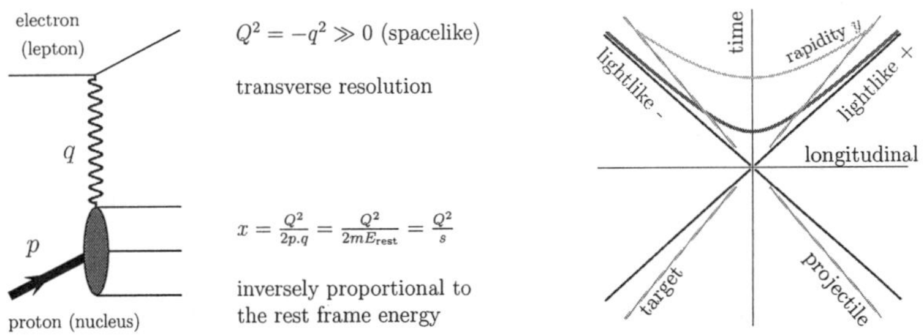

Figure 1. Kinematics in DIS of leptons off protons or nuclei. Projectile and target are separated by a large boost factor $\frac{1}{x} = e^y$

limit of large Q^2 at fixed x. There, twist- and operator product expansion in powers of transverse resolution $1/Q^2$ guarantee diluteness. Constituents of apparent size $1/Q^2$ do not overlap. In this domain all one needs to know to describe the system are single particle properties: quark and gluon distributions. Q^2 dependence of cross sections can be calculated by summing ladder diagrams as is most efficiently done by solving the DGLAP evolution equations. Going to small x at fixed Q^2 leads into an entirely different domain. The reason is soft gluon radiation that gets ever more important the higher the energy (the smaller x) becomes. More and more particles (predominantly gluons) of size $1/Q^2$ populate the same region and will necessarily overlap. Particle correlations will become important and density effects will build up a correlation length $R_s = 1/Q_s$. Consequently one enters a new region characterized by a highly dense gluonic medium the Color Glass Condensate (CGC). The situation is sketched in Fig. 2. I will return to the transition region at the end of Sec. 4.

3. Two nonlinearities

There are two types of nonlinearity which are important in a description of the kinematic domain and they can both be seen as corrections to the wave function of the virtual photon transferring the momentum q to the

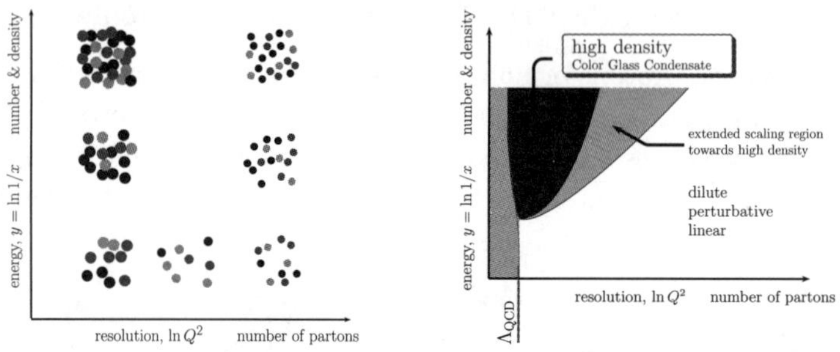

Figure 2. Gluon distributions grow towards small x. To the left: Growth of gluon (parton) numbers with Q^2 and $\ln(1/x)$. Going to sufficiently small x at fixed Q^2 partons start to overlap. To the right: corresponding regions of phase space with qualitatively different behavior.

target. Besides the initial splitting into a $q\bar{q}$ pair one encounters phase space enhanced soft gluon emission:

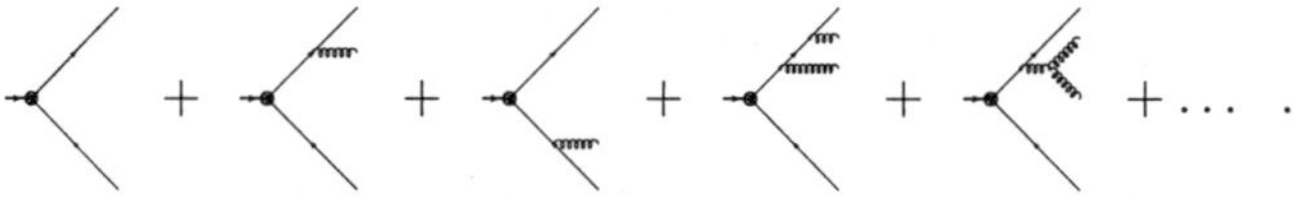
(1)

This complicated photon wave function will then interact with the proton or nuclear target an lead to a cross section that will reflect the presence of the many gluons.

3.1. *Cross sections at fixed x. Nonlinearities I*

The first type of nonlinearity already emerges if one ignores the colored structure of the gluons and thus the last diagram in this equation. It would already be present in a strong electromagnetic field: Due to the high longitudinal momentum of the "hard" $q\bar{q}$ the gauge bosons eikonalize into path ordered exponentials collinear with the projectiles worldline (x^- in Fig. 1):

$$U_{\boldsymbol{x}} = \sum_{\text{gluons}}$$ 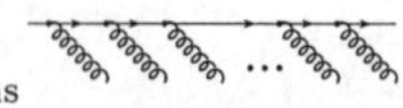 $$= \mathsf{P}\exp -ig \int dz^- b^+(z^-, \boldsymbol{x}, 0) \qquad (2)$$

One may arrive at the same conclusion if one considers this from the targets perspective. Probing this target at small x implies a large rapidity difference $y = \ln(1/x)$ between projectile and target, corresponding to a large relative boost factor e^y of the order of several hundred to several thousand in the experiments listed above. Knowing that the interaction will be dominated by gluonic configurations, one tries to isolate the boost enhanced ones. In terms of the gluon field strength of the target these are the F^{+i} components all others can be set to zero. All components suffer a strong Lorentz contraction in x^- direction, and a time dilation, correspondingly in x^+ direction. With such a field strength tensor, one is left with only one important degree of freedom, which can be taken to be the $+$ component of a gauge field. The gauge field can then be written as

$$A = b + \delta A \qquad \text{with} \qquad b^{i,-} = 0, \quad b^+ = \beta(\boldsymbol{x})\delta(x^-) \tag{3}$$

where the leading contribution b^+ is x^+ independent and Lorentz contracted to a δ-function in x^-. Both of these statements are to be taken to be true with a resolution imposed by the rapidity separation $y = \ln(1/x)$. Note that mathematically one can always trade a gauge field that has only a single component for a path ordered exponential along the direction that picks up this component: $b^+ = i(\partial^+ U)U^\dagger$ with U given by (2). Whether one attributes the gluon fields to the wave function of the target or the projectile is largely a matter of interpretation and choice of frame. As a result cross sections at high energies can be expressed entirely in terms of these eikonal factors. For the DIS case the result schematically reads[a]

$$\frac{d\sigma_{\text{DIS}}}{dx dQ^2} = \text{Im}$$

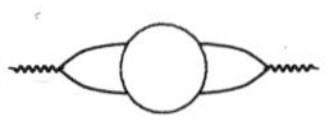

$$= \int d^2r\, |\psi^2|(r^2Q^2) \int d^2b\, \langle \frac{\text{tr}(1 - U_{\boldsymbol{x}} U^\dagger_{\boldsymbol{y}})}{N_c} \rangle_y \tag{4}$$

where $\boldsymbol{r} = \boldsymbol{x} - \boldsymbol{y}$ corresponds to the transverse size of the $q\bar{q}$ dipole and $\boldsymbol{b} = (\boldsymbol{x} + \boldsymbol{y})/2$ to its impact parameter relative to the target. As shown, the leading small x contribution factorizes into a wave function part (the factor $|\psi^2|(r^2Q^2)$) and the dipole cross section part under and including the b integral. The latter contains all the interaction with the targets gluon field via the U factors. The average contains all the information about the perturbative corrections and the nonperturbative features of the target wave function. As such this will be uncalculable, but a few properties can be immediately read off. Contrary to the gluon distributions appearing in the corresponding expression for the Bjorken limit, the dipole function

$$N_{y;\boldsymbol{xy}} = \langle \hat{N}_{y;\boldsymbol{xy}} \rangle_y = \langle \frac{\text{tr}(1 - U_{\boldsymbol{x}} U^\dagger_{\boldsymbol{y}})}{N_c} \rangle_y \tag{5}$$

can only take values between 0 and 1. It will interpolate between these values from small dipoles to large ones and reflects the ideas of color transparency at short distances and saturation at large distances (Fig. 4). In the

[a]An integral over longitudinal momentum fractions has been suppressed.

simplest case, the transition is characterized by a single scale, the saturation scale Q_s. Its inverse R_s is clearly the correlation length of U-fields.

3.2. *JIMWLK: x dependence. Nonlinearities II*

Going to smaller x new gluon emission will cause this correlation length to shrink and the curve for N in Fig. 4 to move towards the left. Calculationally this can be viewed as a Wilson type renormalization group process in which one integrates out fluctuations around b^+ that become indistinguishable from the original ones as one worsens longitudinal (x^-) resolution and freezes additional modes in x^+ by going to higher energy. This forces us to change the averaging procedure over the U fields in (4). If one writes this average as a functional integral with a distribution functional $\hat{Z}_y[U]$ as

$$\langle \ldots \rangle_y = \int \hat{D}[U] \ldots \hat{Z}_y[U] \tag{6}$$

then this amounts to a functional RG equation for $\hat{Z}_y[U]$ which has the form of a Fokker-Planck equation, the Jalilian-Marian+Iancu+McLerran+Weigert+Leonidov+Kovner or JIMWLK equation [1–9]:

$$\partial_y \hat{Z}_y[U] = -H_{\text{JIMWLK}}[U] \hat{Z}_y[U] \ . \tag{7}$$

The JIMWLK Hamiltonian implements the emission of additional gluons shown in Fig. 3 in the projectiles wave function to leading log accuracy

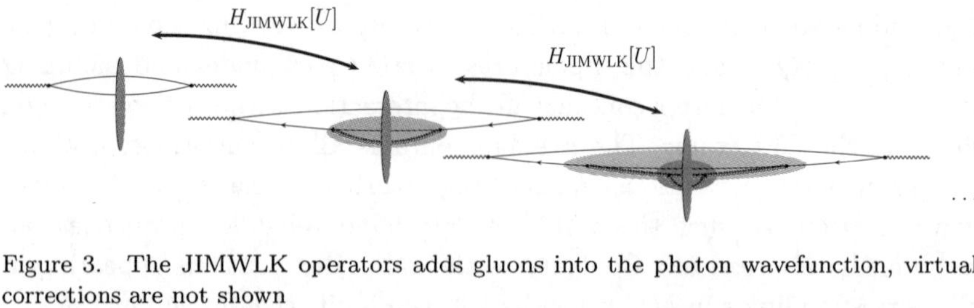

Figure 3. The JIMWLK operators adds gluons into the photon wavefunction, virtual corrections are not shown

(it sums corrections proportional to $\alpha_s \ln(1/x)$) and crucially involves the colored nature of gluons. Without the last diagram in (1) no evolution would take place. To be precise, the JIMWLK equation reinterprets this emission as a change in the average as indicated by the shaded areas in Fig. 3. The emission perspective becomes obvious by rewriting this equation as an infinite coupled hierarchy of evolution equations for correlators of U fields, the Balitsky hierarchy. The simplest of these equations is derived by applying $\int \hat{D}[U]\, \hat{N}_{y;\boldsymbol{xy}}$ to both sides of the equation and letting the Hamiltonian act on $\hat{N}$ instead of $\hat{Z}$. The result is

$$\partial_y \langle \hat{N}_{\boldsymbol{xy}} \rangle_y = \frac{\alpha_s N_c}{2\pi^2} \int d^2z\ \tilde{K}_{\boldsymbol{xzy}} \left\{ \langle \hat{N}_{\boldsymbol{xz}} + \hat{N}_{\boldsymbol{zy}} - \hat{N}_{\boldsymbol{xy}} \rangle_y - \langle \hat{N}_{\boldsymbol{xz}} \hat{N}_{\boldsymbol{zy}} \rangle_y \right\} \quad (8)$$

with $\tilde{K}_{\boldsymbol{xzy}} = \frac{(\boldsymbol{x}-\boldsymbol{y})^2}{(\boldsymbol{x}-\boldsymbol{z})^2(\boldsymbol{z}-\boldsymbol{y})^2}$. The characteristic feature of this equation is the nonlinear term which involves a 3-point function $\langle \hat{N}_{\boldsymbol{xz}} \hat{N}_{\boldsymbol{zy}} \rangle_y$ and thus does not close. The new coordinate $\boldsymbol{z}$ with its associated factors $U_{\boldsymbol{z}}^{(\dagger)}$ is the transverse location of the newly created gluon. This arises from the U-dependence of the JIMWLK Hamiltonian and will in turn couple the evolution equation of this 3-point function to even more complicated 4-point functions and lead to an iteration procedure that will not terminate. The result is the Balitsky hierarchy. A simple factorization assumption $\langle \hat{N}_{\boldsymbol{xz}} \hat{N}_{\boldsymbol{zy}} \rangle_y \to N_{y;\boldsymbol{xz}}\ N_{y;\boldsymbol{zy}}$ decouples the complete hierarchy – the result is a nonlinear equation for $N_{y;\boldsymbol{xy}}$ known as the Balitsky-Kovchegov (BK) equation. That approximation retains the boundedness property crucially depends on the presence of the nonlinearity – dropping it in a small density approximation leads to the well known BFKL equation with its exponentially growing solutions.

4. Q_s scaling of correlators

The crucial feature of both JIMWLK and BK equations is the emergence of solutions which carry all y dependence through dependence on $Q_s(y)$

$$N_{y;\boldsymbol{xz}} = N((\boldsymbol{x}-\boldsymbol{y})^2 Q_s^2(y)) \quad (9)$$

The scaling form is independent of any initial conditions – it is an asymptotic property of the evolution equations. This property is very robust: Although the introduction of running coupling necessarily introduces Λ_{QCD} and thus strictly speaking breaks scaling, the scale breaking is very weak as shown in Fig. 4 (right). A nice illustration on how fast scaling is reached can be obtained by plotting the evolution rate $\lambda(y) := \partial_y \ln Q_s(y)$ against Q_s (the physical scale) or $\alpha_s(Q_s)$ (the characteristic coupling) as in Fig. 5. The perhaps simplest physical interpretation of the evolution process can be cast in terms of BFKL growth as follows: Starting from a dipole function that is saturated at large dipole sizes and perturbatively small at small dipole sizes, evolution is driven by contributions from small sizes. This follows the BFKL pattern of exponential growth. Creation of large objects that would fall into the already saturated domain is prohibited by the nonlinearities, corresponding contributions are absorbed by the medium. Due

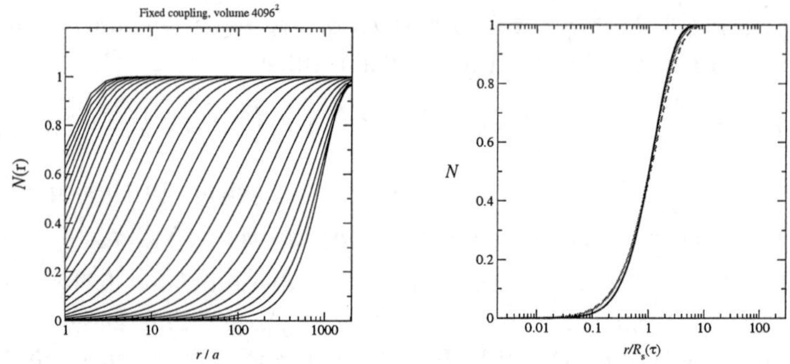

Figure 4. x dependence of dipole functions at fixed coupling shows scaling in Q_s (left) and good approximate scaling at running coupling (right) [10].

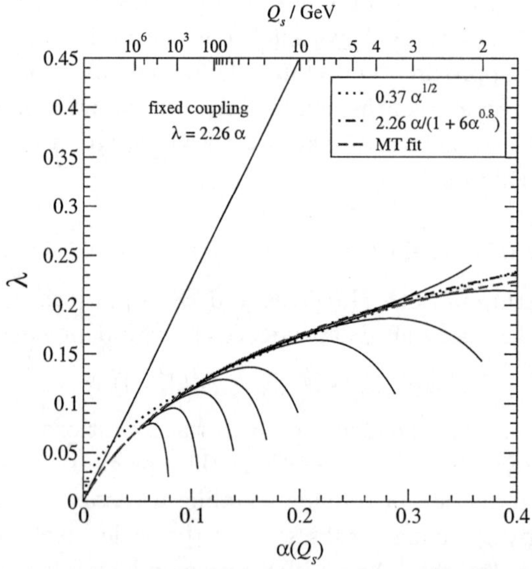

Figure 5. $\lambda(y)$ as a function of $\alpha_s(Q_s(y)^2)$ [10]. A perfect description of the scaling curve is obtained with the MT result [11, 12] by fitting a free parameter. Simple parametrizations emphasize how drastically the functional shape is changed from a simple fixed coupling form $c\ \alpha_s(Q_s^2)$.

to the growth at small distances, densities rise also there and the boundary where the nonlinearities start to act readjust. This happens selfconsistently in JIMWLK and BK. In [11, 12] the picture is taken literally and an even quantitatively comparable result is achieved by combining BFKL evolution with an absorptive boundary which is selfconsistently adjusted to limit growth. This type of analysis leads to scaling forms for the dipole cross section valid at distances smaller than R_s, and predicts the size of the scaling window (the range over which physics is characterized by Q_s despite the fact that saturation is not yet reached) via the corresponding BFKL diffusion radius. This is the transition region between the low density region and the CGC marked in Fig. 2 (right).

5. Phenomenological ideas

The presence of Q_s and more importantly scaling of correlators in terms of $Q_s(y)$ will leave its mark in phenomenological observations. In fact, before the the scaling properties of JIMWLK and BK equations were known, HERA data at small x ($\lesssim 10^{-2}$) had been shown to show "geometric scaling" [13, 14] in terms of the saturation scale. Starting from the scaling of dipole functions $N_{y;\boldsymbol{xy}}$ scaling of the cross section is easily understood. Rescaling the r integral in Eq. (4) and using the fact that the wave function factor scales like $f(\boldsymbol{r}^2 Q^2)/\boldsymbol{r}^2$, this implies

$$\sigma_{\text{tot}}(y, Q^2) = \sigma_{\text{tot}}\Big(y_0, \frac{Q^2}{Q_s^2(y)} Q_s^2(y_0)\Big) \tag{10}$$

and should allow to map all the cross section data onto a single scaling curve. This is shown in Fig. 6 While the explicit parametrizations shown

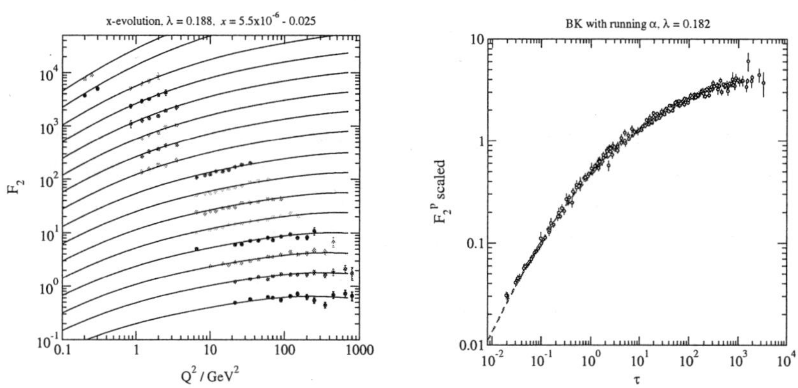

Figure 6. Geometric scaling in HERA data. Shown is the structure function F_2, which at small x is related to the cross section by a rescaling by Q^2: $\sigma(y, Q^2) \sim F_2(y, Q^2) \cdot Q^2$.

in Fig. 6 make use of BK evolution results from [10], this is not essential, at least for the scaling plot on the right.

To tie in also $\gamma^* A$ collisions one can use the McLerran-Vanugopalan (MV) model [15–17] which models correlators of U fields with Gaussian weights and adds the assumption that the energies are high enough that the probes interact with all color charges along their path through the target. Since scattering with gluons from different nuclei should be uncorrelated this leads to a reduction in transverse correlation length (an increase of saturation scale) according to $Q_s^A(y_0)^2 \sim Q_s^A(y_0)^2\ A^{1/3}$. This simple scaling will be changed by evolution, but it should apply at some initial y_0. Based on this idea scaling features have also looked for in older NMC [18] and E665 [19,20] data for the ratios of nuclear to deuterium structure functions $2F_2^A/AF_2^D$. Despite the fact that both x and Q^2 ranges are such that CGC saturation arguments remain marginal, both [21] and [22] find that scaling in both x and A holds surprisingly well.

Due to the limited range in x this can not be taken as an experimental confirmation of CGC physics – such would require a measurement of x dependence and this is clearly a reason to explore this question more closely in experiments at higher energies. One of the options is the Electron Ion Collider (EIC) in planning at BNL [23] as a machine dedicated to such experiments or via pA collisions at the LHC where higher energies are available at the price of a more complicated projectile.

Presently the most intriguing experimental news come from the BRAHMS experiment at RHIC [24] and concern the Cronin effect. Results indicate a disappearance of the Cronin enhancement of particle yields at moderate transverse momenta as one moves from central to forward rapidities. This parallels a prediction that evolution at small x quickly erases a Cronin enhancement originally present at initial y_0 (as modeled in the MV spirit) [25–27]. The rate of disappearance is driven by the respective behavior of saturation scales in protons and appears to be compatible with data. RHIC would appear to probe exactly the onset of the CGC.

One of the most interesting consequences of this is that LHC in the heavy ion program would probe the CGC region already at central rapidities. Even the particle physics program should enter this range if some of the planned forward detectors will be built.

References

1. J. Jalilian-Marian, A. Kovner, L. D. McLerran, and H. Weigert *Phys. Rev.* **D55** (1997) 5414–5428, [hep-ph/9606337].
2. J. Jalilian-Marian, A. Kovner, A. Leonidov, and H. Weigert *Nucl. Phys.* **B504** (1997) 415–431, [hep-ph/9701284].

3. J. Jalilian-Marian, A. Kovner, A. Leonidov, and H. Weigert *Phys. Rev.* **D59** (1999) 014014, [hep-ph/9706377].
4. J. Jalilian-Marian, A. Kovner, and H. Weigert *Phys. Rev.* **D59** (1999) 014015, [hep-ph/9709432].
5. J. Jalilian-Marian, A. Kovner, A. Leonidov, and H. Weigert *Phys. Rev.* **D59** (1999) 034007, [hep-ph/9807462].
6. A. Kovner, J. G. Milhano, and H. Weigert *Phys. Rev.* **D62** (2000) 114005, [hep-ph/0004014].
7. H. Weigert *Nucl. Phys.* **A703** (2002) 823–860, [hep-ph/0004044].
8. E. Iancu, A. Leonidov, and L. D. McLerran *Nucl. Phys.* **A692** (2001) 583–645, [hep-ph/0011241].
9. E. Ferreiro, E. Iancu, A. Leonidov, and L. McLerran *Nucl. Phys.* **A703** (2002) 489–538, [hep-ph/0109115].
10. K. Rummukainen and H. Weigert hep-ph/0309306.
11. A. H. Mueller and D. N. Triantafyllopoulos *Nucl. Phys.* **B640** (2002) 331–350, [hep-ph/0205167].
12. D. N. Triantafyllopoulos *Nucl. Phys.* **B648** (2003) 293–316, [hep-ph/0209121].
13. K. Golec-Biernat and M. Wüsthoff *Phys. Rev.* **D59** (1999) 014017, [hep-ph/9807513].
14. A. M. Stasto, K. Golec-Biernat, and J. Kwiecinski *Phys. Rev. Lett.* **86** (2001) 596–599, [hep-ph/0007192].
15. L. D. McLerran and R. Venugopalan *Phys. Rev.* **D49** (1994) 2233–2241, [hep-ph/9309289].
16. L. D. McLerran and R. Venugopalan *Phys. Rev.* **D49** (1994) 3352–3355, [hep-ph/9311205].
17. L. D. McLerran and R. Venugopalan *Phys. Rev.* **D50** (1994) 2225–2233, [hep-ph/9402335].
18. **New Muon** Collaboration, M. Arneodo *et. al. Nucl. Phys.* **B487** (1997) 3–26, [hep-ex/9611022].
19. **E665** Collaboration, M. R. Adams *et. al. Z. Phys.* **C65** (1995) 225–244.
20. **E665** Collaboration, M. R. Adams *et. al. Z. Phys.* **C67** (1995) 403–410, [hep-ex/9505006].
21. A. Freund, K. Rummukainen, H. Weigert, and A. Schäfer *Phys. Rev. Lett.* **90** (2003) 222002, [hep-ph/0210139].
22. N. Armesto, C. A. Salgado, and U. A. Wiedemann hep-ph/0407018.
23. *The Electron Ion Collider: A white paper*, BNL Report BNL-68933-02/07-REV, Eds. A. Deshpande, R. Milner and R. Venugopalan. .
24. **BRAHMS** Collaboration, I. Arsene *et. al.* nucl-ex/0403005.
25. R. Baier, A. Kovner, and U. A. Wiedemann *Phys. Rev.* **D68** (2003) 054009, [hep-ph/0305265].
26. D. Kharzeev, Y. V. Kovchegov, and K. Tuchin *Phys. Rev.* **D68** (2003) 094013, [hep-ph/0307037].
27. J. Jalilian-Marian, Y. Nara, and R. Venugopalan *Phys. Lett.* **B577** (2003) 54–60, [nucl-th/0307022].

FROM LEADING HADRON SUPPRESSION TO JET QUENCHING AT RHIC AND LHC

URS ACHIM WIEDEMANN

Department of Physics,
CERN, Theory Division,
CH-1211 Geneva 23
Switzerland
E-mail: Urs.Wiedemann@cern.ch

Experiments at the Relativistic Heavy Ion Collider (RHIC) have established that leading high-p_T hadroproduction is suppressed in nucleus-nucleus collisions due to a final state effect. This suppression can be accounted for by models of partonic energy loss ("jet quenching" models). Here, we review the current status of theory and experiment. We emphasize that additional information about the properties of the produced dense QCD matter will become accessible with future measurements of jet-like particle correlations, jet multiplicity distributions and jet shapes.

1. Introduction

High-p_T partons produced in nucleus-nucleus collisions traverse several fm of dense QCD matter. The interaction with this matter was predicted to modify strongly the partonic fragmentation process, mainly due to medium-induced gluon radiation (often referred to as "parton energy loss") [1,2,3,4,5,6]. Motivations for studying this process include i) to use high-p_T partons as a *hard probe* whose modification allows to characterize the properties of the produced dense matter such as its density and its collective flow, ii) to determine the dynamic mechanisms of partonic QCD equilibration processes by studying how and on what time scale high-p_T partons evolve towards thermal equilibrium in dense matter and iii) to get novel access to the dynamics of hadronization by using the medium as a tool to interfere with the hadronization process. In accordance with model predictions based on parton energy loss, experiments at the Relativistic Heavy Ion Collider RHIC have observed that high-p_T leading hadron spectra in Au+Au collisions at $\sqrt{s_{NN}} = 130$ and 200 GeV are strongly suppressed by a factor 4–5 compared to the yield of nucleon-nucleon collisions scaled by the number of equivalent binary scatterings [7,8,9,10,11,12]. However, leading hadron spectra are only of limited use for addressing the main issues mentioned above, since they give only limited access to the fragmentation pattern of hard partons. In what follows, we review their use and we discuss how the measurement of jet-like particle correlations and jets at RHIC and LHC allows to overcome their limitations.

2. Theory of parton energy loss

To account for the medium-modification of high-p_T partons prior to hadronization, we must understand how the parton showers associated to hard partonic production processes interact with a dense medium. This is a difficult problem, which is not solved completely. All existing approaches [2,3,4,5,6] indicate that the leading parton loses additional energy owing to medium-induced gluon radiation and that the parton shower broadens in transverse momentum. Remarkably, only one property of the medium enters these calculations, namely the amount $\hat{q}$ of transverse momentum squared that the medium transfers to the parton per unit path-length. Momentum broadening is characterised by $\langle k_\perp^2 \rangle \sim \hat{q} L_{\rm med}$ [2], the additional radiated energy is determined by the characteristic energy scale $\omega_c = \frac{1}{2}\hat{q} L_{\rm med}^2$ [2], and medium effects regulate the additional gluon radiation in the infrared on a scale $\omega \sim \omega_c / \left(\hat{q} L_{\rm med}^3\right)^{2/3}$ [13]. Also, the expansion of the medium can be controlled in these calculations [14,15,16]. Since the transport coefficient is proportional to the density of scattering centres, it decreases in time, $\hat{q}(\tau)$. The resulting medium-induced radiation spectrum corresponds to the spectrum of an equivalent static medium in which all kinematic variables are rescaled by the linear line-averaged transport coefficient $\bar{\hat{q}} = \frac{2}{L_{\rm med}^2} \int_0^{L_{\rm med}} \tau\, \hat{q}(\tau)\, d\tau$ [16]. In practice, this dynamical scaling law allows a simplified data analysis in terms of static medium properties; a posteriori, one then translates the extracted static average transport coefficient into the realistic dynamical one.

In general, theory makes predictions about the suppression of leading hadron spectra due to partonic energy loss. Moreover, it specifies how this "lost" energy is redistributed in transverse phase space via associated secondary hadroproduction. For comparison to data, one assumes that the hadronization of the partonic fragments occurs outside the medium. Formation time arguments indicate that this assumption holds for partons with $p_T^{\rm parton} \geq 10$ GeV which corresponds to a transverse momentum of the leading hadron of $p_T \geq 7$ GeV [17]. The following discussion is limited to this kinematic range and will detail the implications of parton energy loss for measurements at RHIC and LHC.

3. Leading hadron spectra

Most comparisons of the parton energy loss formalism to data are limited to the leading hadron spectra and to leading back-to-back correlations. In this case, the reduced energy of the leading hadron directly traces the reduced energy of the leading parent parton. The *average* reduction is expected to depend mainly on the parameter combination $\hat{q}\, L^2_{\mathrm{med}}$. However, high-$p_T$ hadronic spectra show a well-known trigger bias: as a consequence of the steep p_T-dependence, their dominant contribution comes from those hadrons which carry much more than the average energy of a hadronic fragment. Thus, what matters is not the average parton energy loss but the smallest energy loss which a significant fraction of all partons can get away with. To this end, one calculates so-called quenching weights [13,18]: these are probability distributions $P(\Delta E, L, \overline{\hat{q}})$ of the energy fraction ΔE radiated from the hard parton in a medium specified by L and $\overline{\hat{q}}$. Current calculations of leading hadron spectra convolute the partonic cross section with these quenching weights and the parton fragmentation function, thus taking into account the trigger bias effect.

We now discuss to what extent leading hadron spectra allow to test predictions for the dependence of medium-effects on in-medium pathlength, matter density and the identity of the parton. The successes and limitations can be illustrated with the help of Fig. 1 which shows the nuclear modification factor

$$R_{AA}(p_T, y) = \frac{d^2N^{AA}/dp_T dy}{\langle T_{AA}\rangle_c d^2\sigma^{NN}/dp_T dy} \,. \tag{1}$$

Here, $\langle T_{AA}\rangle_c$ is the standard nuclear overlap function, calculated as the average in the measured centrality class. In the absence of nuclear or medium effects, $R_{AA} \equiv 1$. Nuclear modifications of the parton distribution functions are taken into account in the calculation [19]. They lead to a reduction of R_{AA} with increasing p_T since the relevant values of Bjorken x move from the anti-shadowing to the so-called EMC region where nuclear parton distribution functions decrease faster than proton ones with increasing x [20]. However, these pdf effects cannot account for the strength of the suppression. In contrast, final state parton energy loss can account for the degree of suppression. Two features are remarkable [21]: First, the calculation indicates an approximately p_T-independent suppression pattern to higher transverse momentum which are accessible in future experiments at

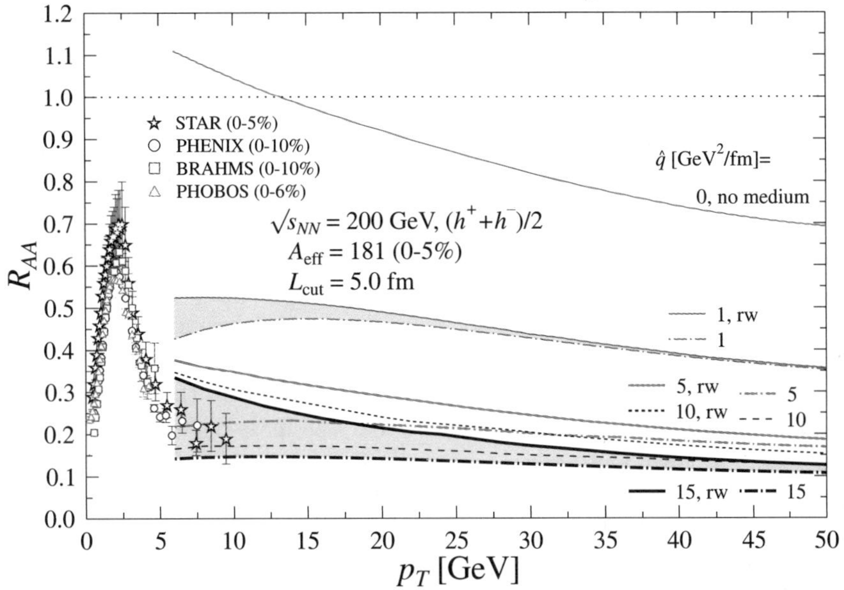

Figure 1. The nuclear modification factor R_{AA} for charged hadrons in the 0-5 % most central Au+Au collisions at $\sqrt{s_{NN}} = 200$ GeV for different values of the time-averaged transport coefficient $\hat{q}$. Differences between solid and dash-dotted curves indicate uncertainties related to finite energy corrections. Figure taken from [21].

RHIC. The reason for this p_T-independence is the slope of the partonic p_T-spectrum which at RHIC increases gradually with increasing p_T. This implies a trigger bias effect where for the same value of R_{AA}, partons at higher p_T have to lose a smaller fraction of their total energy.

Secondly, Fig. 1 shows that R_{AA} allows to establish a lower bound on the density of the produced matter but it is insensitive to the precise value of the transport coefficient if the latter is sufficiently high. The reason is that partons which traverse a larger in-medium pathlength tend to lose more energy. Hence, the dominant fraction of leading partons comes from the outer corona of the two-dimensional transverse overlap of the colliding nuclei, and emission from this low-density corona changes only weakly with increasing density [21,22,23].

It is an important consistency check that the current formalism can account for the centrality dependence of the nuclear modification factor [25,24]. However, information about the centrality dependence does not appear to increase the sensitivity to properties of the medium. In fact, on the basis of the measured centrality dependence, one cannot even discriminate between a linear and a quadratic dependences of parton energy loss on $L_{\rm med}$ [26]. Also, it remains unclear to what extent the dependence of parton energy loss on the parton identity can provide more precise knowledge about the medium. First, gluons are expected to lose more energy than quarks. However, to establish this effect experimentally requires to identify hadrons which receive significantly different contributions from the fragmentation of quarks and gluons (e.g. protons vs. antiprotons). Ideally, this should be done at sufficiently high transverse momentum ($p_T > 7$ GeV) where parton hadronization is expected to occur outside the medium - this requires improved particle identification. Second, the mass dependence of parton energy loss [27,28,29,30] is difficult to establish on the basis of open charm measurements. The problem is that the mass-dependent effects of the medium-dependent parton fragmentation decrease as a function of m/E and are thus expected to be negligible for charm transverse momentum above $p_T = 10$ GeV [28]. For smaller charm transverse momentum, however, the charm may be expected to fragment inside the medium since it moves slower than a massless quark [17]. For open beauty, the perspective is more promising, since mass effects persist at higher transverse momentum.

To sum up, the dependence of parton energy loss on parton identity is a qualitative prediction which remains to be established. While this is feasible in further measurements of leading hadron spectra at RHIC and LHC, the perspectives for gaining a more detailed characterization of properties of the medium appear to be limited. Thus, the main information gained from leading hadron spectra is the value of the time-averaged transport coefficient. So what do we learn from the value $\overline{\hat{q}} \simeq 10$ GeV^2/fm, favoured in Fig. 1 ? In general, the transport coefficient is known to scale with the energy density of the medium like [31]

$$\hat{q}(\tau) = c\, \epsilon^{3/4}(\tau)\,. \tag{2}$$

Here, c is a proportionality constant which can be calculated for specific models of the medium. For an ideal quark gluon plasma, its value is estimated as [31] $c_{\rm QGP}^{\rm ideal} \approx 2$. On the other hand, assuming an expanding medium with expansion parameter α, and $\hat{q}(\tau) = \hat{q}(\tau_0)\ (\tau_0/L)^\alpha$, one finds

$$c = \frac{\hat{q}(\tau_0)}{\epsilon(\tau_0)^{3/4}} = \frac{\overline{\hat{q}}}{\epsilon(\tau_0)^{3/4}}\frac{2-\alpha}{2}\left(\frac{L}{\tau_0}\right)^\alpha . \tag{3}$$

For a typical initial time $\tau_0 \sim 0.2$ fm/c, the average in-medium pathlength can be expected to be $L \sim 10\,\tau_0 = 2$ fm . The energy density averaged over a uniform transverse profile at initial time τ_0 can be as large as $\epsilon(\tau_0) \sim O(100)\frac{\mathrm{GeV}}{\mathrm{fm}^3}$, as suggested by the pQCD+saturation+hydrodynamics model [32]. This implies that realistic scenarios lie in the parameter range $\epsilon(\tau_0) < 100\,\frac{\mathrm{GeV}}{\mathrm{fm}^3}$, $L \sim 10\,\tau_0$, $0.75 < \alpha < 1.5$, for which we find

$$c > 8 \ldots 19\,. \tag{4}$$

Thus, for the entire realistic parameter range, we find a proportionality constant which is a factor $\sim 4-5$ larger than the perturbative estimate. We note that $\bar{\hat{q}}$ is the product of the density of scattering centers times their elastic scattering cross section [31]. Thus, the factor $\sim 4-5$ discrepancy observed here is reminiscent of the opacity problem in elliptic flow, where also scattering cross sections much larger than the perturbative ones are favoured by the data [33].

4. Medium-modification of jets

As discussed in section 2, theory predicts a one-to-one correspondence between the energy degradation of the leading parton (governed by the characteristic gluon energy $\omega_c = \frac{1}{2}\bar{\hat{q}}L^2$) and the transverse-momentum broadening of the parton shower [34,13] (governed by $\bar{\hat{q}}L$). To go beyond leading hadron spectra, one can study jet-like near-side particle correlations associated to high-p_T trigger particles [35,36] or particle distributions associated to trigger particles [37]. Experimentally more challenging is the study of calorimetrically reconstructed "true"jets, which becomes feasible for sufficiently large jet transverse energy ($E_\perp > 50$ GeV, say). In what follows, we focus mainly on the expected modifications of such jets [38,39]. The general problem is to determine how sensitive jet shape observables are to transverse-momentum broadening and how they can be characterized in a high-multiplicity background. One possibility to characterize the transverse phase space distribution of a jet is by measuring the fraction $\rho(R)$ of the total jet energy E_t deposited within a given jet subcone of radius $R = \sqrt{(\Delta\eta)^2 + (\Delta\Phi)^2}$,

$$\rho_{\mathrm{vac}}(R) = \frac{1}{N_{\mathrm{jets}}} \sum_{\mathrm{jets}} \frac{E_t(R)}{E_t(R=1)}\,. \tag{5}$$

Our calculation of the medium-induced broadening of this jet shape resulted in relative small effects [38]. We found a moderately reduced average jet energy fraction inside small jet cones $R = 0.3$ by ~ 5 % for $E_t = 50$ GeV and ~ 3 % for $E_t = 100$ GeV. Despite model-dependent uncertainties of this analysis there is a generic kinematic argument of why medium-modifications of the energy distribution can be expected to remain small:

the most energetic components in the jet are rather collinear and their broadening in azimuthal angle is small even if they receive significant momentum transfer from the medium. In nucleus-nucleus collisions at LHC, the background $E_t^{\rm bg}$ deposited inside the corresponding jet cone (for event multiplicity $dN^{\rm ch}/dy = 2500$, we estimate $E_t^{\rm bg} \sim 100$ GeV for $R = 0.3$ and $E_t^{\rm bg} \sim 250$ GeV for $R = 0.5$) is large. So, it will be possible to disentangle a high-E_t jet from background, but the measurement of a 10 % modification of its shape is expected to be challenging.

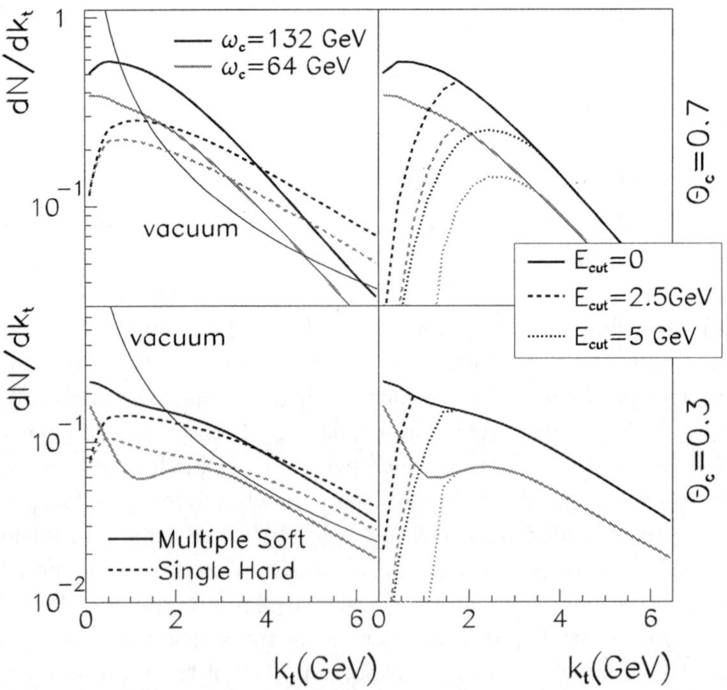

Figure 2. The gluon multiplicity distribution inside a jet cone size $R = \Theta_c$, measured as a function of k_t with respect to the jet axis. Removing gluons with energy smaller than $E_{\rm cut}$ from the distribution (dashed and dotted lines) does not affect the high-k_t tails. Figure taken from [38].

However, if the jet energy distribution does not change significantly while the energy of leading hadrons decreases strongly, then this implies a significant change in the jet multiplicity distributions. Interestingly, one can identify multiplicity measurements which are almost insensitive to the high multiplicity background in nucleus-nucleus collsions. In Fig. 2, we show the medium-induced additional number of gluons with transverse momentum $k_t = |\mathbf{k}|$, produced within a subcone of opening angle θ_c,

$$\frac{dN_{\rm med}}{dk_t} = \int_{k_t/\sin\theta_c}^{E} d\omega\, \frac{dI_{\rm med}}{d\omega\, dk_t}\,. \tag{6}$$

This distribution is compared to the shape of the corresponding vacuum component. The total partonic jet multiplicity is the sum of both components. For realistic values of medium density and in-medium pathlength, medium effects are seen to increase this multiplicity significantly (by a factor > 2) in particular in the high-k_t tails. Also, the shape and width of the distribution (6) changes sensitively with the scattering properties of the medium. Moreover, since gluons must have a minimal energy $\omega > k_t/\sin\Theta_c$ to be emitted inside the jet cone, this high-k_t tail is unaffected by "background" cuts on the soft part of the spectrum, see Fig. 2. These qualitative conclusions are expected to remain unaffected by the uncertainties of our calculation and the effects of subsequent hadronization. The insensitivity of the high-k_t tail to the low E_t background and its sensitivity to the transverse momentum picked up from the medium are both based on kinematic grounds and should not depend on the details of our calculation. On the basis of Fig. 2, we argue that the measurement of the transverse momentum distribution of hadrons with respect to the jet axis is very sensitive to the transverse momentum broadening of the underlying parton shower and should be detectable above background. This allows to test the predicted relation between parton energy loss and transverse momentum broadening.

Jets are not only sensitive to the time-averaged energy density produced in a nucleus-nucleus collision, but also to the collective flow field in the collision region. In Fig. 3 we show a calculation which establishes this effect. Both calorimetric measurements and leading near-side particle correlations can be used to characterize the jet asymmetry introduced by collective flow effects.

In conclusion, the study of jet observables and jet-like particle correlations in heavy ion collisions is still at the very beginning. However, first calculations indicate that they provide novel tests of the partonic energy loss formalism assumed to underly the suppressed high-p_T hadroproduction at RHIC. Moreover, they provide novel information (such as a novel measurement of the collective flow field) about the produced dense QCD matter.

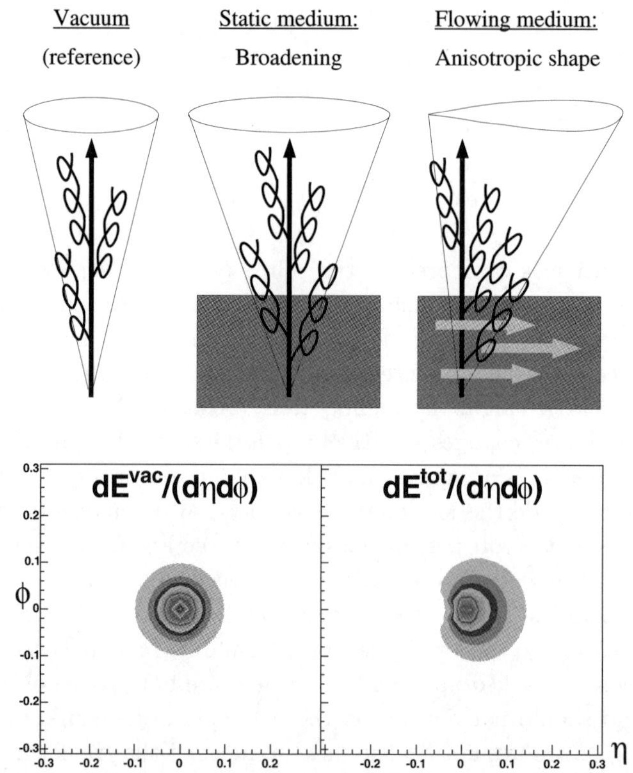

Figure 3. Upper part: sketch of the distortion of the jet energy distribution in the presence of a medium with or without collective flow. Lower part: calculated distortion of the jet energy distribution in the $\eta \times \phi$-plane for a 100 GeV jet. The right hand-side is for an average medium-induced radiated energy of 23 GeV and equal contributions from density and flow effects. Figures taken from [39].

References

1. M. Gyulassy and X. N. Wang, Nucl. Phys. B **420** (1994) 583.
2. R. Baier, Y. L. Dokshitzer, A. H. Mueller, S. Peigne and D. Schiff, Nucl. Phys. B **484** (1997) 265.
3. B. G. Zakharov, JETP Lett. **65** (1997) 615.
4. U. A. Wiedemann, Nucl. Phys. B **588** (2000) 303.
5. M. Gyulassy, P. Levai and I. Vitev, Nucl. Phys. B **594** (2001) 371.
6. X. N. Wang and X. f. Guo, Nucl. Phys. A **696** (2001) 788.
7. K. Adcox *et al.* [PHENIX Collaboration], Phys. Rev. Lett. **88** (2002) 022301.
8. S. S. Adler *et al.* [PHENIX Collaboration], Phys. Rev. C **69** (2004) 034910.
9. C. Adler *et al.* [STAR Collaboration], Phys. Rev. Lett. **89** (2002) 202301.
10. J. Adams *et al.* [STAR Collaboration], Phys. Rev. Lett. **91** (2003) 172302.

11. B. B. Back *et al.* [PHOBOS Collaboration], Phys. Lett. B **578** (2004) 297.
12. I. Arsene *et al.* [BRAHMS Collaboration], Phys. Rev. Lett. **91** (2003) 072305.
13. C. A. Salgado and U. A. Wiedemann, Phys. Rev. D **68** (2003) 014008.
14. R. Baier, Y. L. Dokshitzer, A. H. Mueller and D. Schiff, Phys. Rev. C **58** (1998) 1706.
15. M. Gyulassy, I. Vitev and X. N. Wang, Phys. Rev. Lett. **86** (2001) 2537.
16. C. A. Salgado and U. A. Wiedemann, Phys. Rev. Lett. **89** (2002) 092303.
17. U. A. Wiedemann, J. Phys. G **30** (2004) S649.
18. R. Baier, Y. L. Dokshitzer, A. H. Mueller and D. Schiff, JHEP **0109** (2001) 033 [arXiv:hep-ph/0106347].
19. K. J. Eskola, V. J. Kolhinen and C. A. Salgado, Eur. Phys. J. C **9** (1999) 61.
20. K. J. Eskola and H. Honkanen, Nucl. Phys. A **713** (2003) 167.
21. K. J. Eskola, H. Honkanen, C. A. Salgado and U. A. Wiedemann, arXiv:hep-ph/0406319.
22. B. Muller, Phys. Rev. C **67** (2003) 061901.
23. A. Dainese, C. Loizides and G. Paic, arXiv:hep-ph/0406201.
24. X. N. Wang, arXiv:nucl-th/0305010.
25. X. N. Wang, Phys. Lett. B **579** (2004) 299.
26. A. Drees, H. Feng and J. Jia, arXiv:nucl-th/0310044.
27. Y. L. Dokshitzer and D. E. Kharzeev, Phys. Lett. B **519** (2001) 199.
28. N. Armesto, C. A. Salgado and U. A. Wiedemann, Phys. Rev. D **69** (2004) 114003.
29. M. Djordjevic and M. Gyulassy, arXiv:nucl-th/0310076.
30. B. W. Zhang, E. Wang and X. N. Wang, arXiv:nucl-th/0309040.
31. R. Baier, Nucl. Phys. A **715** (2003) 209 [arXiv:hep-ph/0209038].
32. K. J. Eskola, K. Kajantie, P. V. Ruuskanen and K. Tuominen, Nucl. Phys. B **570** (2000) 379.
33. D. Molnar and M. Gyulassy, Nucl. Phys. A **697** (2002) 495 [Erratum-ibid. A **703** (2002) 893].
34. U. A. Wiedemann, Nucl. Phys. A **690** (2001) 731.
35. S. Pal and S. Pratt, Phys. Lett. B **574** (2003) 21.
36. A. Majumder and X. N. Wang, Phys. Rev. D **70** (2004) 014007.
37. F. Wang [STAR Collaboration], arXiv:nucl-ex/0404010.
38. C. A. Salgado and U. A. Wiedemann, arXiv:hep-ph/0310079.
39. N. Armesto, C. A. Salgado and U. A. Wiedemann, arXiv:hep-ph/0405301.

LATTICE SIMULATIONS WITH CHEMICAL POTENTIAL

C. SCHMIDT

University of Wuppertal, Department of Physics,
Gaussstrasse 20, D-42119 Wuppertal, Germany
E-mail: cschmidt@theorie.physik.uni-wuppertal.de

After giving an overview of recently invented methods for simulating lattice QCD at small μ/T, we discuss some results for bulk thermodynamic quantities of QCD matter coming from those methods. We focus on the transition line and the critical endpoint in the QCD phase diagram.

1. Introduction

Having a profound and quantitative knowledge of the QCD phase diagram and equation of state is of great interest for meany different topics: A few micro seconds after the big bang, the early universe went through the QCD phase transition; Every heigh energy experiment has a certain trajectory in the phase diagram, which may also cross the transition line; Finally the cores of cold neutron and quark stars may be in a color superconducting phase, also part of the QCD phase diagram.

During the last two decades lattice simulations have proven to be the only quantitative approach to the QCD phase transition. At vanishing baryon chemical potential ($\mu_B = 0$), lattice simulations provided detailed informations of several bulk thermodynamic quantities, for instance the transition temperature ($T_c \approx 173$ MeV) and critical energy density ($\varepsilon_c \approx 0.7$ GeV/fm^3)[1]. The sign problem, however, has disabled simulations at non-zero chemical potential. For $\mu > 0$ the determinant of the fermion matrix (detM) becomes complex. Standard Monte Carlo techniques using importance sampling are thus no longer applicable when calculating observables in the grand canonical ensemble according to the partition function

$$Z(\mu) = \int \mathcal{D}U \, \mathrm{detM}(\mu) \exp\{-\mathrm{S_G}\}. \tag{1}$$

Note that all physical observables are real and therefore also the imaginary part of $\mathrm{detM} \equiv |\mathrm{detM}| \exp\{i\theta\}$ vanishes in average. What influences the physics and courses the problems are the fluctuations of the complex phase θ. Several new techniques have been developed to overcome this problem in the region of small μ/T, where the fluctuations are rather mild. For a recent overview see for instance Ref. 2.

2. Methods for small μ/T

2.1. *The Taylor expansion method*

One conceptionally very simple idea is to calculate observables $\langle O\rangle$ at $\mu > 0$ from a Taylor expansion around $\mu = 0$:

$$O(\hat{\mu}) = c_0 + c_1\hat{\mu} + \frac{1}{2}c_2\hat{\mu}^2 + \cdots \tag{2}$$

Since on the lattice all quantities are given in units of the lattice spacing (a), the expansion parameter is $\hat{\mu} = a\mu = N_t^{-1}(\mu/T)$. Here N_t is the number of lattice points in the temporal direction. The idea goes back to the calculation of the quark number susceptibilities in Ref. 3. By this method the response of meson masses[4] as well as the pressure and further bulk thermodynamic quantities[5,6] have been studied. Besides derivatives of the observable itself, the calculation of derivatives of ln detM with respect to $\hat{\mu}$ are required. the first two nontrivial coefficients in Eq. (2) are given by

$$c_1 = \left\langle \frac{\partial O}{\partial \hat{\mu}} \right\rangle + \left\langle O \frac{\partial \ln \mathrm{detM}}{\partial \hat{\mu}} \right\rangle \tag{3}$$

$$c_2 = \left\langle \frac{\partial^2 O}{\partial \hat{\mu}^2} \right\rangle + 2\left\langle \frac{\partial O}{\partial \hat{\mu}} \frac{\partial \ln \mathrm{detM}}{\partial \hat{\mu}} \right\rangle + \left\langle O \frac{\partial^2 \ln \mathrm{detM}}{\partial \hat{\mu}^2} \right\rangle - \left\langle O \right\rangle \left\langle \frac{\partial^2 \ln \mathrm{detM}}{\partial \hat{\mu}^2} \right\rangle .$$

The derivatives have to be taken at $\hat{\mu}_0 = 0$. Note that due to a symmetry of the partition function $(Z(\mu) = Z(-\mu))$ all odd coefficients in Eq. (2) vanish identically. For the same reason we have $\langle \partial \ln \mathrm{detM}/\partial \hat{\mu} \rangle = 0$ at $\hat{\mu}_0 = 0$. We explicitly use this property in Eq. (3) to derive the coefficients. The advantages of this method are:

(i) Expectations values only have to be evaluated at $\hat{\mu} = 0$ (no sign problem).

(ii) All derivatives of the fermion determinant can be expressed in terms of traces by using the identity $\ln \mathrm{detM} = \mathrm{Tr} \ln \mathrm{M}$. This enables the stochastic calculation of the coefficients by the random noise method, which is much faster than the evaluation of the determinant.

(iii) continuum and infinite volume extrapolations are well defined on a coefficient by coefficient basis.

On the other hand it is not a priori clear for how large μ/T the method works and how large the truncation errors are. Furthermore one is strictly limited by phase transitions, since phase transitions are connected with discontinuities or divergencies. An estimation of the convergence radius of the series gives a lower bound on the applicability range and thus also a lower bound to the phase transition line in the (T, μ) plane.

2.2. *The reweighting method*

In principle one can also calculate the exact expectation value $\langle O \rangle (\hat{\mu})$ from an ensemble generated at $\hat{\mu} = 0$. This is possible by giving every configuration an additional weight according to the reweighting factor $R(\hat{\mu}) = \text{detM}(\hat{\mu})/\text{detM}(0)$. We have the identity

$$\langle O \rangle (\hat{\mu}) = \langle O(\hat{\mu}) R(\hat{\mu}) \rangle \, / \, \langle R(\hat{\mu}) \rangle \,, \tag{4}$$

where the expectation values on the right hand side are to be evaluated at $\hat{\mu} = 0$. This method goes back to Ref. 7. For large reweighting distances, however, the generated ensemble at $\hat{\mu} = 0$ shares less and less configurations with the target ensemble. This is the so called overlap problem. The overlap can be improved by the multi-parameter reweighting technique[8], where the reweighting factor is a function of $\hat{\mu}$ and the lattice coupling β. In the $(\beta, \hat{\mu})$ plane one can reweight along lines where the overlap measure[9] (α) is maximal. One of those lines is the transition line. In Ref. 9 the generated ensemble at $(\beta_0, 0)$ was defined to have the overlap α with the ensemble at $(\beta, \hat{\mu})$, if a fraction $\alpha/2$ of the ensemble gives the weight $1 - \alpha/2$ to the expectation value at $(\beta, \hat{\mu})$. In Fig. 1(a) the volume dependence of the overlap measure along the transition line is shown. In Ref. 9 it was empirically found, that the half width $\hat{\mu}_{1/2}$ of α scales with $\hat{\mu}_{1/2} \propto V^{-\gamma}$, with $\gamma \approx 1/3$.

To reduce the numerical effort, which is connected with the calculation of the reweighting factor (R), one can expand R in a Taylor series around $\hat{\mu}_0 = 0$ as it was done in Ref. 10. In lowest order we have

$$\ln R = \text{TrM}^{-1} \left. \frac{\partial \text{M}}{\partial \hat{\mu}} \right|_{\hat{\mu}_0 = 0} \times \hat{\mu} + \mathcal{O}(\hat{\mu}^2) \tag{5}$$

In this expansion all odd terms are purely imaginary, whereas all even terms are real[10]. The complex phase of the determinant is thus given by

$$\theta = \bar{\theta} + \mathcal{O}(\hat{\mu}^3), \qquad \text{with} \qquad \bar{\theta} = \text{ImTrM}^{-1} \left. \frac{\partial \text{M}}{\partial \hat{\mu}} \right|_{\hat{\mu}_0 = 0} \times \hat{\mu}. \tag{6}$$

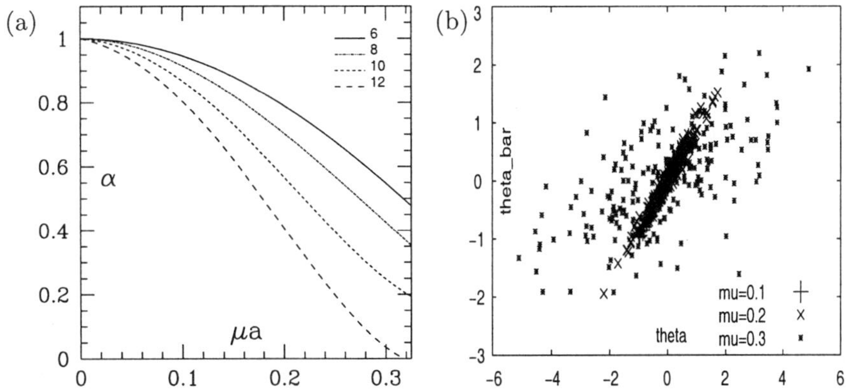

Figure 1. (a) The overlap measure along the transition line for different volumes[9]. Shown lattice sizes are $6^3 \times 4$, $8^3 \times 4$, $10^3 \times 4$ and $12^3 \times 4$ (from top to bottom). (b) The approximated phase $\bar{\theta}$ vs. the exact phase θ (Ref. 11).

Again the problem of the determinant calculations is reduced to calculations of traces. In Ref. 11 it was empirically shown that θ and $\bar{\theta}$ coincide quite well for $\hat{\mu}\lesssim 0.3$ (see Fig. 1(b)). In the region $\hat{\mu}\gtrsim 0.3$, where one also faces the overlap problem, the lowest order expansion seems to break down.

2.3. *Analytical continuations*

At imaginary chemical potentials, the fermion determinant is real and positive, thus simulations by standard Monte Carlo techniques are possible. Results on the imaginary $\hat{\mu}_I$ axis can be analytically continued to the real $\hat{\mu}_R$ axis. It is especially easy to convert a Taylor series in $\hat{\mu}_I$, expanded around $\hat{\mu}_0 = 0$, into a Taylor series in $\hat{\mu}_R$. Since the series has only even powers of $\hat{\mu}$, due to the the symmetry $Z(\hat{\mu}) = Z(-\hat{\mu})$, one only has to switch the sign of every second coefficient ($c_2 \to -c_2, c_6 \to -c_6, \ldots$). There is however another symmetry of the partition function which limits the analytic continuation. Due to the periodicity[12] $Z(\mu_R, \mu_I) = Z(\mu_R, \mu_I + 2\pi T/3)$ simulations with $\mu_I > 0$ will only give excess to the physical region $\mu_R \lesssim \pi T/3$. This corresponds to $\mu_B \lesssim 500$ MeV. This method was also used to map out the phase transition line[13–15]. One should note, that for this method neither an evaluation of the determinant, nor any of its derivatives are required. In order to determine a Taylor series of an observable one needs however many different simulation points for several values of $\hat{\mu}_I$, to interpolate by a certain Ansatz. In future, one should think of combining methods 2.1 and 2.3 to get a better extrapolation along the $\hat{\mu}_R$ axis.

2.4. *The density of states method*

An alternative to the importance sampling technique used in most Monte Carlo simulations is the density of states method. Here one reorders the path integral of the partition function in the following way: first expectation values with a constrained parameter will be calculated. *I.e.* one exposed parameter (ϕ) is fixed. Expectation values according to the usual grand canonical partition function (Z_{GC}) can then be recovered by the integral

$$< O >= \int d\phi \, \langle O f(U) \rangle_\phi \, \rho(\phi) \Big/ \int d\phi \, \langle f(U) \rangle_\phi \, \rho(\phi) \tag{7}$$

where the density of states (ρ) is given by the constrained partition function:

$$\rho(x) \equiv Z_\phi(x) = \int \mathcal{D}U \, g(U) \, \delta(\phi - x). \tag{8}$$

With $\langle \, \rangle_\phi$ we denote the expectation value with respect to the constrained partition function. In addition, the product of the weight functions f, g has to equal the correct measure of Z_{GC}: $fg = \mathrm{detM} \exp\{-\mathrm{S_G}\}$. This idea of reordering the partition functions is rather old and was used for gauge theories[16] and QED with dynamical fermions[17]. For QCD the parameter ϕ is normally set to be the plaquette[18]: $\phi = P$. In Ref. 19, however, the DOS method was constructed for the complex phase ($\phi = \theta$). Within the random matrix model, the authors of Ref. 20 used the quark number density ($\phi = n_q$).

The advantages of this additional integration becomes clear, when choosing $\phi = P$ and $g(U) = 1$. In this case $\rho(\phi)$ is independent of all simulation parameters. The observable can be calculated as a function of all values of the lattice coupling β. If one has stored all eigenvalues of the fermion matrix for all configurations, the observable can also be calculated as a function of quark mass (m) and number of flavors[18] (N_f).

Note that this method does not solve the sign problem. It is, however supposed to solve the overlap problem. Nevertheless, it is also possible to combine the DOS method, with the reweighting method 2.2, by reweighting the constrained expectation values in the case of $g(U) \neq 1$. For large reweighting distances an overlap problem is introduced then once again.

3. The critical temperature

On the lattice, the critical temperature ($T_c(\mu)$) can be calculated from the critical coupling ($\beta_c(\hat{\mu})$). We define the function $\beta_c(\hat{\mu})$ as the line in the ($\beta, \hat{\mu}$) plane, where susceptibilities peak. Together with the lattice

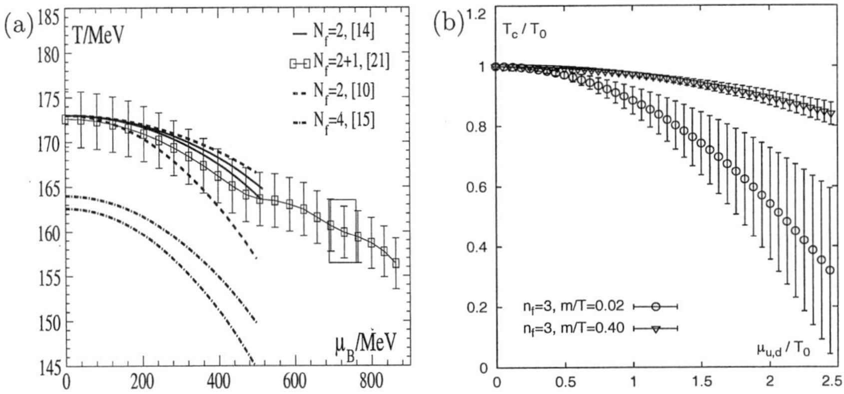

Figure 2. (a) The transition line $T_c(\mu)$ for diferent number of flavors and masses in the range $m/T = 0.1 - 0.4$, for the light quark mass. (b) The mass dependence of the transition line, for $N_f = 3$ (Ref. 23).

beta function ($\beta(a)$) one can calculate Taylor coefficients of the critical temperature, for the first nontrivial coefficient we have

$$\frac{\mathrm{dT_c}}{\mathrm{d}(\mu^2)} = -\frac{1}{N_t^2 T_c(0)} \frac{\partial \beta_c(\hat{\mu})}{\partial(\hat{\mu}^2)} \left(a \frac{\partial \beta}{\partial a} \right)^{-1} . \tag{9}$$

This quantity was calculated by several groups[10,13–15,21,22], with methods 2.1-2.3. Although different quark masses in the range of $m/T = 0.1 - 0.4$, lattice sizes and actions has been used, the results are in good agreement (see Fig. 2(a)). In first order, the transition line for $N_f = 2$ and $m/T = 0.4$ is given by[10]

$$T_c(\mu_q)/T_c(0) = 1 - 0.070(35)\left(\mu_q/T_c(0)\right)^2 . \tag{10}$$

As one can anticipate from Fig. 2(a), the mass dependence of $T_c(\mu)$ is rather small. Comparing $N_f = 3$ results[23] (Fig. 2(b)) for $m/T = 0.4$ and $m/T = 0.02$ one finds however a significant mass dependents. As expected the curvature of the transition line becomes stronger with increasing mass. Using the perturbative asymptotic β-function in Eq. 9 gives

$$\frac{T_c(\mu_q)}{T_c(0)} = \begin{cases} 1 - 0.025(6)\left(\frac{\mu_q}{T_c(0)}\right)^2, \; m/T = 0.4 \\ 1 - 0.114(46)\left(\frac{\mu_q}{T_c(0)}\right)^2, \; m/T = 0.02 \end{cases} . \tag{11}$$

We note that an additional quark mass dependence is hidden here in the transition temperature at $\mu_q = 0$, *i.e.* $T_c(0)$, which is used to normalize the transition temperature at $\mu_q \neq 0$. Furthermore, taking into account

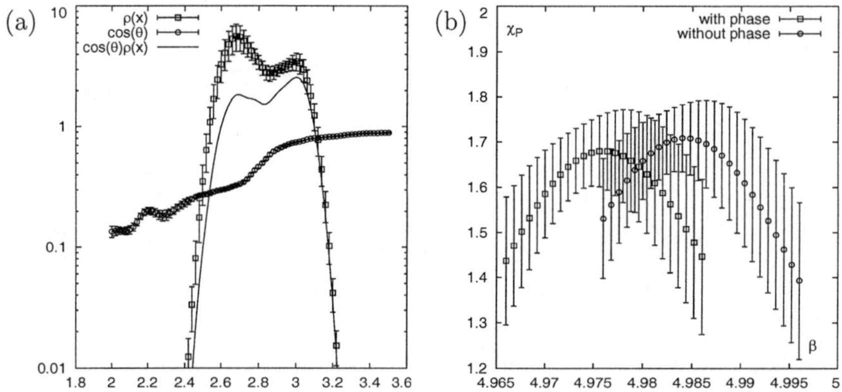

Figure 3. (a) The density of states ρ, the phase factor $\langle\cos\theta\rangle$ and their product [25]. (b) Susceptibilites calculated for the DOS method, with and without the phase factor[25].

violations of asymptotic scaling in the β-function will lead to a further increase of the curvature of $T_c(\mu_q)$.

This increase of the curvature can be understood as a result of much larger phase fluctuations for the smaller quark mass. A detailed analysis of the phase fluctuations can be found in Ref. 24. To make this more clear we have also calculated the transition line as a function of a vector chemical potential ($\mu_V \equiv \mu_u = -\mu_d$). This corresponds to reweighting with the absolute value of the determinant, *i.e.* the influence of the phase is dropped. In this case the mass dependence is much less significant:

$$\frac{T_c(\mu_V)}{T_c(0)} = \begin{cases} 1 - 0.026(5)\left(\frac{\mu_V}{T_c(0)}\right)^2, & m/T = 0.4 \\ 1 - 0.033(6)\left(\frac{\mu_V}{T_c(0)}\right)^2, & m/T = 0.02 \end{cases} . \tag{12}$$

Since for the larger quark mass the two transition lines are in good agreement, we have $T_c(\mu_V) > T_c(\mu_q)$ for $m/T = 0.02$. This result seems to be a bit counter intuitive, since we know that the onset transition of isospin matter at $T = 0$ takes place at $\mu_V = m_\pi/2$, rather than at $\mu_q = m_B/3$. However, this result can also be confirmed by the DOS method. In Fig. 3 we presents results[25] from a 4^4 lattice and $N_f = 4$. Simulations have been performed at finite isospin density with $\hat{\mu}_V = 0.3$, according to Ref. 26. The DOS (ρ) was defined according to Eq. 8 with $g = |\mathrm{detM}|\exp\{-\mathrm{S_G}\}$ and $\phi = P$. In Fig. 3(a) we show ρ, the phase suppression factor $\langle cos(\theta)\rangle$ and their product as a function of the constrained Plaquette P. We find, that the phase fluctuations in the low temperature phase ($P \lesssim 2.8$) are much larger than in the high temperature phase ($P \gtrsim 2.8$). From this we can argue, that including the phase factor, the critical temperature will be lower than without the phase factor. Indeed, after calculating the susceptibilities from those distributions as a function of β, we find a shift in β_c towards lower temperatures (see Fig. 3(b)). Moreover the value of β_c is in good agreement with an earlier result form the reweighting method[8a].

4. The critical end-point

The determination of the critical end-point on the transition line, separating a first order region form a crossover region is a difficult task. At present there are two different strategies used for this calculation. An infinite volume extrapolation of Lee-Yang Zeros[21,22] and a finite size scaling analysis of susceptibilities and Binder Cumulants[27,28,13]. At this stage the most reliable calculation for the critical endpoint and physical quark masses comes form Ref. 22. The result is $\mu_B = 360(40)$ MeV and $T = 162(2)$ MeV. Note that those values are, however not yet free from corrections coming from the finite volume and lattice spacing.

In the case of 3 degenerate quarks the critical mass ($m_c(0)$), where the critical point sits at $\mu = 0$, was determined. In this quantity huge cut-off effects were found. It was determined for standard staggered fermions[27] and $p4$ improved fermions[23], both on $N_t = 4$ lattices. The outcomes for the critical pseudo scalar mass are $m_\pi^{\rm crit} = 190(20)$ MeV, $67(18)$ MeV, respectively. Binder Cumulant method also admits the determination of the universality class of this critical point, which was found to be that of a 3d-Ising model[27].

In addition the mass dependence of the critical point turned out to be rather large. A correct calculation of the critical point may thus only be possible, if the quark masses are known very precisely. The first Taylor coefficient of $m_c(\hat{\mu})$ was calculated in Ref. 13. The result of a combined fit for several lattice sizes and imaginary chemical potentials is

$$\frac{m_c(\mu)}{m_c(\mu=0)} = 1 + 0.84(36)\left(\frac{\mu}{\pi T}\right)^2, \tag{13}$$

where $m_c(\mu)$ is the critical quark mass. This coefficient does not look unnaturally small, since it was expanded in the natural units ($\mu/\pi T$), it will, however give in turn a rather large first coefficient for the critical chemical potential as a function of the quark mass ($\mu_c(m_q)$).

[a] Agreement was found after a linear extrapolation $\lambda \to 0$, with λ being a small parameter in front of an isospin symmetry braking term as defined in Ref. 26

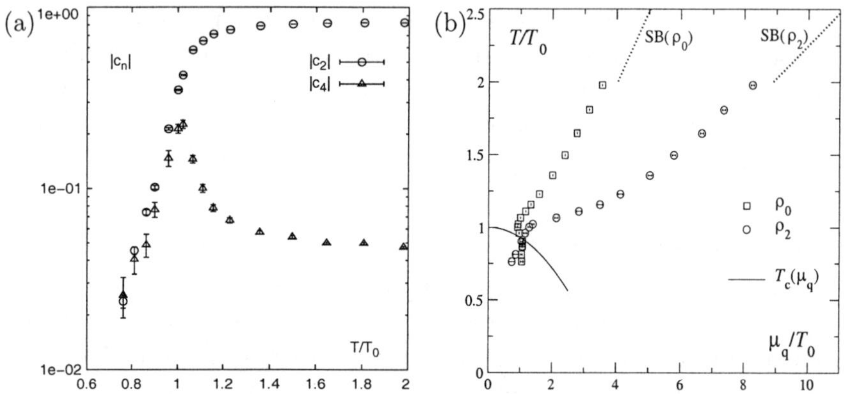

Figure 4. (a) The absolute values of the Taylor coefficients of the pressure[6]. (b) The convergence radius of the pressure[6].

Another interesting method to extract information about the critical end-point is the convergence radius of any bulk thermodynamic observable at $\mu = 0$. As mentioned in Sec. 2.1, the Taylor expansion is limited by a phase transition. Thus for any $T < T_{\text{endpoint}} \equiv T_E$ the convergence radius (ρ) should remain finite, whereas for $T > T_E$ it should be infinite. Since physical observables are even in $\hat{\mu}$, we have

$$\rho = \lim_{n\to\infty} \rho_n \equiv \lim_{n\to\infty} \sqrt{|c_n/c_{n+2}|}. \tag{14}$$

In Fig. 4(a) we show the absolute value of the Taylor coefficients of the pressure[6]. The results are for $N_f = 2$ and $m/T = 0.4$. In Fig. 4(b) the first two terms in the series (Eq. 14) are drawn in a (T, μ)-diagram, in order to compare with the phase transition line. As one can see in Fig. 4, for values of $T/T_c(0) \gtrsim 0.8$ the radius seems to increase. This corresponds to a critical chemical potential of $\mu_E \approx (3 - 3.6)T_0 \approx (600 - 700)$MeV. This is a reasonably large value for $m_q/T = 0.4$.

Note that the pressure and quark number density, which can be constructed from the coefficients in Fig. 4(a) is in good agreement with result from method 2.2 (Ref. 9). Moreover the pressure is like the transition line, almost perfectly described by the leading order in μ/T. For a further discussion of the pressure from Taylor expansion see Ref. 29.

5. Conclusions

For small μ/T several methods have been proposed and tested. The agreement between different methods is good. The phase diagram for small μ/T

is slowly evolving and one can even address the question of locating the critical point. All methods are bases, however, on $\mu_R = 0$ simulations and thus extrapolate. Moreover they give no solution for the sign problem and the results are not yet extrapolated to the continuum and infinite volume.

Acknowledgments

The author would like to thank Z. Fodor, S. Katz and all members of the Bielefeld-Swansea Collaboration for helpful discussions and comments.

References

1. F. Karsch, E. Laermann and A. Peikert, *Nucl. Phys.* **B605**, 579 (2001).
2. S. Muroya *et al.*, *Prog. Theor. Phys.* **110** 615 (2003).
3. S. A. Gottlieb *et al.*, *Phys. Rev.* **D38**, 2888 (1988).
4. S. Choe *et al.* [QCD-TARO Collaboration], *Nucl. Phys. Proc. Suppl.* **106**, 462 (2002); *Phys. Rev.* **D65**, 054501 (2002); *Nucl. Phys.* **A698**, 395 (2002).
5. R. V. Gavai and S. Gupta, *Phys. Rev.* **D68**, 034506 (2003).
6. C. R. Allton *et al.*, *Phys. Rev.* **D68**, 014507 (2003).
7. A. M. Ferrenberg and R. H. Swendsen, *Phys. Rev. Lett.* **61**, 2635 (1988); *Phys. Rev. Lett.* **63** 1195 (1989).
8. Z. Fodor and S. D. Katz, *Phys. Lett.* **B534** 87 (2002).
9. F. Csikor *et al.*, *JHEP* **0405** 046 (2004).
10. C. R. Allton *et al.*, *Phys. Rev.* **D66** 074507 (2002).
11. P. de Forcrand *et al.*, *Nucl. Phys. Proc. Suppl.* **119** 541 (2003).
12. A. Roberge and N. Weiss, *Nucl. Phys.* **B275** 734 (1986).
13. P. de Forcrand and O. Philipsen, *Nucl. Phys.* **B673** 170 (2003);
14. P. de Forcrand and O. Philipsen, *Nucl. Phys.* **B642** 290 (2002).
15. M. D'Elia and M. P. Lombardo, *Phys. Rev.* **D67** 014505 (2003).
16. G. Bhanot *et at.*, *Phys. Lett.* **B187** 381 (1987); *Phys. Lett.* **B188** 246 (1987); M.Karliner *et al.*, *Nucl. Phys.* **B302** 204 (1988).
17. V. Azcoiti, G. di Carlo and A. F. Grillo, *Phys. Rev. Lett.* **65** 2239 (1990).
18. X. Q. Luo, *Mod. Phys. Lett.* **A16** 1615 (2001).
19. A. Gocksch, *Phys. Rev. Lett.* **61** 2054 (1988).
20. J. Ambjorn *et al.*, *JHEP* **0210** 062 (2002).
21. Z. Fodor and S. D. Katz, *JHEP* **0203** 014 (2002).
22. Z. Fodor and S. D. Katz, *JHEP* **0404** 050 (2004).
23. F. Karsch *et al.*, *Nucl. Phys. Proc. Suppl.* **129** 614 (2004).
24. S. Ejiri, *Phys. Rev.* **D69** 094506 (2004).
25. Z. Fodor, S. Katz and C. Schmidt, in preparation.
26. J. B. Kogut and D. K. Sinclair, *Phys. Rev.* **D66** 014508 (2002); *Phys. Rev.* **D66** 034505 (2002).
27. F. Karsch, E. Laermann and C. Schmidt, *Phys. Lett.* **B520** 41 (2001).
28. N. H. Christ and X. Liao, *Nucl. Phys. Proc. Suppl.* **119** 514 (2003).
29. S. Ejiri, this proceedings.

MESONIC CORRELATORS IN HOT QCD

M. LAINE

Faculty of Physics, University of Bielefeld,
D-33501 Bielefeld, Germany
E-mail: laine@physik.uni-bielefeld.de

Certain spacelike mesonic correlation lengths serve as interesting theoretical probes for the reliability of perturbation theory in high-temperature QCD, are directly sensitive to chiral symmetry restoration and to the axial anomaly, and might also have indirect phenomenological signatures in idealised heavy ion collision experiments. I review here the weak-coupling predictions for some of these correlation lengths, to be compared with results from lattice Monte Carlo simulations.

1. Introduction

At a finite temperature T, Lorentz symmetry is broken by the heat bath, so that temporal and spatial directions are, in general, unrelated. Correlation functions in the (Minkowski) time direction can be used to define, after a Fourier transform, spectral functions for various operators, and the spectral functions in turn determine fundamental "real-time" properties of the plasma, such as particle production rates.[1] These are then directly measurable in heavy ion collision experiments (modulo problems with background contamination, etc).

Correlation functions in spatial directions, in contrast, address questions such as: At which length scales are thermal fluctuations correlated? At which length scales are external charges screened? In principle these "static" observables are also physical and may lead to detectable signals, however in practice the relations are indirect and therefore weaker than for real-time observables. On the other hand, static quantities are eminently suited to measurements in *lattice experiments*. Lattice simulations will hence be the most immediate "phenomenological" reference point in the following.

There is naturally a vast variety of different operators that can be correlated. The operators can be classified according to their discrete and continuous global symmetry properties, leading to many independent correlation lengths. The analytic structures of various purely *gluonic* correlators were discussed in Ref. 2, and the corresponding lattice measurements are rather precise by now.[3–8] The purpose of this talk is to analyse the structures of *mesonic* correlators, for which the status has been somewhat less advanced. This talk is based on the original study in Ref. 9.

2. Detailed setup

As just mentioned, we will focus on correlation lengths ξ of mesonic observables, that is operators of the type $\mathcal{O} = \bar\psi \Gamma F^a \psi$, where

$$\Gamma = \{1, \gamma_5, \gamma_\mu, \gamma_\mu\gamma_5\} \,, \tag{1}$$

$$F^a = \{F^s, F^n\}, \; F^s \equiv \mathbb{I}_{N_\mathrm{f}\times N_\mathrm{f}}, \; F^n = \text{traceless } N_\mathrm{f} \times N_\mathrm{f} \text{ matrix} \,, \tag{2}$$

N_f is the number of quark flavours, and all colour, spinor and flavour indices have been suppressed. We may recall that at least some among these operators have direct physical significance, for instance

$$\bar\psi\gamma_5 F^s \psi \propto \eta'\text{-meson} \,, \tag{3}$$

$$\bar\psi\gamma_5 F^n \psi \propto \text{pion} \,, \tag{4}$$

$$\bar\psi\gamma_0 F^s \psi \propto \text{baryon number density} \,, \tag{5}$$

$$\bar\psi\gamma_0 F^n \psi \propto \text{electric charge density} \quad (\text{for } N_\mathrm{f} = 3) \,. \tag{6}$$

The lattice determination of spatial correlation functions for operators of this type was pioneered by DeTar and Kogut[10] and others[11] a long time ago, but quantitatively significant results relevant for the physical infinite volume chiral continuum limit, are only being produced presently.[12–15] In order to streamline the discussion, we will restrict to one of the operators in the following, namely the interpolating field for the pion: $\pi \sim i\bar\psi\gamma_5 F^n\psi$. For $\bar\psi\gamma_0 F^s\psi$, see Sec. 4, and for the other operators, Ref. 9.

The spatial correlation function of the pion field has at large distances the structure

$$C_\mathbf{x} \equiv \int_0^{1/T} \mathrm{d}\tau \, \langle \pi(\tau,\mathbf{x})\pi(0,\mathbf{0})\rangle \sim \frac{1}{|\mathbf{x}|^n} \exp\left(-\frac{|\mathbf{x}|}{\xi(T)}\right) \,, \tag{7}$$

where τ is the Euclidean time coordinate of the imaginary time formulation. In Eq. (7), n is some unspecified power, while $\xi(T)$ defines the correlation length we are interested in. We will denote the inverse of the correlation length by $m \equiv \xi^{-1}$, and call it a "screening mass".

The question now is, how does $m(T)$ behave as the temperature is increased from below to above the critical temperature T_c of the chiral phase transition? At small temperatures, chiral symmetry is broken, and the pion is massless in the chiral limit, resulting in an infinite correlation length, or $m = 0$. At high temperatures, on the other hand, the screening mass approaches $2\pi T$, as we will recall presently. Thus the pion correlation length provides for a finite and gauge-invariant "order parameter" for chiral symmetry restoration.

The reason that the screening mass equals $2\pi T$ at high temperatures, is easily understood. For $T \gg T_c$, asymptotic freedom sets in, and the correlator can be determined in perturbation theory. A computation of the leading order graph,

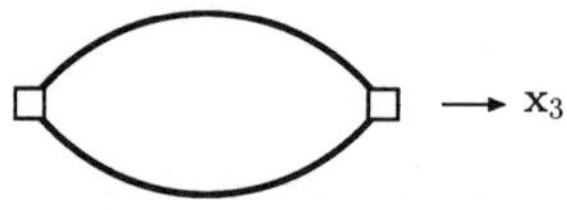

and a subsequent Fourier transform $[C_{\mathbf{P}} = \int_{\tau,\mathbf{x}} \exp(-i\omega_n\tau - i\mathbf{p}\cdot\mathbf{x})C_{\mathbf{x}}]$, yield

$$C_{\mathbf{P}} \sim \mathrm{Tr}[F^n F^n]\, N_c\, T \sum_{n=-\infty}^{\infty} \frac{i|\mathbf{p}|}{8\pi} \ln \frac{2\omega_n - i|\mathbf{p}|}{2\omega_n + i|\mathbf{p}|}\,, \quad \omega_n = 2\pi T(n+\frac{1}{2})\,. \tag{8}$$

Now, the singularity closest to the origin in momentum space (determining the behaviour at the largest distances in configuration space) is seen to be a two-particle threshold at $|\mathbf{p}| = \pm 2i\omega_0$. Referring for a moment to the spatial directions in a (2+1)-dimensional language, with x_3 as the "time" coordinate, both quarks are on-shell at this point and have a minimal "energy" $ip_3 = \pm\omega_0$ [the minimal energy appears for a relative momentum $\mathbf{p}_\perp = \mathbf{0}$, where we denoted $\mathbf{p} \equiv (\mathbf{p}_\perp, p_3)$]. Thus, taking an inverse Fourier transform back to configuration space, we obtain the exponential decay advertised above, $C_{\mathbf{x}} \sim \exp(-2\omega_0|\mathbf{x}|)$.

3. Next-to-leading order for flavour non-singlet correlators

The question we would like to address next is, what is the first weak-coupling correction to $m = 2\pi T$? This computation has been described in Ref. 9, and we only reiterate the main steps here.

For flavour non-singlet operators, the relevant graphs might appear to be

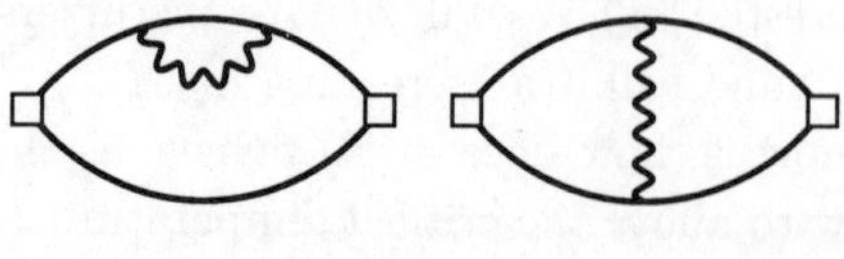

but in fact there are infinitely many higher order graphs that need to be taken into account, symbolically of the types

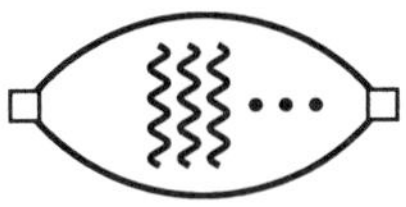

These are not suppressed, because near the two-particle threshold, the quarks are almost on-shell, $p^2 = \omega_n^2 + \mathbf{p}_\perp^2 + p_3^2 \approx 0$, and consequently the dimensionless expansion parameter associated with adding a further line can be seen to be $\mathcal{O}(g^2T/|\not{p}|) \sim \mathcal{O}(1)$. In other words, the situation is analogous to the computation of the energy of a two-quark bound state at zero temperature, where again infinitely many graphs contribute.

The analogy with the bound-state system suggests immediately a tool for organising the computation: As in the study of quarkonia at $T = 0$, energies of bound states can be addressed with an effective theory called "Non-Relativistic QCD" (NRQCD).[16] NRQCD techniques were first employed in the present context by Huang and Lissia.[17]

In order for an effective description to apply, the system must possess a scale hierarchy. Let us recall why one exists here, at high enough temperatures where the QCD coupling constant g is small. The basic point is that because quarks of a definite Matsubara frequency ω_n interact with bosonic Matsubara zero-modes only, we expect that their off-shellness is related to the momentum scales of the latter: $|ip_3 \pm \omega_n| \lesssim gT$. More precisely, we may recall that the static potential has in (2+1) dimensions the structure

$$V(\mathbf{x}_\perp) \sim g^2 T \ln |\mathbf{x}_\perp| \, . \tag{9}$$

Thus the typical (transverse) momentum $\mathbf{p}_\perp$ of the bound-state constituents satisfies

$$\mathbf{p}_\perp^2/\omega_n \sim V(\mathbf{x}_\perp) \Rightarrow \mathbf{p}_\perp^2 \sim (gT)^2 \, , \tag{10}$$

and the binding energy is (dropping again possible logarithms)

$$|ip_3 \pm \omega_n| \sim \mathbf{p}_\perp^2/\omega_n \sim V(\mathbf{x}_\perp) \sim g^2 T \, . \tag{11}$$

We thus indeed find a scale hierarchy

$$|ip_3 \pm \omega_n| \ll |\mathbf{p}_\perp| \ll \omega_n \, . \tag{12}$$

The effective action for this kinematic range can most easily be constructed by choosing a convenient basis for Dirac matrices (making $\gamma_0\gamma_3$

diagonal), and rewriting then the Dirac spinor as

$$\psi \equiv \begin{pmatrix} \chi \\ \phi \end{pmatrix} , \tag{13}$$

where χ, ϕ are two-component spinors. In this basis the pion operator becomes $\pi \sim i(\chi^\dagger\sigma_3\phi - \phi^\dagger\sigma_3\chi)$, where σ_3 is a Pauli matrix. Restricting to the Matsubara mode ω_0 and expanding the QCD action to order $1/\omega_0$, the (on-shell) effective Lagrangian reads

$$\mathcal{L} = i\chi^\dagger\left(\omega_0 - gA_0 + D_3 - \frac{\nabla_\perp^2}{2\omega_0}\right)\chi + i\phi^\dagger\left(\omega_0 - gA_0 - D_3 - \frac{\nabla_\perp^2}{2\omega_0}\right)\phi\,, \tag{14}$$

where D_3 is a covariant derivative. We may note that:

- Both A_0 and A_3 play an important dynamical role.
- The transverse gauge fields A_1, A_2, on the other hand, can be ignored, as long as we are interested in an energy shift of $\mathcal{O}(g^2T)$: they are of higher order than $\nabla_\perp \sim gT$. (They would be of order unity with respect to the magnetic scales $\nabla_\perp \sim g^2T$, but that gives an energy correction of order $\mathcal{O}(g^4T)$ only.)
- To be consistent at $\mathcal{O}(g^2T)$, we should replace ω_0 of the tree-level effective Lagrangian by a matching coefficient $M = \omega_0 + \mathcal{O}(g^2T)$ that needs to be determined.

While the final value of the matching coefficient M is unambiguous, there are many ways to determine it. In order to avoid computing wave function normalisation factors, we determine M by matching on-shell quark self-energies in the original QCD as well as in the effective theory of Eq. (14). On the side of QCD, a simple 1-loop graph,

$p^2 = 0$

produces near the pole the inverse propagator

$$\Sigma(p) = i\not{p} - ig^2C_F \sum\!\!\!\!\!\!\!\int_q \left.\frac{\gamma_\mu(\not{p} - \not{q})_f\gamma_\mu}{(p-q)_f^2(q^2+\lambda^2)_b}\right|_{p^2=0} , \tag{15}$$

where $C_F = (N_c^2 - 1)/2N_c$, N_c is the number of colours, $\sum\!\!\!\!\!\!\int$ is the standard imaginary-time integration measure, $(...)_f$ and $(...)_b$ denote fermionic and bosonic Matsubara four-momenta, respectively, and we have introduced a gluon mass λ as an intermediate infrared (IR) regulator. It can be seen that the integral in Eq. (15) remains finite for $\lambda \to 0$, and then, up to order $\mathcal{O}(g^2)$, solving for the zero of Eq. (15) is equivalent to solving

$$p^2 + m_{\rm eff}^2 = 0 \; , \tag{16}$$

where $m_{\rm eff}^2 = g^2 T^2 C_F / 4$ is nothing but the "hard effective mass" of quarks introduced recently in Ref. 18.

On the side of NRQCD, the computation of the quark self-energy is to be carried out order by order in $1/\omega_0$.[19] It is easy to see that, in fact, 1-loop corrections then vanish,[a] such that the on-shell point is determined directly by the mass scale M appearing in the tree-level propagator, $ip_3 = \pm M$.

Consequently, solving for the zeros in ip_3 of Eq. (16) to order $\mathcal{O}(g^2 T)$, we see that M needs to match $\omega_0 + m_{\rm eff}^2 / 2\omega_0$, leading to

$$M = \omega_0 + g^2 T \frac{C_F}{8\pi} + \mathcal{O}(g^4 T) \; . \tag{17}$$

This value replaces ω_0 in the first terms inside the parentheses in Eq. (14).

The parameters of the effective theory having been determined, it remains to solve its dynamics to order $g^2 T$. We define the correlation function

$$C(\mathbf{r}, x_3) \equiv \int_{\mathbf{R}} \left\langle \phi^\dagger \left(\mathbf{R} + \frac{\mathbf{r}}{2}, x_3 \right) \sigma_3 \chi \left(\mathbf{R} - \frac{\mathbf{r}}{2}, x_3 \right) \, \chi^\dagger(0) \sigma_3 \phi(0) \right\rangle \; , \tag{18}$$

and integrate out A_0, A_3. To order $\mathcal{O}(g^2)$, $C(\mathbf{r}, x_3)$ satisfies the partial differential equation

$$\left[\partial_{x_3} + 2M - \frac{1}{\omega_0} \nabla_{\mathbf{r}}^2 + V(\mathbf{r}) \right] C(\mathbf{r}, x_3) \propto \delta(x_3) \delta(\mathbf{r}) \; , \tag{19}$$

where

$$V(\mathbf{r}) = g^2 T C_F \int \frac{\mathrm{d}^{2-2\epsilon} q}{(2\pi)^{2-2\epsilon}} \left\{ \frac{1 - e^{i\mathbf{q}\cdot\mathbf{r}}}{q^2 + \lambda^2} - \frac{1 + e^{i\mathbf{q}\cdot\mathbf{r}}}{q^2 + m_D^2} \right\} \tag{20}$$

$$= g^2 T \frac{C_F}{2\pi} \left[\ln \frac{m_D r}{2} + \gamma_E - K_0(m_D r) \right] \; . \tag{21}$$

On the last line we took the continuum limit ($\epsilon \to 0$) and removed the IR regulator from the propagator of A_3 ($\lambda \to 0$). Moreover, K_0 is a modified Bessel function, and m_D is the Debye mass appearing in the propagator of A_0, $m_D^2 = g^2 T^2 (N_{\rm c}/3 + N_{\rm f}/6)$. Solving for the lowest energy level of the Schrödinger equation following from Eq. (19) at $x_3 \neq 0$, we finally obtain

$$m = 2\pi T + g^2 T \frac{C_F}{2\pi} \left(\frac{1}{2} + \hat{E}_0 \right) , \quad \hat{E}_0^{N_{\rm f}=0} = 0.164 \; , \tag{22}$$

[a]To be precise, it is important to note that at this point we are using un-resummed gluon propagators both on the QCD and on the effective theory side.

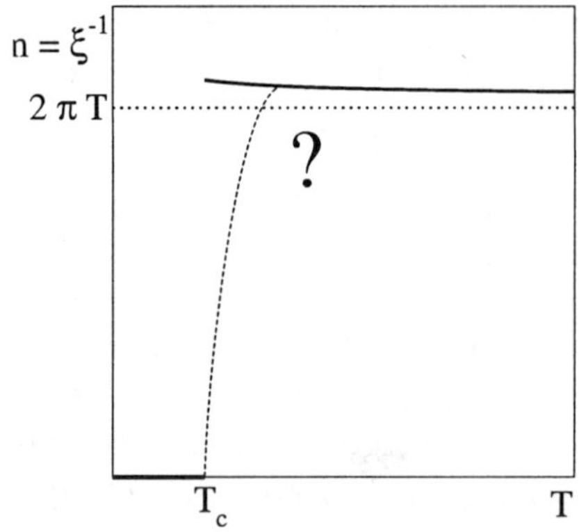

Figure 1. The pion screening mass as a function of the temperature. The dashed line is a guess for how the low-T and high-T limits might be connected, and in the continuous form drawn assumes a second order transition (i.e. $N_f = 2$). Note, however, that no sign of the indicated overshooting of $2\pi T$ has been observed in lattice simulations.

where $\hat{E}_0$ depends mildly on N_f (cf. Ref. 9), and we have shown explicitly the value relevant for the quenched theory.[b]

Note that in Eq. (22), the factor 1/2 inside the parentheses comes from the "constituent mass" correction in Eq. (17), while $\hat{E}_0$ comes from the solution of the Schrödinger equation. If the solution corresponds to a bound state, why then is $\hat{E}_0$ positive? The answer lies simply in the form of the potential in Eq. (21): it is a logarithmically "confining" potential, with $V(\mathbf{r} \to \mathbf{0}) \to -\infty$, $V(\mathbf{r} \to \infty) \to +\infty$, whose overall normalisation just happens to be such that the lowest energy eigenvalue is positive.

The result of Eq. (22) is plotted, with a certain choice of renormalisation scale for g^2 (cf. Ref. 3), in Fig. 1. We note that the correction appears small even at realistic T, but positive in sign. The change of m across T_c is thus very rapid — it is not smoothed by the $\mathcal{O}(g^2)$ term like for the pressure,[20] but rather made more pronounced. Close enough to T_c the mass of course has to decrease, either continuously (the case for $N_f = 2$, indicated with a dashed line in Fig. 1), or with a small discontinuity (for $N_f = 0, 3$).

For the pressure, the next-to-next-to-leading order correction, $\mathcal{O}(g^3)$, comes with an opposite sign to $\mathcal{O}(g^2)$, and appears very large.[21] (At the same time the ultimate result, including higher orders still, resembles more

[b]To be precise, the quenched theory value applies in the "perturbative" broken $Z(N_c)$ vacuum, i.e. the one where the phase of the Polyakov loop is trivial. In the unquenched theory this vacuum constitutes in any case the global ground state of the system.

the $\mathcal{O}(g^2)$ approximation.[22]) It would be interesting to find out whether there is an $\mathcal{O}(g^3)$ correction in the pion correlation length as well.

4. Flavour-singlet correlators

For flavour singlets the pattern is very different from the flavour non-singlets, as the singlet operators couple to purely gluonic "glueball" states. For baryon density, for instance, one finds that[8,23]

$$\bar{\psi}\gamma_0 F^s \psi \longleftrightarrow -\frac{iN_{\rm f}}{3\pi^2} g^3 \,{\rm Tr}[A_0^3] \,, \tag{23}$$

or graphically, that the correlation function is dominated by graphs like

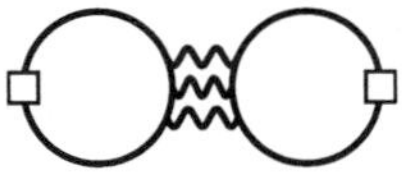

The corresponding screening mass has been determined some time ago,[8] and evaluates to $m({\rm Tr}[A_0^3]) \approx 5T$ at $T \approx 2T_c$. The inverse, $\xi \approx 1/5T$, is then the distance scale at which fluctuations in baryon number density are correlated. The glueball operators mediating the correlations of a number of other mesonic flavour-singlets have been worked out in Ref. 9.

5. On $\text{U}_A(1)$ symmetry at high temperatures

The $\text{U}_A(1)$ axial symmetry is broken explicitly by the anomaly. It has been suggested,[24] however, that it might get "effectively restored" somewhat above the critical temperature, say at $T \gtrsim 2T_c$. What is meant by this is that the topological susceptibility, $\chi = \langle \nu^2 \rangle / V$, where ν is the topological charge of the gauge configuration, or the net number of zero-modes of the Dirac operator, and V is the volume, decreases rapidly at temperatures above T_c. This becomes quite obvious particularly if the number of colours is increased above $N_{\rm c} = 3$; see, e.g., Ref. 25 and references therein. Consequently, it could be expected that anomalous effects cease to operate.

It may be appropriate to remark at this point, though, that it is possible to find observables for which the $\text{U}_A(1)$ symmetry appears not to get restored. Consider, for instance, the screening masses for the operators $V_3^s = \bar{\psi}\gamma_3\psi$, $A_3^s = \bar{\psi}\gamma_3\gamma_5\psi$, measured from correlation functions in the x_3-direction. Now, $\text{U}_A(1)$ symmetry restoration would imply that the correlation functions for V_3^s, A_3^s are identical and, therefore, that the screening masses are identical. Yet, a perturbative analysis can be used to indicate[9] that A_3^s couples to the glueball-like state $g^2\epsilon_{ij3}\,\mathrm{Tr}[A_0 F_{ij}]$ of the dimensionally reduced theory,[26] and has thus a non-vanishing screening mass $m = m_D + \mathcal{O}(g^2T)$,[2,3] where the next-to-leading order correction is also known numerically.[5] At the same time, the correlator related to V_3^s is exactly conserved, and has no non-trivial screening mass associated with it. Therefore, the $\mathrm{U}_A(1)$ symmetry appears broken even at high temperatures.

6. Conclusions

Various mesonic correlation lengths are well-defined gauge-invariant physical observables that are computable, for flavour non-singlets, up to next-to-leading order in the gauge coupling g^2 with quite a manageable amount of work. Higher orders could also be reached in principle. For flavour singlets, on the other hand, the correlation lengths can be related to those of purely gluonic states, which have been determined previously from the dimensionally reduced theory with good numerical precision.

The comparison of these analytic predictions with results from four-dimensional lattice simulations poses an interesting puzzle. Indeed, lattice results so far are consistent with the leading order prediction $m = 2\pi T$, as soon as the temperature is just above the critical one,[12–15] while the next-to-leading order correction advertised above is positive, and would require an "overshooting" of $m = 2\pi T$. Lattice results involve systematic errors, though, so that the discrepancy cannot be considered too serious at present.

On the side of analytic efforts, one could envisage a number of computations which might facilitate the comparison with more precise future lattice simulations. For instance, $\mathcal{O}(a)$ discretization effects could be determined analytically, even though this of course is specific to the fermion action used. (Different groups have rather different preferences here.) Another point of possible relevance is that some of the correlation functions involve an x_3-dependent prefactor in front of the exponential, and it might be useful to take it into account in the fitting procedure.

To conclude, let us recall that the ultimate theoretical goal of the exercise discussed here is to estimate the reliability of perturbation theory at finite T, in order to then use perturbation theory with more confidence for other observables for which lattice is not well suited. Given that good QCD predictions can eventually be obtained for static correlation lengths and that there is a remarkably stark difference between the two phases, one is however also lead to wonder, once again, whether some interesting phenomenological signatures might also be found for these quantities.

Acknowledgments

I wish to thank M. Vepsäläinen for collaboration on the work presented in this talk, and S. Datta, E. Laermann, G.D. Moore and R.D. Pisarski for useful comments related to topics covered in it.

References

1. L.D. McLerran and T. Toimela, *Phys. Rev.* **D31**, 545 (1985).
2. P. Arnold and L.G. Yaffe, *Phys. Rev.* **D52**, 7208 (1995).
3. K. Kajantie *et al*, *Nucl. Phys.* **B503**, 357 (1997).
4. M.J. Teper, *Phys. Rev.* **D59**, 014512 (1999).
5. M. Laine and O. Philipsen, *Phys. Lett.* **B459**, 259 (1999); A. Hart and O. Philipsen, *Nucl. Phys.* **B572**, 243 (2000).
6. O. Kaczmarek *et al*, *Phys. Rev.* **D62**, 034021 (2000).
7. S. Datta and S. Gupta, *Phys. Lett.* **B471**, 382 (2000); *Phys. Rev.* **D67**, 054503 (2003).
8. A. Hart, M. Laine and O. Philipsen, *Nucl. Phys.* **B586**, 443 (2000); *Phys. Lett.* **B505**, 141 (2001).
9. M. Laine and M. Vepsäläinen, *JHEP* **02**, 004 (2004).
10. C. DeTar and J.B. Kogut, *Phys. Rev. Lett.* **59**, 399 (1987); *Phys. Rev.* **D36**, 2828 (1987).
11. K.D. Born *et al*, *Phys. Rev. Lett.* **67**, 302 (1991).
12. P. de Forcrand *et al.*, *Phys. Rev.* **D63**, 054501 (2001).
13. E. Laermann and P. Schmidt, *Eur. Phys. J.* **C20**, 541 (2001).
14. R.V. Gavai, S. Gupta and R. Lacaze, *Phys. Rev.* **D65**, 094504 (2002); R.V. Gavai and S. Gupta, *Phys. Rev.* **D67**, 034501 (2003).
15. For a review and references, see E. Laermann and O. Philipsen, *Ann. Rev. Nucl. Part. Sci.* **53**, 163 (2003), and E. Laermann, these proceedings.
16. W.E. Caswell and G.P. Lepage, *Phys. Lett.* **B167**, 437 (1986).
17. S. Huang and M. Lissia, *Nucl. Phys.* **B480**, 623 (1996).
18. P. Arnold, G.D. Moore and L.G. Yaffe, *JHEP* **01**, 030 (2003).
19. A.V. Manohar, *Phys. Rev.* **D56**, 230 (1997).
20. E.V. Shuryak, *Sov. Phys. JETP* **47**, 212 (1978); S.A. Chin, *Phys. Lett.* **B78**, 552 (1978).
21. J.I. Kapusta, *Nucl. Phys.* **B148**, 461 (1979).
22. K. Kajantie *et al*, *Phys. Rev.* **D67**, 105008 (2003); JHEP **04**, 036 (2003); A. Vuorinen, *Phys. Rev.* **D68**, 054017 (2003).
23. D. Bödeker and M. Laine, JHEP **09**, 029 (2001).
24. E.V. Shuryak, *Comm. Nucl. Part. Phys.* **21**, 235 (1994); G. Boyd *et al*, hep-lat/9607046; J.B. Kogut, J.F. Lagae and D.K. Sinclair, *Phys. Rev.* **D58**, 054504 (1998); S. Chandrasekharan *et al*, *Phys. Rev. Lett.* **82**, 2463 (1999).
25. B. Lucini, M. Teper and U. Wenger, hep-lat/0401028; L. Del Debbio, H. Panagopoulos and E. Vicari, hep-th/0407068.
26. P. Ginsparg, *Nucl. Phys.* **B170**, 388 (1980); T. Appelquist and R.D. Pisarski, *Phys. Rev.* **D23**, 2305 (1981).

THERMALIZATION AND PLASMA INSTABILITIES

PETER ARNOLD

Department of Physics, University of Virginia
P.O. Box 400714, Charlottesville, VA 22901-4714

In this talk, I review the important role played by plasma instabilities in the thermalization of quark-gluon plasmas at very high energy.

1. Introduction

Here is a basic question: What is the (local) thermalization time for quark-gluon plasmas (QGPs) in heavy ion collisions? That's a difficult question, so let's ask a simpler one: What is it for *arbitrarily high energy* collisions, where the running coupling constant is small, $\alpha_s \ll 1$? Even that turns out to be complicated, so let me focus on an even simpler version, first posed by Baier, Mueller, Schiff and Son:[1] How does that time depend on α_s? In particular, in the saturation picture of heavy-ion collisions, what is the exponent n in the relation

$$t_{eq} \sim \frac{\alpha_s^n}{Q_s}\,? \tag{1}$$

Before discussing this question in more detail, I wish to make a general side comment about plasma physics. Plasma physics is complicated! This is made abundantly clear simply by looking at pictures of various plasma phenomena, such as the image in Fig. 1 of a solar coronal filament from NASA's TRACE satellite. Theoretical discussions of quark-gluon plasmas, however, are generally much less complicated. There are several reasons why such discussions can usually avoid the full complication of traditional plasma physics. First, much of the theoretical literature discusses QGPs that are at or very close to thermal equilibrium. The physics of plasmas near global thermal equilibrium is much less complicated than the physics of non-equilibrium situations. Because electromagnetic interactions are long-ranged, traditional electromagnetic plasmas can be very complicated even when they are in *local* thermal equilibrium. Macroscopic currents in one region of the plasma can interact magnetically with other currents in other regions, over tremendous distance scales, creating complicated structures like Fig. 1. Non-Abelian plasmas, however, are somewhat different. From theoretical studies of the equilibrium properties of such plasmas, we know that the non-Abelian interactions cause magnetic *confinement* over distances of order $1/(g^2T)$. It is reasonable to assume that, even dynamically, color magnetic fields cannot exists on distance scales larger than the confinement length. So, unlike traditional electromagnetic plasmas, there are no large-distance magnetic fields. As far as the color degrees of freedom are concerned, the long-distance effective theory of a non-Abelian plasma is hydrodynamics rather than magneto-hydrodynamics.

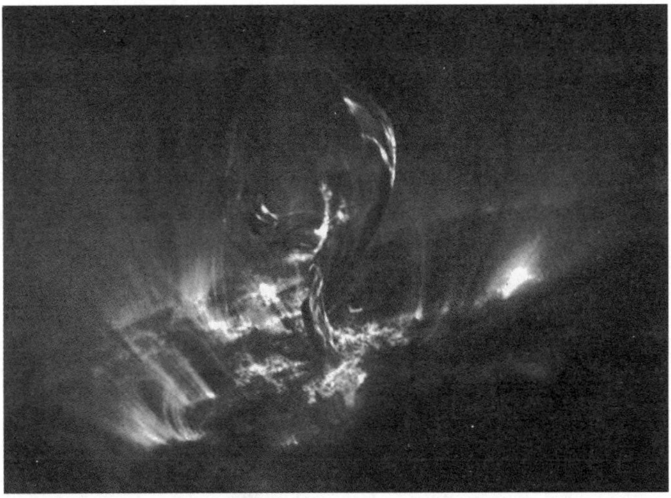

Figure 1. Image of a solar coronal filament from NASA's TRACE satellite, from ⟨http://antwrp.gsfc.nasa.gov/apod/ap000809.html⟩. (It's much more dramatic in color—see the web page.)

The color magnetic fields *can* play a role on small scales. But the full complications of plasma physics might be ignored on small distance scales if the relevant physics on those scales is weakly interacting. This was the proposal of the original bottom-up scenario for thermalization of quark-gluon plasmas.[1] However, as we shall discuss, even at small distance scales, there can be plasma instabilities. These instabilities cause the growth of non-perturbatively large magnetic fields, bringing in some of the complicated non-linear physics of traditional plasma physics.

In the original bottom-up thermalization scenario of Baier *et al.*, the starting assumption is the saturation scenario, which assumes that the formation of the quark-gluon plasma starts at times $t \sim Q_s^{-1}$ dominated by gluons with momenta $p \sim Q_s$, where the scale Q_s is known as the saturation scale. In this talk, these gluons will be referred to as "hard" gluons, since we will soon be discussing even softer momentum scales. This situation is depicted in Fig. 1a. The occupation numbers of each gluon mode are initially non-perturbative, with $f(\boldsymbol{x},\boldsymbol{p}) \sim 1/\alpha_s$, where f is the phase-space density. As the system expands 1-dimensionally immediately after the collision, the density per unit volume decreases, and one might therefore expect the hard gluon interactions to become more perturbative. To understand what happens next, let's ignore these perturbative interactions for the moment and think about free expansion. As the nuclear pancakes fly apart after the collision, the gluons, which started in the center, will separate themselves in space according to the z components of their velocities, as $\Delta z \sim v_z t$. This is depicted in Fig. 1b. The gluons left in the central region will be those with small v_z, as shown in Fig. 1c. As a result, the momentum distribution in that central region will have an anisotropic pancake shape, as shown in Fig. 1d. In other regions, the momentum distribution is simply a boosted version of this—that is, it looks the same if one works in the local rest frame.

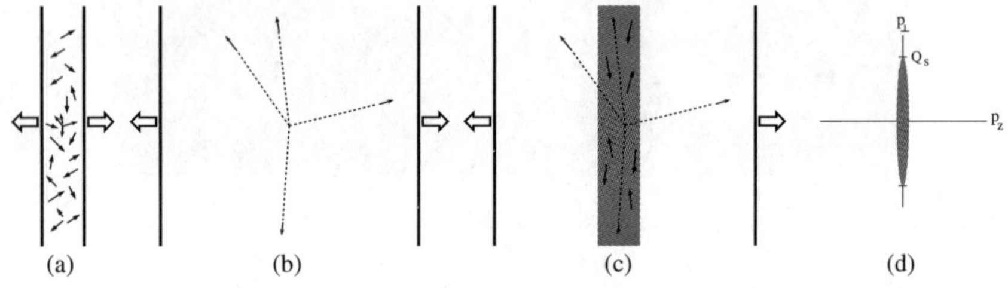

Figure 2. Approximately free expansion at relatively early stages of the bottom-up scenario.

It should be emphasized that this anisotropic momentum distribution has nothing to do with the usual elliptic flow distribution measured in non-central heavy ion collisions. Fig. 1d is a statement of the local momentum distribution in the local fluid rest frame at very early times, before collisions have brought the system to local equilibrium. Elliptic flow is instead a measure of the *net* fluid flow on large scales of a system that has quite possibly come to local thermal equilibrium (and so locally has isotropic momentum distributions in the local fluid rest frame).

In the original bottom-up scenario, equilibration of the plasma was assumed to occur through individual 2-body collisions between particles (with some LPM effect thrown in). In the first stage of the scenario, $1 \ll Q_s t \ll \alpha^{-5/2}$, the important processes were small angle scattering, which slightly widens the hard particle distribution in p_z, and soft Bremsstrahlung from colliding hard particles, which creates soft gluons with momenta $k \ll Q_s$. In the second stage, $\alpha^{-5/2} \ll Q_s t \ll \alpha^{-13/5}$, these soft gluons come to dominate the number density of particles, but the hard gluons still dominate the energy density. Collisions between the soft gluons cause the soft gluons to thermalize. Finally, the hard particles begin to lose energy by Bremsstrahlung plus cascading, as shown in Fig. 1. In this scenario, local thermalization became complete at $Q_s t \sim \alpha^{-13/5}$.

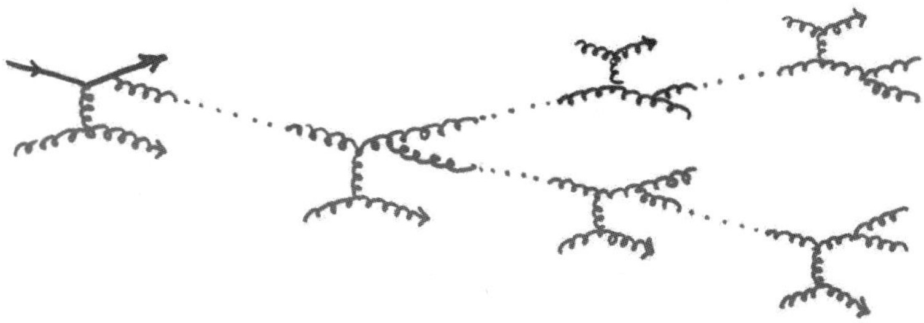

Figure 3. Cascading process for energy transfer from hard particles to soft sector in the original bottom-up scenario. Straight lines denote hard gluons, $p \sim Q_s$.

The original bottom-up scenario overlooked the possibility that *collective* processes (as opposed to sequences of individual collisions) could play an important role in the equilibration of the plasma. In the case at hand, these collective processes are related to the appearance of plasma instabilities in the analysis of the equilibration of the quark-gluon plasma.

2. Plasma Instabilities

The hero of this story is Stan Mrówczyński,[2,3,4] who over the years has been the major proponent of the idea that plasma instabilities are important for the equilibration of the quark-gluon plasma. The application of this idea to bottom-up thermalization was made by myself, Jonathan Lenaghan and Guy Moore.[5] A selection of other folks who have considered the idea past and present include Refs. 6–9 and the work by Romatschke and Strickland,[10] which was reported on at this conference.

Let me start with a slightly formal explanation of the origin of plasma instabilities, and I will give a more physical picture afterward. Imagine calculating the self-energy $\Pi(\omega, k)$ for a particle moving through the plasma. The self-energy represents the effect on the particle of forward-scattering off of other particles in the medium, as in Fig. 2a, which one can alternatively calculate as in Fig. 2b using one's favorite formalism for field theory in a medium, or more simply calculate using linearized kinetic theory. (Here, the straight lines represent hard particles, which are gluons in the bottom-up scenario.) The result is the same either way. Generically, if the momentum distribution $f(\boldsymbol{p})$ of hard particles is anisotropic,[a] one finds that there are negative eigenvalues of $\Pi^{ij}(0, \boldsymbol{k})$. (See, for example, the general arguments in Ref. 5.) Such negative eigenvalues indicate instabilities at small k, which means exponentially growing soft gauge fields.

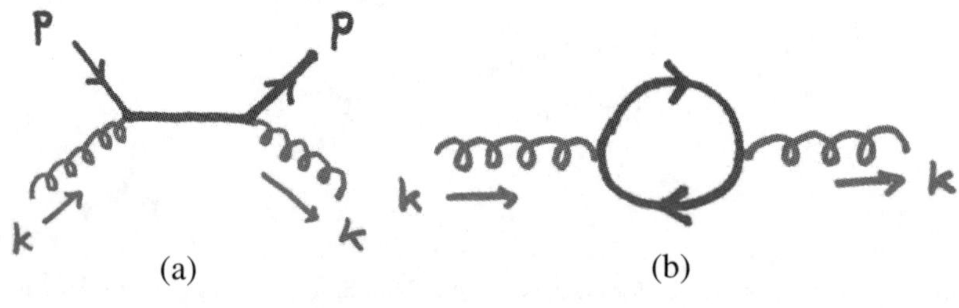

Figure 4. Self-energy of soft modes due to hard particles.

Let me give an analogy from scalar ϕ^4 theory at finite temperature. Imagine integrating out the hard particles to get an effective thermal self-energy Π of order λT^2. The effective linearized equation of motion is then

$$\omega^2\phi = (k^2 + m_0^2 + \Pi)\phi \equiv (k^2 + m_{\text{eff}}^2)\phi. \tag{2}$$

Let's set $m_0 = 0$ to improve the analogy, since gluons do not have any intrinsic mass. If we were in a situation where $k^2 + \Pi$ were less than zero, then there would be solutions with $\omega = \pm i\Gamma$ pure imaginary, which would lead to exponentially growing solutions to the linearized equation.

[a]Here and throughout, I always assume f is parity symmetric: $f(-\boldsymbol{p}) = f(\boldsymbol{p})$.

Alternatively, think about the effective potential,

$$V_{\text{eff}}(\phi) = m_{\text{eff}}^2 \phi^2 + \lambda\phi^4 = \Pi\phi^2 + \lambda\phi^4. \tag{3}$$

If $\Pi < 0$, as happens in some multi-scalar theories,[11] then the potential looks like a double-well potential: the naive vacuum $\phi = 0$ is unstable; there is exponential growth from $\phi = 0$ of modes with $k < (-\Pi)^{1/2}$; and the growth stops once ϕ becomes *non-perturbatively* large.

Now let's turn to a physical picture of the instability, which I will adapt from Refs. 2,5. For simplicity, imagine two inter-penetrating, homogeneous streams of charged hard particles, one going up the page and one going down, which I'll call the $\pm x$ directions. Now also imagine that, due to some fluctuation, there is a very tiny seed magnetic field of the form $\boldsymbol{B} = B\boldsymbol{e}_y \cos(kz)$, as shown in Fig. 2a. Here, crosses denote magnetic fields pointing into the page, and dots fields pointing out of the page. Using the right-hand rule, you can check that the magnetic fields bend the trajectories of positively charged particles in the directions shown. This then focuses the net downward and upward currents into different channels, as shown in Fig. 2b. Again using the right-hand rule, one finds that these currents in turn create magnetic fields that *add* to the original seed field. With bigger fields, the effect becomes more pronounced, and the fields continue to grow through this mechanism. This is the Weibel instability.

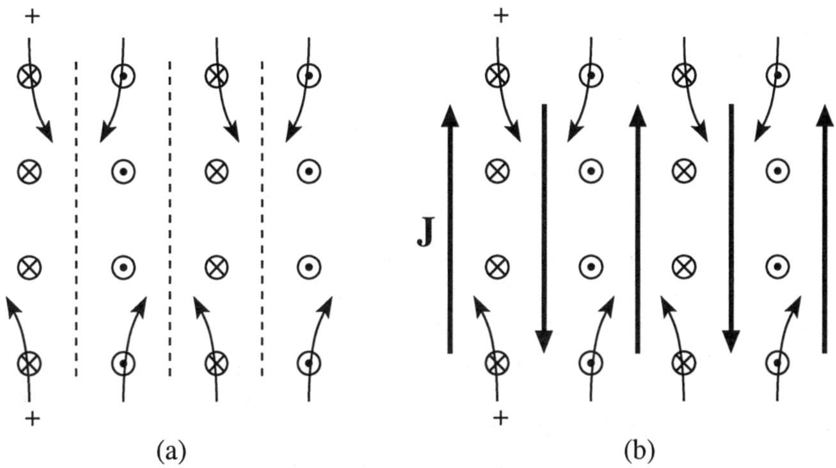

Figure 5. Origin of Weibel instability.

We get a seemingly contradictory picture of what happens if we considers hard particles which move in other directions, such as in Fig. 2. Following how these particles are bent by the seed magnetic field, we find they are directed a little more upward in some regions and a little more downward in others. This results in a net current as shown in the picture, but these currents create magnetic fields which *oppose* the seed field. Let $\boldsymbol{p}$ be the original momentum of the hard particle and $\boldsymbol{k}$ the wave number of the soft magnetic field fluctuation. What happens is that particles with $\boldsymbol{p}\cdot\boldsymbol{k} \simeq 0$ get trapped in valleys, as shown in Fig. 2b, and give the de-stabilizing effect discussed earlier. Other particles, with $\boldsymbol{p}\cdot\boldsymbol{k} \not\simeq 0$, are "untrapped" as in Fig. 2a, and give a stabilizing effect. For *isotropic* $f(\boldsymbol{p})$, these two contributions turn out to cancel, giving $\Pi^{ij}(0,\boldsymbol{k}) = 0$.

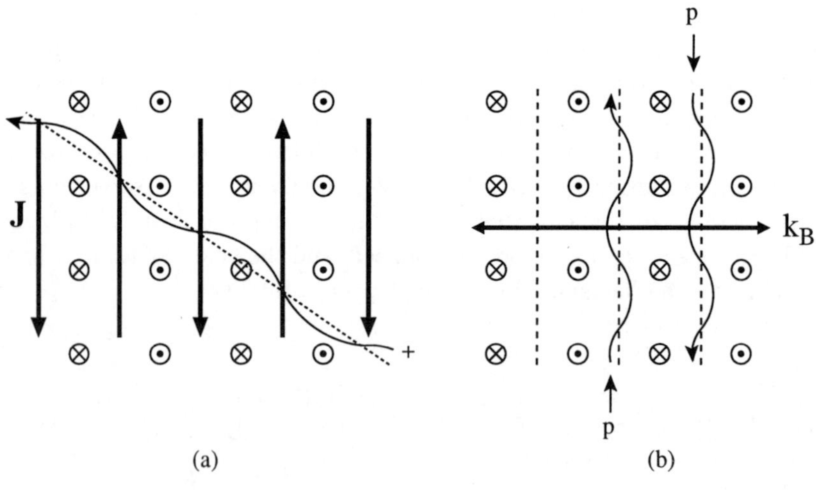

Figure 6. (a) untrapped particles; (b) trapped particles.

Now, instead of an isotropic hard particle distribution $f(\boldsymbol{p})$, think of the one depicted earlier in Fig. 1d. For $\boldsymbol{k}$ in the z direction, we will get a relatively smaller percentage of particles with $\boldsymbol{p}\cdot\boldsymbol{k} \simeq 0$ than we would with an isotropic distribution, and so the net effect will be stabilizing. On the other hand, for $\boldsymbol{k}$ in the $\perp$ direction, we will get a relatively larger percentage of particles with $\boldsymbol{p}\cdot\boldsymbol{k} \simeq 0$ than we would with an isotropic distribution, and so the net effect will be de-stabilizing. For $\boldsymbol{k}$ at significant angles to the axis, most of the particles with have $\boldsymbol{p}\cdot\boldsymbol{k} \not\simeq 0$, and we will tend to get stability. The moral of this story, to which we will later return, is that the unstable modes associated with this distribution have $\boldsymbol{k}$ pointing very close to the z axis.

3. When does the growth stop?

As in the scalar analogy discussed earlier, the growth of instabilities should stop when the soft fields become non-perturbatively large. From considering covariant derivatives $D = \partial - igA$, the effects of soft fields A become non-perturbative when $A \sim \partial/g$. There are two possibilities for the momentum scale associated with the derivatice ∂. The first, a possibility for both QCD and QED, is that growth stops when the effects of the soft fields on *hard* particles becomes non-perturbative, which will happen when $A \sim p_{\rm hard}/g$ and corresponds to the trajectories of hard particles being bent dramatically from straight lines. The second possibility, which cannot occur in QED, is that growth stops when the non-Abelian self-interactions of the soft fields become non-perturbative. This corresponds to $A \sim k_{\rm soft}/g$. Whenever there is a significant separation $k \ll p$ between soft and hard physics (as in the bottom-up scenario), these two possibilities correspond to parametrically different scales for A.

Jonathan Lenaghan and I conjecture[12] that growing QCD instabilities "abelianize." That is, the growth stops when $A \sim k_{\rm soft}/g \sim |\Pi|^{1/2}/g$, just as in QED. That in turn suggests that the complicated stuff that happens afterward is closely related to the mainstream plasma physics of (collisionless) relativistic QED plasmas. What follows is a summary of suggestive arguments that we make for abelianization.

Start with the general HTL effective action for anisotropic $f(\boldsymbol{p})$, which I adapt from Mrówczyński, Rebhan, and Strickland.[13] Shcematically,

$$S_{\rm eff} = -\int_x F^2 - g^2 \int_{x\boldsymbol{p}} f(\boldsymbol{p})\, W^2, \tag{4}$$

$$W \equiv W_\alpha(x, \boldsymbol{p}) \equiv \frac{p^\mu}{p \cdot D} F_{\mu\alpha}. \tag{5}$$

Now imagine finding the effective potential $V_{\rm eff}$ by looking at $S_{\rm eff}$ for time-independent configurations in $A_0{=}0$ gauge. There is a problem, which is that $(p \cdot D)^{-1}$ is a complicated, non-local operator. But now recall that, for the hard particle distribution of Fig. 1d, the typical unstable modes have $\boldsymbol{k}$ pointing very close the z direction. Inspired by this, let's ignore $k_\perp$ altogether and consider configurations $\boldsymbol{A} = \boldsymbol{A}(z)$ depending on z only. There is then an amazing simplification, noted by Blaizot and Iancu:[14] W

given by (5) then turns out to be *linear* in A. As a result, the HTL term $g^2 \int fW^2$ in the effective action (4) is then *quadratic* in A. That means that it consists of nothing but the HTL self-energy, so that

$$S_{\text{eff}}[\boldsymbol{A}(z)] = -\int_x F^2 - \int_x A\Pi A. \tag{6}$$

The Π here is for $\boldsymbol{k}$'s proportional to $\hat{z}$. It is also transverse, corresponding to $\Pi_\perp(0, k\hat{z})$. This turns out not to depend on the magnitude k of $\boldsymbol{k}$. I will define the constant $-\mu^2 \equiv \Pi_\perp(0, k\hat{z})$. The effective potential from (6) is then

$$V_{\text{eff}}[A(z)] = B^2 - \mu^2 A_\perp^2, \tag{7}$$

where B is the full, non-Abelian magnetic field, which includes non-linear terms in A. Now consider the arbitrarily soft limit $k \to 0$, where the derivative term in B vanishes and only the non-Abelian commutator term survives:

$$V_{\text{eff}} \to [A_x, A_y]^2 - \mu^2(A_x^2 + A_y^2). \tag{8}$$

What does this potential look like? As an example, let's consider just two of the degrees of freedom: color 1 of A_x and color 2 of A_y. A plot of the resulting potential is shown in Fig. 3. As one moves away from $A = 0$, the potential bends down due to the $-\mu^2 A^2$ term. If both colors are present, it then later bends back up again due to the quartic $[A_x, A_y]^2$ term. However, if the configuration is single-colored ($A_x^{(1)} = 0$ or $A_y^{(2)} = 0$), the commutator vanishes, and the potential continues to run away. If we were to roll a ball from the origin in such a potential, it would eventually roll away down into one of the Abelian directions $[A_x, A_y] = 0$.

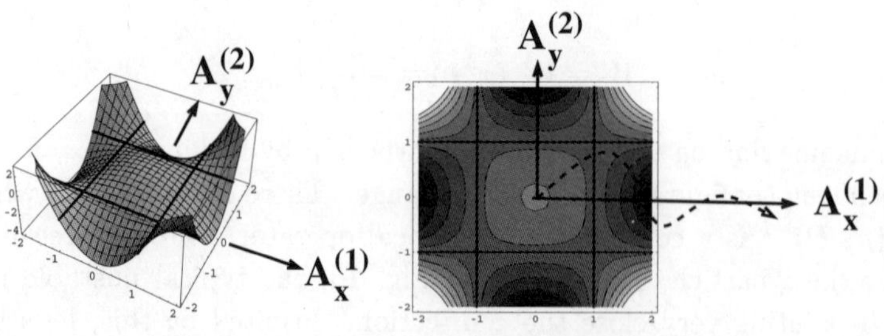

Figure 7. A picture of the effective potential.

The above argument suggests the following conjecture, which is the conclusion of my talk: The growth of Weibel instabilities eventually "abelianizes" the soft gauge fields into the maximal Abelian subgroup of the gauge group. For SU(2) gauge theory, this would give traditional U(1) plasma physics. For SU(3) gauge theory, it would give U(1)×U(1) plasma physics, corresponding to two copies of Abelian electromagnetism. Further discussion, including corroboration from numerical simulation results of a related toy model, may be found in Ref. 12.

References

1. R. Baier, A. H. Mueller, D. Schiff and D. T. Son, *Phys. Lett.* **B502**, 51 (2001) [hep-ph/0009237].
2. S. Mrówczyński, *Phys. Lett.* **B214**, 587 (1988); *ibid.* **B314**, 118 (1993); *Phys. Rev.* **C49**, 2191 (1994); *Phys. Lett.* **B393**, 26 (1997) [hep-ph/9606442].
3. S. Mrówczyński and M. H. Thoma, *Phys. Rev.* **D62**, 036011 (2000) [hep-ph/0001164].
4. J. Randrup and S. Mrówczyński, *Phys. Rev.* **C68**, 034909 (2003) [nucl-th/0303021].
5. P. Arnold, J. Lenaghan and G. D. Moore, *JHEP* **0308**, 002 (2003) [hep-ph/0307325].
6. U. W. Heinz, *Nucl. Phys.* **A418**, 603C (1984).
7. Y. E. Pokrovsky and A. V. Selikhov, *JETP Lett.* **47**, 12 (1988) [*Pisma Zh. Eksp. Teor. Fiz.* **47**, 11 (1988)]; *Sov. J. Nucl. Phys.* **52**, 146 (1990) [*Yad. Fiz.* **52**, 229 (1990)]. *ibid.* **52**, 385 (1990) [*Yad. Fiz.* **52**, 605 (1990)].
8. O. P. Pavlenko, *Sov. J. Nucl. Phys.* **55**, 1243 (1992) [*Yad. Fiz.* **55**, 2239 (1992)].
9. M. C. Birse, C. W. Kao and G. C. Nayak, *Phys. Lett.* **B570**, 171 (2003) hep-ph/0304209.
10. P. Romatschke and M. Strickland, *Phys. Rev.* **D68**, 036004 (2003) [hep-ph/0304092].
11. S. Weinberg, *Phys. Rev.* **D9**, 3357 (1974).
12. P. Arnold and J. Lenaghan, hep-ph/0408052.
13. S. Mrowczynski, A. Rebhan and M. Strickland, *Phys. Rev.* **D70**, 025004 (2004) [hep-ph/0403256].
14. J. P. Blaizot and E. Iancu, *Phys. Lett.* **B326**, 138 (1994) [hep-ph/9401323].

TRANSPORT COEFFICIENTS IN HOT QCD

GUY D. MOORE

Department of Physics, McGill University,
3600 rue University, Montreal, QC H3A 2T8, Canada

I give a physical explanation of what shear viscosity is, and what physics determines its value. Then I explain why determining the shear viscosity of the Quark-Gluon Plasma is interesting. I outline the leading-order calculation of the QGP shear viscosity (and baryon number diffusion constant), explaining why the quite complicated physics of parton splitting and Landau-Pomeranchuk-Migdal interference effects are required for its calculation. Then I briefly explore the range of applicability, emphasizing the importance of plasma instabilities.

1. Introduction

In this talk I will discuss work with my collaborators, Peter Arnold and Larry Yaffe, on transport coefficients in QCD, particularly shear viscosity.

Interest in transport coefficients in the Quark-Gluon Plasma (QGP) started in the 1980's, and several rough estimates were made at that time [1]. The first really mature work was by Baym *et al.* [2] in 1990. They showed that the dominant physics involved is small angle scattering, and they demonstrated how to perform a calculation accurate to leading order in the logarithm of the strong coupling, $\ln(1/g_s)$. We started work around 1999 under the mistaken impression that not much was required to improve their work to leading order in the coupling g_s. After discovering that even the leading-log calculation was wrong in the literature, we decided to write two papers, rather than one–a quick paper [3] computing the leading log correctly, and a longer one presenting a leading order calculation. The quick paper took 8 months and the "second" paper turned out to be six papers [4,5,6,7,8,9], spanning the next two and a half years.

My goal is to explain how the physics of transport coefficients turns out to be much richer than we had anticipated. In Section 2, I will explain what shear viscosity is and what physics is involved; and I will briefly explain why it is an interesting property of the QGP. Section 3 will outline our calculation, emphasizing the roles played by soft scatterings, identity changing scatterings, and collinear processes. Then, Section 4 will briefly discusses how limited the range of the calculation is; not only does it depend on weak coupling, but on a near-equilibrium assumption outside of which the physics is completely different.

2. The physics of shear viscosity

The velocity of an ideal fluid evolve according to Euler's equations,

$$\frac{\partial \boldsymbol{v}}{\partial t} = -\boldsymbol{v}\cdot\nabla\,\boldsymbol{v} - \frac{1}{\rho}\nabla p\,, \qquad p = p(\rho)\,. \tag{1}$$

Furthermore, the velocity at a boundary equals the velocity of that boundary (no slipping), and the stress tensor is determined by the pressure alone, that is, in the $\boldsymbol{v} = 0$ frame it is given by

$$T^{\mu\nu} = (\rho + p)u^{\mu}u^{\nu} + p\eta^{\mu\nu}\,, \tag{2}$$

(with $\eta^{\mu\nu} = \mathrm{Diag}[-,+,+,+]$). In fact, Euler's equations follow simply from the assumption that the stress tensor will satisfy this form, from an equation of state, and from stress-energy conservation, $\partial_{\mu}T^{\mu\nu} = 0$.

Local equilibrium ensures that these conditions are met. What enforces local equilibrium is interactions between degrees of freedom; free theories never equilibrate. Therefore, the ideal description is a better and better approximation as a theory is more and more strongly coupled; in this case ideality is the opposite of free behavior.

To see this, consider a fluid undergoing shear flow. Shear flow means that, in the local rest frame, the traceless part of $\partial_i v_j$ is nonzero. The simplest example of shear flow is a fluid trapped between two plates, one fixed and one moving laterally with velocity v, as in Fig. 1; under ideal

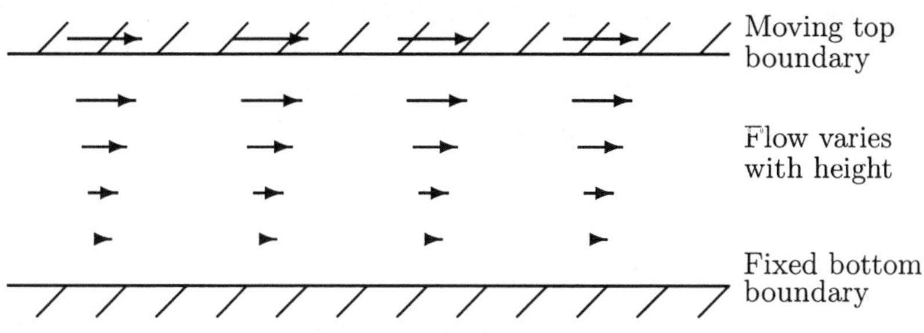

Figure 1. a system undergoing shear flow

Eulerian flow, the fluid at the top is moving with the upper boundary, with $T_{xz} = 0$; therefore there is no lateral force on the top boundary.

In a real fluid, there will be a "drag" force on the top plate; the shear viscosity is defined by the relation that the force per unit area is,

$$\frac{F_x}{A} = T_{xz} = \eta \frac{dv_x}{dz}\,, \quad \text{implying } T_{ij} = p\delta_{ij} + \eta \left(\partial_i v_j + \partial_j v_i - \frac{2}{3}\delta_{ij}\partial_k v_k \right) . \quad (3)$$

Shear occurs because particles can fly freely. The particles near the top of the fluid region are primarily moving to the right; those near the bottom are moving right or left with equal likelihood. If the particles fly freely, the upward moving particles near the bottom, and the downward moving particles from the top, will meet in the middle, as shown in Fig. 2. However, the

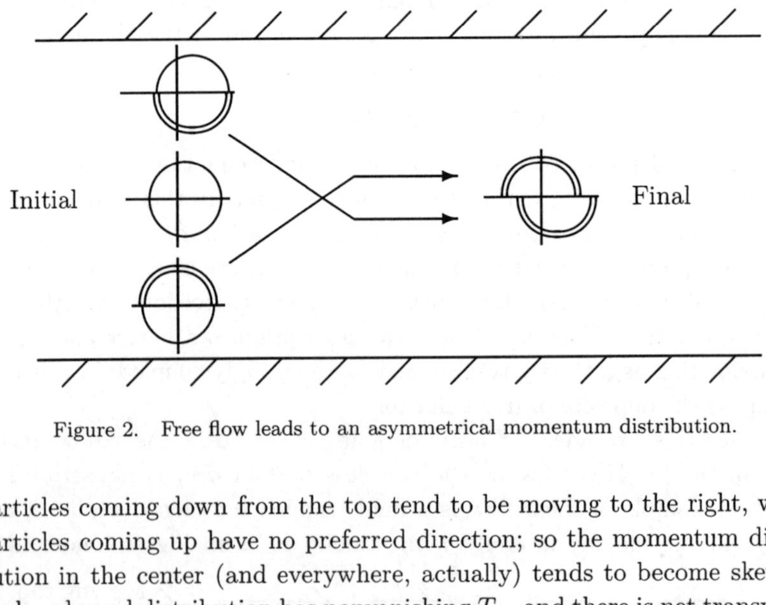

Figure 2. Free flow leads to an asymmetrical momentum distribution.

particles coming down from the top tend to be moving to the right, while particles coming up have no preferred direction; so the momentum distribution in the center (and everywhere, actually) tends to become skewed. Such a skewed distribution has nonvanishing T_{xz} and there is net transverse momentum flow and a net drag on the top surface. Scatterings restore the distribution towards equilibrium, evening out the momentum distribution. The fewer scatterings, the more distorted the momentum distribution will be. Therefore, the shear viscosity will be *smaller* if the scattering rate is *larger*; $\eta \propto 1/\Gamma$ with Γ the mean scattering rate. [a] The problem of deter-

[a]Familiar viscous fluids like honey are viscous because the molecules get tangled in each other, and are not a good analogy with any system which behaves approximately like a collection of freely flying objects. A better example is the air; as the pressure is lowered, there are fewer particles available to carry momentum, $T_{ij} \propto p$, so one would naively expect $\eta \propto p$. However, at lower pressure, collisions are rarer, $\Gamma \propto p$; and so the shear viscosity of air is actually independent of air pressure.

mining the shear viscosity will be the problem of determining how collision processes re-arrange the momentum distribution of particles from skewed towards isotropic.

Why is the shear viscosity of the QGP interesting? One of the main focuses of this conference is the understanding of nonequilibrium dynamics; QCD is the most interesting theory to study this and viscosity and other transport coefficients (baryon number diffusion and electrical conductivity, which have very similar physics) are theoretically clean and well defined nonequilibrium dynamical quantities. That makes them ideal objects for "cutting our teeth" on calculational strategies for nonequilibrium field theories.

The shear viscosity of QCD is also directly interesting because it gives a handle on the quality of the hydrodynamical description of heavy ion collisions; this has been considered at some length recently by Teaney [10], who found the corrections to flow parameters like v_2 due to viscosity at RHIC. Viscosity and particularly diffusion coefficients are also important in electroweak baryogenesis.

3. The shear viscosity of the QGP

Shear viscosity describes momentum transfer, and most of the momentum of a weakly coupled, near-equilibrium QGP is in quasiparticle excitations with momentum $p \sim 3T$. These should be adequately described by kinetic theory, as I implicitly assumed in the last section's discussion. In fact, the validity of a kinetic description has essentially been derived diagrammatically, by Jeon for $\lambda\phi^4$ theory [11] and by others for gauge theories [12,13].

The physics of viscosity is the physics of direction changing scattering, and this is dominated by $2 \leftrightarrow 2$ t-channel gluon exchange [1]. In terms of the spatial momentum q exchanged in the scattering process, such scatterings have a cross section of form,

$$d\sigma \propto \frac{d^3q}{q}\frac{1}{q^4}, \tag{4}$$

which is quadratically small q divergent—the familiar Coulomb divergence. However, what is important for shear viscosity is how quickly the particles change direction. Momentum changes from different scatterings add incoherently; so

$$\int q^2 d\sigma \propto \int q^3 \frac{d^3q}{q}\frac{1}{q^4} \sim \int \frac{dq}{q}, \tag{5}$$

is the relevant quantity and enough collisions must occur for it to add up to $|\boldsymbol{p}|^2$, for a particle to change direction. This integral is log IR divergent. There are two key lessons.

(1) Small q collisions dominate, but only by a logarithm.
(2) Large momentum particles take the longest time to change direction; since $(\delta \boldsymbol{p})^2 \sim |\boldsymbol{p}|^2$ is required, the time scale for direction change goes as $\tau \propto |\boldsymbol{p}|^2$. Therefore high energy particles contribute disproportionally to η.

Baym *et al.*[2] provided the first quantitative treatment which accounted for both of these points. To find the shear flow, they wrote the Boltzmann equation for the nonequilibrium distribution function $f(x,p)$,

$$\frac{\partial f[p,x,t]}{\partial t} = -\boldsymbol{v} \cdot \nabla_x f[p,x,t] - \mathcal{C}[f][p,x,t] \,, \tag{6}$$

and linearized it. That is, they took $f = f_0 + \delta f$ and expanded the collision integral $\mathcal{C}$ to first order in δf. To this order, one can write

$$\delta f = f_0(1 \pm f_0)\partial_i v_j (\hat{p}_i \hat{p}_j - \frac{\delta_{ij}}{3})\chi(p) \,. \tag{7}$$

The collision integral is nonlocal in momentum space, so we have an integral equation to solve, which in general is nasty. They showed how to approach this variationally. They also showed that the physics of dynamical screening by the plasma, which is contained within the Hard Thermal Loops (HTL's) of Braaten and Pisarski [14], is sufficient to render the Coulomb divergence finite; the would be logarithmic divergence becomes a large logarithm $\ln(1/g)$. Further, at leading order in this logarithm $\mathcal{C}$ is semilocal in momentum, which allowed Baym *at al.* to solve for the shear viscosity of QCD at leading logarithmic order.

In our first paper [3], we showed that, even at leading logarithmic order, one must also include Compton type processes,

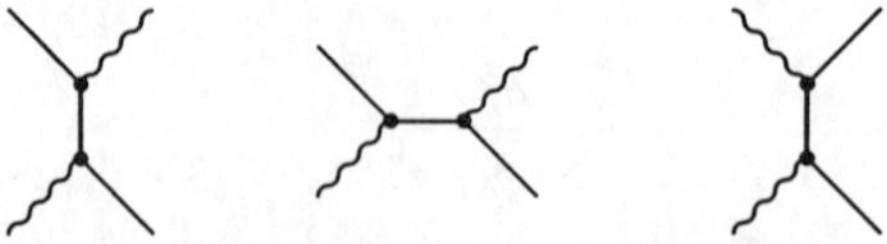

which are logarithmically IR divergent. The role of these processes is not to change particle direction, but to change particle *identity*; a quark on scattering becomes a gluon and *vice versa*. Because of their different group Casimir, gluons equilibrate more quickly than quarks; so the conversion of hard to equilibrate quarks into easy to equilibrate gluons accelerates thermalization. This is needed at leading-log order, though it is only numerically important for electrical conductivity in QED. With these processes, a leading-log treatment is possible [3], yielding, for instance,

$$\sigma = \frac{12^4 \zeta^2(3) \pi^{-3} N_{lept}}{3\pi^2 + 32 N_{spec}} \frac{T}{e^2 \ln 1/e}, \tag{8}$$

$$D_{\rm quark} = \frac{2^4 3^6 \zeta^2(3) \pi^{-3}}{24 + 4N_{\rm f} + \pi^2} \frac{1}{g^4 T \ln 1/g}. \tag{9}$$

The result for η is more cumbersome and can be found in our paper [3]. The above results are only accurate to 0.5% because they were made with the *Ansatz* $\xi(p) \propto p^2$. More accurate results are presented in the paper.

What is required to convert these to a leading-order computation? One only needs to treat the collision integral more carefully, both in the $q \sim T$ momentum region where the collision integral is nonlocal, and in the $q \sim m_{\rm D} \sim g_{\rm s} T$ region where screening effects must be dealt with. For heavy quark diffusion, this is sufficient; Derek Teaney and I recently derived a complete leading order result;

$$D = \frac{72\pi}{C_{\rm f} g_{\rm s}^4 T} \left[\left(N_{\rm c} + \frac{N_{\rm f}}{2} \right) \left(\ln \frac{2T}{m_D} + \frac{1}{2} - \gamma_E + \frac{\zeta'(2)}{\zeta(2)} \right) + \frac{N_{\rm f} \ln 2}{2} \right]^{-1}. \tag{10}$$

However, viscosity in the QGP and light quark diffusion turn out to involve more physics, and therefore more work; one must also include inelastic processes with collinear external states.

Particles in the plasma undergo the left process at an $O(\alpha_{\rm s} T)$ rate, not $\alpha_{\rm s}^2 T$, because of the Coulomb divergence already discussed. The right process is slower by a factor of $\alpha_{\rm s}$; this is true even when the radiated gluon has momentum $O(T)$, provided it is collinear with the particle which emits it. The rate for this process is therefore as large as the large angle scattering rate.

We saw previously that the particles furthest from equilibrium were those with the most energy, $\delta f \propto f |\boldsymbol{p}|^2$. Collinear processes like the one just presented tend to split these high energy particles into lower energy ones, which are more easily bent; therefore they lead to equilibration and lower η, even though they do not change particle direction. Numerically, we find that they reduce η by about 10% [9].

But life is even more complicated; one cannot ignore interference between emission amplitudes of form

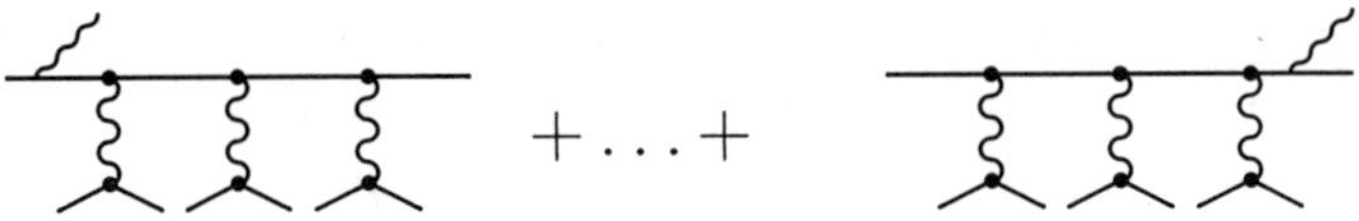

The effect of such interference is referred to as the Landau-Pomeranchuk-Migdal (LPM) effect [16], and it suppresses the efficiency of collinear emission. To see the physical origin of the suppression, consider a particle undergoing multiple scatterings, and remember that it possesses a finite transverse wave packet. That wave packet continues to physically overlap the wave packet of the emitted radiation for some time, called the formation time; if there is a second scattering in that time, the two radiation fields overlap and interfere;

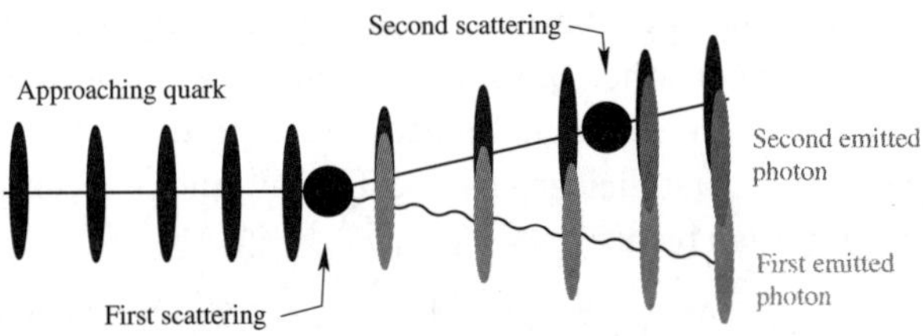

The radiations sum in amplitude; on average the interference is destructive.

Dealing with this complication takes a lot of work, and has been the topic of several studies [17,5,7]. The upshot is that the rate at which a particle of momentum p undergoes a soft scattering which induces a bremsstrahlung of a gluon of energy k, is

$$\frac{d\Gamma(p,k)}{dkdt} = \frac{C_s g_s^2}{16\pi p^7} \frac{1}{1 \pm e^{-k/T}} \frac{1}{1 \pm e^{-(p-k)/T}} \times \left\{ \begin{array}{ll} \frac{1+(1-x)^2}{x^3(1-x)^2} & q \to qg \\ N_{\rm f} \frac{x^2+(1-x)^2}{x^2(1-x)^2} & g \to qq \\ \frac{1+x^4+(1-x)^4}{x^3(1-x)^3} & g \to gg \end{array} \right\} \times$$
$$\times \int \frac{d^2\boldsymbol{h}}{(2\pi)^2} 2\boldsymbol{h} \cdot {\rm Re}\, \boldsymbol{F}(\boldsymbol{h},p,k)\,, \tag{11}$$

where $\boldsymbol{F}$ is the solution of the following integral equation:

$$2\boldsymbol{h} = i\delta E \boldsymbol{F}(\boldsymbol{h}) + g^2 \int \frac{d^2\boldsymbol{q}_\perp}{(2\pi)^2} C(\boldsymbol{q}_\perp) \Big\{ \frac{2C_s - C_{\rm A}}{2} [\boldsymbol{F}(\boldsymbol{h}) - \boldsymbol{F}(\boldsymbol{h} - k\,\boldsymbol{q}_\perp)] +$$
$$+ \frac{C_{\rm A}}{2} [\boldsymbol{F}(\boldsymbol{h}) - \boldsymbol{F}(\boldsymbol{h} + p\,\boldsymbol{q}_\perp) + \boldsymbol{F}(\boldsymbol{h}) - \boldsymbol{F}(\boldsymbol{h} - (p-k)\,\boldsymbol{q}_\perp)] \Big\},$$
$$\delta E \equiv \frac{\boldsymbol{h}^2}{2pk(p-k)} + \frac{m_k^2}{2k} + \frac{m_{p-k}^2}{2(p-k)} - \frac{m_p^2}{2p}\,, \tag{12}$$

with $C(\boldsymbol{q}_\perp) = \frac{m_{\rm D}^2}{\boldsymbol{q}_\perp^2(\boldsymbol{q}_\perp^2 + m_{\rm D}^2)}$, $m_{\rm D}^2 = \frac{g_s^2 T^2}{6}(2N_{\rm c} + N_{\rm f})$. Determining $d\Gamma/dkdt$

therefore requires the solution of an integral equation; and this for a collision term which, to solve the Boltzmann equation, must appear as a coefficient in an integral equation which is to be solved. However, the numerics turn out not to be very demanding.

These ingredients have been combined into a kinetic theory for the QGP [8], which has been linearized and solved to determine the transport coefficients at leading order in the strong coupling [9]. Results for shear viscosity are presented in Fig. 3. At $\alpha_s = 1/3$, η is quite small, and hydrodynamics should be nearly ideal; but the perturbative computation is breaking down at this point and the results cannot be trusted (factor of 3 uncertainty?).

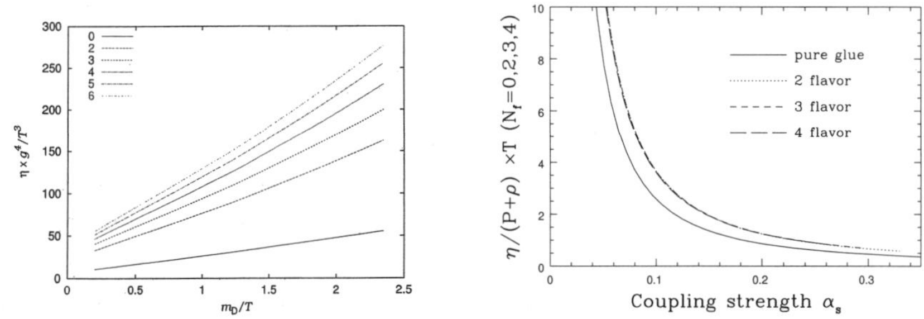

Figure 3. Shear viscosity, scaled by α_s^{-2} (left) and by enthalpy density (right).

4. Range of applicability of QGP transport coefficients

Two key assumptions underlie the above calculation. The first is that the coupling is actually weak, at the scale relevant to the physics in question. For electroweak phase transition applications this is probably a good approximation; for RHIC it is dubious, though most of the theory community is guilty of this assumption. I will not discuss this more here.

The other assumption underpinning our treatment is, that the system is near equilibrium. One can check self-consistency by comparing the time scale of relaxation with the age of the plasma. The ratio $\eta/(\rho+p)$ is a time or length scale, called the sound attenuation length; it differs by a geometric factor $\sim 1/5$ from the relaxation time of the plasma. At $\alpha_s = 1/3$, Fig. 3 shows that this is $\sim 2/3T$, so $\tau \sim 3/T$, which is a few fermis at RHIC. Therefore we expect hydrodynamic treatments to be valid after ~ 3 fermi at RHIC. One might question a hydro description for times earlier than that.

How different could the early time physics be? The answer turns out to be, pretty different. For systems which are far from equilibrium but approximately isotropic, kinetic theory should be applicable under fairly weak conditions [8]. But for systems with strong momentum space anisotropy, this turns out not to be the case. A treatment involving soft gauge fields becomes essential, because these soft gauge fields show exponential growth due to plasma instabilities [18]. It has recently been noted that this may be important for QGP physics [19,20,21], and it is the topic of two other talks at this conference [22,23]. The short story is that, for plasmas which are far from isotropic, soft gauge field modes can grow at a rate suppressed by only a single power of g_s, and may [24] randomize the directions of other particles on this time scale.

5. Conclusions

Transport coefficients in gauge theories can be computed within kinetic theory, but they involve rich physics; hard thermal loops, multiple scattering processes, collinear splitting. The kinetic description is only possible after resumming certain multiple scattering emission processes to account for LPM interference. It has taken nearly 20 years to go from the first attempts at a quantitative analysis to one which is complete at leading order in the coupling g_s.

The calculation of transport coefficients has been directly useful to the heavy ion community [10], but it has also been a good object lesson in non-equilibrium field theory in QCD. Most importantly, we have learned just how narrow the domain of validity of kinetic theory is, and how much more complicated the physics of highly anisotropic plasmas may be. The problem of understanding plasma instabilities in QCD is only beginning to be addressed and should be the subject of much future work.

Acknowledgements

I particularly thank my co-workers, Peter Arnold and Larry Yaffe. I also thank Mike Strickland, Stan Mrówczyński, and Francois Gelis for conversations which have been useful in this work over the past years. My work is currently supported by grants from the Natural Sciences and Engineering Research Council of Canada and by le Fonds Québécois de la Recherche sur la Nature et les Technologies.

References

1. A. Hosoya, M. Sakagami and M. Takao, Annals Phys. **154**, 229 (1984); A. Hosoya and K. Kajantie, Nucl. Phys. **B250**, 666 (1985).
2. G. Baym, H. Monien, C. J. Pethick and D. G. Ravenhall, Phys. Rev. Lett. **64**, 1867 (1990); Nucl. Phys. **A525**, 415C (1991).
3. P. Arnold, G. D. Moore and L. G. Yaffe, JHEP **0011**, 001 (2000) [hep-ph/0010177].
4. G. D. Moore, JHEP **0105**, 039 (2001) [hep-ph/0104121].
5. P. Arnold, G. D. Moore and L. G. Yaffe, JHEP **0111**, 057 (2001) [hep-ph/0109064].
6. P. Arnold, G. D. Moore and L. G. Yaffe, JHEP **0112**, 009 (2001) [hep-ph/0111107].
7. P. Arnold, G. D. Moore and L. G. Yaffe, JHEP **0206**, 030 (2002) [hep-ph/0204343].
8. P. Arnold, G. D. Moore and L. G. Yaffe, JHEP **0301**, 030 (2003) [hep-ph/0209353].
9. P. Arnold, G. D. Moore and L. G. Yaffe, JHEP **0305**, 051 (2003) [hep-ph/0302165].
10. D. Teaney, Phys. Rev. C **68**, 034913 (2003) [nucl-th/0301099].
11. S. Jeon, Phys. Rev. D **52**, 3591 (1995) [hep-ph/9409250]; S. Jeon and L. G. Yaffe, Phys. Rev. D **53**, 5799 (1996) [hep-ph/9512263].
12. G. Aarts and J. M. Martinez Resco, JHEP **0211**, 022 (2002) [hep-ph/0209048]; G. Aarts and J. M. Martinez Resco, Phys. Rev. D **68**, 085009 (2003) [hep-ph/0303216].
13. M. A. Valle Basagoiti, Phys. Rev. D **66**, 045005 (2002) [hep-ph/0204334].
14. E. Braaten and R. D. Pisarski, Nucl. Phys. B **337**, 569 (1990).
15. D. Teaney and G. D. Moore, in preparation.
16. L. D. Landau and I. Pomeranchuk, Dokl. Akad. Nauk Ser. Fiz. **92** (1953) 535; Dokl. Akad. Nauk Ser. Fiz. **92** (1953) 735; A. B. Migdal, Dokl. Akad. Nauk S.S.S.R. **105**, 77 (1955); Phys. Rev. **103**, 1811 (1956).
17. See for instance, R. Blankenbecler and S. D. Drell, Phys. Rev. D **53**, 6265 (1996); R. Baier, Y. L. Dokshitzer, A. H. Mueller, S. Peigne and D. Schiff, Nucl. Phys. B **478**, 577 (1996) [hep-ph/9604327]; B. G. Zakharov, Phys. Atom. Nucl. **61**, 838 (1998) [Yad. Fiz. **61**, 924 (1998)] [hep-ph/9807540].
18. E. S. Weibel, Phys. Rev. Lett. **2**, 83 (1959).
19. S. Mrówczyński, Phys. Lett. B **214**, 587 (1988); Phys. Lett. B **314**, 118 (1993); Phys. Lett. B **393**, 26 (1997) [hep-ph/9606442].
20. P. Romatschke and M. Strickland, Phys. Rev. D **68**, 036004 (2003) [hep-ph/0304092]; hep-ph/0406188.
21. P. Arnold, J. Lenaghan and G. D. Moore, JHEP **0308**, 002 (2003) [hep-ph/0307325].
22. P. Romatschke and M. Strickland, these proceedings, hep-ph/0408314.
23. Peter Arnold, these proceedings.
24. P. Arnold and J. Lenaghan, hep-ph/0408052.

CLASSICAL FIELDS AND HEAVY ION COLLISIONS

T. LAPPI

Department of Physical Sciences, Theoretical Physics Division and Helsinki Institute of Physics
P.O. Box 64, FIN-00014 University of Helsinki, FINLAND
E-mail: tuomas.lappi@helsinki.fi

This is a review of numerical applications of classical gluodynamics to heavy ion collisions. We recall some results from calculations of gluon production, discuss their implications for heavy ion phenomenology, and outline a strategy to calculate the number of quark pairs produced by these classical fields.

1. Introduction

To understand what is happening in relativistic heavy ion colliders, currently RHIC in Brookhaven and in the future the LHC at CERN, one needs to understand not only hard probes, but also bulk particle production and thermalisation. It can be argued that the large phase space densities of partons in the small-x wavefunction of the nuclei generate a hard enough momentum scale to allow weak coupling methods to be used. We shall discuss first some general ideas concerning weak coupling, classical field methods in this context. Then we will go on to discuss first gluon and then quark pair production in the McLerran-Venugopalan model.

2. Relativistic heavy ion collisions

We shall be interested in studying the case where two nuclei move at the speed of light along the $x^{\pm} = 0$-axes, i.e. at $\tau = 0$[a]. These nuclei then collide and leave behind them, at finite values of η and τ, some matter which is then observed in detectors located in some region, varying between different experiments, around $\eta = 0$. We can divide the collision process in different stages:

[a]The light cone coordinates are defined as $x^{\pm} = (t \pm z)/\sqrt{2}$ and the proper time and spacetime rapidity as: $\tau = \sqrt{t^2 - z^2}$ and $\eta = \frac{1}{2} \ln \frac{x^+}{x^-}$.

(1) The initial condition at $\tau = 0$ depends on the properties of the nuclear wavefunction at small x.
(2) The thermal and chemical equilibration of the matter formed at $\tau \lesssim \tau_0$ requires understanding of time dependent, nonequilibrium Quantum Field Theory.
(3) The Quark Gluon Plasma, surviving for some fermis around $\tau_0 \lesssim \tau \lesssim 10\text{fm}$. If the system reaches local thermal equilibrium, finite temperature field theory and relativistic hydrodynamics can be used to describe its behaviour.
(4) Finally, for $\tau \gtrsim 10\text{fm}$ the system hadronises and decouples.

The question of the thermalisation timescale τ_0 remains poorly understood. Hydrodynamical calculations have been very successful in explaining the experimental observations, but their success depends on the assumption of a very early thermalisation time[1]. Most perturbative estimates, i.e. the bottom-up scenario[2], generically produce quite a large thermalisation time, $\tau_0 \gtrsim 3\text{fm}$. On the other hand, one could argue that if the behaviour of the system is characterised by some quite large momentum scale, i.e. the saturation scale $Q_\mathrm{s} \sim 1 \dots 2\text{GeV}$, thermalisation should occur already at times $\tau_0 \sim 1/Q_\mathrm{s} \sim 0.2\text{fm}$. It has been pointed out recently (see e.g. Ref. 3) that plasma instabilities could provide the rapid thermalisation that hydrodynamical models require. In this context classical field models of the nuclear wavefunction and particle production can provide some insight into understanding the collision process.

3. Saturation and the classical field model

The general idea of parton saturation is that the small-x components of the nuclear wavefunction are dominated by a transverse momentum scale, the *saturation scale* Q_s. The scale Q_s is supposed to grow for decreasing x as $Q_\mathrm{s}^2(x) \sim x^{-\lambda}$ with $\lambda \approx 0.3$ giving a good fit to HERA data on deep inelastic scattering[4]. Thus for small enough x or, equivalently, large enough $\sqrt{s}$ we have $Q_\mathrm{s} \gg \Lambda_\mathrm{QCD}$ and we can use weak coupling methods. The presence of the scale Q_s is supposed to be caused by the color fields in the nuclear wave function becoming so strong that the gluonic interactions are dominated by the nonlinearities in the Yang-Mills Lagrangian.

This idea can also be thought of as parton percolation. Let us attach, arguing by the Heisenberg uncertainty principle, to an individual parton with transverse momentum p_T a transverse area $\sim 1/p_\mathrm{T}^2$. One expects a qualitative change in the behaviour of the system at momentum scales where the partons overlap and percolate the transverse plane: $N/p_{\rm T}^2 \sim \pi R_A^2$, where N is the number of partons and πR_A^2 the nuclear transverse area.

Because the color fields are strong, the occupation numbers of quantum states of the system are large, and it is natural to use a classical field approximation. The saturation model has been dubbed the "Color Glass Condensate" and even called "a new form of matter". In theory the term "color glass condensate" refers to the small-x wavefunction of the nucleus, characterised by the saturation scale $Q_{\rm s}$. In practice, most applications of these ideas to heavy ion collisions have so far not been classical field computations, but perturbative calculations with some phenomenological gluon $k_{\rm T}$-distribution depending on $Q_{\rm s}^2(x)$.

3.1. *Heavy ion collisions in the classical field model*

To study particle production in the classical field model we have to solve the Yang-Mills equations of motion:

$$[D_\mu, F^{\mu\nu}] = J^\nu, \tag{1}$$

with a current of two infinitely Lorentz-contracted nuclei moving along the two light cones:

$$J^\mu = \delta^{\mu +}\rho_{(1)}(x_{\rm T})\delta(x^-) + \delta^{\mu -}\rho_{(2)}(x_{\rm T})\delta(x^+). \tag{2}$$

The model part of this approach comes when one must take some form for the charge density $\rho(x_{\rm T})$. The suggestion of McLerran and Venugopalan[5] was to take a random stochastic color source with a Gaussian distribution:

$$\langle \rho^a(x_{\rm T})\rho^b(y_{\rm T})\rangle = g^2\mu^2\delta^{ab}\delta^2(x_{\rm T} - y_{\rm T}) \tag{3}$$

and then average all quantities calculated from $\rho(x_{\rm T})$ with this distribution. An important development has been to consider a more general x-dependent probability distribution $W_x[\rho(x_{\rm T})]$ and derive a renormalisation group equation for this distribution as a function of x, the JIMWLK equation[6].

3.2. *Note on different saturation scales*

Different ways of understanding saturation have led to different definitions of the saturation scale in the literature, which can be a source of considerable confusion.

- The p_{sat} in the EKRT[7] model is a final state saturation scale, not a property of the nuclear wave function (initial state), and thus conceptually a bit different from the saturation scale we have discussed so far.
- We are using the old notation of Krasnitz et. al. In their newer work[8] they define $\Lambda_s = g^2\mu$.
- By calculating the gluon distribution in the McLerran-Venugopalan model one can relate the strength of the color source to the saturation scale Q_s defined by A. Mueller and Yu. Kovchegov (see e.g. Refs. 8, 9)[b]:

$$Q_s^2 = \frac{g^4\mu^2 C_A}{4\pi} \ln\left(\frac{g^4\mu^2}{\Lambda_{QCD}^2}\right) \tag{4}$$

- The saturation scale, or radius, $Q_s = 1/R_s$ in the work of Golec-Biernat and Wusthoff[4] or Rummukainen and Weigert[10] is the same except with C_A replaced by C_F.[c]
- For a comment on the relation to the saturation scale used by E. Iancu et. al., see Ref. 11.

4. Gluon production in the classical field model

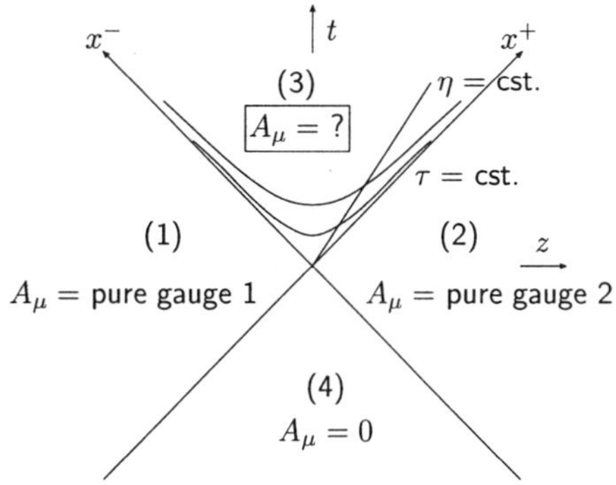

The solution of the Yang-Mills equations in regions (1) and (2) is an analytically known pure gauge field[12], and gives the initial condition for the numerical solution in the forward light cone (3).
The numerical method for solving the Yang-Mills equations in the forward light cone was developed by Krasnitz, Nara and Venugopalan[13,14]. The

---

[b]Note the dependence on an infrared cutoff Λ_{QCD} that gives a large numerical uncertainty.
[c]The techical reason for this is that they consider correlators of Wilson lines in the fundamental representation, whereas the gluon distribution that Mueller and Kovchegov consider involves the same correlator in the adjoint representation.

correct result for the transverse energy was found in Ref. 15, (see also the erratum to Ref. 14 in Ref. 16).

The numerical computation is done in the Hamiltonian formalism. Due to the boost invariance of the initial conditions the Yang-Mills equations can be dimensionally reduced to a 2+1 dimensional gauge theory with an adjoint scalar field. With the assumption of boost invariance one is explicitly neglecting the longitudinal momenta of the gluons; a restriction that should be relaxed in future computations. In the Hamiltonian formalism one obtains directly the (transverse) energy. By decomposing the fields in Fourier modes one can also define a gluon multiplicity corresponding to the classical gauge fields.

4.1. *Numerical results*

Let us define the dimensionless ratios describing the energy and multiplicity:

$$f_E = \frac{\mathrm{d}E/\mathrm{d}\eta}{g^4\mu^3\pi R_A^2} \quad f_N = \frac{\mathrm{d}N/\mathrm{d}\eta}{g^2\mu^2\pi R_A^2} \tag{5}$$

These dimensionless ratios depend, apart from the lattice spacing a, only on one dimensionless parameter characterising the field strength, $g^2\mu R_A$. As can be seen from Fig. 1, for strong enough fields ($g^2\mu R_A \gtrsim 50$) both f_E and f_N are approximately independent of $g^2\mu R_A$ and the lattice spacing. Figure 2 shows the energy as a function of time in different field components and the spectrum of the produced gluons.

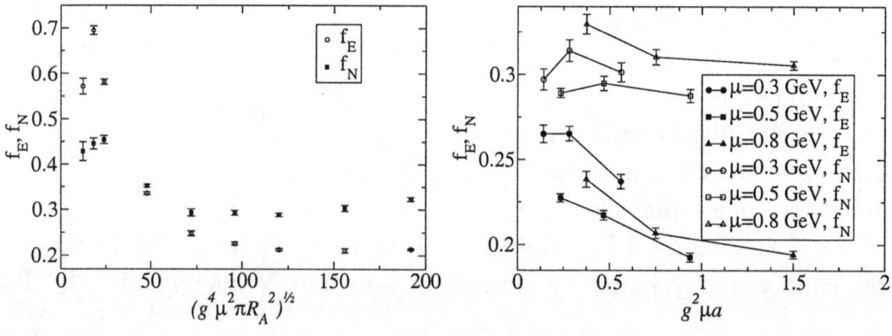

Figure 1. Left: The dependence of f_E and f_N on the field strength paramater $g^2\mu R_A$. Right: The dependence of f_E and f_N on lattice spacing for different values of μ.

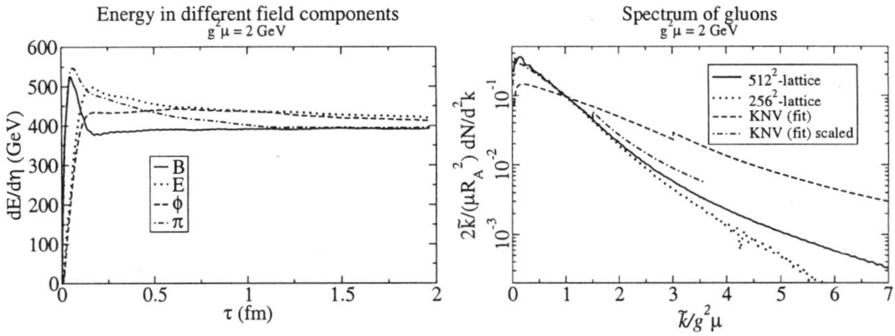

Figure 2. Left: the energy per unit rapidity is evenly distributed between different field components and almost constant after a very short time of the order of $\frac{1}{g^2\mu}$. This means that the 3+1 dimensional energy density decreases as $\varepsilon \sim \tau^{-1}$ Right: The differential multiplicity for two different transverse lattice sizes. The curves labeled "KNV" are a fit to the numerical result of Ref. 14, and KNV "scaled" after these results have been corrected (see Ref. 16).

4.2. *Phenomenology*

At the level of the present discussion $g^2\mu$ is still a free parameter that needs to be fixed from the experimental data. One can distinguish between two broad scenarios for relating the measured results to the calculated initial state quantities.

Hydro scenario If the system thermalises fast, one can use ideal hydrodynamics to follow its subsequent evolution. In this scenario entropy and thus multiplicity are approximately conserved, but the transverse energy decreases by a factor of ~ 3 due to pdV-work going down the beampipe. The corresponding value for the saturation scale is $g^2\mu \approx 1.9$GeV.

Free streaming scenario Assuming that the gluons interact very weakly and do not thermalise, one can argue that longitudinal pressure is negligible and thus the transverse energy is conserved. Assuming energy conservation one gets a lower value for the saturation scale: $g^2\mu \sim 1.4$GeV. This lower value means that the multiplicity must increase during thermalisation or hadronisation of the system approximately by a factor of 2.

At least in the first case a prediction for the multiplicity at central rapidities at the LHC is straightforward; one expects to see $\left(\frac{5500\text{GeV}}{130\text{GeV}}\right)^{0.3} \times 1000 \approx 3000$ particles per unit rapidity[d].

[d]Total, including neutral particles. At this level of approximation, $N_{\rm ch} \approx \frac{2}{3} N_{\rm tot}$.

5. Quark pair production

Given the classical fields corresponding to gluon production a natural question to ask is: how are quark-antiquark pairs produced by these color fields? Formally quark production is suppressed by α_s and group theory factors compared to gluons, so in a first approximation we should be able to treat the quarks as a small perturbation and neglect their backreaction on the color fields. Heavy quark production is calculable already in perturbation theory[e], but one can ask how much the strong color fields change the result. Understanding light quark production would address the question of chemical equilibration; turning the color glass condensate into a *quark* gluon plasma.

The calculation of quark pair production, outlined in more detail in Ref. 18, proceeds by solving the Dirac equation in the background color field of the two nuclei. This can be done analytically for QED[f]. The initial condition for $t \to -\infty$ is a negative energy plane wave. Similarly to the QED case, one can find analytically the solution for the regions $x^\pm > 0$, $x^\mp < 0$. These then give the initial condition at $\tau = 0$ for a numerical solution of the Dirac equation in the forward light cone $\tau > 0$. To find the number of quark pairs one then projects the numerically calculated wave function to a positive energy plane wave at some sufficiently large time τ.

A major technical challenge in this calculation is the coordinate system. In order to include the hard sources of the color fields, the colliding nuclei, only in the initial condition of the numerical calculation, one wants to use the proper time τ instead of the Minkowski time t. Unlike the gluon production case, where one was able to assume strict boost invariance, one now has a nontrivial correlation between the rapidities of the quark and the antiquark. Thus, although the background gauge field is boost invariant, one must solve the Dirac equation 3+1 dimensions.

5.1. *1+1-d toy model*

The longitudinal direction is the one posing the most technical problems. To understand how to handle the longitudinal dimension we can construct a 1+1-dimensional toy model without the transverse dimensions. In 1+1-

---

[e]In the weak field limit quark pair production from this classical field model reduces to a known result in k_T-factorised perturbation theory[17].

[f]The QED calculation, of interest for ultraperipheral collisions, is done e.g. by Baltz and McLerran[19] and others. Baltz, Gelis, McLerran and Peshier[20] discuss the theory in more detail.

dimensions Dirac matrices are 2-dimensional. In the temporal gauge $A_\tau = 0$ that we have been using throughout the calculation there is only one component, A_η, in the external gauge field. The mass m_{eff}^2 in the 1+1-dimensional model corresponds to the transverse mass $k_{\text{T}}^2 + m^2$ of the full theory.

The initial condition at $\tau = 0$ involves a longitudinal momentum scale (the longitudinal momentum of the incoming antiquark) and thus we must use a dimensionful longitudinal variable to be able to represent the initial condition. We choose to take as our coordinates τ and z. The Dirac equation becomes

$$\partial_\tau \psi = \frac{\sqrt{\tau^2 + z^2} + \gamma^0\gamma^3 z}{\tau} \left(-\gamma^0\gamma^3\partial_z\psi - i m_{\text{eff}}\gamma^0\psi\right) - i\gamma^0\gamma^3 g \frac{A_\eta}{\tau}\psi. \qquad (6)$$

Because of the way the coefficients depend explicitly on the coordinates the discretisation of this equation is potentially very unstable and we have to use an explicit discretisation method.

Let us take the background field as[g]:

$$A_\eta = c Q_{\text{s}} \tau J_1(Q_{\text{s}}\tau). \qquad (7)$$

For weak fields $c \ll 1$ we can compute the amplitude for quark pair production using the first order in perturbation theory (diagram (a) in Fig. 3). The result is a peak at

$$2k^+k^- = (p+q)^2 = 2m_{\text{eff}}^2\left(1 + \cosh(\Delta y)\right) = Q_{\text{s}}^2. \qquad (8)$$

As can be seen from Fig. 4, for weak fields our numerical computation reproduces this perturbative peak. For stronger fields the numerical solution sums over all the diagrams in Fig. 3 and the position of the peak is shifted. In the full 3+1-dimensional case this peak is washed out by integration over the relative transverse momentum of the pair.

A rapid back-of-the envelope estimate suggests that with transverse lattices of the order of 256^2 points (lattices from 128^2 to 256^2 were used in the gluon production computation) one could just manage to have a large enough lattice in the z-direction with the computers available to us. Then the computation should in principle be repeated for all the different values of p_{T} (of the antiquark). As this would be prohibitively expensive in terms of CPU-time one will, however, probably have to do with an interpolation from a reasonable amount of points on the p_{T} lattice.

[g]This is the correct time dependence of the perturbative solution to the gauge field equations of motion[12].

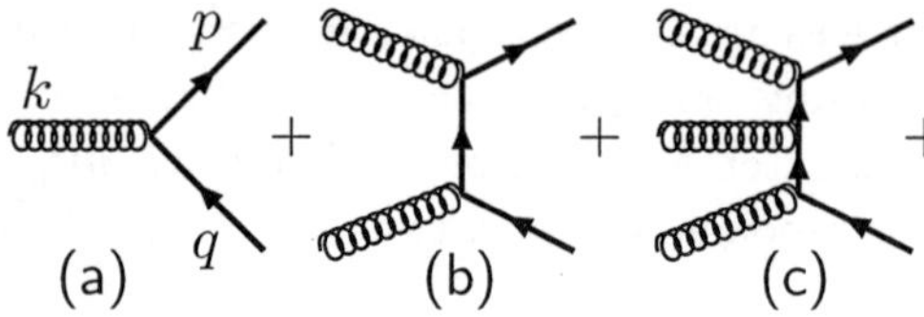

Figure 3. Diagrams contributing to the quark pair prodction amplitude in the 1+1-dimensional toy model.

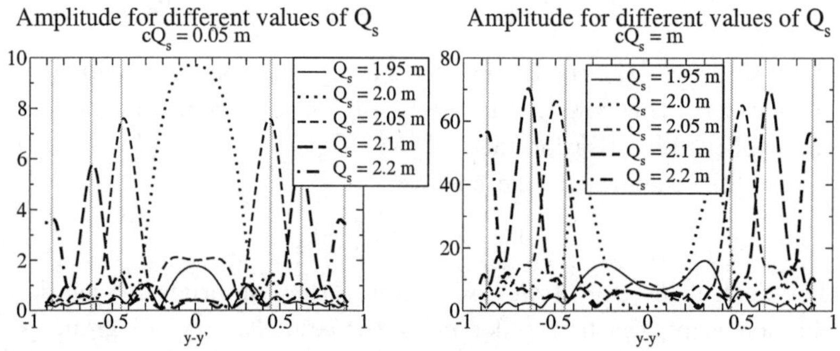

Figure 4. Absolute value of the quark pair production amplitude for different values of the oscillation scale Q_{s}. Left: weak fields, the peaks are at the location explained by Eq. (8). Right: strong fields.

6. Conclusions

Production of gluons has been computed numerically in the classical field model for heavy ion collisions, and the results are in rough agreement with RHIC observations, although not very conclusive due to the uncertainty in the numerical value of the saturation scale. A numerical calculation of the number of quark pairs produced from these classical fields is under way. It is hoped that this computation will tell us something about the how a purely gluonic system can transform into a plasma of both quarks and gluons. Understanding kinetic thermalisation in this context will require treatment of the longitudinal dimension and implementing initial conditions from the JIMWLK equation.

Acknowledgments

The author wishes to thank K. Kajantie and F. Gelis for collaboration in the work presented here and K. Rummukainen, K. Tuominen, A. Krasnitz, Y. Nara and R. Venugopalan for discussions. This work was supported by the Finnish Cultural Foundation and the Magnus Ehrnrooth Foundation.

References

1. P. Kolb, These proceedings, nucl-th/0407066.
2. R. Baier, A. H. Mueller, D. Schiff and D. T. Son, Phys. Lett. **B502**, 51 (2001), [hep-ph/0009237].
3. P. Arnold, These proceedings, hep-ph/0409002.
4. K. Golec-Biernat and M. Wusthoff, Phys. Rev. **D59**, 014017 (1999), [hep-ph/9807513].
5. L. D. McLerran and R. Venugopalan, Phys. Rev. **D49**, 2233 (1994), [hep-ph/9309289].
6. H. Weigert, These proceedings.
7. K. J. Eskola, K. Kajantie, P. V. Ruuskanen and K. Tuominen, Nucl. Phys. **B570**, 379 (2000), [hep-ph/9909456].
8. A. Krasnitz, Y. Nara and R. Venugopalan, Nucl. Phys. **A717**, 268 (2003), [hep-ph/0209269].
9. Y. V. Kovchegov, Nucl. Phys. **A692**, 557 (2001), [hep-ph/0011252].
10. K. Rummukainen and H. Weigert, Nucl. Phys. **A739**, 183 (2004), [hep-ph/0309306].
11. C. S. Lam, G. Mahlon and W. Zhu, Phys. Rev. **D66**, 074005 (2002), [hep-ph/0207058].
12. A. Kovner, L. D. McLerran and H. Weigert, Phys. Rev. **D52**, 3809 (1995), [hep-ph/9505320].
13. A. Krasnitz and R. Venugopalan, Nucl. Phys. **B557**, 237 (1999), [hep-ph/9809433].
14. A. Krasnitz, Y. Nara and R. Venugopalan, Phys. Rev. Lett. **87**, 192302 (2001), [hep-ph/0108092].
15. T. Lappi, Phys. Rev. **C67**, 054903 (2003), [hep-ph/0303076].
16. A. Krasnitz, Y. Nara and R. Venugopalan, Nucl. Phys. **A727**, 427 (2003), [hep-ph/0305112].
17. F. Gelis and R. Venugopalan, Phys. Rev. **D69**, 014019 (2004), [hep-ph/0310090].
18. F. Gelis, K. Kajantie and T. Lappi, hep-ph/0409058.
19. A. J. Baltz and L. D. McLerran, Phys. Rev. **C58**, 1679 (1998), [nucl-th/9804042].
20. A. J. Baltz, F. Gelis, L. D. McLerran and A. Peshier, Nucl. Phys. **A695**, 395 (2001), [nucl-th/0101024].

PROGRESS IN NONEQUILIBRIUM QUANTUM FIELD THEORY II *

JÜRGEN BERGES† AND JULIEN SERREAU‡

Institut für Theoretische Physik, Universität Heidelberg
Philosophenweg 16, 69120 Heidelberg, Germany

We review recent developments for the description of nonequilibrium quantum fields, continuing hep-ph/0302210.[1]

1. Motivations

Nonequilibrium quantum fields play a central role in various areas of physics, such as cosmology, ultra-relativistic heavy-ion collisions, or condensed matter physics. A timely example in cosmology concerns the issue of reheating[2] of the early universe: At the end of inflation the universe is left in a frozen albeit very high-energy state, very different from the subsequent hot universe observed from the cosmic microwave background. This amounts to an enormous entropy growth due to particle production. It is important progress that questions such as the far-from-equilibrium particle production at the end of the inflationary universe, including the subsequent process of thermalization, can be addressed in quantum field theory from first principles.[3] The late-time approach to quantum thermal equilibrium represents a complex collective phenomenon, which can occur on time scales dramatically larger than typical relaxation times.[4,5,6] For phenomenological applications it is therefore often crucial that different quantities effectively thermalize on different time scales and a complete thermalization of all quantities may not be necessary. For instance, an approximately time-independent equation of state $p = p(\epsilon)$, characterized by an almost fixed relation between pressure p and energy density ϵ, can form very early – even

*Prepared for SEWM2004. Based on an invited talk by J.S. and a contributed talk by J.B.

†email: j.berges@thphys.uni-heidelberg.de

‡email: serreau@thphys.uni-heidelberg.de

though the system is still far from equilibrium. Such prethermalized quantities approximately take on their final thermal values already at a time when the occupation numbers of individual momentum modes still show strong deviations from the late-time Bose-Einstein or Fermi-Dirac distribution. Prethermalization is a universal far-from-equilibrium phenomenon which occurs on time scales dramatically shorter than the thermal equilibration time.[7] In order to establish such a behavior it is crucial to be able to compare between the time scales of prethermalization and thermal equilibration. Approaches based on small deviations from equilibrium, or on a sufficient homogeneity in time underlying kinetic descriptions, are not applicable in this case to describe the required "link" between the early and the late-time behavior.

Similar issues arise in the study of high-energy nuclear collisions, where highly excited matter is produced far from equilibrium. The experiments seem to indicate early thermalization whereas the present theoretical understanding of QCD suggests a much longer thermal equilibration time,[8] thereby questioning for instance the picture of a weakly interacting partonic gas.[9] This also points to the need to go beyond kinetic approaches, which do not provide a valid description on very short time scales. Efforts to go beyond kinetic descriptions include studies of the non-equilibrium dynamics of a chiral quark-meson model,[5,7] or photon production[10] in heavy ion collisions.

The recent progress in the theoretical description of nonequilibrium quantum fields is based on efficient functional integral techniques, so-called n-particle irreducible effective actions, for which powerful nonperturbative approximation schemes are available.[11] The techniques provide an efficient approximation scheme also for situations close to, or in equilibrium where resummations are needed. This includes e.g. the calculation of transport coefficients[12] and thermodynamic quantities[13] in hot field theories or the study of critical phenomena near second-order phase transitions.[14] These are promising systematic approaches to cure the convergence problems of perturbative calculations of thermodynamic quantities at high temperature such as pressure or entropy.[15]

Though we focus on particle physics and cosmology applications, we emphasize that these techniques can be equally applied to other nonequilibrium phenomena in complex many body systems. A prominent example is the study of Bose-Einstein condensates. The interpretation of various recent experiments, for instance concerning Bose condensation of molecules,[16] indicate important contributions beyond standard mean-field type approximations for the real-time dynamics of correlations.[17] These can be systematically taken into account using nPI effective action techniques.[18]

Apart from these applications, there has been much progress in understanding formal aspects of approximation schemes based on nPI effective actions, such as renormalization and the validity of Ward identities for gauge theories. In the following we review these developments.

2. Non-equilibrium quantum field theory: methods

2.1. *Standard approximations fail out of equilibrium*

Nonequilibrium dynamics requires the specification of an initial state at some time $t = 0$, characterized by an initial density matrix ρ_D. All the information about the time-evolution of the system is encoded in the correlation functions of field operators $\hat{\varphi}$, which may be obtained from the generalized generating functional:

$$Z[R_1, R_2, \ldots; \rho_D] = \mathrm{Tr}\left\{\rho_D \, \mathrm{e}^{i \sum_n R_n \cdot \hat{\varphi}^n}\right\}, \tag{1}$$

with n-point classical sources $R_n \equiv R_n(x_1, \ldots, x_n)$. Here, $R_n \cdot \hat{\varphi}^n \equiv \int_{x_1 \ldots x_n} R_n(x_1, \ldots, x_n)\, \hat{\varphi}(x_1) \cdots \hat{\varphi}(x_n)$, with $\int_x \equiv \int_{\mathcal{C}} dx^0 \int d^3x$. The closed time contour $\mathcal{C}$,[19] runs from the initial time $t = 0$, where the system has been prepared, to an arbitrarily large time, and backward to the initial time. The functional (1) can be given the following path-integral representation:

$$Z[R_1, R_2, \ldots; \rho_D] = \int d\varphi_0 \, d\varphi_0' \, \langle \varphi_0 | \rho_D | \varphi_0' \rangle \int_{\varphi_0}^{\varphi_0'} \mathcal{D}\varphi \, \mathrm{e}^{i\left(\mathcal{S}[\varphi] + \sum_n R_n \cdot \varphi^n\right)} \tag{2}$$

where $\mathcal{S}[\varphi]$ is the classical action and $\int_{\varphi_0}^{\varphi_0'} \mathcal{D}\varphi$ represents the integral over classical paths along the contour $\mathcal{C}$, with fixed boundaries on both ends at $t = 0$ corresponding to the field configurations $\varphi_0(\mathbf{x})$ and $\varphi_0'(\mathbf{x})$. One finally performs the appropriate average over possible initial conditions. All correlation functions can be given similar expressions by functional differentiation. A direct evaluation of the corresponding path integrals in real time is currently not possible with standard Monte-Carlo techniques. Equivalently, the solution of the real-time dynamics of the functional Schrödinger equation is prohibitively difficult. A frequently employed strategy is to concentrate on classical statistical field theory instead, which can be simulated. This can give important insights when the number of field quanta per mode is sufficiently large such that quantum fluctuations are suppressed compared to statistical fluctuations. However, classical Rayleigh-Jeans divergences and the lack of genuine quantum effects – such as the approach to quantum thermal equilibrium characterized by Bose-Einstein or Fermi-Dirac statistics – limit their use. A coherent understanding of the time evolution in quantum field theory is required – a program which has made substantial progress in recent years with the development of powerful theoretical techniques.

There exist various powerful approximation methods for vacuum field theory as well as for thermal equilibrium at nonzero temperature. However, most standard field theoretical techniques are not applicable for out-of-equilibrium situations, where one encounters additional complications. The first new aspect concerns secularity: The perturbative time evolution suffers from the presence of spurious, so-called secular terms, which grow with time and invalidate the expansion even in the presence of a weak coupling. Here it is important to note that the very same problem appears as well for nonperturbative approximation schemes such as standard $1/N$ expansions, where N denotes the number of field components.[a] Typically, secularity is a not a very difficult problem and for a given approximation there can be various ways to resolve it. There is a requirement, however, which poses very strong restrictions on the possible approximations: Universality, i.e. the insensitivity of the late-time behavior to the details of the initial conditions. If thermal equilibrium is approached then the late-time result is universal in the sense that it becomes uniquely determined by the energy density and further conserved charges. To implement the necessary nonlinear dynamics is demanding. Both requirements of a non-secular and universal behavior can indeed be fulfilled using efficient functional integral techniques: so-called n-particle irreducible effective actions, where the simplest example is given by the two-particle-irreducible (2PI) effective action,[20] which is described in the following.

2.1.1. *Secularity*

Here, we discuss the well-known secularity problem in the context of a relativistic real scalar field theory with potential $m^2\varphi^2/2 + \lambda\varphi^4/4!$. For the argument it is irrelevant that we assume spatial homogeneity and isotropy as well as Z_2-symmetry, so that $\mathrm{Tr}\{\rho_D\, \hat{\varphi}(x)\} = 0$, with Gaussian initial density matrix ρ_D.

[a]Note that restrictions to mean-field type approximations such as leading-order large-N are insufficient. They suffer from the presence of an infinite number of spurious conserved quantities and are known to fail to describe thermalization. Secularity enters the required next-to-leading order corrections and beyond.

The time-ordered propagator along the contour $\mathcal{C}$ is defined as:

$$G(x,y) = \mathrm{Tr}\{\rho_D\, T_{\mathcal{C}}(\hat{\varphi}(x)\,\hat{\varphi}(y))\} = \Theta_{\mathcal{C}}^{+}\, G^{>}(x,y) + \Theta_{\mathcal{C}}^{-}\, G^{<}(x,y)\,, \qquad (3)$$

where $\Theta_{\mathcal{C}}^{\pm} \equiv \Theta_{\mathcal{C}}(\pm(x^0 - y^0))$. One obtains to $\mathcal{O}(\lambda^2)$:

$$G_{2-\mathrm{loop}}(x,y) = G_0(x-y) + \int_{uv} G_0(x-u)\,\Sigma_2(u-v)\,G_0(v-y)\,, \qquad (4)$$

where the bare inverse propagator is given by $G_0^{-1}(x,y) = i(\Box + m^2)\delta_{\mathcal{C}}(x-y)$ and where Σ_2 denotes the self-energy to two-loop order: $\Sigma_2(x-y) = -i\lambda G_0(x,x)\delta_{\mathcal{C}}(x-y)/2 - \lambda^2 G_0(x-y)/6$. The first (one-loop) contribution corresponds to a simple mass shift and plays no role for the present argument. We ignore it as well as all local insertions in the following discussion. We therefore write:

$$\Sigma_2(x-y) = -\frac{\lambda^2}{6} G_0^3(x-y)\,. \qquad (5)$$

For illustrational purposes, we introduce an effective particle number $n_p(t)$, defined as:

$$\omega_p\,(2n_p(t)+1) = \left[\partial_t \partial_{t'} G^{<}(t,t';p) + \omega_p^2 G^{<}(t,t';p)\right]_{t'=t}\,, \qquad (6)$$

where $\omega_p = \sqrt{p^2+m^2}$ and where $G(t,t';p)$ denotes the spatial Fourier transform of the propagator. Writing explicitly the time integrals in (4), one easily obtains, for the two-loop contribution to the particle number:

$$n_p^{2-\mathrm{loop}}(t) - n_p^0 = (1 + n_p^{(0)})\mathcal{R}_p^{<}(t) - n_p^{(0)}\mathcal{R}_p^{>}(t)\,, \qquad (7)$$

where, introducing the Fourier transform in frequency space $\bar{\Sigma}_2^{<,>}(\omega,p)$:

$$\mathcal{R}_p^{<,>}(t) = -\frac{1}{\omega_p}\int \frac{d\omega}{2\pi}\,\bar{\Sigma}_2^{<,>}(\omega,p)\,\frac{1-\cos(\omega-\omega_p)\,t}{(\omega-\omega_p)^2}\,. \qquad (8)$$

In the large-time limit, one gets:

$$\mathcal{R}_p^{<,>}(t) \simeq \bar{\sigma}_p^{<,>} \times t\,, \qquad (9)$$

with the on-shell rates: $\bar{\sigma}_p^{<,>} = -\bar{\Sigma}_2^{<,>}(\omega_p,p)/2\omega_p$. We recover the usual secular behavior of perturbation theory, which occurs whenever the rates in (9) do not vanish on-shell. For instance, in the present two-loop approximation, the production rate $\bar{\sigma}_p^{<}$ receives contributions from in-medium two-body scatterings $\mathbf{q} + \mathbf{k} \to \mathbf{r} + \mathbf{p}$:

$$\bar{\sigma}_p^{<} = \frac{\lambda^2}{4\omega_p}\int \tilde{dq}\tilde{dk}\tilde{dr}\,(2\pi)^4\delta^{(4)}(Q+K-R-P) \times n_q^0\, n_k^0\,(1+n_r^0)\,. \qquad (10)$$

where we introduced four-momenta as *e.g.* $Q^\mu \equiv (\omega_q, \mathbf{q})$ and $\tilde{dq} \equiv \frac{d^3q}{(2\pi)^3 2\omega_q}$.

A similar analysis reveals that higher-order contributions to Eq. (4) are secular as well. The standard solution to this problem corresponds to summing the infinite series of self-energy insertions:

$$\begin{aligned} G_2 &= G_0 + G_0 \cdot \Sigma_2 \cdot G_0 + G_0 \cdot \Sigma_2 \cdot G_0 \cdot \Sigma_2 \cdot G_0 + \ldots \\ &= G_0 + G_0 \cdot \Sigma_2 \cdot G_2 \, , \end{aligned} \tag{11}$$

where we have used an obvious schematic notation. This corresponds to the usual resummation in terms of one-particle-irreducible (1PI) proper vertices. In the two-loop example discussed here, this leads to a non-secular time-evolution. However, the problem reappears for higher order contributions to the self-energy. For instance, the following $\mathcal{O}(\lambda^4)$ contribution

$$\frac{\lambda^4}{36} G_0^2(x-y) \times \int_{uv} G_0(x-u)\, G_0^3(u-v)\, G_0(v-y) \tag{12}$$

contains the same time-integral as analyzed above and thus diverges at large times. Therefore, the systematic resolution of the secularity problem requires infinite resummations beyond the standard 1PI resummation scheme.

To further elaborate the discussion, we consider the sum of contributions (5) and (12), which can be written as: $\Sigma_4 = -(\lambda^2/6) G_0^2 \, \{G_0 + G_0 \cdot \Sigma_2 \cdot G_0\}$. Our previous analysis suggests to resum the infinite geometric series of self-energy (Σ_2) insertions in the term in brackets: $-(\lambda^2/6)\, G_0^2 \, \{G_0 + G_0 \cdot \Sigma_2 \cdot G_0 + \ldots\} = -(\lambda^2/6)\, G_0^2 \, G_2$. Similarly, one can resum self-energy insertions on the other two lines of the original two-loop diagram, to arrive at the following simple – and suggestive – expression:

$$\Sigma_{[2]} = -\frac{\lambda^2}{6}\, G_2^3 \, . \tag{13}$$

To improve on this, we note that the simple expression (13), suggests the following iterative procedure to include higher orders without encountering new secular terms: From the propagator at the n-th iteration $G_{[2n]}$, compute the resummed two-loop self-energy $\Sigma_{[2n]} = -\lambda^2 G_{[2n]}/6$. The next iteration $G_{[2(n+1)]}$ is then computed from $G^{-1}_{[2(n+1)]} = G_0^{-1} - \Sigma_{[2n]}$. Starting from $G_{[0]} = G_0$, this converges, after infinitely many iterations, to the following equation for $G \equiv G_{[\infty]}$:[b]

$$G^{-1} = G_0^{-1} - \Sigma[G] \tag{14}$$

with

$$\Sigma[G] = -\frac{\lambda^2}{6} G^3 . \tag{15}$$

We arrive at a non-linear equation for the two-point function G. The procedure described above guarantees that possible secular terms are automatically resummed at each step of the iteration. Equation (15) is remarkably simple and suggests to reorganize the original perturbative expansion in terms of the full propagator. Before we discuss below that this is efficiently described in terms of the 2PI effective action, we describe in the following that this procedure also automatically respects late-time universality.

2.1.2. *Universality*

The price to be paid for the above non-secular scheme is that at some order one necessarily has to solve nonlinear equations without further approximations. However, as we will see next, it is precisely the nonlinearity which is required to be able to obtain universality in the sense mentioned above. We consider first Eq. (11), which is more conveniently rewritten for initial-value problems as: $G_0^{-1} \cdot G_2 = 1 + \Sigma_2 \cdot G_2$. Using this equation, evaluated along the contour $\mathcal{C}$, one obtains for the time-derivative of the effective particle number $n_p(t)$:

$$\omega_p \, \dot{n}_p(t) = \mathrm{Re} \int_0^t dt' \Big\{ \Sigma_2^{<}(t,t';p)(i\partial_t - \omega_p) G_2^{>}(t',t;p) \\ - \Sigma_2^{>}(t,t';p)(i\partial_t - \omega_p) G_2^{<}(t',t;p) \Big\} . \tag{16}$$

To illustrate the content of this equation, we employ the following free-field like ansatz: $(i\partial_t - \omega_p) G_2^{<}(t',t;p) \to -n_p(t)\, \mathrm{e}^{-i\omega_p(t'-t)}$, and similarly for $G_2^{>}$ with $n_p(t)$ replaced by $1 + n_p(t)$. Using (8)) one finds:

$$\dot{n}_p(t) = (1 + n_p(t))\, \dot{\mathcal{R}}_p^{<}(t) - n_p(t)\, \dot{\mathcal{R}}_p^{>}(t) . \tag{17}$$

[b]This procedure is a direct generalization of the successive summations of so-called daisy and super-daisy diagrams, which correspond to the standard Hartree approximation. With present notations, the latter corresponds to iterating the one-loop self-energy: $\Sigma_{[n]}(x,x') = -i\lambda G_{[n]}(x,x)\, \delta_{\mathcal{C}}(x-x')/2$. In the discussion presented here, we have discarded these (trivial) one-loop tadpole insertions for simplicity.

Using the late-time behavior (9), one finally gets:

$$n_p(t) \simeq n_p^0 \, \mathrm{e}^{-\gamma_p t} + \frac{\bar{\sigma}_p^<}{\gamma_p} \left(1 - \mathrm{e}^{-\gamma_p t}\right) \to \frac{\bar{\sigma}_p^<}{\gamma_p} , \tag{18}$$

where $\gamma_p = \bar{\sigma}_p^> - \bar{\sigma}_p^<$. The late-time limit, indicated on the RHS, is completely determined by the initial particle number n_p^0, see (10). This dependence on initial conditions is rooted in the fact that Eq. (11) is a linear equation for the propagator. Late-time universality – and the associated effective loss of information – requires nonlinear equations.

To illustrate the importance of the resummation procedure discussed in Sec. 2.1.1 and, in particular, of the non-linear character of the resulting equations, we consider the time evolution of the particle number (6) for the approximation (15). We emphasize that these particle numbers are only employed for illustrational purposes. The quantum field theoretical description for the propagator contains off-shell as well as memory effects and is not limited to late-times.[4] Moreover, the functional formulation of this approach, based on the 2PI effective action described below provides a systematic framework for higher-order calculations. Using Eq. (14), it is easy to check that, in the present approximation, the time derivative of $n_p(t)$ has a similar form as in (16) with the replacements $G_2^{<,>} \to G^{<,>}$ and $\Sigma_2^{<,>} \to \Sigma^{<,>}$. Employing a similar free-field like ansatz as above, one obtains the following Boltzmann-like equation at sufficiently large times:

$$\begin{aligned} \dot{n}_p(t) = \frac{\lambda^2}{4\omega_p} \int \tilde{d}q \tilde{d}k \tilde{d}r \, (2\pi)^4 \, \delta^{(4)}(Q + K - R - P) \\ \times \Big\{ n_q(t) n_k(t) (1 + n_r(t))(1 + n_p(t)) \\ - n_p(t) n_r(t)(1 + n_k(t))(1 + n_q(t)) \Big\} . \end{aligned} \tag{19}$$

The late-time equilibrium solution - corresponding to $\dot{n}_p(t) = 0$ - is given by the Bose-Einstein distribution:

$$n_p^{\mathrm{eq}} = \frac{1}{\mathrm{e}^{(\omega_p - \mu)/T} - 1} , \tag{20}$$

and is only characterized by the temperature T and chemical potential μ. The latter are uniquely determined by the total energy and particle number densities, which are conserved in the present approximation.[c] As expected,

[c]While the former is an exactly conserved quantity, the latter is only conserved in the approximation (19), which describes on-shell two-body elastic scatterings. In particular, it is not conserved by the original equation (14), which includes off-shell, particle-number-

the late-time result is insensitive to the details of the initial condition. Note that, in contrast, resummation schemes based on local insertions, such as *e.g.* the so-called 2PPI loop expansion, fail to describe late-time thermalization at finite loop order.[21] This demonstrates that it is crucial to go beyond local (mass) resummations.

2.2. *The 2PI effective action*

The previous analysis illustrates that a systematic way of respecting both non-secularity as well as universality is to reorganize the perturbative expansion in terms of the full propagator. This can be done in an efficient way by performing a double Legendre transform of the functional (1) with respect to the linear and bilinear sources $R_1(x)$ and $R_2(x,y)$. One then obtains the so-called two-particle-irreducible (2PI) effective action, a functional of the one and two-point functions $\phi(x)$ and $G(x,y)$, which can be conveniently parametrized as:[20]

$$\Gamma[\phi,G] = S[\phi] + \frac{i}{2}\,\mathrm{Tr}\ln G^{-1} + \frac{i}{2}\,\mathrm{Tr}\,G_0^{-1}\cdot G + \Gamma_2[\phi,G]\,, \qquad (21)$$

where Γ_2 can be written as an infinite series of closed 2PI diagrams with lines corresponding to G and vertices given by the shifted action $\mathcal{S}[\phi+\varphi]$. Physical solutions corresponds to the stationarity conditions:

$$\frac{\delta\Gamma[\phi,G]}{\delta\phi} = 0 \quad , \quad \frac{\delta\Gamma[\phi,G]}{\delta G} = 0\,. \qquad (22)$$

Using (21), the second of these equations is equivalent to the equation of motion (14) for G, with self-energy given by:

$$\Sigma[\phi,G] = 2i\frac{\delta\Gamma_2[\phi,G]}{\delta G}\,. \qquad (23)$$

Thus, one automatically obtains non-linear, self-consistent equations, as required to describe late-time thermalization. This generalizes the resummation procedure described previously.[d] Systematic approximations include a loop or a $1/N$-expansion of the 2PI effective action beyond leading order.[22]

Finally, the effective action $\Gamma[\phi]$, which encodes all the n-point functions of the theory, is obtained as:

$$\Gamma[\phi] = \Gamma[\phi,G[\phi]]\,, \qquad (24)$$

---

changing processes. When the latter are taken into account, the equilibrium distribution is given by (20) with a vanishing chemical potential.[4]

[d]For instance, the resummed two-loop self-energy (15) is obtained from the three-loop contribution to the functional Γ_2: $i(\lambda^2/48)\int_{xy} G^4(x,y)$.

where $G[\phi]$ is the solution of the second of Eqs. (22). Equation (24) is an exact relation between the 1PI and 2PI functionals which merely states that both are equivalent representations of the theory in the absence of higher-than-linear sources $R_{n\geq 2}$. However, systematic (*e.g.* loop or $1/N$) expansions of the 1PI and 2PI functionals do not coincide order by order in general. Equation (24) therefore defines a non-trivial approximation scheme for the computation of n-point functions. We call it the 2PI resummation scheme. It is non-perturbative in nature and, as we have argued previously, it includes all the ingredients needed to overcome both the secularity and universality problems of non-equilibrium quantum field theory.

2.3. *Symmetries*

Linear symmetries of the classical action are directly promoted at the level of the 2PI effective action. For example, for a $O(N)$-symmetric scalar field theory, $\Gamma[\phi, G]$ is invariant under the transformation:

$$\phi(x) \to \mathcal{R}\phi(x) \quad ; \quad G(x,y) \to \mathcal{R}G(x,y)\mathcal{R}^\dagger \,, \tag{25}$$

where $\mathcal{R}$ denotes a $O(N)$ rotation. This ensures that the resummed effective action (24) has the same symmetry properties as the classical action: $\Gamma[\phi] = \Gamma[\mathcal{R}\phi]$. Notice that this holds for any 2PI approximation which respect the symmetry. In turn, this directly implies that Ward identities are automatically satisfied by the n-point functions calculated from the resummed effective action (24). For instance, in the case of spontaneously broken symmetry, the two-point function, evaluated at the solution $\phi = \bar{\phi}$ of the first of Eqs. (22):

$$\Gamma^{(2)}_{ab}(x,y) = \left.\frac{\delta^2\Gamma[\phi]}{\delta\phi_a(x)\delta\phi_b(y)}\right|_{\phi=\bar{\phi}} \,, \tag{26}$$

where a, b are $O(N)$ indices, exactly satisfies Goldstone's theorem[e].[22,24]

We emphasize that similar arguments can be applied to arbitrary linear and/or affine symmetry transformations including local ones. For instance, for QED in linear gauges, one can show that the 2PI resummed effective action satisfies the usual Ward identities. In particular, one immediately

[e]In contrast, the solution $G_{ab}[\bar{\phi}]$ of the second of Eqs. (22) violates Goldstone's theorem in general. However, it can be shown that for systematic expansion schemes, such as a 2PI loop or $1/N$-expansion at a given order, the functions (26) and $iG^{-1}_{ab}[\bar{\phi}]$ only differ by higher-order terms and, therefore, these violations are of higher order as well.[23]

concludes that the photon polarization tensor calculated from it is transverse in momentum space at any order of a 2PI loop-expansion.[23] The generalization to nonabelian gauge theories is technically more involved and needs to be further investigated.

2.4. *Renormalization*

A systematic renormalization procedure of the resummed propagator at vanishing field, $G[\phi = 0]$, has been proposed in Refs. [24,25] for the 2PI loop-expansion in scalar field theories.[f] This is based on applying a BPHZ subtraction procedure to diagrams with resummed propagators. In particular, it has been shown that the subtleties arising from self-consistent resummations reduce in this case to a proper renormalization of the following Bethe-Salpeter like equation (here, $G \equiv G[\phi = 0]$):

$$\bar{V}(x,y,z,t) = \bar{\Lambda}(x,y,z,t) + \frac{i}{2}\int_{uv\bar{u}\bar{v}} \bar{\Lambda}(x,y,u,v)G(u,\bar{u})G(v,\bar{v})\bar{V}(\bar{u},\bar{v},z,t)\,, \tag{27}$$

where

$$\bar{\Lambda}(x,y,z,t) \equiv 4\frac{\delta^2\Gamma_2[\phi=0,G]}{\delta G(x,y)\delta G(z,t)}\Big|_{G=G[\phi=0]}. \tag{28}$$

This analysis has been recently generalized in [13,23] to include the field-dependence of the 2PI resummed effective action (24). This allows one to determine all the counterterms of the theory at a given level of approximation from suitable renormalization conditions and to obtain finite results for arbitrary n-point functions. As an explicit example, we consider the one-component scalar field theory discussed in the previous sections at order λ_R^2 in the 2PI coupling-expansion, where λ_R denotes the renormalized coupling. We employ renormalization conditions for the two-point function, $\Gamma^{(2)}$, and four-point function, $\Gamma^{(4)}$, given by:

$$\Gamma^{(2)}(x,y) \equiv \frac{\delta^2\Gamma[\phi]}{\delta\phi(x)\delta\phi(y)}\Big|_{\phi=0}, \tag{29}$$

$$\Gamma^{(4)}(x,y,z,t) \equiv \frac{\delta^4\Gamma[\phi]}{\delta\phi(x)\delta\phi(y)\delta\phi(z)\delta\phi(t)}\Big|_{\phi=0}. \tag{30}$$

Without loss of generality we use renormalization conditions for $\phi = 0$

[f]For a recent application to the 2PI $1/N$-expansion, see [26].

which in Fourier space read:

$$Z\,\Gamma^{(2)}(p^2)|_{p=0} = -m_R^2\,, \tag{31}$$

$$Z\,\frac{d}{dp^2}\Gamma^{(2)}(p^2)|_{p=0} = -1\,, \tag{32}$$

$$Z^2\,\Gamma^{(4)}(p_1,p_2,p_3)|_{p_1=p_2=p_3=0} = -\lambda_R\,, \tag{33}$$

with the wave function renormalization Z. Defining the renormalized fields as: $\phi_R = Z^{-1/2}\phi$ and $G_R[\phi_R] = Z^{-1}G[\phi]$, and introducing the usual counterterms: $Zm^2 = m_R^2 + \delta m^2$, $Z^2\lambda = \lambda_R + \delta\lambda$ and $\delta Z = Z - 1$, the classical contribution to the effective action (24) reads:

$$S[\phi] = \int_x \left(\frac{1+\delta Z}{2}\partial_\mu\phi_R\partial^\mu\phi_R - \frac{m_R^2+\delta m^2}{2}\phi_R^2 - \frac{\lambda_R+\delta\lambda}{24}\phi_R^4 \right). \tag{34}$$

Similarly, one can write for the one-loop part: $\mathrm{Tr}\ln G^{-1}[\phi] = \mathrm{Tr}\ln G_R^{-1}[\phi_R]$ up to an irrelevant constant, and:

$$\frac{i}{2}\mathrm{Tr}\,G_0^{-1}[\phi]G[\phi] = -\frac{1}{2}\int_x \left[(1+\delta Z_1)\Box_x + m_R^2 + \delta m_1^2\right] G_R(x,y;\phi_R)|_{x=y} \\ -\frac{\lambda_R+\delta\lambda_1}{4}\int_x \phi_R^2(x)G_R(x,x;\phi_R)\,. \tag{35}$$

Here, δZ_1, δm_1^2 and $\delta\lambda_1$ denote the same counterterms as above, however approximated to the given order. Finally, the contribution from the 2PI functional Γ_2 in Eq. (21) reads, at order λ_R^2:

$$\Gamma_2[\phi,G[\phi]] = -\frac{\lambda_R+\delta\lambda_2}{8}\int_x G_R^2(x,x;\phi_R) + i\frac{\lambda_R^2}{48}\int_{xy} G_R^4(x,y;\phi_R) \\ +i\frac{\lambda_R^2}{12}\int_{xy}\phi_R(x)G_R^3(x,y;\phi_R)\phi_R(y)\,. \tag{36}$$

One first has to calculate the solution $G_R[\phi_R]$ of the stationarity condition for the 2PI effective action. For this, one has to impose the same renormalization conditions as for the propagator (31)-(32) in Fourier space:

$$iG_R^{-1}(p^2;\phi_R)|_{p=0,\phi_R=0} = -m_R^2\,, \tag{37}$$

$$\frac{d}{dp^2}iG_R^{-1}(p^2;\phi_R)|_{p=0,\phi_R=0} = -1\,, \tag{38}$$

together with the condition:

$$\bar{V}_R(p_1,p_2,p_3)|_{p_1=p_2=p_3=0} = -\lambda_R\,, \tag{39}$$

for the renormalized "four-point" field, obtained from Eq. (27). It is easy to check that conditions (37)-(39) fix all the counterterms of the effective action at vanishing field, namely δZ_1, δm_1^2 and $\delta\lambda_2$. This corresponds to the case considered in [24,25].

The remaining counterterms are to be determined from the renormalization conditions (31)-(33). A similar analysis to the one performed in [25] shows that the coupling divergences appearing in the two-point function (29) can be absorbed in a proper renormalization of the following four-point field:[23]

$$V_R(x,y,z,t) \equiv \frac{\delta^2 iG_R^{-1}(x,y;\phi_R)}{\delta\phi_R(z)\phi_R(t)}\bigg|_{\phi_R=0}, \tag{40}$$

which satisfies a Bethe-Salpeter like equation similar to (27). The corresponding condition reads, in momentum space:

$$V_R(p_1,p_2,p_3)|_{p_1=p_2=p_3=0} = -\lambda_R\,. \tag{41}$$

The latter, together with Eqs. (31)-(32) fix the counterterms associated with the quadratic field-dependence in Eqs. (34)-(36), namely δZ, δm^2 and $\delta\lambda_1$.[g] Finally, the remaining counterterm $\delta\lambda$, associated with the quartic field-dependence, is determined from Eq. (33).

It is important to emphasize that Eqs. (37) and (38), as well as (39) and (41) are not independent conditions. Instead, they are fixed by the renormalization conditions (31)-(33). This express the fact that in the exact theory, the following relations between renormalized quantities hold: $\Gamma_R^{(2)} = iG_R^{-1}(\phi_R = 0)$ and $\Gamma_R^{(4)} = \bar{V}_R = V_R$. Thus, the renormalization procedure described here provides an efficient fixing of all the above counterterms, including those associated with the explicit field-dependence of the effective action. The latter play a crucial role *e.g.* in the broken phase and are essential for the determination of the effective potential.

2.5. *Equivalence hierarchy for nPI effective actions*

The 2PI resummation scheme provides a self-consistent dressing of the field ϕ and the propagator G. This can be generalized to higher n-point functions as well: The nPI effective action $\Gamma[\phi, G, V_3, \ldots, V_n]$ provides a self-consistent description for the dressed proper vertices $V_3, \ldots, V_n$, which are obtained the stationarity conditions: $\delta\Gamma/\delta V_3 = 0, \ldots, \delta\Gamma/\delta V_n = 0$. As

[g] In the present approximation, one has $\delta Z_1 = \delta Z$ and $\delta m_1^2 = \delta m^2$, which follow from the identity $\Gamma^{(2)}(x,y) = iG^{-1}(x,y;\phi = 0)$, as well as $\delta\lambda_2 = \delta\lambda_1$, which follows from $V(x,y;z,t) = \bar{V}(x,y,z,t)$.[13] Notice that, although these identities are true in the exact theory, they may not be satisfied in general at the level of approximations.

before, non-secular approximation respecting late-time universality can be obtained from systematic expansion schemes of higher effective actions. In general this leads to non-linear integro-differential equations for the n-point functions, which are now independent self-adjusting variables. As described previously, this is a crucial ingredient for late-time universality.

The use of nPI effective actions with $n > 2$ can be important to describe non-Gaussian initial density matrix for non-equilibrium systems. They are also relevant in the context of high-temperature gauge theories for a quantitative description of transport coefficients beyond the "leading-log" approximation,[27] or in the context of critical phenomena near second-order phase transitions. For instance, the quantitative description of the critical exponents of scalar ϕ^4 theory goes beyond any finite order 2PI loop-expansion[h].[14] It requires taking into account vertex corrections that start with the 4PI effective action to four-loop order.

For practical purposes, a crucial observation is that there exists an equivalence hierarchy between nPI effective actions. For instance, in the case of a loop expansion one has:[27]

$$
\begin{aligned}
&\Gamma^{(1\mathrm{loop})}[\phi] \neq \Gamma^{(1\mathrm{loop})}[\phi, G] = \Gamma^{(1\mathrm{loop})}[\phi, G, V_3] = \ldots\,, \\
&\Gamma^{(2\mathrm{loop})}[\phi] \neq \Gamma^{(2\mathrm{loop})}[\phi, G] = \Gamma^{(2\mathrm{loop})}[\phi, G, V_3] = \Gamma^{(2\mathrm{loop})}[\phi, G, V_3, V_4] = \ldots \\
&\Gamma^{(3\mathrm{loop})}[\phi] \neq \Gamma^{(3\mathrm{loop})}[\phi, G] \neq \Gamma^{(3\mathrm{loop})}[\phi, G, V_3] = \Gamma^{(3\mathrm{loop})}[\phi, G, V_3, V_4] = \ldots \\
&\qquad\vdots
\end{aligned}
$$

where $\Gamma^{(n\mathrm{loop})}$ denotes the approximation of the respective effective action to n-loop order in the absence of external sources. E.g. for a two-loop approximation all nPI descriptions with $n \geq 2$ are equivalent and the 2PI effective action captures already the complete answer for the self-consistent description up to this order. In contrast, a self-consistently complete result to three-loop order requires at least the 3PI effective action in general, etc.[i] Typically the 2PI, 3PI or maybe the 4PI effective action captures already the complete answer for the self-consistent description to the desired/computationally feasible order of approximation.

We thank the organizers for this stimulating meeting. We also thank Sz. Borsányi, U. Reinosa and C. Wetterich for fruitful collaborations.

---

[h] Critical phenomena can be described using the 2PI $1/N$-expansion beyond LO.[14]

[i] There exist further simplifications which can decrease the hierarchy even further such that lower nPI effective actions can be sufficient in practice. For instance, for a vanishing field ϕ, one has, in the absence of sources: $\Gamma^{(3\mathrm{loop})}[\phi = 0, G] = \Gamma^{(3\mathrm{loop})}[\phi = 0, G, V_3 = 0, V_4] = \ldots$.

References

1. J. Berges, J. Serreau, "Progress in nonequilibrium quantum field theory", in Strong and Electroweak Matter 2002, ed. M.G. Schmidt , hep-ph/0302210.
2. J.H. Traschen, R.H. Brandenberger, Phys. Rev. D **42** (1990) 2491. L. Kofman, A.D. Linde, A.A. Starobinsky, Phys. Rev. Lett. **73** (1994) 3195.
3. J. Berges, J. Serreau, Phys. Rev. Lett. **91** (2003) 111601. A. Arrizabalaga, J. Smit, A. Tranberg, hep-ph/0409177. A. Tranberg, these proceedings.
4. J. Berges, J. Cox, Phys. Lett. B **517** (2001) 369. J. Berges, Nucl. Phys. A **699** (2002) 847.
5. J. Berges, S. Borsányi, J. Serreau, Nucl. Phys. B **660** (2003) 51.
6. G. Aarts, J. Berges, Phys. Rev. D **64** (2001) 105010. S. Juchem, W. Cassing, C. Greiner, Phys. Rev. D **69** (2004) 025006. D.J. Bedingham, Phys. Rev. D **68** (2003) 105007.
7. J. Berges, S. Borsányi, C. Wetterich, Phys. Rev. Lett. **93** (2004) 142002 [hep-ph/0403234]. S. Borsányi, these proceedings, hep-ph/0409184.
8. For a review see J. Serreau, Nucl. Phys. A **715** (2003) 805.
9. E. Shuryak, J. Phys. G **30** (2004) S1221.
10. J. Serreau, JHEP **0405** (2004) 078.
11. For an introductory review see J. Berges, "Introduction to nonequilibrium quantum field theory," hep-ph/0409233.
12. G. Aarts, J.M.M. Resco, Phys. Rev. D **68** (2003) 085009. JHEP **0402** (2004) 061. These proceedings, hep-ph/0409090.
13. J. Berges, S. Borsányi, U. Reinosa, J. Serreau, hep-ph/0409123.
14. M. Alford, J. Berges, J.M. Cheyne, appears in Phys. Rev. D [hep-ph/0404059].
15. For a review see J. P. Blaizot, E. Iancu, A. Rebhan, in "Quark Gluon Plasma 3", eds. R.C. Hwa and X.N. Wang, World Scientific, Singapore, 60-122 [hep-ph/0303185].
16. N.R. Claussen, E.A. Donley, S.T. Thompson, C.E. Wieman, Phys. Rev. Lett. **89** (2002) 010401.
17. T. Köhler, T. Gasenzer, K. Burnett, Phys. Rev. A **67** (2003) 013601.
18. J. Berges, T. Gasenzer, M.G.Schmidt, M. Seco, in preparation.
19. J.S. Schwinger, J. Math. Phys. **2** (1961) 407.
20. J.M. Cornwall, R. Jackiw, E. Tomboulis, Phys. Rev. D **10** (1974) 2428.
21. J. Baacke, A. Heinen, Phys. Rev. D **67** (2003) 105020.
22. G. Aarts, D. Ahrensmeier, R. Baier, J. Berges, J. Serreau, Phys. Rev. D **66** (2002) 045008.
23. J. Berges, S. Borsányi, U. Reinosa, J. Serreau, in preparation. U. Reinosa, these proceedings.
24. H. van Hees, J. Knoll, Phys. Rev. D **65** (2002) 025010; Phys. Rev. D **65** (2002) 105005; Phys. Rev. D **66** (2002) 025028.
25. J.P. Blaizot, E. Iancu, U. Reinosa, Phys. Lett. B **568** (2003) 160. Nucl. Phys. A **736** (2004) 149.
26. F. Cooper, B. Mihaila, J.F. Dawson, hep-ph/0407119.
27. J. Berges, appears in Phys. Rev. D, hep-ph/0401172.

A GENERAL EFFECTIVE THEORY FOR DENSE QUARK MATTER

P. T. REUTER, Q. WANG AND D. H. RISCHKE*

Institute für Theoretische Physik,
Johann Wolfgang Goethe – Universität,
Robert-Mayer-Str. 10,
D-60054 Frankfurt am Main, Germany

A general effective action for quark matter at nonzero temperature and/or nonzero density is derived. Irrelevant quark modes are distinguished from relevant quark modes, and hard from soft gluon modes, by introducing two separate cut-offs in momentum space, one for quarks, Λ_q, and one for gluons, Λ_g. Irrelevant quark modes and hard gluon modes are then exactly integrated out in the functional integral representation of the QCD partition function. Depending on the specific choice for Λ_q and Λ_g, the resulting effective action contains well-known effective actions for hot and/or dense quark matter, for instance the "Hard Thermal Loop" (HTL) or the "Hard Dense Loop" (HDL) action, as well as the high-density effective theory proposed by Hong and others.

1. Introduction

Sufficiently cold and dense quark matter is a color superconductor [1]. Due to asymptotic freedom, at asymptotically large quark chemical potential, $\mu \gg \Lambda_{\rm QCD}$, the strong coupling constant $g \ll 1$, and one can apply weak-coupling methods to compute the color-superconducting gap parameter. For N_f massless quark flavors which participate in the screening of gluon fields, to subleading order in g the result is [2,3,4,6,7]

$$\phi = 512\,\pi^4 \left(\frac{2}{N_f\, g^2} \right)^{5/2} \mu \exp\left(-\frac{3\pi^2}{\sqrt{2}g} - \frac{\pi^2+4}{8} \right) [1+O(g)] \ . \qquad (1)$$

Sub-subleading corrections enter as $O(g)$ corrections to the prefactor of the exponential. For a reliable extrapolation of this result to quark chemical potentials of physical interest, $\mu \sim 500$ MeV, it is mandatory to know the order of magnitude of these sub-subleading corrections. However, in full

*Invited speaker

QCD their calculation appears prohibitively difficult. In order to proceed, note that in weak coupling, $g \ll 1$, there is an ordering of scales $\phi \sim \mu \exp(-1/g) \ll g\mu \ll \mu$. This ordering of scales implies that the modes near the Fermi surface, which participate in the formation of Cooper pairs and are therefore of primary relevance in the gap equation, can be considered to be independent of the detailed dynamics of the modes deep within the Fermi sea. This suggests that the most efficient way to compute properties such as the color-superconducting gap parameter is via an *effective theory for quark modes near the Fermi surface.* Such an effective theory has been originally proposed by Hong [8,9] and was subsequently refined by others [10,11,12,13].

In this paper, we derive a general effective theory for hot and/or dense quark matter from first principles (see also Ref. [14]). We introduce cut-offs in momentum space for quarks, Λ_q, and gluons, Λ_g. These cut-offs separate relevant from irrelevant quark modes and soft from hard gluon modes. We then explicitly integrate out irrelevant quark and hard gluon modes. We show that the standard HTL and HDL effective actions are contained in our general effective action for a certain choice of the quark and gluon cut-offs Λ_q, Λ_g. It can also be shown [14] that the action of the high-density effective theory derived by Hong and others [8,9,10,11,12,13] is a special case of our general effective action. We also outline the computation of the color-superconducting gap parameter with this effective action (for more details, see Ref. [14]).

Our units are $\hbar = c = k_B = 1$. 4-vectors are denoted by $K^\mu = (k_0, \mathbf{k})$, with $\mathbf{k}$ being a 3-vector of modulus $|\mathbf{k}| \equiv k$. For the summation over Lorentz indices, we use the familiar Minkowski metric $g^{\mu\nu} = \mathrm{diag}(+,-,-,-)$, although we exclusively work in compact Euclidean space-time with volume V/T, where V is the 3-volume and T the temperature of the system. Space-time integrals are denoted as $\int_0^{1/T} d\tau \int_V d^3\mathbf{x} \equiv \int_X$. Energy-momentum sums are denoted as $(T/V)\sum_K \equiv T\sum_n (1/V)\sum_{\mathbf{k}}$. The sum over n runs over the Matsubara frequencies, $\omega_n^{\mathrm{b}} = 2n\pi T$ for bosons and $\omega_n^{\mathrm{f}} = (2n+1)\pi T$ for fermions. The 4-dimensional delta-function is conveniently defined as $\delta^{(4)}(X) \equiv \delta(\tau)\,\delta^{(3)}(\mathbf{x})$.

2. Deriving the Effective Theory

The partition function for QCD in the absence of external sources reads

$$Z = \int [DA] \exp\{S_A[A]\}\, Z_q[A] \,. \tag{2}$$

Here the (gauge-fixed) gluon action is

$$S_A[A] = \int_X \left[-\frac{1}{4} F_a^{\mu\nu}(X)\, F^a_{\mu\nu}(X) \right] + S_{\rm gf}[A] + S_{\rm ghost}[A] \ , \qquad (3)$$

where $F^a_{\mu\nu} = \partial_\mu A^a_\nu - \partial_\nu A^a_\mu + g f^{abc} A^b_\mu A^c_\nu$ is the gluon field strength tensor, $S_{\rm gf}$ is the gauge-fixing part, and $S_{\rm ghost}$ the ghost part of the action.

The partition function for quarks in the presence of gluon fields is

$$Z_q[A] = \int [D\bar{\Psi}]\, [D\Psi]\, \exp\left\{ S_q[A, \bar{\Psi}, \Psi] \right\} \ , \qquad (4)$$

where the quark action is

$$S_q[A, \bar{\Psi}, \Psi] = \frac{1}{2} \int_{X,Y} \bar{\Psi}(X)\, G_0^{-1}(X,Y)\, \Psi(Y) + \frac{g}{2} \int_X \bar{\Psi}(X)\, \hat{\Gamma}^\mu_a A^a_\mu(X)\, \Psi(X) \ . \qquad (5)$$

Here,

$$\Psi \equiv \begin{pmatrix} \psi \\ \psi_C \end{pmatrix} \ , \qquad \bar{\Psi} \equiv (\bar{\psi}, \bar{\psi}_C) \ , \qquad (6)$$

are Nambu-Gor'kov quark spinors, where $\psi_C \equiv C\bar{\psi}^T$, $\bar{\psi}_C \equiv \psi^T C$ are charge-conjugate quark spinors; $C \equiv i\gamma^2\gamma_0$ is the charge conjugation matrix. The free inverse quark propagator in the Nambu-Gor'kov basis is

$$G_0^{-1}(X,Y) \equiv \begin{pmatrix} [G_0^+]^{-1}(X,Y) & 0 \\ 0 & [G_0^-]^{-1}(X,Y) \end{pmatrix} \ , \qquad (7)$$

with the free inverse propagator for quarks and charge-conjugate quarks

$$[G_0^\pm]^{-1}(X,Y) \equiv (i\partial\!\!\!/_X \pm \mu\gamma_0 - m)\, \delta^{(4)}(X-Y) \ . \qquad (8)$$

The quark-gluon vertex in the Nambu-Gor'kov basis is defined as

$$\hat{\Gamma}^\mu_a \equiv \begin{pmatrix} \gamma^\mu T_a & 0 \\ 0 & -\gamma^\mu T_a^T \end{pmatrix} \ . \qquad (9)$$

We now transform all fields into energy-momentum space. The partition function for quarks becomes

$$Z_q[A] = \int [D\bar{\Psi}]\, [D\Psi]\, \exp\left[\frac{1}{2}\, \bar{\Psi}\, (G_0^{-1} + gA)\, \Psi \right] \ . \qquad (10)$$

Here, we employ a compact matrix notation,

$$\bar{\Psi}\, (G_0^{-1} + gA)\, \Psi \equiv \sum_{K,Q} \bar{\Psi}(K)\, \left[G_0^{-1}(K,Q) + gA(K,Q) \right]\, \Psi(Q) \ , \qquad (11)$$

with

$$G_0^{-1}(K,Q) = \frac{1}{T} \begin{pmatrix} [G_0^+]^{-1}(K) & 0 \\ 0 & [G_0^-]^{-1}(K) \end{pmatrix} \delta^{(4)}_{K,Q} \, , \tag{12}$$

where $[G_0^\pm]^{-1}(K) \equiv \not{K} \pm \mu\gamma_0 - m$, and

$$A(K,Q) \equiv \frac{1}{\sqrt{VT^3}} \hat{\Gamma}^\mu_a A^a_\mu(K-Q) \, . \tag{13}$$

The next step is to integrate out irrelevant quark modes. We define projection operators P_1, P_2 for relevant and irrelevant quark modes, respectively,

$$\Psi_1 \equiv P_1 \Psi \, , \quad \Psi_2 \equiv P_2 \Psi \, , \quad \bar{\Psi}_1 \equiv \bar{\Psi}\, \gamma_0 P_1 \gamma_0 \, , \quad \bar{\Psi}_2 \equiv \bar{\Psi}\, \gamma_0 P_2 \gamma_0 \, . \tag{14}$$

The Grassmann integration over the irrelevant quark fields $\bar{\Psi}_2$, Ψ_2 can be done exactly. The result for the QCD partition function is

$$Z = \int [DA]\, [D\bar{\Psi}_1]\, [D\Psi_1]\, \exp\left\{ S[A, \bar{\Psi}_1, \Psi_1] \right\} \, , \tag{15a}$$

$$S[A,\bar{\Psi}_1,\Psi_1] \equiv S_A[A] + \frac{1}{2}\bar{\Psi}_1 \left(G^{-1}_{0,11} + gB[A] \right) \Psi_1 + \frac{1}{2}\mathrm{Tr}_q \ln G^{-1}_{22}[A] , \tag{15b}$$

where

$$gB[A] \equiv gA_{11} - gA_{12}\, G_{22}[A]\, gA_{21} \, , \tag{16a}$$

$$G^{-1}_{22}[A] \equiv G^{-1}_{0,22} + gA_{22} \, . \tag{16b}$$

The quantities $G^{-1}_{0,11}$, $G^{-1}_{0,22}$, A_{11}, A_{12}, A_{21}, A_{22} carry as argument a pair of quark 4-momenta, (K,Q). The set of subscripts $(nm), n, m = 1, 2$, at these quantities indicates whether the corresponding 3-momenta $\mathbf{k}$, $\mathbf{q}$ belong to the subspace of relevant ($n = 1$) or irrelevant ($n = 2$) quark modes.

Similarly to the treatment of fermions we now define projectors Q_1, Q_2 for soft and hard gluon modes, respectively,

$$A_1 \equiv Q_1\, A \, , \quad A_2 \equiv Q_2\, A \, . \tag{17}$$

Inserting $A \equiv A_1 + A_2$ into Eq. (15b), we sort the result with respect to powers of the hard gluon field, A_2,

$$\begin{aligned} S[A,\bar{\Psi}_1,\Psi_1] &= S[A_1,\bar{\Psi}_1,\Psi_1] + A_2\, J[A_1,\bar{\Psi}_1,\Psi_1] \\ &\quad - \frac{1}{2}\, A_2\, \Delta^{-1}_{22}[A_1,\bar{\Psi}_1,\Psi_1]\, A_2 + S_I[A_1,A_2,\bar{\Psi}_1,\Psi_1] \, . \end{aligned} \tag{18}$$

The first term in this expansion, containing no hard gluon fields at all, is simply the action (15b), with A replaced by the soft gluon field A_1. The second term, $A_2 J$, contains a single power of the hard gluon field, where $J[A_1,\bar{\Psi}_1,\Psi_1] \equiv J_B[A_1,\bar{\Psi}_1,\Psi_1] + J_{\rm loop}[A_1] + J_V[A_1]$. The first contribution,

J_B, arises from the coupling of the relevant fermions to the "modified" gluon field $B[A]$, i.e., from the second term in Eq. (15b). The second contribution, $J_{\rm loop}$, arises from the term $\mathrm{Tr}_q \ln G_{22}^{-1}$ in Eq. (15b), plus an analogous contribution from the Fadeev-Popov determinant contained in $S_A[A]$. Finally, the third contribution, J_V, arises from the non-Abelian vertices in QCD.

The third term in Eq. (18) is quadratic in A_2, where $\Delta_{22}^{-1}[A_1, \bar{\Psi}_1, \Psi_1] \equiv \Delta_{0,22}^{-1} + \Pi_{22}[A_1, \bar{\Psi}_1, \Psi_1]$. Here, $\Delta_{0,22}^{-1}$ is the free inverse propagator for hard gluons. Similarly to the "current" J the "self-energy" Π_{22} of hard gluons consists of three different contributions, $\Pi_{22}[A_1, \bar{\Psi}_1, \Psi_1] = \Pi_B[A_1, \bar{\Psi}_1, \Psi_1] + \Pi_{\rm loop}[A_1] + \Pi_V[A_1]$.

Finally, we collect all terms with more than two hard gluon fields A_2 in Eq. (18) in the "interaction action" for hard gluons, $S_I[A_1, A_2, \bar{\Psi}_1, \Psi_1]$. In order to perform the functional integration over the hard gluon fields A_2 one adds the source term $A_2 J_2$ to the action (15b) and then replaces the fields A_2 in S_I by functional differentiation with respect to J_2, at $J_2 = 0$. Moving the factor $\exp\{S_I[A_1, \delta/\delta J_2, \bar{\Psi}_1, \Psi_1]\}$ in front of the functional A_2-integral, the latter is Gaussian and can be readily performed (after a suitable shift of A_2). We refrain from giving the explicit result, because we will simplify it anyway by employing the following two approximations. The first is based on the principle assumption in the construction of any effective theory, namely that soft and hard modes are well separated in momentum space. Consequently, momentum conservation does not allow a hard gluon to couple to any (finite) number of soft gluons. Under this assumption, the diagrams generated by $A_2(J_{\rm loop} + J_V)$ will not occur in the effective theory. In the following, we shall therefore omit these terms, such that $J \equiv J_B$. Our second approximation is that we approximate $S_I \simeq 0$. With these approximations, the partition function reads

$$Z = \int [DA_1]\, [D\bar{\Psi}_1]\, [D\Psi_1]\, \exp\{S_{\rm eff}[A_1, \bar{\Psi}_1, \Psi_1]\}\,, \tag{19}$$

with the effective action

$$\begin{aligned} S_{\rm eff}[A_1, \bar{\Psi}_1, \Psi_1] \equiv\; & S_A[A_1] + \frac{1}{2}\, \bar{\Psi}_1 \left\{G_{0,11}^{-1} + gB[A_1]\right\} \Psi_1 \\ & + \frac{1}{2}\, \mathrm{Tr}_q \ln G_{22}^{-1}[A_1] - \frac{1}{2}\, \mathrm{Tr}_g \ln \Delta_{22}^{-1}[A_1, \bar{\Psi}_1, \Psi_1] \\ & + \frac{1}{2}\, J_B[A_1, \bar{\Psi}_1, \Psi_1]\, \Delta_{22}[A_1, \bar{\Psi}_1, \Psi_1]\, J_B[A_1, \bar{\Psi}_1, \Psi_1]. \end{aligned} \tag{20}$$

This is the desired action for the effective theory describing the interaction of relevant quark modes, $\bar{\Psi}_1, \Psi_1$, and soft gluons, A_1. The vertices arising in the effective action (20) are displayed in Figs. 1 – 5.

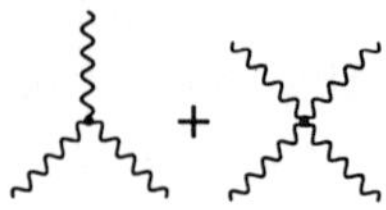

Figure 1. The three- and four-gluon vertices in $S_A[A_1]$.

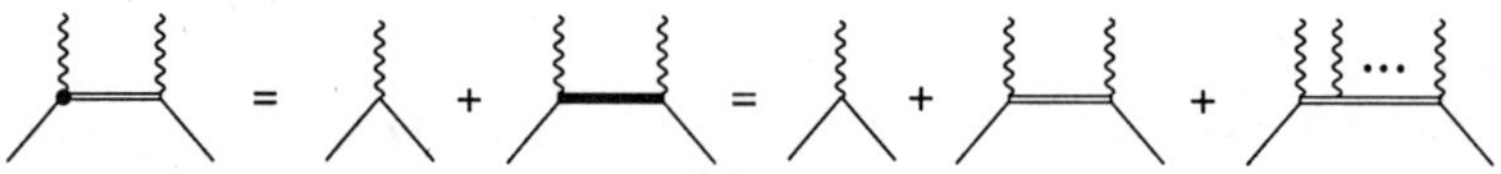

Figure 2. The term $\bar{\Psi}_1 g B[A_1] \Psi_1$ in the effective action (20). The left-hand side is a graphical definition for this term. The right-hand side is obtained from the definition of $gB[A_1]$, Eq. (16a), and subsequently expanding the full propagator for irrelevant quarks, G_{22} (thick solid line), in this equation in powers of the soft gluon field, A_1 (wavy line), according to Eq. (16b). A double line stands for the free propagator for irrelevant quarks, $G_{0,22}$. Thin lines stand for relevant quark fields $\Psi_1, \bar{\Psi}_1$.

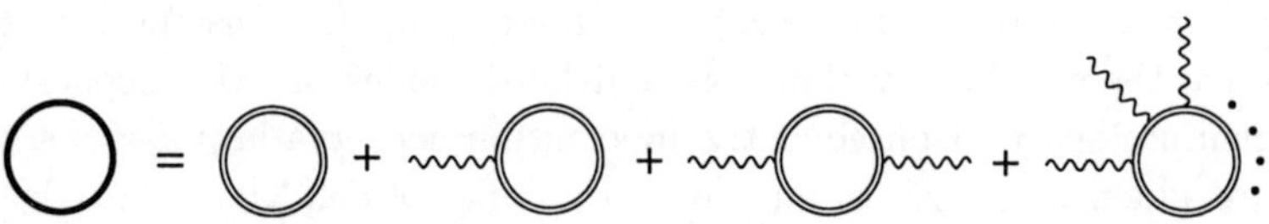

Figure 3. The term $\mathrm{Tr}_q \ln G_{22}^{-1}[A_1]$ in the effective action (20). The right-hand side is obtained upon expanding the full propagator G_{22} in powers of the soft gluon field, A_1, according to Eq. (16b).

3. Examples

In all examples discussed in this section, the gluon projectors are chosen as

$$Q_1(P_1, P_2) \equiv \Theta(\Lambda_g - p_1)\, \delta^{(4)}_{P_1,P_2}\,, \tag{21a}$$

$$Q_2(P_1, P_2) \equiv \Theta(p_1 - \Lambda_g)\, \delta^{(4)}_{P_1,P_2}\,. \tag{21b}$$

The gluon cut-off momentum Λ_g separates soft gluon fields, $A_1 \equiv Q_1 A$, from hard gluon fields, $A_2 \equiv Q_2 A$.

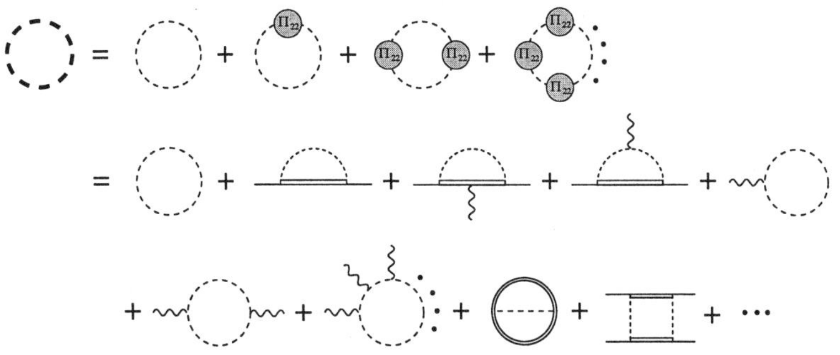

Figure 4. The term $\text{Tr}_g \ln \Delta_{22}^{-1}[A_1, \bar{\Psi}_1, \Psi_1]$ in the effective action (20). The first line is obtained from the expression for Δ_{22}; a dashed line denotes the free propagator for hard gluons $\Delta_{0,22}$, a bubble denotes the full self-energy Π_{22}. The second and third lines contain some examples for diagrams generated when explicitly inserting Π_{22}.

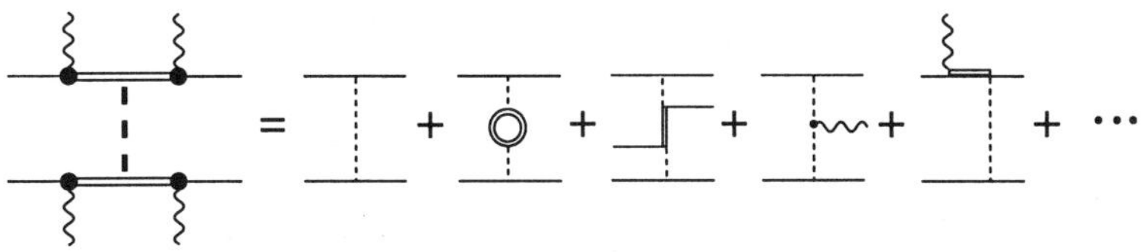

Figure 5. The term $J_B \Delta_{22} J_B$ in the effective action (20). The left-hand side serves as a graphical definition for this term, the right-hand side shows some examples for diagrams when explicitly inserting the expressions for J_B and Δ_{22}.

3.1. *HTL/HDL Effective Action*

This action defines an effective theory for massless quarks and gluons with small momenta in a system at high temperature T (HTL), or large chemical potential μ (HDL). Consequently, the projectors $P_{1,2}$ for quarks are given by

$$P_1(K,Q) = \Theta(\Lambda_q - k)\, \delta^{(4)}_{K,Q} \,, \tag{22a}$$

$$P_2(K,Q) = \Theta(k - \Lambda_q)\, \delta^{(4)}_{K,Q} \,, \tag{22b}$$

The essential assumption to derive the HTL/HDL effective action is that there is a single momentum scale, $\Lambda_q = \Lambda_g \equiv \Lambda$, which separates hard modes with momenta $\sim T$, or $\sim \mu$, from soft modes with momenta $\sim gT$, or $\sim g\mu$. In order to show that the HTL/HDL effective action is contained in the effective action (20), we first note that a soft particle cannot become hard by interacting with another soft particle. This has the consequence that a soft quark cannot turn into a hard one by soft-gluon scattering.

Therefore,

$$gB[A_1] \equiv gA_{11} \, . \tag{23}$$

Another consequence is that the last term in Eq. (20), $J_B \Delta_{22} J_B$, vanishes since J_B is identical to a vertex between a soft quark and a hard gluon, which is kinematically forbidden. The resulting action then reads

$$S_{\text{high}\,T/\mu}[A_1, \bar{\Psi}_1, \Psi_1] \equiv S_A[A_1] + \frac{1}{2}\,\bar{\Psi}_1 \left(G_{0,11}^{-1} + gA_{11}\right) \Psi_1 \\ + \frac{1}{2}\text{Tr}_q \ln G_{22}^{-1}[A_1] - \frac{1}{2}\text{Tr}_g \ln \Delta_{22}^{-1}[A_1, \bar{\Psi}_1, \Psi_1] . \tag{24}$$

We realize that the term $\text{Tr}_q \ln G_{22}^{-1}$ generates all one-loop diagrams, where n soft gluon legs are attached to a hard quark loop. This is precisely the quark-loop contribution to the HTL/HDL effective action.

Now consider the last term, $\text{Tr}_g \ln \Delta_{22}^{-1}$. For hard gluons with momentum $\sim T$ or $\sim \mu$, the free inverse gluon propagator is $\Delta_{0,22}^{-1} \sim T^2$ or $\sim \mu^2$, while the contribution Π_{loop} to the hard gluon "self-energy" Π_{22} is at most of the order $\sim g^2T^2$ or $\sim g^2\mu^2$. Consequently, Π_{loop} can be neglected and Π_{22} only contains tree-level diagrams, $\Pi_{22} \equiv \Pi_B + \Pi_V$. Now expand $\text{Tr}_g \ln \Delta_{22}^{-1}$ in powers of Π_{22}. The terms which contain only insertions of Π_V correspond to one-loop diagrams where n soft gluon legs are attached to a hard gluon loop. This is precisely the pure gluon loop contribution to the HTL effective action. For the HDL effective action, this contribution is $\sim g^2T^2$, and thus negligible.

The "self-energy" Π_B contains only two soft quark legs attached to a hard quark propagator (via emission and absorption of hard gluons). Consequently, in the expansion of $\text{Tr}_g \ln \Delta_{22}^{-1}$, the terms which contain insertions of Π_V and Π_B correspond to one-loop diagrams where an arbitrary number of soft quark and gluon legs is attached to the loop. It was shown in Ref. [15] that of these diagrams, only the ones with two soft quark legs and no four-gluon vertices are kinematically important and thus contribute to the HTL/HDL effective action. We have thus shown that the latter is contained in the effective action (24), and constitutes its leading contribution.

3.2. *High-density Effective Theory*

In an effective theory for cold, dense quark matter the projectors are chosen as

$$P_1(K,Q) \equiv \begin{pmatrix} \Lambda_{\mathbf{k}}^+ & 0 \\ 0 & \Lambda_{\mathbf{k}}^- \end{pmatrix} \Theta(\Lambda_q - |k - k_F|)\, \delta^{(4)}_{K,Q} \,, \tag{25a}$$

$$P_2(K,Q) \equiv \left[\begin{pmatrix} \Lambda_{\mathbf{k}}^- & 0 \\ 0 & \Lambda_{\mathbf{k}}^+ \end{pmatrix} + \begin{pmatrix} \Lambda_{\mathbf{k}}^+ & 0 \\ 0 & \Lambda_{\mathbf{k}}^- \end{pmatrix} \Theta(|k - k_F| - \Lambda_q) \right] \delta^{(4)}_{K,Q} \,. \tag{25b}$$

Here, $\Lambda_{\mathbf{k}}^e \equiv [E_{\mathbf{k}} + e\gamma_0 (\boldsymbol{\gamma} \cdot \mathbf{k} + m)] / (2E_{\mathbf{k}})$ are projection operators onto states with positive ($e = +$) or negative ($e = -$) energy, where $E_{\mathbf{k}} = \sqrt{\mathbf{k}^2 + m^2}$ is the relativistic single-particle energy. The momentum cut-off Λ_q controls how many quark modes (with positive energy) are integrated out. Thus, all quark modes within a layer of width $2\Lambda_q$ around the Fermi surface are considered as relevant, while all antiquark modes and quark modes outside this layer are considered as irrelevant, cf. Fig. 6. It can be shown [14] that the effective theory with the projection operators (25) is equivalent to the high-density effective theory proposed by Hong [8,9] and refined by others [10,11,12,13].

Armed with the effective action (20), one can compute the color-superconducting gap parameter in weak coupling [14]. One first derives the 2PI effective action [16], Γ, with Eq. (20) as the underlying *tree-level* action. The stationary points of Γ yield Dyson-Schwinger equations for the propagators of relevant quarks and soft gluons, which can be solved in a particular many-body approximation scheme. The gap equation for the color-superconducting gap parameter is equivalent to the equation for the off-diagonal component of the quark propagator in the Nambu-Gor'kov basis. The introduction of the quark and gluon cut-offs Λ_q and Λ_g allows for a rigorous power-counting of various contributions to the gap equation. It turns out that, in order to obtain the subleading-order result (1), one has to assume $\Lambda_q \lesssim g\mu \ll \Lambda_g \lesssim \mu$.

The situation is illustrated in Fig. 6. Relevant quark modes lie in a narrow "shell" of width $2\Lambda_q$ around the Fermi surface, irrelevant quarks lie outside this shell. The leading contribution to the QCD gap equation arises from soft magnetic gluon exchange, which mediates between quark states inside a "patch" of diameter $2\Lambda_g$ located on this shell. Subleading contributions arise from soft electric gluon exchange, as well as from hard magnetic and electric gluon exchange. The latter mediates between quark states in- and outside this patch, inside the shell of relevant quark modes.

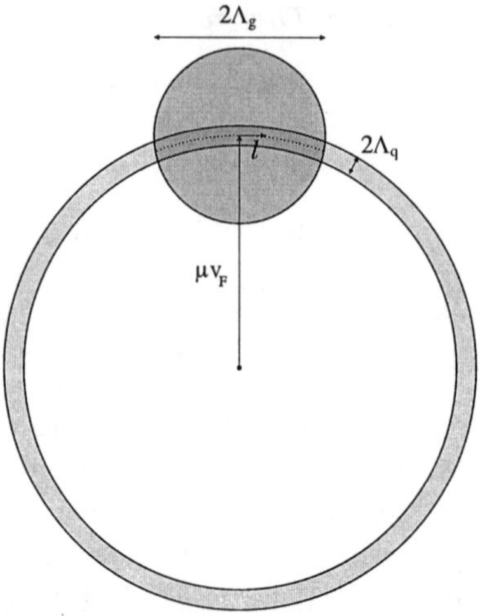

Figure 6. The shell of relevant quark modes of width $2\Lambda_q$ around the Fermi surface, and the patch on this shell, characterizing the range of soft gluon interactions.

Acknowledgments

The work of Q.W. is supported by the Virtual Institute VH-VI-041 of the Helmholtz Association.

References

1. D. Bailin and A. Love, *Phys. Rept.* **107**, 325 (1984).
2. D.T. Son, *Phys. Rev.* **D59**, 094019 (1999).
3. T. Schäfer and F. Wilczek, *Phys. Rev.* **D60**, 114033 (1999).
4. R.D. Pisarski and D.H. Rischke, *Phys. Rev.* **D61**, 051501, 074017 (2000).
5. D.K. Hong, V.A. Miransky, I.A. Shovkovy, and L.C.R. Wijewardhana, *Phys. Rev.* **D61**, 056001 (2000) [Erratum-ibid. **D62**, 059903 (2000)].
6. W.E. Brown, J.T. Liu, and H.c. Ren, *Phys. Rev.* **D61**, 114012 (2000); *ibid.* **62**, 054013, 054016 (2000).
7. Q. Wang and D.H. Rischke, *Phys. Rev.* **D65**, 054005 (2002).
8. D.K. Hong, *Phys. Lett.* **B473**, 118 (2000); *Nucl. Phys.* **B582**, 451 (2000).
9. D.K. Hong, *Prog. Theor. Phys. Suppl.* **153**, 241 (2004).
10. T. Schäfer, *Nucl. Phys.* **A728**, 251 (2003); *eConf* **C030614**, 038 (2003).
11. T. Schäfer, arXiv:hep-ph/0402032.
12. T. Schäfer and K. Schwenzer, arXiv:hep-ph/0405053.
13. G. Nardulli, *Riv. Nuovo Cim.* **25N3**, 1 (2002).
14. P.T. Reuter, Q. Wang, and D.H. Rischke, arXiv:nucl-th/0405079.
15. E. Braaten and R.D. Pisarski, *Nucl. Phys.* **B337**, 569 (1990).
16. J.M. Cornwall, R. Jackiw, and E. Tomboulis, *Phys. Rev.* **D10**, 2428 (1974).

THERMAL LEPTOGENESIS

M. PLÜMACHER
Department of Physics, CERN Theory Division
1211 Geneva 23, Switzerland

Properties of neutrinos may be the origin of the matter-antimatter asymmetry of the universe. In the seesaw model for neutrino masses this leads to important constraints on the properties of light and heavy neutrinos. In particular, an upper bound on the light neutrino masses of 0.1 eV can be derived. We review the present status of thermal leptogenesis and briefly comment on the 'leptogenesis conspiracy', the remarkable fact that neutrino masses may lie in the range where leptogenesis works best.

1. Introduction

The observed baryon asymmetry of the universe is one of the most intriguing problems of particle physics and cosmology. This baryon asymmetry is conveniently expressed as the ratio of baryon minus anti-baryon density to the photon density and has recently been measured to a high degree of accuracy by observations of the cosmic microwave background (CMB) [1] combined with measurements of large scale structures of the universe [2]:

$$\eta_B^{\rm CMB} = (6.3 \pm 0.3) \times 10^{-10} \, , \tag{1}$$

which is consistent with the results of standard primordial nucleosynthesis (BBN), obtained from the measurements of relic nuclear abundances.

A simple and elegant explanation of the observed baryon asymmetry is offered by neutrino physics. During the past years data on atmospheric and solar neutrinos have provided strong evidence for neutrino masses and mixings. In the seesaw mechanism [3] the smallness of these neutrino masses is naturally explained by the mixing of the left-handed neutrinos with heavy Majorana neutrinos. Further, the connection between baryon and lepton number in the high-temperature, symmetric phase of the standard model due to rapid sphaleron transitions [4] is by now firmly established [5]. As in classical GUT baryogenesis [6], out-of-equilibrium decays of the heavy Majorana neutrinos can then generate a lepton asymmetry which, by sphaleron processes, is partially transformed into a baryon asymmetry [7].

A beautiful aspect of this 'leptogenesis' mechanism is the connection between the cosmological baryon asymmetry and neutrino properties. This connection is established by standard kinetic calculations [8,9,10], very much like in big bang nucleosynthesis [6], where light nuclei play the role analogous to leptons in leptogenesis. The requirement of 'successful baryogenesis', i.e. the existence of neutrino masses and mixings for which the predicted and the observed value of the baryon asymmetry are in agreement, constitutes a severe test for models of neutrino masses, which has been extensively explored during the past years [11].

The dynamical generation of a baryon asymmetry requires that the particle interactions do not conserve baryon number, C and CP. In leptogenesis these conditions are realized by the couplings of the heavy Majorana neutrinos N_i. Their decays can generate an asymmetry in the number of leptons and antileptons, and therefore in $B-L$. The crucial departure from thermal equilibrium is provided by the expansion of the universe. At temperatures $T \sim \mathcal{O}(M_i)$ the abundance of heavy neutrinos exceeds the thermal abundance due to their weak interactions with the thermal bath.

In the following we shall assume that the dominant contribution to the baryon asymmetry is given by decays of N_1, the lightest of the heavy Majorana neutrinos. This assumption is well justified in the case of a mass hierarchy among the heavy neutrinos, i.e. $M_1 \ll M_2, M_3$, and it is also known to be a good approximation, if $M_{2,3} - M_1 = \mathcal{O}(M_1)$.

2. Leptogenesis

A quantitative description of the generation of a baryon asymmetry is obtained by means of two kinetic equations, one for the number of heavy neutrinos N_1 and one for the $B-L$ asymmetry [8,9,10]:

$$\frac{dN_{N_1}}{dz} = -(D+S)\,(N_{N_1} - N_{N_1}^{\mathrm{eq}})\,, \qquad (2)$$

$$\frac{dN_{B-L}}{dz} = -\varepsilon_1\, D\,(N_{N_1} - N_{N_1}^{\mathrm{eq}}) - W\, N_{B-L}\,, \qquad (3)$$

where $z \equiv M_1/T$ and ε_1 is the CP asymmetry in the decay of the right-handed neutrinos. N_i is the particle number in a comoving volume element $R_\star^3(t)$, chosen to contain one photon at a time $t_\star$ before the onset of leptogenesis. For a massless boson with g_B degrees of freedom in thermal equilibrium one has $N_B = g_B/2$, whereas for a fermion with g_F degrees of freedom $N_F = 3g_F/8$. The final baryon asymmetry is conveniently expressed in terms of the baryon-to-photon ration η_B, to be compared with the observed value $\eta_B^{\rm CMB}$ (cf. (1)). This is related to the $B-L$ asymmetry in a co-moving volume element by

$$\eta_B \simeq 0.013 N_{B-L} \,. \tag{4}$$

The relevant processes in the thermal plasma are: decays and inverse decays of N_1 into leptons and Higgs bosons $N_1 \leftrightarrow l\phi$, and anti-leptons and anti-Higgs bosons, $N_1 \leftrightarrow \bar{l}\bar{\phi}$, $\Delta L = 1$ scatterings of N_1 off top-quarks and gauge bosons and $\Delta L = 2$ processes mediated by the exchange of all heavy right-handed neutrinos. The first three contribute all together to modify the N_1 abundance. Denoting thermally averaged reaction rates by Γ_i, the term $D \equiv \Gamma_D/(H\,z)$ accounts for the decays and inverse decays, where H is the Hubble parameter. Similarly, the term $S \equiv \Gamma_S/(H\,z)$ accounts for the $\Delta L = 1$ scatterings. The decays are also the source term for the generation of the $B-L$ asymmetry, the first term in the second equation, while all the other processes contribute to the washout term $W \equiv \Gamma_W/(H\,z)$ that competes with the decay source term.

In order to understand the dependence of the solutions of the rate equations (2) and (3) on the neutrino parameters, it is crucial to note that the interaction terms D and S as well as the contribution from N_1 exchange to W are all dimensionless functions of the inverse temperature z, proportional to

$$D,\ S,\ W_1 \ \propto\ \frac{M_{\rm Pl}\widetilde{m}_1}{v^2}\ , \tag{5}$$

where v is the vacuum expectation value of the standard model Higgs field. The effective neutrino mass $\widetilde{m}_1$ [9] is given by

$$\widetilde{m}_1 = \frac{(m_D^\dagger m_D)_{11}}{M_1}\ , \tag{6}$$

where m_D is the Dirac neutrino mass matrix. The strength of the remaining contribution to the washout term, ΔW, is determined by $\overline{m}^2 = m_{\nu_1}^2 + m_{\nu_2}^2 + m_{\nu_3}^2$, the sum over the squares of light neutrino masses,

$$\Delta W \ \propto\ \frac{M_{\rm Pl} M_1\,\overline{m}^2}{v^4}\ . \tag{7}$$

This implies that, as long as ΔW can be neglected, the generated lepton asymmetry is independent of M_1.

If one assumes a vanishing initial $B-L$ asymmetry before the onset of leptogenesis, i.e. at $z \ll 1$, the solution for N_{B-L} has the simple form

$$N_{B-L}(z) = -\frac{3}{4}\,\varepsilon_1\,\kappa(z;\widetilde{m}_1, M_1\overline{m}^2)\ , \tag{8}$$

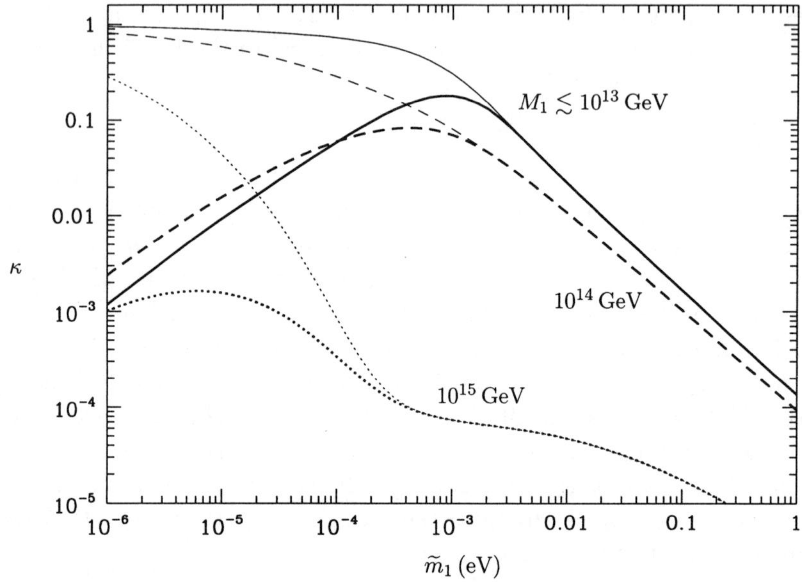

Figure 1. The final efficiency factor κ_f as a function of $\widetilde{m}_1$, for the indicated values of M_1 assuming either an initially vanishing N_1 abundance (thick lines) or a thermal initial N_1 abundance, $N_{N_1}^{\rm in} = 3/4$ (thin lines). Here we have assumed a hierarchical spectrum of light neutrino masses, i.e. $\overline{m} = 0.05\,$eV.

where we have introduced the *efficiency factor* κ [12] which does not depend on the *CP* asymmetry ε_1 and parametrizes the effect of scattering and decay processes. It is given by the following integral expression:

$$\kappa(z) = \frac{4}{3} \int_{z_i}^{z} dz'\, D\, \left(N_{N_1} - N_{N_1}^{\rm eq}\right) e^{-\int_{z'}^{z} dz''\, W(z'')} \,. \qquad (9)$$

It is normalized in such a way that its final value $\kappa_f = \kappa(\infty)$ approaches one in the limit of thermal initial abundance of heavy neutrinos, $N_{N_1}(z \ll 1) = N_{N_1}^{\rm eq} = 3/4$ and no washout, $W = 0$. Hence, determining the amount of baryon asymmetry produced in leptogenesis requires the calculation of both the final efficiency factor κ_f and the CP asymmetry ε_1.

The results for the final efficiency factor are summarized in Fig. 1. For values $M_1 \ll 10^{14}\,{\rm GeV}\,[\overline{m}/(0.05\,{\rm eV})]^2$ the washout term ΔW can be neglected and the final efficiency factor depends only on the effective neutrino mass $\widetilde{m}_1$.

Two different regimes can clearly be distinguished. In the *weak washout regime*, $\widetilde{m}_1 \ll 10^{-3}\,\mathrm{eV}$, the suppression of the efficiency factor in the case of an initially vanishing N_1 abundance is due to the fact that Yukawa interactions are not strong enough to create an initial thermal distribution of right-handed neutrinos. On the other hand, assuming an initial thermal abundance, the neutrinos decay far out of equilibrium and one approaches the limit $\kappa \to 1$. In the weak washout regime the results therefore strongly depend on the initial conditions and on theoretical assumptions, i.e. in that case predictions are strongly model dependent and affected by large theoretical uncertainties.

In the *strong washout regime*, $\widetilde{m}_1 \gg 10^{-3}\,\mathrm{eV}$, inverse decays and scatterings rapidly thermalize the heavy neutrinos N_1 and the washout due to decays and inverse decays is strong enough to destroy an initial asymmetry that may have been present before the onset of leptogenesis [13], leading to a negligible dependence on initial conditions.

The only model independent information we have on $\widetilde{m}_1$ is that it has to be larger than the smallest neutrino mass m_{ν_1} [14]. However, a situation where $m_{\nu_1} < \widetilde{m}_1 \ll m_{\mathrm{sol}}$ is rather artificial within neutrino mass models and the leptogenesis predictions are then very model dependent in the weak washout regime. In typical neutrino mass models values of $\widetilde{m}_1$ are usually in the mass range suggested by neutrino oscillations. It is remarkable that both the scale of solar neutrino oscillations, $m_{\mathrm{sol}} \equiv \sqrt{\Delta m^2_{\mathrm{sol}}} \simeq 8 \times 10^{-3}\,\mathrm{eV}$, and the scale of atmospheric neutrino oscillations, $m_{\mathrm{atm}} \equiv \sqrt{\Delta m^2_{\mathrm{atm}}} \simeq 0.05\,\mathrm{eV}$, are much larger than $10^{-3}\,\mathrm{eV}$. Hence, the parameter range suggested by neutrino oscillations, $m_{\mathrm{sol}} \lesssim \widetilde{m}_1 \lesssim m_{\mathrm{atm}}$, lies entirely in the strong washout regime where theoretical uncertainties are small and the efficiency factor is still large enough to allow for successful leptogenesis. In this parameter range the final efficiency factor is given, within theoretical uncertainties, by the simple power law [15]

$$\kappa_{\mathrm{f}} = (2 \pm 1)\, 10^{-2} \left(\frac{0.01\,\mathrm{eV}}{\widetilde{m}_1}\right)^{1.1\pm 0.1} . \tag{10}$$

3. CP asymmetry

The *CP* asymmetry ε_1 is the second crucial ingredient needed to calculate the baryon asymmetry. To leading order in the Yukawa coupling, the *CP* asymmetry is determined by the interference between tree level and vertex plus self-energy one-loop diagrams [16,17] and can be consistently extracted from the $l\phi \to l\phi$ scattering processes [18].

We will be interested in the maximal CP asymmetry for given neutrino masses. Assuming a hierarchy of the heavy neutrinos, $M_{2,3} \gg M_1$, $\varepsilon_1^{\max}$ depends on M_1, $\widetilde{m}_1$, m_{ν_1} and m_{ν_3} [13]. It can be expressed as

$$\varepsilon_1^{\max}(M_1, \widetilde{m}_1, m_{\nu_1}, m_{\nu_3}) = \varepsilon_1^{\max}(M_1)\, \beta(\widetilde{m}_1, m_{\nu_1}, m_{\nu_3}) \;, \quad \beta \leq 1 \;. \tag{11}$$

The maximal asymmetry, i.e. $\beta = 1$, is reached for $m_{\nu_1} = 0$ and, with $m_{\nu_3} = \sqrt{m_{\rm atm}^2 + m_{\nu_1}^2}$, it is given by [19]

$$\varepsilon_1^{\max}(M_1) = \frac{3}{16\pi} \frac{M_1\, m_{\rm atm}}{v^2} \simeq 10^{-6} \left(\frac{M_1}{10^{10}\,{\rm GeV}}\right) \left(\frac{m_{\rm atm}}{0.05\,{\rm eV}}\right) \;. \tag{12}$$

An improved bound is obtained in the limit $m_{\nu_1}/\widetilde{m}_1 \to 0$, where one has [20],

$$\beta(m_{\nu_1}) = \frac{m_{\nu_3} - m_{\nu_1}}{m_{\rm atm}} \;. \tag{13}$$

Also important is the case of quasi-degenerate neutrinos, $m_{\nu_3} \simeq m_{\nu_1}$. In this region one finds [21,22],

$$\beta(\widetilde{m}_1, m_{\nu_1}) \simeq \frac{m_{\nu_3} - m_{\nu_1}}{m_{\rm atm}} \sqrt{1 - \frac{m_{\nu_1}^2}{\widetilde{m}_1^2}} \;. \tag{14}$$

For all neutrino mass models with moderately hierarchical heavy neutrinos, such that Eq. (11) applies, the observed baryon asymmetry yields constraints on the three neutrino parameters $\widetilde{m}_1, m_{\nu_1}$ and M_1. In the following section we follow mostly the discussion in Ref. [15].

4. Constraints on neutrino parameters

4.1. *Lower bounds on heavy neutrino masses and reheating temperature*

Knowledge of the maximal CP asymmetry $\varepsilon_1^{\max}$ allows to compute the maximal baryon asymmetry that can be produced in leptogenesis:

$$\eta_B^{\max}(M_1, \widetilde{m}_1, m_{\nu_1}, m_{\nu_3}) = 0.96 \times 10^{-2}\, \varepsilon_1^{\max}(M_1, \widetilde{m}_1, m_{\nu_1}, m_{\nu_3})\, \kappa_f \;. \tag{15}$$

Lines of constant baryon asymmetry in the $\widetilde{m}_1$-M_1-plane are shown in Fig. 2, where we have assumed a hierarchical spectrum of light neutrinos, i.e. $m_{\nu_1} = 0$ and $m_{\nu_3} = m_{\rm atm}$.

Together with the CMB constraint, $\eta_B^{\max} \geq \eta_B^{\rm CMB}$, eq. (15) yields interesting constraints on the space of the three neutrino parameters $\widetilde{m}_1, m_{\nu_1}$ and M_1. We have seen that the absolute maximum of the CP asymmetry is obtained for $m_{\nu_1} = 0$. For $m_{\nu_1} \neq 0$ the function β suppresses the CP

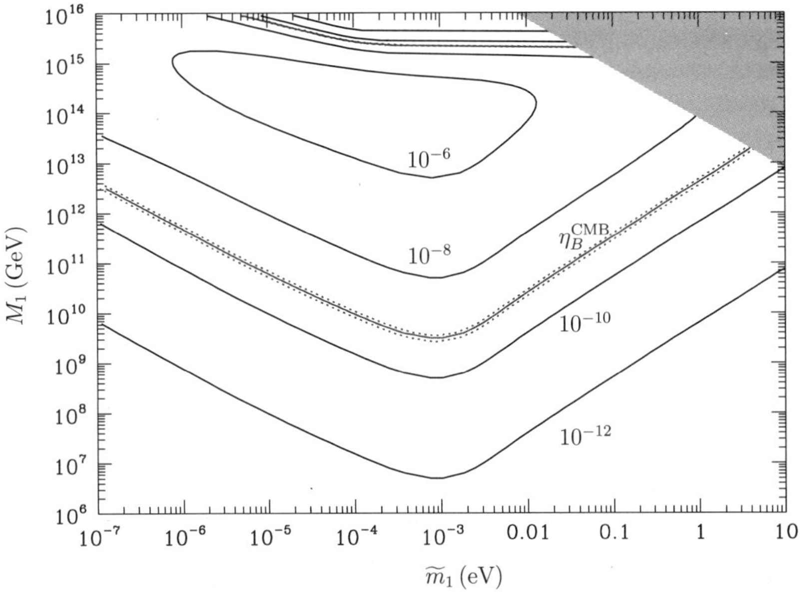

Figure 2. Iso-$\eta_B^{\rm max}$ curves for $N_{N_1}^{\rm in} = 0$ for the case of hierarchical light neutrino masses, i.e. $m_{\nu_1} = 0$ and $m_{\nu_3} = m_{\rm atm}$. The region within the dashed lines is currently allowed by the CMB constraint (cf. eq. (1)) for the (3σ) lower η_B^{CMB} value.

asymmetry [20]. Furthermore, the $\Delta L = 2$ washout term becomes stronger when the absolute neutrino mass scale increases (cf. (7)). Therefore, the maximal baryon asymmetry $\eta_B^{\rm max}(M_1, \widetilde{m}_1, m_{\nu_1})$ is maximized for $m_{\nu_1} = 0$, and in this case the allowed region in the space of the parameters M_1 and $\widetilde{m}_1$ is maximal [10].

In this way one finds an important lower bound on the value of M_1 [20,10], by inserting the expression (12) into eq. (15) and using the CMB constraint $\eta_B^{\rm max} \geq \eta_B^{\rm CMB}$,

$$M_1 \geq M_1^{\rm min} \simeq 6.4 \times 10^8\,{\rm GeV} \left(\frac{\eta_B^{CMB}}{6 \times 10^{-10}}\right) \left(\frac{0.05\,{\rm eV}}{m_{\rm atm}}\right) \kappa_{\rm f}^{-1}\,. \qquad (16)$$

For *thermal initial abundance*, and in the limit $\widetilde{m}_1 \to 0$, one has by definition $\kappa_f = 1$, and therefore

$$M_1 \geq (6.6 \pm 0.8) \times 10^8\,{\rm GeV} \gtrsim 4 \times 10^8\,{\rm GeV}\,. \qquad (17)$$

Here the last inequality is the $3\,\sigma$ bound, with the experimental value of Eq. (1) for η_B^{CMB}, and $m_{\rm atm} = (0.051 \pm 0.004)\,{\rm eV}$. In the case of a *dynam-*

ically generated N_1 abundance the maximal efficiency factor is $\kappa_{\rm f} \simeq 0.18$, which yields the more stringent bound

$$M_1 \geq (3.6 \pm 0.4) \times 10^9\,\text{GeV} \gtrsim 2 \times 10^9\,\text{GeV}\,. \tag{18}$$

The most interesting case corresponds to the range $m_{\rm sol} \lesssim \widetilde{m}_1 \lesssim m_{\rm atm}$, for which the power law Eq. (10) for $\kappa_{\rm f}$ can be used. Using the central value of κ_f and neglecting the theoretical uncertainty, one obtains

$$M_1 \gtrsim (3.3 \pm 0.4) \times 10^{10}\,\text{GeV} \left(\frac{\widetilde{m}_1}{10^{-2}\,\text{eV}}\right)^{1.1}\,, \tag{19}$$

which yields the $3\,\sigma$ bound

$$M_1 \gtrsim (1.5 - 10) \times 10^{10}\,\text{GeV}\,. \tag{20}$$

The lower bound on M_1 is particularly interesting since it can be translated into a lower bound on the initial temperature $T_{\rm i}$ which, within inflationary models, corresponds to the reheating temperature. In the interesting parameter range $m_{\rm sol} \lesssim \widetilde{m}_1 \lesssim m_{\rm atm}$ it has been shown [15], that the final baryon asymmetry is produced at a temperature that is a factor 4-6 lower than M_1, corresponding to the somewhat less restrictive bound on the reheating temperature

$$T_{\rm i} \gtrsim (4 \times 10^9 - 2 \times 10^{10})\,\text{GeV}\,. \tag{21}$$

4.2. *Upper bound on light neutrino masses*

For larger values of the absolute neutrino mass scale the ΔW washout term cannot be neglected. The final efficiency factor then reads

$$\kappa_{\rm f}(\widetilde{m}_1, M_1\overline{m}^2) \simeq \kappa_{\rm f}(\widetilde{m}_1) \exp\left\{-\frac{\omega}{z_B}\left(\frac{M_1}{10^{10}\,\text{GeV}}\right)\left(\frac{\overline{m}}{\text{eV}}\right)^2\right\}\,, \tag{22}$$

where $\omega \simeq 0.186$, $\kappa_{\rm f}(\widetilde{m}_1)$ is the efficiency factor calculated in the regime of small neutrino masses, neglecting the ΔW term, and z_B is the temperature at which the baryon asymmetry is being produced[15]. Because of the assumption of strong washout one can use the simple power law Eq. (10) for κ_f. This finally yields the leptogenesis bound on light neutrino masses[23,15]

$$m_{\nu_i} \lesssim 0.12\,\text{eV}\,. \tag{23}$$

This can clearly be seen in Fig. 3, where allowed regions in the plane of the parameters $\widetilde{m}_1$ and M_1 are plotted for different choices of light neutrino mass spectra, i.e. different choices of $\overline{m}$.

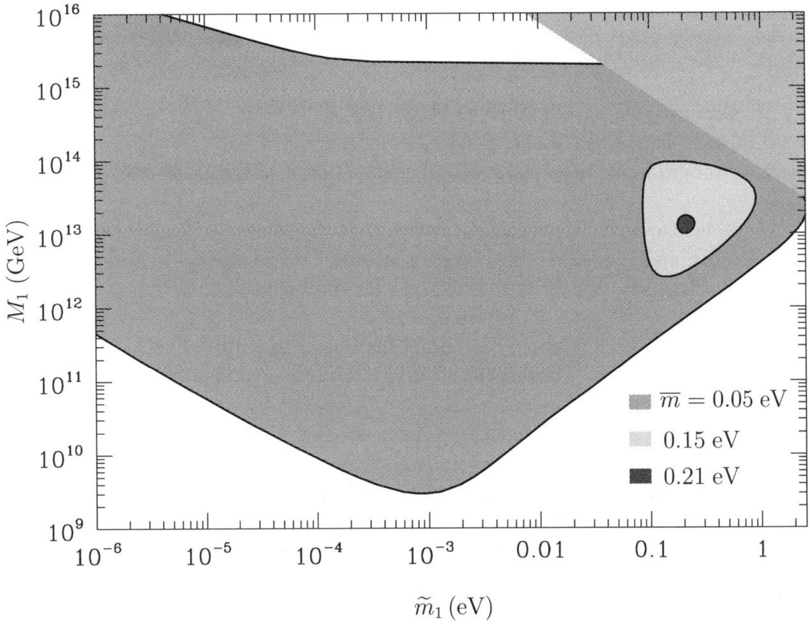

Figure 3. Regions in the $\widetilde{m}_1$-M_1-plane allowed by baryogenesis for different values of $\overline{m}$. There is no allowed region for $\overline{m} = 0.22\,\text{eV}$.

5. Conclusions

We have shown that in the minimal version of thermal leptogenesis, where the lightest of the heavy Majorana neutrinos N_1 is the source of the baryon asymmetry, the final baryon asymmetry is determined by only four parameters: the CP asymmetry ε_1, the heavy neutrino mass M_1, the effective light neutrino mass $\widetilde{m}_1$ and the quadratic mean $\overline{m}$ of light neutrino masses.

The observed baryon asymmetry can only be explained if M_1, and therefore the baryogenesis temperature T_i, satisfy stringent lower bounds and the light neutrinos masses are smaller than about $0.1\,\text{eV}$.

Acknowledgments

I would like to thank W. Buchmüller and P. Di Bari for an enjoyable collaboration on the topic of this review. Further, I would like to thank the organizers of SEWM'04 for organizing a stimulating and fruitful meeting.

References

1. WMAP Collaboration, D. N. Spergel *et al.*, Astrophys. J. Suppl. **148** (2003) 175
2. M. Tegmark et al., Phys. Rev. **D 69** (2004) 103501
3. P. Minkowski, Phys. Lett. B **67** (1977) 421;
   T. Yanagida, in *Workshop on Unified Theories*, KEK report 79-18 (1979) p. 95;
   M. Gell-Mann, P. Ramond, R. Slansky, in *Supergravity* (North Holland, Amsterdam, 1979) eds. P. van Nieuwenhuizen, D. Freedman, p. 315
4. V. A. Kuzmin, V. A. Rubakov, M. E. Shaposhnikov, Phys. Lett. **B 155** (1985) 36
5. D. Bödeker, G. D. Moore, K. Rummukainen, Phys. Rev. **D 61** (2000) 056003
6. E. W. Kolb, M. S. Turner, *The Early Universe*, Addison-Wesley, New York, 1990
7. M. Fukugita, T. Yanagida, Phys. Lett. **B 174** (1986) 45
8. M. A. Luty, Phys. Rev. **D 45** (1992) 455
9. M. Plümacher, Z. Phys. **C 74** (1997) 549
10. W. Buchmüller, P. Di Bari and M. Plümacher, Nucl. Phys. B **643** (2002) 367
11. For a review and references, see
    W. Buchmüller, M. Plümacher, Int. J. Mod. Phys. **A 15** (2000) 5047
12. R. Barbieri, P. Creminelli, A. Strumia, N. Tetradis, Nucl. Phys. **B 575** (2000) 61
13. W. Buchmüller, P. Di Bari, M. Plümacher, Nucl. Phys. **B 665** (2003) 445
14. M. Fujii, K. Hamaguchi and T. Yanagida, Phys. Rev. D **65** (2002) 115012
15. W. Buchmüller, P. Di Bari, M. Plümacher, `hep-ph/0401240`, accepted for publication in Ann. of Phys.
16. M. Flanz, E. A. Paschos, U. Sarkar, Phys. Lett. **B 345** (1995) 248; *ibid.* **B 384** (1996) 487 (E)
17. L. Covi, E. Roulet, F. Vissani, Phys. Lett. **B 384** (1996) 169
18. W. Buchmüller, M. Plümacher, Phys. Lett. **B 431** (1998) 354
19. K. Hamaguchi, H. Murayama, T. Yanagida, Phys. Rev. **D 65** (2002) 043512
20. S. Davidson and A. Ibarra, Phys. Lett. B **535** (2002) 25
21. T. Hambye, Y. Lin, A. Notari, M. Papucci, A. Strumia, Nucl. Phys. B **695** (2004) 169
22. P. Di Bari, Proceedings of the Second International Woorkshop on Neutrino Oscillations in Venice, ed. M. Baldo Ceolin, December 3-5, 2003
23. W. Buchmüller, P. Di Bari and M. Plümacher, Phys. Lett. B **547** (2002) 128

COLD ELECTROWEAK BARYOGENESIS

JAN SMIT

Institute for Theoretical Physics, University of Amsterdam
Valckenierstraat 65, 1018 XE Amsterdam, the Netherlands

We present arguments that the CKM CP-violation in the standard model may be sufficient for the generation of the baryon asymmetry, if the electroweak transition in the early universe was of the cold, tachyonic, type after electroweak-scale inflation. A model implementing this is described which complies with the CMB constraints and which is falsifiable with the LHC. Numerical simulations of the tachyonic transition with an effective CP-bias indicate that the observed baryon asymmetry can be generated this way.

1. Introduction

There are three families in the minimal Standard Model (SM), why? Such why-questions lie outside the scope of the model, but the thought 'to allow for CP-violation and baryogenesis' comes up easily. What is currently known experimentally about CP violation[1] can be attributed to the CKM matrix being complex, which is only possible for three or more families. This provides a strong motivation for finding a scenario for baryogenesis that makes use of this fact. SM electroweak baryogenesis[2], is believed by many to be impossible, because of (*i*) the weakness of CKM CP-violation, and (*ii*) the smoothness of the finite-temperature electroweak transition[3].

Of course, the SM assumes neutrinos to be massless, and a better framework is its extension that includes right-handed neutrino fields with renormalizable Yukawa couplings and a Majorana mass matrix, which may be called the Extended Standard Model (ESM). Leptogenesis is an attractive possibility within this framework[4]. However, it makes use of physics at the scale of 10^{10} GeV and it seems worth putting effort into scenarios closer to what we know on the scale $\lesssim 200$ GeV. So we must face issues (*i*) and (*ii*).

In the following a scenario[5,6,7,8,9] will be reviewed based on a cold ('tachyonic') electroweak transition caused by the coupling of a gauge-singlet inflaton to the ESM. First, we shall argue that CP violation is much stronger at *zero* temperature than often stated.

2. Magnitude of CKM CP-violation

The usual estimate for the magnitude of CP violation is[10,2]

$$\delta_{\rm CP} = J\,(m_u^2-m_c^2)(m_c^2-m_t^2)(m_t^2-m_u^2)(m_d^2-m_s^2)(m_s^2-m_b^2)(m_b^2-m_d^2)/T^{12}, \tag{1}$$

where $m_u, \ldots, m_b$ are the quark masses and[1]

$$J = |{\rm Im}(V_{fg}V_{hi}V^*_{fi}V^*_{hg})| = (2.88 \pm 0.33) \times 10^{-5} \tag{2}$$

is the simplest rephasing-invariant combination of the CKM matrix V. For $T = 100$ GeV, of order of the finite-temperature electroweak phase transition, (1) gives the discouraging value $\delta_{\rm CP} \approx 10^{-19}$. At zero temperature one would replace m_j/T by the Yukawa coupling $\lambda_j = \sqrt{2}\,m_j/v$, with $v = 246$ GeV the vev of the Higgs field, giving

$$\delta_{\rm CP} = J\,(\lambda_u^2-\lambda_c^2)(\lambda_c^2-\lambda_t^2)(\lambda_t^2-\lambda_u^2)(\lambda_d^2-\lambda_s^2)(\lambda_s^2-\lambda_b^2)(\lambda_b^2-\lambda_d^2) \approx 10^{-22}, \tag{3}$$

even smaller than (1). However, even with the Higgs field settled in its vev, the measured CP violating effects in accelerator experiments are at a much higher level than 10^{-23}, e.g. the magnitude of the decay asymmetry ϵ' in the $K^0 - \bar{K}^0$ system is about[1] 4×10^{-6}. This suggests that the above order of magnitude estimates of $\delta_{\rm CP}$ are misleading, at least at zero temperature.

To make the discussion more concrete, consider the effective action obtained by integrating our the fermions, $\Gamma_{\rm eff} = -{\rm Tr}(\ln D)$, with D the Dirac operator. In studies and discussions of electroweak baryogenesis, CP violation has been taken into account by approximating the CP asymmetry in $\Gamma_{\rm eff}$ by a leading dimension-six term[11]

$$-\mathcal{L}_{\rm CP} = \frac{3\delta_{\rm CP}}{16\pi^2 M^2}\,\varphi^\dagger\varphi\;{\rm tr}(F^{\mu\nu}\tilde{F}_{\mu\nu}), \tag{4}$$

where $F_{\mu\nu}$ is the SU(2) field strength tensor, $\tilde{F}_{\mu\nu}$ its dual, and M is a mass depending on the scale of the problem. It could be the mass scale of an extension of the ESM. *Within* the ESM, at finite temperature, one could take $M = T$, but what to use when $T = 0$? The fermion masses are $\propto \langle\varphi\rangle$, but $\varphi^\dagger\varphi/\langle\varphi^\dagger\varphi\rangle$ does not seem to make sense. The ordering of the terms in $\Gamma_{\rm eff}$ according to increasing dimension is questionable in a zero-temperature transition in which $\langle\varphi^\dagger\varphi\rangle$ increases from zero to some finite value.

Luckily, there exists a very detailed calculation of the zero-temperature effective action in a general chiral gauge model, based on a gauge-covariant derivative-expansion that is completely non-perturbative in the Higgs field, by L.L. Salcedo[12,13]. We have applied his results to the ESM (without Majorana mass terms), and found that (4) is incorrect for this case: the rephasing invariant J does not appear as a coefficient of $\mathrm{tr}(F^{\mu\nu}\tilde{F}_{\mu\nu})$ times a function of the un-differentiated Higgs field[14]. The first CKM CP-violating contribution lies unfortunately still beyond the scope of Salcedo's results, but a typical term is expected to have the form, in *unitary gauge* $A_\mu \to W_\mu$, $\varphi = (0, h)^T$,

$$\epsilon^{\kappa\lambda\mu\nu} W^a_{\kappa\lambda} W^b_{\mu\nu} W^c_\rho W^{d\rho}\, n'_{abcd}(\lambda h),$$

where a, b, c, d are $SU(2)$ indices and $n'_{abcd}(\lambda h)$ is a non-trivial function of the Yukawa couplings times the Higgs field, $d_j \equiv \lambda_j h$. For e.g. $a = b = c = d = 1$ the above expression violates CP and is expected to contain the invariant J. The crucial point is now that n'_{abcd} is expected to be a *homogeneous* function of the d_j. This is the case for the explicitly calculated coefficient functions at fourth order of the gauge-covariant derivative-expansion (involving four Lorentz indices), which are homogeneous of degree zero[12,13], which is why (4) cannot occur. For the higher order term we expect that, when we rescale the Yukawa couplings $\lambda_j \to s\lambda_j$, the coefficient functions $n(\lambda h)$ scale by some negative power of s (perhaps up to logarithms). This strongly suggests that we should not include the product of Yukawa couplings (3) in estimating $\delta_{\rm CP}$, leaving only J, which is a factor of about 10^{17} larger. For example,

$$\frac{(\lambda_u^2 - \lambda_c^2)(\lambda_c^2 - \lambda_t^2)(\lambda_t^2 - \lambda_u^2)(\lambda_d^2 - \lambda_s^2)(\lambda_s^2 - \lambda_b^2)(\lambda_b^2 - \lambda_d^2)}{(\lambda_u^2 + \lambda_c^2)(\lambda_c^2 + \lambda_t^2)(\lambda_t^2 + \lambda_u^2)(\lambda_d^2 + \lambda_s^2)(\lambda_s^2 + \lambda_b^2)(\lambda_b^2 + \lambda_d^2)} \simeq 0.99.$$

The argument applies only at zero temperature, since at finite T a new scale appears, e.g. the thermal QCD quark mass $m_{\rm th} = g_{\rm s}T/\sqrt{6}$.

3. A falsifiable model of electroweak-scale inflation

Here we take seriously a minimal phenomenological extension of the ESM with an additional gauge-singlet inflaton that couples only to the Higgs field[9]. Its inflaton-Higgs potential is constructed after [7],

$$V(\sigma, \phi) = V_0 - \frac{1}{p}\alpha_p\sigma^p + \frac{1}{q}\alpha_q\sigma^q - \frac{1}{2}\lambda_{\sigma\phi}\sigma^2\phi^2 + \frac{1}{2}\mu_\phi^2\phi^2 + \frac{1}{4}\lambda_\phi(\phi^2)^2, \quad (5)$$

where σ is the inflaton field and $\phi^2 \equiv 2\varphi^\dagger\varphi$. During inflation, σ slow-rolls away from the origin. After inflation has ended it accelerates, and the effective Higgs mass

$$\mu_{\rm eff}^2 = \mu_\phi^2 - \lambda_{\sigma\phi}\sigma^2 \quad (6)$$

changes sign from positive to negative, and the tachyonic electroweak transition starts. During the transition the baryon asymmetry is to be generated through the electroweak anomaly and the CKM CP-bias. To avoid sphaleron washout of the asymmetry the reheating temperature $T_{\rm rh}$ has to be small enough. Taking $V_0^{1/4} = 100$ GeV, the Hubble rate is tiny, $H \approx 10^{-14}$ GeV, and $T_{\rm rh} = [30V_0/(\pi^2 g_*)]^{1/4} \simeq 43$ GeV ($g_* = 86.25$ is the effective number of SM degrees of freedom below the W mass), at which temperature the sphaleron rate is negligible.

As a phenomenological model it has to comply with what is currently known. Firstly, we have the CMB constraints[15]: the large-scale normalization $|\delta_{\mathbf{k}}|^2 = (3.8 \pm 0.5) \times 10^{-5}$, and the scalar spectral-index $n_S - 1 = -0.03 \pm 0.03$. The latter is given by $n_S - 1 \simeq -(2/N)(p-1)/(p-2)$, with N the number of e-folds before the end inflation, and $p > 2$ for slow-roll to end naturally. For electroweak-scale inflation $N \simeq 24$, which is much lower than usually mentioned ≈ 60, and the resulting $n_S - 1 \lesssim -0.083$ is too low. This is fixed[9] as mentioned below.

Secondly, in the minimum of the potential we have (approximating the cosmological constant by zero)

$$\frac{dV}{d\sigma}(v_\sigma, v_\phi) = 0, \quad \frac{dV}{d\phi}(v_\sigma, v_\phi) = 0, \quad V(v_\sigma, v_\phi) = 0, \quad v_\phi = 246 \text{ GeV},$$

$$\mu_\phi^2 - \lambda_{\sigma\phi} v_\sigma^2 = -\frac{1}{2} m_\phi^2, \;\; m_\phi^2 = 2\lambda_\phi v_\phi^2, \tag{7}$$

where m_ϕ would be the SM Higgs mass, were it not that $\lambda_{\sigma\phi}$ leads to a considerable inflaton-Higgs mixing, to be addressed shortly. For definiteness we continue with $V_0^{1/4} = 100$ GeV, $\mu_\phi = 100$ GeV, $m_\phi = 200$ GeV, for which the Higgs self-coupling $\lambda_\phi = 0.33$, and we expect a particle mass above the current lower bound of 114 GeV[1].

Thirdly, the transition has to be reasonably rapid to allow for successful baryogenesis, which leads[7,9] to the constraint $\lambda_{\sigma\phi} \gtrsim 0.01$. Essentially this constraint now fixes $p = 5$ (the choices for V_0, μ_ϕ and m_ϕ are not critical in this respect). The basic reason is that for $p < 5$ the potential goes down rather slowly, giving a large σ-vev v_σ and consequently a too small $\lambda_{\sigma\phi}$ (cf. (7)), whereas for $p > 5$ the result is a ridiculously large $\lambda_{\sigma\phi}$. So we end up with non-renormalizable couplings, and to keep the powers of σ as low as possible we have chosen $q = 6$.

Returning to the minimum of the potential, the inflaton-Higgs mixing follows from the mass matrix

$$\partial^2 V/\partial(\sigma, \phi)_{v_\sigma, v_\phi} \Rightarrow m_1 = 385 \text{ GeV}, \; m_2 = 125 \text{ GeV}, \; \xi = 0.44 \;,$$

where ξ is the mixing angle. The model thus predicts (only) two scalar particles with couplings to the rest of the SM determined by the mixing.

The CMB scalar spectral-index can be accommodated by adding terms $-\frac{1}{2}\alpha_2\sigma^2 + \frac{1}{3}\alpha_3\sigma^3 - \frac{1}{4}\alpha_4\sigma^4$ to the potential that were artificially set to zero in (5). For example, still with $\alpha_3 = 0$ we found that we could fit n_S easily, see figure 1. Furthermore, in a one-loop study we verified that the

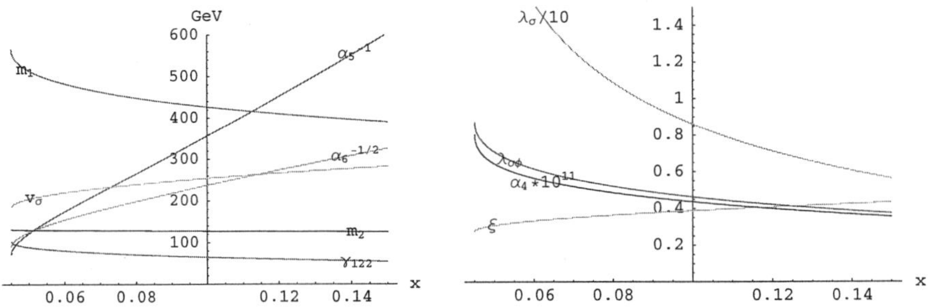

Figure 1. Dimension-full (left) and dimension-less (right) parameters as a function of $x \equiv \alpha_2 m^2_{\text{Planck}}/V_0$, for the central WMAP value $n_S - 1 = -0.03$.

potential can be kept flat enough so as not to upset slow-roll inflation; the non-renormalizable couplings lead to the conclusion that the model breaks down at scales around a TeV or so.

The flatness of the potential is finely tuned, but we stress that there is nothing wrong with this in a *phenomenological* model. It is a minimal extension of the SM that incorporates inflation as well as the tachyonic electroweak transition desired for baryogenesis, with signatures due to inflaton-Higgs mixing that allow it to be *falsified* by the LHC.

4. Tachyonic way-out-of-equilibrium transition

After inflation and before the transition, the universe is cold and approximately in the ground state corresponding to the symmetric phase of the SM. Then μ^2_{eff} changes sign and the tachyonic transition develops. The initial part of such transitions with given 'speed' $\partial\mu^2_{\text{eff}}/\partial t$ has been studied analytically as well as numerically[16,17,18,19,20]. We used the infinitely-fast quench,

$$\begin{aligned}\mu^2_{\text{eff}} &= +\mu^2, \quad t < 0\\ &= -\mu^2, \quad t > 0,\end{aligned}$$

in our simulations of the $SU(2)$-Higgs model described by

$$-\mathcal{L} = \frac{1}{2g^2}\,\mathrm{tr}(F_{\mu\nu}F^{\mu\nu}) + (D_\mu\varphi)^\dagger D^\mu\varphi + \mu_{\text{eff}}^2\varphi^\dagger\varphi + \lambda(\varphi^\dagger\varphi)^2 \qquad (8)$$

as an explorative approximation for studying baryogenesis[8]. In the gaussian approximation the Fourier modes of a real component $\phi_{\mathbf{k}}$ of the Higgs field can be solved in the form $\alpha_{\mathbf{k}}\, e^{-i\omega_k^- t} + \beta_{\mathbf{k}}\, e^{i\omega_k^- t}$, with $\omega_k^- = \sqrt{-\mu^2 + k^2}$. The modes with $k < \mu$ are unstable and grow exponentially $\propto e^{|\omega_k^-|t}$. After some time $\mu t \gg 1$, the equal-time correlators of $\phi_{\mathbf{k}}$ and $\pi_{\mathbf{k}} = \partial_t\phi_{\mathbf{k}}$ are generically well approximated by the *classical distribution*[21]

$$\exp\left[-\frac{1}{2}\sum_{|\mathbf{k}|<\mu}\left(\frac{|\xi_{\mathbf{k}}^+|^2}{n_k + 1/2 + \tilde{n}_k} + \frac{|\xi_{\mathbf{k}}^-|^2}{n_k + 1/2 - \tilde{n}_k}\right)\right], \qquad (9)$$

where $\xi_{\mathbf{k}}^\pm = \frac{1}{\sqrt{2\omega_k}}(\omega_k\,\phi_{\mathbf{k}} \pm \pi_{\mathbf{k}})$, and $n_k + 1/2 + \tilde{n}_k \gg 1$ and $n_k + 1/2 - \tilde{n}_k \ll 1$. So the typical ξ_k^+ grow large, while $\xi_k^- \to 0$; the distribution gets *squeezed.* In the classical approximation[a], realizations of the distribution (9) are used as initial conditions for time evolution using the classical equations of motion. For a weakly coupled system such as the SU(2)-Higgs sector of the SM we don't have to wait till the occupation numbers have grown large, as the quantum and classical evolutions are identical in the gaussian approximation. Our 'just a half' initial conditions start directly at $t = 0$, for which the occupation numbers $n_k = \tilde{n}_k = 0$. In temporal gauge, $A_0 = 0$, the gauge field $\mathbf{A}$ is initialized to zero, and the initial $\partial_t\mathbf{A}$ is found from Gauss' law[8].

In [22] we studied the particle-number distribution of the Higgs and W. For example, Fig. 2 shows how the W-particle n_k in Coulomb gauge change in time. The rapid rise of the low momentum modes end at about $tm_H = 7$ and after $tm_H = 30$ the distribution does not change much anymore. These n_k differ initially very much from the corresponding ones in the unitary gauge, but by time $tm_H = 100$ they are practically the same. The particle numbers of the longitudinal polarization are initially similar to those of the Higgs but later appear to be slower in adjusting to the same distribution. The distributions at $tm_H = 100$ can be fitted to Bose-Einstein distributions, separately in the more important low momentum region and for the higher momentum tails, giving effective temperatures and (at that time still considerable) chemical potentials. The overall lesson learned from

[a]See [23] for a comparison in the $O(4)$ model with the quantum case using Φ-derivable approximation.

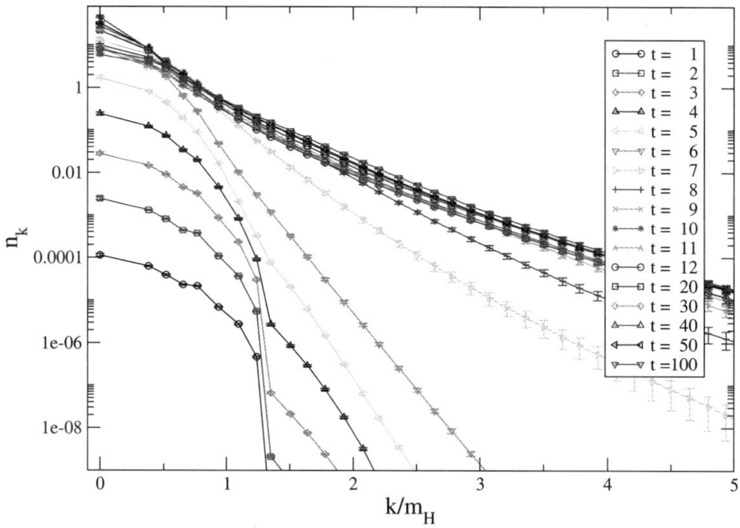

Figure 2. Coulomb-gauge W-particle numbers vs. momentum, time in units of m_H^{-1}.

these studies is that the transition is strongly out of equilibrium, and over after a few tens of m_H^{-1}. The finer details of thermalization will be different when the other d.o.f. from the quarks, gluons, etc. have been taken into account.

5. Baryogenesis

To estimate the asymmetry generated by the transition under influence of CP violation we simulated[8] the $SU(2)$-Higgs model with the effective CP violating term (4) added to the lagrangian (8). Although it was mentioned in section 2 that this term does not apply to the ESM, it is still interesting to see how large its parameter $\delta_{\rm CP}$ has to be in order to obtain the observed asymmetry. For the simulation we relabelled

$$\frac{3\delta_{\rm CP}}{16\pi^2 M^2} \equiv \kappa, \quad k \equiv 16\pi^2 m_W^2\, \kappa. \tag{10}$$

The baryon asymmetry is given by the anomaly equation

$$B(t) = 3N_{\rm CS}(t) = 3\frac{1}{16\pi^2}\int_0^t dt' \int d^3x \left\langle {\rm tr}\left(F_{\mu\nu}(\mathbf{x},t')\tilde{F}^{\mu\nu}(\mathbf{x},t')\right)\right\rangle,$$

where we assumed $B(0) = 0$ (after inflation, just before the transition) and initial Chern-Simon number $N_{\rm CS}(0) = 0$ (choice of gauge). Another useful observable is the winding number of the Higgs field,

$$N_{\rm w} = \frac{1}{24\pi^2}\int d^3x\, \epsilon_{klm}{\rm tr}(\partial_k UU^\dagger \partial_l UU^\dagger \partial_m UU^\dagger),$$
$$U = (i\tau_2\varphi^*,\, \varphi)\,/(\varphi^\dagger\varphi)^{1/2} \in SU(2).$$

At relatively low energy the gauge and Higgs fields are close to being pure-gauge and then $N_{\rm CS}(A) \approx N_{\rm w}(\varphi)$.

We simulated with parameters such that $m_H/m_W = 1$ and $\sqrt{2}$, and various values of k defined in (10). The initial conditions were 'just a half' (cf. Sect. 4) and 'thermal'. The latter are simply Higgs-field realizations drawn from a free BE ensemble at low temperature $T = 0.1\,m_H$, thermal noise that we used to get an impression of the sensitivity to the initial conditions. Figure 3 shows an example of how at first $\langle\varphi^\dagger\varphi\rangle$ rises exponentially

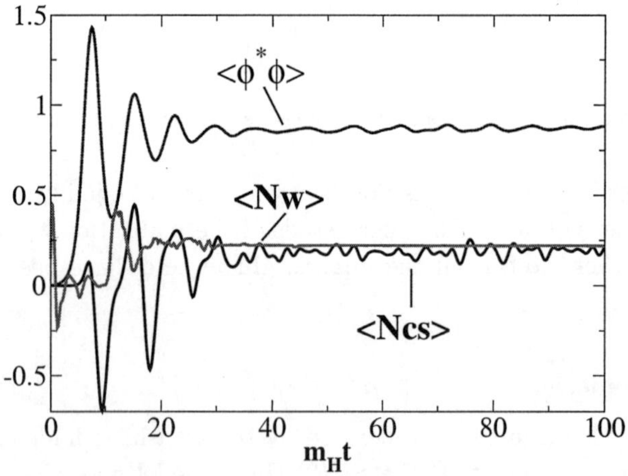

Figure 3. Averages: Higgs field $2\langle\varphi^\dagger\varphi\rangle/v^2$, Chern-Simons number $\langle N_{\rm CS}\rangle$ and Higgs winding number $\langle N_{\rm w}\rangle$, for $m_H = m_w$, $k = 16$, 'just a half' initial conditions.

and then settles into a damped oscillation near the vev. The Chern-Simons number has a small bump near $tm_H = 7$, which we can understand analytically, after which it appears to resonate with $\langle\varphi^\dagger\varphi\rangle$ and settle near the winding number. The latter is at first erratic but then appears to stabilize earlier than $N_{\rm CS}$.

Fig. 4 shows results for the Chern-Simons number-density at time $tm_H = 100$. There is no evidence for a sensitive dependence on the initial ensemble, and apparently the dependence on the magnitude of CP violation becomes non-linear for $k \gtrsim 5 - 10$. The dependence on the mass ratio

m_H/m_W appears limited. This is in contrast to what we found[21] in the 1+1D abelian-Higgs toy model, for which there is a sensitive dependence on m_H/m_W; it also showed nice linear behavior in the 1+1D analog κ for small enough values. The dashed lines are linear fits through the origin,

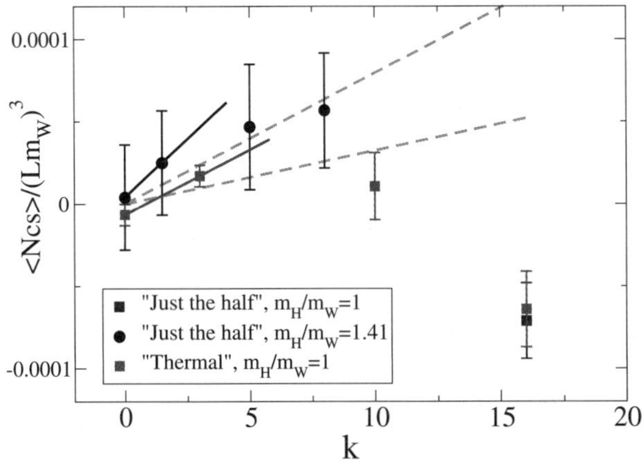

Figure 4. Chern-Simons number density vs. k.

ignoring the data at $k = 0$ and 16. The straight full lines go through the two data points with lowest k. Including the data at $k = 0$ is meaningful because these were generated with the same sequence of pseudo-random numbers as for $k = 3$ ($m_H = m_W$) and $k = 1.5$ ($m_H = \sqrt{2}m_W$). (Note that the $k = 0$ data is shifted somewhat away from zero for $m_H = m_W$). Using the dashed-line fit for the case $m_H = \sqrt{2}\,m_W \simeq 114$ GeV gives a baryon asymmetry

$$\frac{n_B}{n_\gamma} = (0.46 \pm 0.21) \times 10^{-2} \kappa\, m_W^2.$$

To fit the observed $n_B/n_\gamma \simeq 6.5 \times 10^{-10}$, requires κ to be

$$\kappa \approx \frac{1.4 \times 10^{-7}}{m_W^2} \approx \frac{2.2 \times 10^{-5}}{1\text{TeV}^2}.$$

The 1 TeV scale is reasonable if we interpret $\mathcal{L}_{\text{CP}}$ as being due to physics beyond the ESM. On the other hand, if we boldly assume that the results make sense for the ESM by taking the scale $M = m_W$, then the required δ_{CP} turns out to be of the order of J in (2):

$$\delta_{\text{CP}} \approx 0.7 \times 10^{-5}.$$

We consider this as very encouraging for further developing the scenario of cold electroweak baryogenesis.

Acknowledgments

I would like to thank Anders Tranberg, Jon-Ivar Skullerud and Bartjan van Tent, for a very fruitful and pleasant collaboration. This work was supported by FOM/NWO.

References

1. S. Eidelman *et al.* [Particle Data Group Collaboration], Phys. Lett. B **592** (2004) 1.
2. V. A. Rubakov and M. E. Shaposhnikov, Usp. Fiz. Nauk **166**, 493 (1996) [Phys. Usp. **39**, 461 (1996)] [arXiv:hep-ph/9603208].
3. K. Kajantie, M. Laine, K. Rummukainen and M. E. Shaposhnikov, Nucl. Phys. B **466** (1996) 189 [arXiv:hep-lat/9510020].
4. M. Plümacher, these proceedings.
5. J. García-Bellido, D. Y. Grigoriev, A. Kusenko and M. E. Shaposhnikov, Phys. Rev. D **60**, 123504 (1999) [arXiv:hep-ph/9902449].
6. L.M. Krauss and M. Trodden, Phys. Rev. Lett. **83**, 1502 (1999).
7. E. J. Copeland, D. Lyth, A. Rajantie and M. Trodden, Phys. Rev. D **64**, 043506 (2001) [arXiv:hep-ph/0103231].
8. A. Tranberg and J. Smit, JHEP **0311**, 016 (2003) [arXiv:hep-ph/0310342].
9. B. van Tent, J. Smit and A. Tranberg, JCAP **0407**, 003 (2004).
10. C. Jarlskog, Phys. Rev. Lett. **55**, 1039 (1985).
11. M. E. Shaposhnikov, Nucl. Phys. B **287**, 757 (1987).
12. L. L. Salcedo, Eur. Phys. J. C **20**, 147 (2001) [arXiv:hep-th/0012166].
13. L. L. Salcedo, Eur. Phys. J. C **20**, 161 (2001) [arXiv:hep-th/0012174].
14. J. Smit, arXiv:hep-ph/0407161.
15. D. N. Spergel *et al.*, Astrophys. J. Suppl. **148**, 175 (2003).
16. T. Asaka, W. Buchmuller and L. Covi, Phys. Lett. B **510**, 271 (2001).
17. E. J. Copeland, S. Pascoli and A. Rajantie, Phys. Rev. D **65**, 103517 (2002).
18. J. García-Bellido, M. García-Pérez, A. Gonzáles-Arroyo, Phys. Rev. D **67**, 103501 (2003) [arXiv:hep-ph/0208228].
19. J. García-Bellido, M. García-Pérez, A. Gonzáles-Arroyo, Phys. Rev. D **69**, 023504 (2004) [arXiv:hep-ph/0304285].
20. S. Borsanyi, A. Patkos and D. Sexty, Phys. Rev. D **68**, 063512 (2003).
21. J. Smit and A. Tranberg, JHEP **0212**, 020 (2002) [arXiv:hep-ph/0211243].
22. J. I. Skullerud, J. Smit and A. Tranberg, JHEP **0308**, 045 (2003).
23. A. Tranberg, these proceedings.

PROTON-NUCLEUS COLLISIONS IN THE COLOR GLASS CONDENSATE FRAMEWORK

J.-P. BLAIZOT, F. GELIS, R. VENUGOPALAN
CEA/DMS/SPhT
91191, Gif-sur-Yvette cedex, France

We discuss proton-nucleus collisions in the framework of the color glass condensate. By assuming that the proton can be described as a low density color source, we solve exactly the Yang-Mills equation corresponding to this type of collision, and then use this solution in order to calculate inclusive gluon production or quark-antiquark production. Our result shows that $k_\perp$-factorization, while valid for gluon production, is violated for quark pair production in proton-nucleus collisions.

1. Introduction

When one increases the energy of a hadron (or equivalently when one looks at partons with a smaller and smaller momentum fraction x), its parton density increases. At moderate energies, this increase is governed by the linear BFKL equation. This equation leads to a growth of the gluon distribution $xG(x, Q^2)$ as a positive power of $1/x$ which would lead to violations of unitarity. However, it has been known for quite some time that processes which are of higher order in the gluon density – like the recombination of two gluons – become important at small x. These non-linear effects in the evolution limit the growth of the gluon density [1].

Note that these non-linear effects will arise sooner in a large nucleus than in a proton, because its parton density per unit of transverse area is enhanced by a factor $A^{1/3}$ where A is the atomic number of the nucleus. Therefore, there is a window in energy for which the proton can be considered to a be a dilute projectile while the nucleus is a dense projectile. In this talk, we address this situation in the framework provided by the Color Glass Condensate[2]. In this description, the projectiles are effectively described as static random sources of color charge on the light-cone [3]. In order to evaluate a given process in this framework, one should first calculate the classical color field (i.e. solve the classical Yang-Mills equations) generated by these sources of color, evaluate the matrix element of interest in the presence of this classical field, and average over the random color sources of both projectiles.

The distribution of the random sources ρ of a projectile is some functional $W_x[\rho]$ that depends on the typical momentum fraction x probed in the projectile by the process one considers. The Color Glass Condensate predicts the evolution of this distribution when x decreases, via the JIMWLK equation [4,2]. This equation is a Wilson renormalization group equation that reduces to the BFKL equation in the low density limit. Note however that the initial condition for this evolution cannot in general be predicted from perturbative QCD, although for a large nucleus it is expected that a Gaussian distribution might be a good ansatz (as in the McLerran-Venugopalan model [3]).

2. Solution of the classical Yang-Mills equations

The first step in using the color glass condensate framework in order to calculate a given process is to solve the classical Yang-Mills equations for two sources representing the colliding hadrons. This cannot be done analytically in the general case, but it turns out that an exact solution exists in the limit where the classical source representing the proton is treated to first order while all orders in the source representing the nucleus are kept. More precisely, one needs to solve the equation $[D_\mu, F^{\mu\nu}] = J^\nu$ where, at lowest order in the sources,

$$J_a^\nu = g\delta^{\nu+}\delta(x^-)\rho_{p,a}(\boldsymbol{x}_\perp) + g\delta^{\nu-}\delta(x^+)\rho_{A,a}(\boldsymbol{x}_\perp)\,. \tag{1}$$

Corrections to the current J^μ are determined from the requirement that it must be covariantly conserved: $[D_\nu, J^\nu] = 0$. Finally, in order to close the set of equations, one must also impose a gauge condition to the field A^μ. In this work, we use the Lorenz gauge, $\partial_\mu A^\mu = 0$. Indeed, in this gauge, the solution has a very simple expression which turns out to be very convenient when we later calculate the pair production cross-section.

It is fairly straightforward to solve this set of equations to first order in the source ρ_p of the proton, and the Fourier transform of the solution reads [5]:

$$A^\mu(q) = \frac{ig}{q^2 + iq^+\epsilon}\int\frac{d^2\boldsymbol{k}_{1\perp}}{(2\pi)^2}\Big\{C_U^\mu(q,\boldsymbol{k}_{1\perp})\left[U(\boldsymbol{k}_{2\perp}) - (2\pi)^2\delta(\boldsymbol{k}_{2\perp})\right] + C_V^\mu(q)\left[V(\boldsymbol{k}_{2\perp}) - (2\pi)^2\delta(\boldsymbol{k}_{2\perp})\right]\Big\}\frac{\rho_p(\boldsymbol{k}_{1\perp})}{k_{1\perp}^2}\,, \tag{2}$$

where implicitly $\boldsymbol{k}_{2\perp} \equiv \boldsymbol{q}_\perp - \boldsymbol{k}_{1\perp}$. The 4-vectors $C^\mu_{_{U,V}}$ are,

$$C^+_{_U}(q, \boldsymbol{k}_{1\perp}) \equiv -\frac{k^2_{1\perp}}{q^- + i\epsilon}\,,\ C^-_{_U}(q, \boldsymbol{k}_{1\perp}) \equiv \frac{k^2_{2\perp} - q^2_\perp}{q^+}\,,\ C^i_{_U}(q, \boldsymbol{k}_{1\perp}) \equiv -2k^i_1\,,$$

$$C^+_{_V}(q) \equiv 2q^+ \quad , \quad C^-_{_V}(q) \equiv 2\frac{q^2_\perp}{q^+} - 2q^- \quad , \quad C^i_{_V} \equiv 2q^i\,, \tag{3}$$

and the quantities U, V are Fourier transforms of Wilson lines:

$$U(\boldsymbol{k}_\perp) \equiv \int d^2\boldsymbol{x}_\perp e^{-i\boldsymbol{k}_\perp\cdot\boldsymbol{x}_\perp} \mathcal{P} e^{ig\int dz^+ A^-_{_A}(z^+,\boldsymbol{x}_\perp)\cdot T}\,,$$

$$V(\boldsymbol{k}_\perp) \equiv \int d^2\boldsymbol{x}_\perp e^{-i\boldsymbol{k}_\perp\cdot\boldsymbol{x}_\perp} \mathcal{P} e^{i\frac{g}{2}\int dz^+ A^-_{_A}(z^+,\boldsymbol{x}_\perp)\cdot T}\,. \tag{4}$$

In these Wilson lines, the field $A^\mu_{_A}$ is the field generated by the source $\rho_{_A}$ of the nucleus. All the higher-order effects in the strong source of the nucleus enter via these two Wilson lines.

3. Inclusive gluon production

From this solution, one can obtain the gluon production amplitude at leading order as $\mathcal{M}_\lambda(\boldsymbol{q}) = q^2 A^\mu(q)\epsilon^{(\lambda)}_\mu(\boldsymbol{q})$ where $\epsilon^{(\lambda)}_\mu(\boldsymbol{q})$ is the polarization vector for a gluon of 3-momentum $\boldsymbol{q}$ in the polarization state λ. Squaring, summing over the polarizations of the produced gluon, averaging over the sources ρ_p and $\rho_{_A}$ and integrating over the impact parameter, one obtains the following expression for the inclusive gluon production cross-section[a]:

$$\omega_{\boldsymbol{q}}\frac{d\sigma}{d^3\boldsymbol{q}} = \frac{\alpha_s N}{\pi^4 d_{_A} q^2_\perp}\int \frac{d^2\boldsymbol{k}_\perp}{(2\pi)^2}\,\varphi_p(\boldsymbol{k}_\perp)\phi_{_A}(\boldsymbol{q}_\perp - \boldsymbol{k}_\perp)\,. \tag{5}$$

In this equation, the functions φ_p and $\phi_{_A}$ respectively describe the proton and the nucleus, and are expressed as follows:

$$\varphi_p(\boldsymbol{k}_\perp) \equiv \pi^2 R^2_p g^2 k^2_\perp \int d^2\boldsymbol{x}_\perp e^{i\boldsymbol{k}_\perp\cdot\boldsymbol{x}_\perp}\left\langle \rho_{p,a}(0)\rho_{p,a}(\boldsymbol{x}_\perp)\right\rangle\,,$$

$$\phi_{_A}(\boldsymbol{k}_\perp) \equiv \frac{\pi^2 R^2_{_A} k^2_\perp}{g^2 N}\int d^2\boldsymbol{x}_\perp e^{i\boldsymbol{k}_\perp\cdot\boldsymbol{x}_\perp}\,\mathrm{tr}\left\langle U(0)U^\dagger(\boldsymbol{x}_\perp)\right\rangle\,. \tag{6}$$

This formula clearly indicates that $k_\perp$-factorization remains valid for single gluon production in collisions involving a dilute and a dense projectile.

---

[a]Our result for gluon production is equivalent to the result of ref. [6].

4. Quark pair production in pA collisions

One can also use the solution of the Yang-Mills equations in order to calculate the cross-section for the production of quark-antiquark pairs. This calculation basically amounts to finding the quark propagator in the color field generated by the colliding projectiles [7]. For the sake of illustration, let us write here the cross-section for single quark production[b] [8]:

$$\frac{d\sigma_q}{d^2\boldsymbol{q}_\perp dy_q} = \frac{\alpha_s^2 N}{8\pi^4 d_A}\int\frac{dp^+}{p^+}\int_{\boldsymbol{k}_{1\perp},\boldsymbol{k}_{2\perp}}\frac{1}{\boldsymbol{k}_{1\perp}^2\boldsymbol{k}_{2\perp}^2}$$
$$\times\Big\{\mathrm{tr}\Big[(\not{q}+m)T_{q\bar{q}}(\boldsymbol{k}_{1\perp},\boldsymbol{k}_{2\perp})(\not{p}-m)\gamma^0 T^\dagger_{q\bar{q}}(\boldsymbol{k}_{1\perp},\boldsymbol{k}_{2\perp})\gamma^0\Big]\frac{C_F}{N}\phi_A^{q,q}(\boldsymbol{k}_{2\perp})$$
$$+\int_{\boldsymbol{k}_\perp}\mathrm{tr}\Big[(\not{q}+m)T_{q\bar{q}}(\boldsymbol{k}_{1\perp},\boldsymbol{k}_\perp)(\not{p}-m)\gamma^0 T^\dagger_g(\boldsymbol{k}_{1\perp})\gamma^0\Big]\phi_A^{q\bar{q},g}(\boldsymbol{k}_{2\perp};\boldsymbol{k}_\perp) + \mathrm{h.c.}$$
$$+\mathrm{tr}\Big[(\not{q}+m)T_g(\boldsymbol{k}_{1\perp})(\not{p}-m)\gamma^0 T^\dagger_g(\boldsymbol{k}_{1\perp})\gamma^0\Big]\phi_A^{g,g}(\boldsymbol{k}_{2\perp})\Big\}\varphi_p(\boldsymbol{k}_{1\perp})\,, \tag{7}$$

where we use the following shorthands:

$$T_{q\bar{q}}(\boldsymbol{k}_{1\perp},\boldsymbol{k}_\perp) \equiv \frac{\gamma^+(\not{q}-\not{k}+m)\gamma^-(\not{q}-\not{k}-\not{k}_1+m)\gamma^+}{2p^+[(\boldsymbol{q}_\perp-\boldsymbol{k}_\perp)^2+m^2]+2q^+[(\boldsymbol{q}_\perp-\boldsymbol{k}_\perp-\boldsymbol{k}_{1\perp})^2+m^2]}$$
$$T_g(\boldsymbol{k}_{1\perp}) \equiv \frac{\not{C}_U(p+q,\boldsymbol{k}_{1\perp})+\frac{1}{2}\not{C}_V(p+q)}{(p+q)^2}\,. \tag{8}$$

The distribution φ_p that appears in this formula is the same as the one given in eq. (6) and the function $\phi_A^{g,g}$ is identical to the function ϕ_A of eq. (6). However, this function is not the only piece of information we need about the nucleus in order to compute quark production. It turns out that we need also two other functions, denoted here $\phi_A^{q,q}$ and $\phi_A^{q\bar{q},g}$. The first one, $\phi_A^{q,q}$, has a definition identical to that of $\phi_A^{g,g}$, except that the Wilson line must be evaluated in the fundamental representation. On the contrary, $\phi_A^{q\bar{q},g}$ is defined in terms of a three-point correlator:

$$\frac{\phi_A^{q\bar{q},g}(\boldsymbol{l}_\perp;\boldsymbol{k}_\perp)}{\pi R_A^2} = \frac{2\pi l_\perp^2}{g^2 N}\int_{\boldsymbol{x}_\perp,\boldsymbol{y}_\perp} e^{i(\boldsymbol{k}_\perp\cdot\boldsymbol{x}_\perp+(\boldsymbol{l}_\perp-\boldsymbol{k}_\perp)\cdot\boldsymbol{y}_\perp)}\mathrm{tr}\left\langle\widetilde{U}(\boldsymbol{x}_\perp)t^a\widetilde{U}^\dagger(\boldsymbol{y}_\perp)t^b U_{ab}(0)\right\rangle\,,$$

where $\widetilde{U}$ is a Wilson line in the fundamental representation while U is the adjoint Wilson line already encountered. Eq. (7) is a generalization to the case of a dense target of a well formula for proton-proton collisions[10].

[b] A similar formula has been obtained in [9] for the special case of sources having Gaussian correlations.

Clearly, the need for three different functions in order to describe the nucleus in this process implies that $k_\perp$-factorization is broken. In particular, we see that it is not correct to simply replace the distribution function that describes the nucleus by some saturation inspired ansatz in the formula derived in [10].

5. Conclusions

In this talk, we have seen that for proton-nucleus collisions described in the color glass condensate framework, physical cross-sections can be expressed in terms of correlators of Wilson lines. It appeared that $k_\perp$-factorization holds for gluon production, while it is broken for quark production. In order to go further than this analytical work and attack the case of nucleus-nucleus collisions, one will have to perform a numerical analysis of the Dirac equation in an external color field [11].

References

1. L.V. Gribov, E.M. Levin, M.G. Ryskin, Phys. Rept. **100**, 1 (1983), A.H. Mueller, J-W. Qiu, Nucl. Phys. **B 268**, 427 (1986).
2. E. Iancu, A. Leonidov, L.D. McLerran, Nucl. Phys. **A 692**, 583 (2001), Phys. Lett. **B 510**, 133 (2001), E. Ferreiro, E. Iancu, A. Leonidov, L.D. McLerran, Nucl. Phys. **A 703**, 4 89 (2002).
3. L.D. McLerran, R. Venugopalan, Phys. Rev. **D 49**, 2233 (1994), *ibid* **49**, 3352 (1994), *ibid* **50**, 2225 (1994), *ibid* **59**, 094002 (1999).
4. J. Jalilian-Marian, A. Kovner, A. Leonidov, H. Weigert, Nucl. Phys. **B 504**, 415 (1997), Phys. Rev. **D 59**, 014014 (1999), I. Balitsky, Nucl. Phys. **B 463**, 99 (1996), Yu.V. Kovchegov, Phys. Rev. **D 54**, 5463 (1996), *ibid* **61**, 074018 (2000), J. Jalilian-Marian, A. Kovner, L.D. McLerran, H. Weigert, Phys. Rev. **D 55**, 5414 (1997), H. Weigert, Nucl. Phys. **A 703**, 823 (2002).
5. J.P. Blaizot, F. Gelis, R. Venugopalan, hep-ph/0402256.
6. Yu.V. Kovchegov, A.H. Mueller, Nucl. Phys. **B 529**, 451 (1998), A. Dumitru, L.D. McLerran, Nucl. Phys. **A 700**, 492 (2002).
7. A.J. Baltz, F. Gelis, L.D. McLerran, A. Peshier, Nucl. Phys. **A 695**, 395 (2001), F. Gelis, A. Peshier, Nucl. Phys. **A 697**, 879 (2002), *ibid* **707**, 175 (2002).
8. J.P. Blaizot, F. Gelis, R. Venugopalan, hep-ph/0402257.
9. K. Tuchin, hep-ph/0401022.
10. J.C. Collins, R.K. Ellis, Nucl. Phys. **B 360**, 3 (1991), S. Catani, M. Ciafaloni, F. Hautmann, Nucl. Phys. **B 366**, 135 (1991), F. Gelis, R. Venugopalan, Phys. Rev. **D 69**, 014019 (2004).
11. D. Dietrich, Phys. Rev. **D 68**, 105005 (2003), hep-th/0402026, hep-ph/0408343, F. Gelis, K. Kajantie, T. Lappi, hep-ph/0409058, T. Lappi, these proceedings.

FROM CLASSICAL TO QUANTUM SATURATION IN THE NUCLEAR WAVEFUNCTION

D. N. TRIANTAFYLLOPOULOS
Service de Physique Théorique, Saclay, CEA
91191 Gif sur Yvette, France
E-mail: dionysis@spht.saclay.cea.fr

We study the gluon content of a large nucleus (i) in the semi-classical McLerran-Venugopalan model and (ii) in the high energy limit as given by the quantum evolution of the Color Glass Condensate. We give a qualitative description of the Cronin effect and high-$p_{\perp}$ suppression in proton-nucleus collisions.

1. Introduction

At high energies and/or for large atomic number A, the wavefunction of a hadron is expected to be dominated by a high density gluonic system. Gluons having occupation numbers φ of order $1/\alpha_s$, which is the maximal density allowed by their mutual interactions, overlap in phase space and saturate[1,2]. A strong classical field is associated with the wavefunction and assumes a value $\mathcal{A} \sim \sqrt{a^{\dagger}a} \sim \sqrt{\varphi} \sim 1/g$ at saturation. At the same time scattering amplitudes become of order 1 and unitarity limits are reached[3]. The problem can be attacked by weak coupling methods, since the non-linear phenomena "push" gluons to occupy higher momenta, and the saturation momentum Q_s, which is defined as the scale where $\varphi(Q_s) \sim 1/\alpha_s$, is a hard scale increasing as a power of energy in the small Bjorken-x limit.

The modern and most complete approach to saturation is the effective theory of the Color Glass Condensate (CGC)[4]. Fast moving partons have a large lifetime due to time dilation and act as "frozen" sources ρ for the virtual emission of softer gluons. One solves the classical Yang-Mills equations to obtain the field $\mathcal{A}(\rho)$ and an observable $O(\mathcal{A})$ is determined by averaging over the possible color sources, with a probability distribution $W_Y[\rho]$. Increasing the rapidity $Y = \ln(1/x)$, more gluons need to be included in the source, and a resummation of $\alpha_s Y$ enhanced terms in the presence of a background field leads to a functional RGE for $W_Y[\rho]$. This gives a hierarchy of coupled equations, and the first one, describing the evolution of the scattering of a dipole off the CGC, can be closed by a mean field approximation and reduced to the BFKL equation in the weak field limit.

2. Classical Saturation

Classical saturation, where there is no small-x evolution, can be realized only in a large nucleus. The $A \times N_c$ valence quarks are the sources for the emission of gluons, and in the McLerran-Venugopalan (MV) model[2] they are assumed to be uncorrelated for transverse separations $\Delta x_\perp \lesssim \Lambda_{\rm QCD}^{-1}$, so that the probability distribution is given by the Gaussian

$$W_{\rm MV}[\rho] \propto \exp\left[-\frac{1}{2}\int^{1/\Lambda} d^2x_\perp \, \frac{\rho_a(x_\perp)\rho_a(x_\perp)}{\mu_A^2}\right], \qquad (1)$$

where $\mu_A^2 = 2\alpha_s A/R_A^2 \sim A^{1/3}\Lambda^2$ is the color charge density squared, with R_A the nuclear radius. Even though the sources are uncorrelated, the created field $\mathcal{A}$ is obtained from a non-linear equation. Thus, starting from its canonical definition, the gluon occupation number φ_A is not the one that we would obtain from a simple superposition of the sources. When $A \gg 1$, the density can be high, sets the magnitude of the saturation scale as $Q_s^2(A) \approx \Lambda^2 A^{1/3} \ln A \gg \Lambda^2$, and we find[5]

$$\varphi_A = \frac{1}{\alpha_s}\Gamma(0,z) + \varphi_A^{\rm twist}(z), \qquad z \equiv k^2/Q_s^2(A). \qquad (2)$$

The first term, enhanced by $1/\alpha_s$, dominates for all $z \lesssim 1$. We interpret this compact distribution, falling exponentially at large z, as the occupation number in the CGC phase. The twist term contains the bremsstrahlung spectrum (BS) $\varphi_{\rm BS} \sim 1/z$, and is important for the large-z behavior. Due to the lack of correlations among the valence quarks, a sum rule exists[5,6]

$$\int^Z dz\, [\varphi_A(z) - \varphi_{\rm BS}(z)] \xrightarrow[Z\to\infty]{} 0; \qquad (3)$$

the integrated distribution is obtained by "summing" over the nucleons when $Q^2 \equiv Z Q_s^2(A) \gg Q_s^2(A)$. The repulsive interactions in the nucleus redistribute the gluons in momenta; the spectra φ_A and $\varphi_{\rm BS}$ become equal at a scale $Q_c(A)$ such that $\Lambda^2 \ll Q_c^2(A) \approx \alpha_s Q_s^2(A) \ll Q_s^2(A)$ and gluons in excess in the BS spectrum are located at $k \sim Q_s(A)$ in the MV model; *the MV spectrum is enhanced around the saturation scale*, as shown in Fig. 1.

As an immediate consequence let us consider the Cronin ratio

$$\mathcal{R}_{\rm pA} \equiv \frac{\varphi_A}{A^{1/3}\varphi_p} = \frac{\varphi_A}{\varphi_{\rm BS}} = z\,\varphi_A, \qquad (4)$$

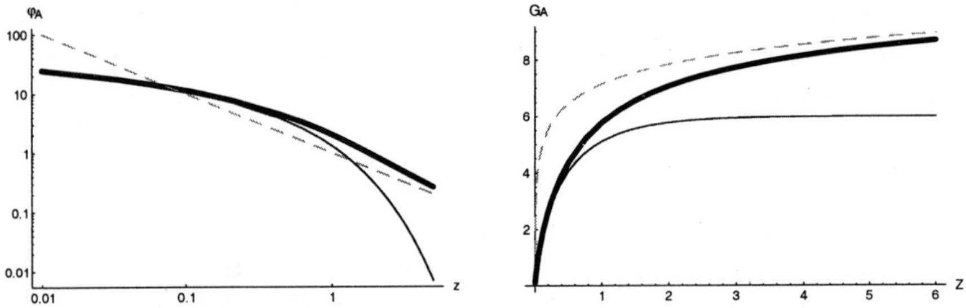

Figure 1. Left: Gluon occupation number in MV model. Right: Integrated distribution. Thick, solid and dashed lines show the total, CGC and BS quantities respectively.

with φ_p the spectrum of the proton, when this is obtained from a simple superposition of the gluons emitted by its valence quarks. It behaves as

$$\begin{aligned} &\bullet\ \mathcal{R}_{\rm pA} \ll 1 && \text{if } z \ll 1 \\ &\bullet\ \mathcal{R}_{\rm pA} \sim \mathcal{O}(1/\alpha_s) \gg 1 && \text{if } z \sim 1 \\ &\bullet\ \mathcal{R}_{\rm pA} \to 1^+ && \text{if } z \gg 1. \end{aligned} \tag{5}$$

The ratio, shown in Fig. 2, has a maximum at $z_m = 0.435 + \mathcal{O}(\alpha_s)$. The maximal value $\mathcal{R}_{\rm max} = 0.281/\alpha_s + \mathcal{O}(\text{const})$ corresponds to a pronounced peak[5,7] originating from the compact nature of the nuclear wavefunction at saturation. This value increases with A (since $1/\alpha_s \equiv \ln Q_s^2(A)/\Lambda^2$).

3. Quantum Saturation

Now consider the evolution of a hadron to higher energies. Its wavefunction contains more and more soft gluons, due to the $\alpha_s Y$ increase in the available longitudinal phase space, and correlations among the color sources are induced. Only a "piecewise" expression for $\alpha_s Y \gtrsim 1$ is known and reads

$$\varphi(k,Y) = \begin{cases} \dfrac{1}{\alpha_s} \ln \dfrac{Q_s^2(Y)}{k^2} & \text{if } k^2 \ll Q_s^2(Y)\text{: CGC} \\[2ex] \dfrac{1}{\alpha_s} \left(\dfrac{Q_s^2(Y)}{k^2}\right)^{\gamma_s} \left(\ln \dfrac{k^2}{Q_s^2(Y)} + \Delta\right) & \text{if } k^2 \gtrsim Q_s^2(Y)\text{: BFKL} \\[2ex] \dfrac{Q_0^2}{k^2}\, \mathrm{I}_0 \left(\sqrt{4\bar{\alpha}_s Y \ln \dfrac{k^2}{Q_0^2}}\right) & \text{if } k^2 \gg Q_s^2(Y)\text{: DLA,} \end{cases} \tag{6}$$

where the saturation momentum is $Q_s^2(Y) = \# Q_s^2(0) \exp[\bar{\alpha}_s \chi(\gamma_s) Y/\gamma_s]$, with $\bar{\alpha}_s = \alpha_s N_c/\pi$, $\chi(\gamma)$ the eigenvalue of the BFKL equation and $\gamma_s = 0.628$ the associated anomalous dimension. The solution exhibits geometrical scaling below and in a certain wide region above Q_s; it depends

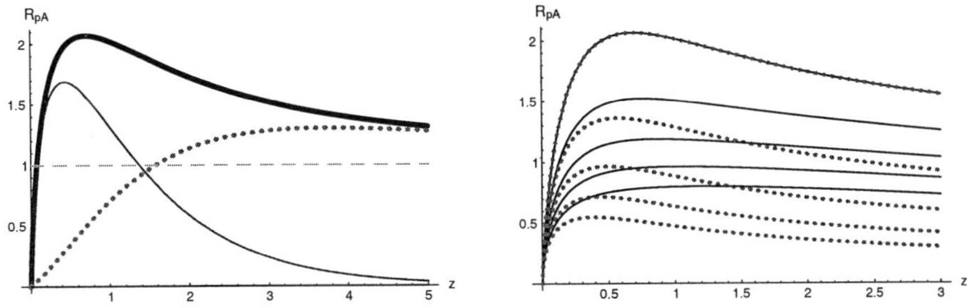

Figure 2. Left: Cronin ratio in MV model. Thick, solid and dotted lines show the total, CGC and twist contributions respectively. Right: The ratio for $Y = 0, 1/2, ..., 2$. Solid lines correspond to an evolved nuclear wavefunction and dotted to an unevolved one.

on k and Y only through the combination $k^2/Q_s^2(Y)$. It is instructive to do a first step in the non-linear evolution, valid so long as $Y \ll 1/\alpha_s$. To the order of accuracy and for momenta $k \lesssim Q_s(A,Y)$, it is enough to evolve only the compact piece in Eq. (2) which will add a correction of order Y. This correction contains power-law tails which are generated from the tails of the evolution kernel. It is clear that, when $Y \gtrsim 1/\alpha_s$ all the components will be "mixed" and, unlike the classical case, *in the quantum case there is no compact distribution for* $k \lesssim Q_s(Y)$ *and no parametric separation between the solutions above and below* $Q_s(Y)$, as can be seen in Eq. (6).

The analysis of the Cronin ratio is not trivial since we do not know the solution in the whole k-Y plane. Furthermore, given a "point" in this plane, the proton and the nucleus can be in different phases, e.g. the nucleus could be saturated while the proton is still dilute. However, we can understand the generic important features.

• The proton is "less saturated" than the nucleus, since the initial proton scale $\sim \Lambda^2$ is much smaller than the initial nuclear one $Q_s^2(A)$, and therefore the available transverse space for the proton is larger. Thus, *the proton evolves faster than the nucleus and the ratio* $\mathcal{R}_{\rm pA}$ *decreases.* For example, along the particular line $k = Q_s(A,Y)$, one has

$$\mathrm{d}\mathcal{R}_{\rm pA}/\mathrm{d}Y < 0 \quad \& \quad \mathcal{R}_{\rm pA} \xrightarrow[Y\to\infty]{} (\alpha_s A^{-1/3})^{1-\gamma_s}. \tag{7}$$

• For fixed Y and for extremely high momenta both systems are dilute, described by the DLA solution, and the above mentioned difference in transverse space is unimportant. The ratio approaches 1 from *below*, i.e.

$$\mathrm{d}\mathcal{R}_{\rm pA}/\mathrm{d}k^2 > 0 \quad \text{for } k^2 \gg Q_s^2(A,Y) \quad \& \quad \mathcal{R}_{\rm pA} \xrightarrow[k^2\to\infty]{} 1^-. \tag{8}$$

• The sum rule breaks down for any $Y > 0$ because of the correlations induced among the sources. The peak remains for a while, since *the sum rule is a sufficient but not a necessary condition for the existence of a peak.*

• One can follow analytically the evolution of the peak, as shown in Fig. 2, until it becomes of order 1, since this happens very fast due (again) to the large separation between the scales Λ^2 and $Q_s^2(A)$. The nuclear wavefunction is almost unevolved, while the proton is still dilute. One finds

$$\mathcal{R}_{\rm pA} = \mathcal{O}(1) \quad \text{when} \quad Y = (1/4)\ln^2(1/\alpha_s) \ll 1/\alpha_s. \tag{9}$$

• Even though smaller than 1, a peak persists under further evolution until $Y \sim 1/\alpha_s$, when the power-law tails will have "washed-out" the compact piece in the nuclear wavefunction; *the peak flattens out due to the nuclear evolution* and the ratio becomes a monotonic function of k^2, that is

$$\mathrm{d}\mathcal{R}_{\rm pA}/\mathrm{d}k^2 > 0 \quad \text{when} \quad Y \gtrsim 1/\alpha_s. \tag{10}$$

These features of saturation and the Cronin ratio[5] extend previous discussions[6], agree with the results obtained in numerical solutions[8], and remain qualitatively unaltered under a running coupling treatment[5].

4. Epilogue

Saturation phenomena can play a significant role in determining the produced particle spectra in high-energy heavy ion collisions. In d-Au collisions at RHIC at BNL, final state interactions are not important, and with $k_\perp$ and η the transverse momentum and the (pseudo)rapidity of the produced particle, one probes the nuclear wavefunction at a value $x_{\rm Au} \simeq 2|k_\perp|\exp[-\eta]/\sqrt{s_{NN}}$, where $\sqrt{s_{NN}}$ is the COM energy per nucleon. At current energies one expects to reveal classical saturation properties in the mid-rapidity region, while quantum saturation should be realized in the forward one. Inded, the CGC predictions (and postdictions) are in reasonable qualitative and quantitative agreement with the RHIC data[9].

References

1. L. Gribov, E. Levin and M. Ryskin, *Phys. Rep.* **100**, 1 (1983).
2. L. McLerran and R. Venugopalan, *Phys. Rev.* **D49**, 2233 (1994).
3. A. H. Mueller, *Nucl. Phys.* **B437**, 107 (1995).
4. E. Iancu, A. Leonidov and L. McLerran, *Nucl. Phys.* **A692**, 583 (2001).
5. E. Iancu, K. Itakura and D. Triantafyllopoulos, *Nucl. Phys.* **A742**, 182 (2004).
6. D. Kharzeev, Y. Kovchegov and K. Tuchin, *Phys. Rev.* **D68**, 094013 (2003).
7. F. Gelis and J. Jalilian-Marian, *Phys. Rev.* **D67**, 074019 (2003).
8. J. Albacete *et al.*, *Phys. Rev. Lett.* **92**, 082001 (2004).
9. I. Arsene *et al.* [BRAHMS Collaboration], arXiv:nucl-ex/0403005.

CHARGE CORRELATIONS IN HEAVY ION COLLISIONS

A. RAJANTIE
DAMTP, University of Cambridge
Wilberforce Road, Cambridge CB3 0WA, UK
E-mail: a.k.rajantie@damtp.cam.ac.uk

When hot quark gluon plasma expands and cools down after an heavy ion collision, charge conservation leads to non-trivial correlations between the charge densities at different rapidities. If these correlations can be measured, they will provide information about dynamical properties of quark gluon plasma.

1. Introduction

The view to the quark gluon plasma phase (QGP) in heavy ion collisions is obscured by events taking place later on. One way around this is to focus on conserved charges, such as electric charge, baryon number or strangeness.[1,2] Later evolution can only change these charges locally, through pair creation or annihilation, but charge fluctuations with long enough wavelengths will remain unchanged.

One possibility is to consider the fluctuation $\langle \Delta Q^2 \rangle$ of the charge Q in a given rapidity window $\Delta\eta$.[1,2] Because of charge conservation, this quantity can only change when charges move in and out of the window, but if the window is wide enough, this effect should be small. Therefore, the measured charge fluctuation should reflect the initial value and be different for QGP and hadronic phases.

Ignoring long-range Coulomb forces, one can estimate that in the QGP phase, where the electric charges of the quarks are either $2/3$ or $-1/3$, the fluctuation is $\langle \Delta Q^2 \rangle \approx 0.2 N_{\rm ch}$. In the hadronic phase, elementary charges are ± 1, and as a consequence the fluctuation is much larger $\langle \Delta Q^2 \rangle \approx 0.7 N_{\rm ch}$.[2] In principle, this difference could be used to find out whether QGP formed in the early stages of the collision. However, current experimental results are consistent with the hadronic value.[3,4,5] They are also very close to $\langle \Delta Q^2 \rangle = N_{\rm ch}$, which would correspond to a purely random distribution of ± 1 charges.

In this talk, I will further explore the information obtainable from charge fluctuations. In particular, I will consider correlations between charge densities at different rapidity values, and show that they carry quantitative information about different stages of the collision. It is hopefully easier to subtract the contributions due to phenomena taking place in the later stages of the collision, such as decays of hadronic resonances, from these correlations than from the charge fluctuation signal $\langle \Delta Q^2 \rangle$.

2. Diffusion in an expanding system

In the first stages of a heavy ion collision, the expansion in the direction of the beam is much faster than that in the orthogonal directions. The effects I will be discussing are all due to this expansion, and therefore I will not consider the orthogonal directions. I will also assume that the two nuclei are moving at the speed of light so that the collision event is boost invariant. In that case, it is convenient to use the Bjorken coordinates τ and η defined by $t = \tau \cosh \eta$, $z = \tau \sinh \eta$. In these coordinates, the Minkowski metric becomes $ds^2 = d\tau^2 - \tau^{-2} dz^2$, which is the metric of a 1+1 dimensional FRW universe with scale factor $a(\tau) = 1/\tau$. This means that very similar considerations apply to charge considerations in the early universe, as well.

I will assume that the evolution of the charge density is purely diffusive. Charge annihilations in that case have been studied for a long time,[6] but often without coupling the system to a heat bath. In the presence of a thermal bath, the diffusion equation in Bjorken coordinates for the comoving charge density $\tilde{\rho} = dQ/d\eta$ is

$$\partial_\tau \tilde{\rho} = \frac{D(\tau)}{\tau^2} \partial_\eta^2 \tilde{\rho} + \partial_\eta \xi_\eta, \tag{1}$$

where ξ_η is a stochastic variable that describes thermal noise. It has a symmetric Gaussian distribution with a two-point function

$$\langle \xi_\eta(\tau, \eta) \xi_\eta(\tau', \eta') \rangle = 2D(\tau) G_{\rm eq}(\tau) \delta(\tau - \tau') \delta(\eta - \eta'). \tag{2}$$

The amplitude of the noise is given by $G_{\rm eq}(\tau)$, which depends on the temperature and can be written as $G_{\rm eq}(\tau) = q^2 \tau n_{\rm eq}(\tau)$, where q is the elementary charge and $n_{\rm eq}(\tau)$ is the equilibrium particle density.

The stochastic term in Eq. (1) can be eliminated by considering the two-point function $G(\tau, \eta - \eta') = \langle \tilde{\rho}(\tau, \eta) \tilde{\rho}(\tau, \eta') \rangle$. It satisfies the equation of motion

$$\partial_\tau G(\eta) = \frac{2D(\tau)}{\tau^2} \partial_\eta^2 \left[G(\eta) - G_{\rm eq}(\tau) \delta(\eta) \right]. \tag{3}$$

The Fourier modes $G(k_\eta) = \int d\eta e^{ik_\eta \eta} G(\eta)$ satisfy

$$\partial_\tau G(k_\eta) = -\frac{2D(\tau)k_\eta^2}{\tau^2}\left[G(k_\eta) - G_{\rm eq}(\tau)\right]. \tag{4}$$

If we assume that initially at $\tau = \tau_{\rm ini}$, the system is in equilibrium, i.e., $G(\tau_{\rm ini}) = G_{\rm eq}(\tau_{\rm ini})$, then the solution is

$$G(\tau, k_\eta) = G_{\rm eq}(\tau) - \int_{\tau_{\rm ini}}^{\tau} d\tau' e^{-\frac{1}{2}\Delta^2(\tau')k_\eta^2}\dot{G}_{\rm eq}(\tau'), \tag{5}$$

where $\Delta^2(\tau') = 4\int_{\tau'}^{\tau} d\hat{\tau}[D(\hat{\tau})/\hat{\tau}^2]$.

Back in coordinate space, we have

$$G(\tau, \eta) = G_{\rm eq}(\tau)\delta(\eta) - \int_{\tau_{\rm ini}}^{\tau} d\tau' \frac{e^{-\eta^2/2\Delta^2(\tau')}}{\sqrt{2\pi\Delta^2(\tau')}}\dot{G}_{\rm eq}(\tau') \tag{6}$$

Assuming that the charged particles are ultrarelativistic, their equilibrium density depends on the temperature as $n_{\rm eq} \propto T^3$. Entropy density scales in the same way, and therefore the temperature decreases with the expansion as $T \propto \tau^{-1/3}$, meaning that $G_{\rm eq}$ remains constant. This means that to a good approximation, $\dot{G}_{\rm eq}$ is simply a sum of delta functions at phase transitions and other non-adiabatic events. Consequently, $G(\tau, \eta)$ becomes a sum of Gaussians with different heights and widths.

As an example, let us assume that the system is initially in thermal equilibrium in the QGP phase so that $G_{\rm eq}(\tau_{\rm ini}) = G_{\rm QGP}$. When the system enters the hadronic phase at $\tau = \tau_{\rm tr}$, $G_{\rm eq}$ jumps discontinuously to $G_{\rm had}$, which is higher because the elementary charge q is ± 1 instead of $2/3$ or $-1/3$. Thus, $\dot{G}_{\rm eq}(\tau) = (G_{\rm had} - G_{\rm QGP})\delta(\tau - \tau_{\rm tr})$, and the final two-point function is

$$G(\tau, \eta) = G_{\rm had}\delta(\eta) - (G_{\rm had} - G_{\rm QGP})\frac{e^{-\eta^2/2\Delta^2(\tau_{\rm tr})}}{\sqrt{2\pi\Delta^2(\tau_{\rm tr})}}. \tag{7}$$

The charge fluctuation signal $\langle \Delta Q^2 \rangle$ can be written in terms of $G(\tau, \eta)$ as

$$\langle \Delta Q^2 \rangle = \frac{1}{\Delta\eta^2}\int_{-\Delta\eta/2}^{\Delta\eta/2} d\eta d\eta' G(\tau, \eta - \eta'). \tag{8}$$

It is easy to see that it, indeed, gives the QGP result if $\Delta\eta \gg \Delta(\tau_{\rm tr})$ and the hadronic result otherwise.

If we then assume that the hadrons become non-relativistic and start to annihilate at a later time $\tau_{\rm nr}$, the delta function peak spreads into a Gaussian

$$G(\tau, \eta) = G_{\rm had}\frac{e^{-\eta^2/2\Delta^2(\tau_{\rm nr})}}{\sqrt{2\pi\Delta^2(\tau_{\rm nr})}} - (G_{\rm had} - G_{\rm QGP})\frac{e^{-\eta^2/2\Delta^2(\tau_{\rm tr})}}{\sqrt{2\pi\Delta^2(\tau_{\rm tr})}}. \tag{9}$$

It is interesting to note that at short enough distances ($\eta \lesssim \Delta(\tau_{\rm nr})$), the correlator is positive.[6,7]. This also means that for small $\Delta\eta$, the charge fluctuation should behave as $\langle \Delta Q^2 \rangle \propto \Delta\eta^2$.

3. Long-range forces

The discussion in Section 2 applies to global charges such as baryon number or strangeness, but for electric charges, one has to take into account the long-range Coulomb interaction, and the motion of the charges is not purely diffusive. By combining Ohm's law $\vec{\jmath} = \sigma \vec{E}$ with Gauss's law $\vec{\nabla} \cdot \vec{E} = \rho$, the diffusion equation gets modified and becomes

$$\partial_\tau \tilde{\rho} = \frac{D(\tau)}{\tau^2} \partial_\eta^2 \tilde{\rho} - \sigma(\tau) \tilde{\rho} + \partial_\eta \xi_\eta. \tag{10}$$

It is instructive to note that when all the coefficients are time-independent, one finds that the equilibrium two-point function is

$$G(k) = \frac{G_{\rm eq} k^2}{k^2 + \sigma/D}, \tag{11}$$

which shows that the Debye screening mass m_D is given by $m_D^2 = \sigma/D$.

Analogously to Eq. (3), we can derive an equation of motion for the two-point function

$$\partial_\tau G(k_\eta) = -\frac{2D(\tau)k_\eta^2}{\tau^2} \left[G(k_\eta) - G_{\rm eq}(\tau) \right] - 2\sigma(\tau) G(k_\eta). \tag{12}$$

Assuming that initially $G(k_\eta)$ vanishes, the solution is

$$G(\tau, k_\eta) = G_{\rm eq}(\tau) - \int_0^\tau d\tau' e^{-\frac{1}{2}\Delta^2(\tau')k_\eta^2 - 2\Sigma(\tau)} \left[2\sigma(\tau') G_{\rm eq}(\tau') + \dot{G}_{\rm eq}(\tau') \right], \tag{13}$$

where $\Sigma(\tau') = \int_{\tau'}^{\tau} d\hat{\tau} \sigma(\hat{\tau})$. Again, this is simply a superposition of Gaussians, and can therefore be easily Fourier transformed back to coordinate space.

As an example, let us consider a simple case in which $G_{\rm eq}$ jumps from 0 to 1 at $\tau_{\rm tr} = 1$. Furthermore, we assume that $D(\tau) = \beta\tau$, where β is constant, and $\sigma(\tau) = m_D^2 D(\tau)$ with constant m_D^2. The correlator $G(\tau, k_\eta)$ has been plotted in the left panel of Fig. 1 for $m_D^2 = 0$ [global charge, from Eq. (7)] and for $m_D^2 = 0.01$ [local charge with long-range forces]. One can see that for a local charge, the correlator is more strongly peaked around zero.

We then imagine that $G_{\rm eq}$ drops instantaneously to zero at $\tau_{\rm nr} = 10$ as the charged particles become non-relativistic. The correlators for global

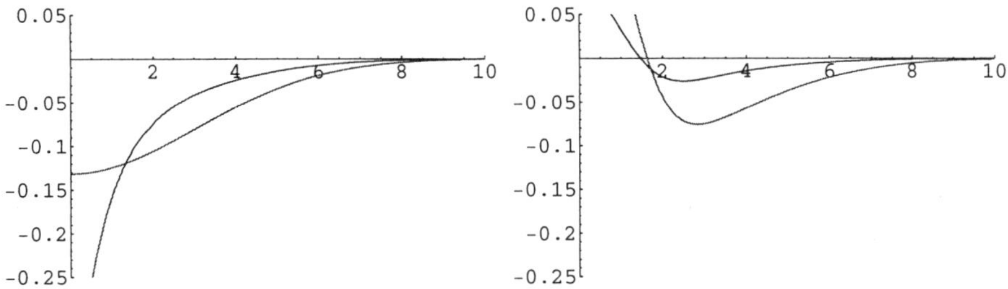

Figure 1. $G(\tau, k_\eta)$ as a function of k_η at $\tau = 10$ (left) and at $\tau = 13$ (right). The two curves correspond to $m_D^2 = 0$ and $m_D^2 = 0.01$.

and local charges at $\tau = 13$ are shown in the right panel of Fig. 1. Because of the long-range forces, annihilation is faster and the signal significantly weaker for local charges.

4. Conclusions

We have seen that the charge correlators measured at late times carry detailed quantitative information about the properties of the system at different stages of the collision. Particle-antiparticle pairs produced later on by neutral resonances etc. will add an extra contribution to these correlations, but if it is properly understood it can be subtracted, at least in principle.

The signals will be much stronger in global charges, i.e., baryon number and strangeness than in the electric charge, but unfortunately they are also much more difficult to measure. It is possible, though, that some of the interesting features survive even in the electric charge correlators, but a more detailed calculation is needed to find out if that is actually the case.

Acknowledgments

This work was supported by Churchill College, Cambridge.

References

1. M. Asakawa, U. W. Heinz and B. Muller, Phys. Rev. Lett. **85**, 2072 (2000).
2. S. Jeon and V. Koch, Phys. Rev. Lett. **85**, 2076 (2000).
3. K. Adcox *et al.* [PHENIX Collaboration], Phys. Rev. Lett. **89**, 082301 (2002).
4. J. Adams *et al.* [STAR Collaboration], Phys. Rev. C **68**, 044905 (2003).
5. C. Alt *et al.* [NA49 Collaboration], arXiv:nucl-ex/0406013.
6. D. Toussaint and F. Wilczek, J. Chem. Phys. **78**, 2642 (1982).
7. T. A. DeGrand, Phys. Rev. D **30**, 2001 (1984).

WHITENING OF THE QUARK-GLUON PLASMA

STANISLAW MRÓWCZYŃSKI
Sołtan Institute for Nuclear Studies
ul. Hoża 69, PL - 00-681 Warsaw, Poland
and Institute of Physics, Świętokrzyska Academy
ul. Świętokrzyska 15, PL - 25-406 Kielce, Poland
E-mail: mrow@fuw.edu.pl

It is discussed how the local color charges of quark-gluon plasma are neutralized.

1. Introduction

This talk is based on two works done in collaboration with Cristina Manuel[1,2]. In the first paper[1], we have analyzed the local equilibrium of the quark-gluon plasma which is defined as a state of maximal local entropy. Using the kinetic equations with the collision terms of the Waldmann-Snider form, we have proved that such a state is generically *colorful*, *i.e.* the color four-current is non-vanishing. Thus, the collisions, which are responsible for equilibration of the parton momenta, do *not* neutralize the local color density. Since the color current is (covariantly) conserved in every collision process, the inter-parton collisions redistribute the color charges among various momentum modes but they do not change a local macroscopic color charge. Consequently, if the color charges are not homogeneously distributed in the process of the plasma production due to, say, statistical fluctuations, the inter-parton collisions will not neutralize the system. On the other hand, the global equilibrium of the quark-gluon plasma is locally colorless because of the maximum entropy principle. We assume here that the system does not curry a global color charge and that it does not experience an external chromodynamic field. Once the inter-parton collisions are not responsible for the neutralization of the local charges, one has to invoke other collective mechanism to whiten the plasma. This is the subject of the second article[2].

Local charges are neutralized due to the currents that flow in the system. We consider the diffusive currents generated by the charge density gradient (Fick's law) and the ohmic currents caused by the chromoelectric field (Ohm's law) which is, in turn, induced by the charge density. The color conductivity of the quark-gluon plasma has been studied for long time[3,4,5,6,7,8,9,10], but only recently the problem has been well understood[11,12,13,14,15,16,17,18]. As far as we know, the color diffusion was only briefly discussed by Heiselberg[10]. In all these papers, the plasma near the colorless global equilibrium was studied. We are, however, interested in the plasma that is locally colorful. Thus, in Sec. 2 we derive the diffusion and conductivity coefficients in such a plasma, and then, in Sec. 3 the temporal evolution of the color charge density is considered.

2. Color diffusion and conductivity coefficients

Using the transport theory, we derive here the diffusive and ohmic currents in a plasma that is locally colorful. The transport equations of quarks, antiquarks and gluons read

$$\big(D^0 + \mathbf{v}\cdot\mathbf{D}\big)Q(\mathbf{p},x) - \frac{g}{2}\{\mathbf{E}+\mathbf{v}\times\mathbf{B}, \nabla_p Q(\mathbf{p},x)\} = C[Q,\bar{Q},G]\,, \quad (1)$$

$$\big(D^0 + \mathbf{v}\cdot\mathbf{D}\big)\bar{Q}(\mathbf{p},x) + \frac{g}{2}\{\mathbf{E}+\mathbf{v}\times\mathbf{B}, \nabla_p \bar{Q}(\mathbf{p},x)\} = \bar{C}[Q,\bar{Q},G]\,, \quad (2)$$

$$\big(D^0 + \mathbf{v}\cdot\mathbf{D}\big)G(\mathbf{p},x) - \frac{g}{2}\{\mathbf{E}+\mathbf{v}\times\mathbf{B}, \nabla_p G(\mathbf{p},x)\} = C_g[Q,\bar{Q},G]\,. \quad (3)$$

The (anti-)quark on-mass-shell distribution functions $Q(\mathbf{p},x)$ and $\bar{Q}(\mathbf{p},x)$, which are $N_c \times N_c$ hermitean matrices, belong to the fundamental representation of the SU(N_c) group, while the gluon distribution function $G(\mathbf{p},x)$, which is a $(N_c^2-1)\times(N_c^2-1)$ matrix, belongs to the adjoint representation. The covariant derivative $D^\mu \equiv \partial^\mu - ig[A^\mu(x),\cdots]$, the chromoelectric and chromomagnetic fields, $\mathbf{E}$ and $\mathbf{B}$, which enter the transport equations also belong to either the fundamental or adjoint representation, correspondingly. To simplify the notation, we use the same symbols D^0, $\mathbf{D}$, $\mathbf{E}$ and $\mathbf{B}$ to denote a given quantity in the fundamental or adjoint representation. $x \equiv (t,\mathbf{x})$ denotes the four-position while $\mathbf{p}$ is the three-momentum. Because the partons are assumed to be massless, the velocity $\mathbf{v}$ equals $\mathbf{p}/|\mathbf{p}|$. The collision terms C, $\bar{C}$ and C_g will be discussed later on.

We are interested in a state close to the colorful local equilibrium. When the effects of quantum statistics are neglected, the (on-mass-shell) local equilibrium distribution functions are[1]

$$Q^{\rm eq}(\mathbf{p},x) = \exp\Big[-\beta(x)\big(u_\nu(x)p^\nu - \mu_b(x) - \widetilde{\mu}(x)\big)\Big] \,, \tag{4}$$

$$\bar{Q}^{\rm eq}(\mathbf{p},x) = \exp\Big[-\beta(x)\big(u_\nu(x)p^\nu + \mu_b(x) + \widetilde{\mu}(x)\big)\Big] \,, \tag{5}$$

$$G^{\rm eq}(\mathbf{p},x) = \exp\Big[-\beta(x)\big(u_\nu(x)p^\nu - \widetilde{\mu}_g(x)\big)\Big] \,, \tag{6}$$

where $p^\mu = (E_p, \mathbf{p})$, and $E_p = |\mathbf{p}|$ for massless quarks and antiquarks, and for gluons; $\beta = 1/T$, u^μ, μ_b denote, respectively, the inverse temperature, hydrodynamic velocity and baryon chemical potential; the colored chemical potentials of quarks ($\widetilde{\mu}$) and of gluons ($\widetilde{\mu}_g$) obey the relationship

$$\widetilde{\mu}_g(x) = 2T^a {\rm Tr}[\tau^a \widetilde{\mu}(x)] = \mu_a(x) T^a \,, \tag{7}$$

where τ^a, T^a with $a = 1, \ldots, N_c^2 - 1$ are the SU(N_c) group generators in the fundamental and adjoint representations, normalized as ${\rm Tr}[\tau^a\tau^b] = \frac{1}{2}\delta^{ab}$ and ${\rm Tr}[T^a T^b] = N_c \delta^{ab}$. In the local equilibrium state there is a non-vanishing color charge density and current. However, the current vanishes in the local rest frame because the distribution functions are locally isotropic.

We further consider small deviations from local equilibrium, and we write down the quark distribution function as $Q(\mathbf{p},x) = Q^{\rm eq}(\mathbf{p},x) + \delta Q(\mathbf{p},x)$. Assuming that

$$|Q^{\rm eq}| \gg |\delta Q| \,, \qquad |D^\mu Q^{\rm eq}| \gg |D^\mu \delta Q| \,, \qquad |\nabla_p Q^{\rm eq}| \gg |\nabla_p \delta Q| \,, \tag{8}$$

and taking into account the local isotropy of the equilibrium state, the transport equation (1) can be approximated as

$$\left(D^0 + \mathbf{v}\cdot\mathbf{D}\right) Q^{\rm eq} + \frac{g}{2}\{\mathbf{E}, \nabla_p Q^{\rm eq}\} = L[\delta Q] \,, \tag{9}$$

where $L[\delta Q]$ is the collision term linearized around the local equilibrium distribution function. Analogous equations hold for antiquarks and gluons. Let us recall here that the collision terms evaluated for the local equilibrium distribution functions (2) vanish[1].

To get the transport coefficients of color diffusion and conductivity we assume that the linearized collision terms L, $\bar{L}$ and L_g satisfy the relationship

$$\int \frac{d^3p}{(2\pi)^3}\, \mathbf{v}\, L[\delta Q] = -\gamma \int \frac{d^3p}{(2\pi)^3}\, \mathbf{v}\, \delta Q \,, \tag{10}$$

where $1/\gamma$ is the characteristic relaxation time which, for simplicity, is assumed to be the same for quarks, antiquarks and gluons. The relation is obeyed[15] by the Waldmann-Snider collision term linearized around global (colorless) equilibrium. The relationship (10) is also trivially satisfied by the collision term in the relaxation time approximation, which, unfortunately, violates the covariant current conservation. However, the collision term can be easily improved. In analogy to the so-called BGK collision term[19], one constructs the expression[2] which obeys both the covariant current conservation and the relation (10).

Using Eq. (10), the color current (in the adjoint representation), which is generated by deviations from equilibrium, gets the form

$$\mathbf{j}^a(x) = -d\,\mathbf{D}^{ab}\rho^b(x) + \sigma^{ab}\mathbf{E}^b(x)\;, \tag{11}$$

where $\mathbf{D}^{ab} = \delta^{ab}\nabla + gf^{acb}\mathbf{A}^c$, the diffusion constant equals $d = 1/3\gamma$ and the color conductivity tensor reads

$$\sigma^{ab} = \frac{g^2}{3\gamma}\int \frac{d^3p}{(2\pi)^3}\frac{1}{E_p}\Big(\mathrm{Tr}[\{\tau^a,\tau^b\}(Q^{\mathrm{eq}}+\bar{Q}^{\mathrm{eq}})] + \mathrm{Tr}[\{T^a,T^b\}G^{\mathrm{eq}}]\Big)\,. \tag{12}$$

When the equilibrium is colorless, the conductivity σ is proportional to the unit matrix in color space.

3. Temporal evolution of the color density

The color current generated by both the gradient of color density and the chromoelectric field is given by Eq. (11). The covariant current conservation $D^0\rho + \mathbf{Dj} = 0$ combined with the Gauss law $\mathbf{DE}(x) = \rho(x)$, provides the equation

$$\big[D^0 - d\,\mathbf{D}^2 + \sigma\big]\rho(x) = 0\;, \tag{13}$$

where the term $[\mathbf{D},\sigma]\,\mathbf{E}$ has been neglected. It can be shown[2] that $[\mathbf{D},\sigma]\,\mathbf{E} \ll \sigma\mathbf{DE}$ in the small coupling limit. Since the commutator $[D^0,\sigma]$ can be also shown to be small, Eq. (13) is solved by

$$\rho(x) = e^{-\sigma t}\,n(x) \tag{14}$$

with n satisfying the diffusion equation

$$\big[D^0 - d\,\mathbf{D}^2\big]n(x) = 0\,. \tag{15}$$

$\rho(x)$ and $n(x)$ obey the initial condition $\rho(0,\mathbf{x}) = n(0,\mathbf{x}) = \rho_0(\mathbf{x})$. As expected, the Ohmic currents appear to dominate over the diffusive ones[2].

4. Discussion

The question arises what is the characteristic time of plasma whitening. The only reliable estimates of the time scales of interest have been found for the perturbative quark-gluon plasma close to global (colorless) equilibrium. The three characteristic times: of the color conductivity, of the parton-parton collisions at momentum transfers of order T and of order g^2T are[8,15]

$$\frac{1}{\sigma} \sim \frac{\ln(1/g)}{T}, \quad t_{\rm hard} \sim \frac{1}{g^4 \ln(1/g)\, T}, \quad t_{\rm soft} \sim \frac{1}{g^2 \ln(1/g)\, T}. \tag{16}$$

Our derivation of the color conductivity and further analysis apply for $t \geq 1/\gamma$. Taking into account the above estimate of color conductivity, the charge density is found to vanish at $t \geq 1/\gamma$. Since, the characteristic time scale of color dissipation cannot be shorter than $t_{\rm soft}$, the whitening of the quark-gluon plasma occurs at $t \sim t_{\rm soft}$. At shorter times scales the color density is expected to oscillate.

References

1. C. Manuel and St. Mrówczyński, *Phys. Rev.* **D68**, 094010 (2003).
2. C. Manuel and St. Mrówczyński, arXiv:hep-ph/0403024.
3. U. W. Heinz, *Annals Phys.* **168**, 148 (1986).
4. W. Czyz and W. Florkowski, *Acta Phys. Polon.* **B17**, 819 (1986).
5. A. Dyrek and W. Florkowski, *Phys. Rev.* **D36**, 2172 (1987).
6. St. Mrówczyński, *Acta Phys. Polon.* **B19** (1988) 91.
7. St. Mrówczyński, *Adv. Ser. Direct. High Energy Phys.* **6**, 185 (1990).
8. A. Selikhov and M. Gyulassy, *Phys. Lett.* **B316**, 373 (1993).
9. A. V. Selikhov and M. Gyulassy, *Phys. Rev.* **C49** (1994) 1726.
10. H. Heiselberg, *Phys. Rev. Lett.* **72**, 3013 (1994).
11. D. Bödeker, *Phys. Lett.* **426**, 351 (1998).
12. D. Bödeker, *Nucl. Phys.* **B559**, 502 (1999).
13. D. F. Litim and C. Manuel, *Phys. Rev. Lett.* **82**, 4981 (1999).
14. D. F. Litim and C. Manuel, *Nucl. Phys.* **B562**, 237 (1999).
15. P. Arnold, D. T. Son and L. G. Yaffe, *Phys. Rev.* **D59**, 105020 (1999).
16. J. P. Blaizot and E. Iancu, *Nucl. Phys.* **B557**, 183 (1999).
17. P. Arnold and L. G. Yaffe, *Phys. Rev.* **D62**, 125014 (2000).
18. J.M. Martinez Resco and M.A. Valle Basagoiti, *Phys. Rev.* **D63**, 056008 (2001).
19. P.L. Bhatnagar, E.P. Gross, and M.Krook, *Phys. Rev.* **94**, 511 (1954).

PROGRESS IN ANISOTROPIC PLASMA PHYSICS

PAUL ROMATSCHKE

Institut für Theoretische Physik
Technische Universität Wien
Wiedner Hauptstrasse 8-10
A-1040 Vienna, Austria

MICHAEL STRICKLAND

Institut für Theoretische Physik
Technische Universität Wien
Wiedner Hauptstrasse 8-10
A-1040 Vienna, Austria

In 1959 Weibel demonstrated that when a QED plasma has a temperature anisotropy there exist unstable transverse magnetic excitations which grow exponentially fast. In this paper we will review how to determine the growth rates for these unstable modes in the weak-coupling and ultrarelativistic limits in which the collective behavior is describable in terms are so-called "hard-loops". We will show that in this limit QCD is subject to instabilities which are analogous to the Weibel instability in QED. The presence of such instabilities dominates the early time evolution of a highly anisotropic plasma; however, at longer times it is expected that these instabilities will saturate (condense). We will discuss how the presence of non-linear interactions between the gluons complicates the determination of the saturated state. In order to discuss this we present the generalization of the Braaten-Pisarski isotropic hard-thermal-loop effective action to a system with a temperature anisotropy in the parton distribution functions. The resulting hard-loop effective action can be used to determine the time and energy scales associated with the possible saturation (condensation) of the gluonic modes. We will also discuss the effects of anisotropies on observables, in particular on the heavy quark energy loss.

1. Introduction

In the last few years a more or less standard picture of the early stages of a relativistic heavy ion collision has emerged. In its most simplified form there are three assumptions: (1) that the system is boost invariant along the beam direction, (2) that it is homogenous in the directions perpendicular to the beam direction, and (3) that the physics at early times is dominated by gluons with momentum at a "saturation" scale Q_s which have occupation numbers of order $1/\alpha_s$. The first two assumptions are reasonable for describing the central rapidity region in relativistic heavy-ion collisions. The third assumption relies on the presence of gluonic "saturation" of the nuclear wavefunction at very small values of the Bjorken variable x [1]. In this regime one can determine the growth of the gluon distribution by requiring that the cross section for deep inelastic scattering at fixed Q^2 does not violate unitarity bounds. As a result the gluon distribution function saturates at a scale Q_s changing from $1/k_\perp^2 \to \log(Q_s^2/k_\perp^2)/\alpha_s$. Luckily, despite this saturation, due to the factor of $1/\alpha_s$ in the second scaling relation the occupation number of small-x gluonic modes in the nuclear wavefunction is still large enough to determine their distribution function analytically using classical nonlinear field theory [1].

These distribution functions are used as input for the subsequent thermalization of the quark-gluon plasma. At short times, $\tau_0 \sim Q_s^{-1}$, the gluon distribution function is isotropic, however, after the system begins to expand it rapidly develops a large anisotropy between the transverse and beam directions with $p_\perp \gg p_z$ due to the fact that initially $p_z \sim \tau^{-1}$. In the weak-coupling limit the assumptions above have been used by Baier et al. in an attempt to systematically describe the early stages of quark-gluon plasma evolution in a framework called "bottom-up" thermalization [2]. Using collisional mechanisms they are able to show that during the first stage of evolution hard gluons scatter out-of-plane counteracting the effect of the expansion of the system reducing the rate at which the longitudinal momentum decreases instead to $p_z \sim Q_s^{2/3}\tau^{-1/3}$. Although less extreme in terms of the rate at which the longitudinal momenta decreases this scenario still results in considerable momentum-space anisotropies in the gluon distribution function. In such anisotropic systems it has been shown that the physics of the QCD collective modes changes dramatically compared to the isotropic case and instabilities are present which can accelerate the thermalization and isotropization of the plasma [3,4,5,6,8,9,10,11].

In addition to their role in isotropization and thermalization it is also important to have an understanding of what the impact of such instabilities will be on observables at RHIC and LHC energies. Here we will discuss one test observable – the heavy fermion energy loss [12,13]. One might be worried that there is a fundamental problem with perturbation theory since the presence of instabilities naively causes the calculation of the soft contribution to the heavy quark energy loss to be divergent; however, there is protection mechanism dubbed "dynamical shielding" which renders the collisional energy loss finite for QED and QCD [12,13]. However, at realistic couplings the presence of instabilities and associated poles on neighboring Riemann sheets [10] causes significant changes in the soft energy loss contribution for both QED and QCD. In fact, as we will discuss, the unphysical poles can even change the sign of the heavy quark energy loss at low momentum turning it instead into energy *gain*.

2. Anisotropic Gluon Polarization Tensor

We consider a quark-gluon plasma with a parton distribution function which is decomposed as

$$f(\mathbf{p}) \equiv 2N_f\left(n(\mathbf{p}) + \bar{n}(\mathbf{p})\right) + 4N_c n_g(\mathbf{p}) \,, \tag{1}$$

where n, $\bar{n}$, and n_g are the distribution functions of quarks, anti-quarks, and gluons, respectively, and the numerical coefficients collect all appropriate symmetry factors. Using the result of Ref. [6] the spacelike components of the high-temperature gluon self-energy for gluons with soft momentum ($k \sim gT$) can be written as

$$\Pi^{ij}(K) = -\frac{g^2}{2}\int \frac{d^3\mathbf{p}}{(2\pi)^3} v^i \partial^l f(\mathbf{p}) \left(\delta^{jl} + \frac{v^j k^l}{K\cdot V + i\epsilon}\right) \,, \tag{2}$$

where the parton distribution function $f(\mathbf{p})$ is, in principle, completely arbitrary. In what follows we will assume that $f(\mathbf{p})$ can be obtained from an isotropic distribution function by the rescaling of only one direction in momentum space. In practice this means that, given any isotropic parton distribution function $f_{\rm iso}(p)$, we can construct an anisotropic version by changing the argument of the isotropic distribution function

$$f(\mathbf{p}) = \sqrt{1+\xi}\, f_{\rm iso}\left(\sqrt{\mathbf{p}^2 + \xi(\mathbf{p}\cdot\hat{\mathbf{n}})^2}\right) \,, \tag{3}$$

where the factor of $\sqrt{1+\xi}$ is a normalization constant which ensures that the same parton density is achieved regardless of the anisotropy introduced, $\hat{\mathbf{n}}$ is the direction of the anisotropy, and $\xi > -1$ is an adjustable anisotropy parameter with $\xi = 0$ corresponding to the isotropic case. Here we will concentrate on $\xi > 0$ which corresponds to a contraction of the distribution along the $\hat{\mathbf{n}}$ direction since this is the configuration relevant for heavy-ion collisions at early times, namely two hot transverse directions and one cold longitudinal direction.

Making a change of variables in (2) it is possible to integrate out the $|p|$-dependence giving [6]

$$\Pi^{ij}(\omega/k,\theta_n) = \mu^2 \int \frac{d\Omega}{4\pi} v^i \frac{v^l + \xi(\mathbf{v}\cdot\hat{\mathbf{n}})\hat{n}^l}{(1+\xi(\mathbf{v}\cdot\hat{\mathbf{n}})^2)^2} \left(\delta^{jl} + \frac{v^j k^l}{K\cdot V + i\epsilon}\right), \quad (4)$$

where $\cos\theta_n \equiv \hat{\mathbf{k}}\cdot\hat{\mathbf{n}}$ and $\mu^2 \equiv \sqrt{1+\xi}\, m_D^2 > 0$. The isotropic Debye mass, m_D, depends on $f_{\rm iso}$. In the case of pure-gauge QCD with an equilibrium $f_{\rm iso}$ we have $m_D = gT$.

The next task is to construct a tensor basis for the spacelike components of the gluon self-energy and propagator. We therefore need a basis for symmetric 3-tensors which depend on a fixed anisotropy 3-vector $\hat{n}^i$ with $\hat{n}^2 = 1$. This can be achieved with the following four component tensor basis: $A^{ij} = \delta^{ij} - k^i k^j/k^2$, $B^{ij} = k^i k^j/k^2$, $C^{ij} = \tilde{n}^i\tilde{n}^j/\tilde{n}^2$, and $D^{ij} = k^i\tilde{n}^j + k^j\tilde{n}^i$ with $\tilde{n}^i \equiv A^{ij}\hat{n}^j$. Using this basis we can decompose the self-energy into four structure functions α, β, γ, and δ as $\mathbf{\Pi} = \alpha\,\mathbf{A} + \beta\,\mathbf{B} + \gamma\,\mathbf{C} + \delta\,\mathbf{D}$. [a]

3. Collective Modes

As shown in Ref. [6] this tensor basis allows us to express the propagator in terms of the following three functions

$$\Delta_\alpha^{-1}(K) = k^2 - \omega^2 + \alpha\,,$$
$$\Delta_\pm^{-1}(K) = \omega^2 - \Omega_\pm^2\,,$$

where $2\Omega_\pm^2 = \bar{\Omega}^2 \pm \sqrt{\bar{\Omega}^4 - 4((\alpha+\gamma+k^2)\beta - k^2\tilde{n}^2\delta^2)}$ and $\bar{\Omega}^2 = \alpha + \beta + \gamma + k^2$.

Taking the static limit of these three propagators we find that there are three mass scales: $m_\pm$ and m_α. In the isotropic limit, $\xi \to 0$, $m_\alpha^2 = m_-^2 = 0$ and $m_+^2 = m_D^2$. However, for $\xi > 0$ we find that $m_\alpha^2 < 0$ for all $|\theta_n| \neq \pi/2$ and $m_-^2 < 0$ for all $|\theta_n| \leq \pi/4$. Note also that for $\xi > 0$ both m_α^2 and m_-^2 have there largest negative values at $\theta_n = 0$ where they are equal.

The fact that for $\xi > 0$ both m_α^2 and m_-^2 can be negative indicates that the system is unstable to both magnetic and electric fluctuations with the fastest growing modes focused along the beamline ($\theta_n = 0$). In fact it can be shown that there are two purely imaginary solutions to each of the dispersions relations $\Delta_\alpha^{-1}(K) = 0$ and $\Delta_-^{-1}(K) = 0$ with the solutions in the upper half plane corresponding to unstable modes. We can determine

[a] Explicit analytic integral expressions for α, β, γ, and δ can be found in Ref. [6]. Additionally, analytic expressions in the small-ξ limit and for propagation along the anisotropy direction can be found in Refs. [6] and [10], respectively.

the growth rate for these unstable modes by taking $w \to i\Gamma$ and then solving the resulting dispersion relations for $\Gamma(k)$.

In Fig. 1 we plot the instability growth rates, Γ_α and Γ_-, as a function of wave number for $\xi = 10$ and $\theta_n = \pi/8$. Note that both growth rates vanish at $k = 0$ and have a maximum $\Gamma_* \sim \mu/10$ at $k_* \sim \mu/3$. The fact that they have a maximum means that at early times the system will be dominated by unstable modes with spatial frequency $1/k_*$.

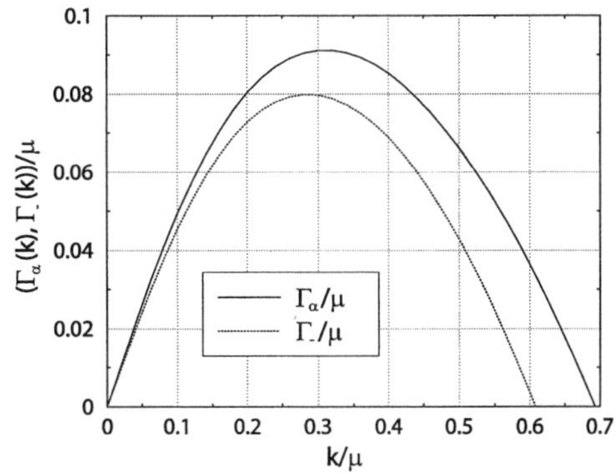

Fig. 1: Instability growth rates.

4. The Hard-Loop Effective Action

In the previous section by only considering the polarization tensor we implicitly used a linear analysis which suffices only during very early times. As the instabilities grow non-linear interactions become more important and can halt their growth. In QED simulations such a saturated state does emerge and is called a Bernstein-Greene-Kruskal wave; however, in QCD any corresponding saturated state would necessarily be a much more complex beast. Answering these questions requires knowledge of the full hard-loop effective Lagrangian in an anisotropic system. Luckily, it is possible to obtain a simple expression which generates all hard-loop (HL) vertex functions [9]

$$\mathcal{L} = -\frac{1}{4}F_{\mu\nu}F^{\mu\nu} + \frac{g^2}{2}\int_{\mathbf{p}} \frac{f(\mathbf{p})}{|\mathbf{p}|} F_{\mu\nu}(x)\frac{p^\rho p^\nu}{(p\cdot D)^2}\, F_\rho{}^{\mu}(x) \tag{5}$$

For example, from this we can obtain the hard-loop 3-gluon vertex

$$\Gamma^{\mu\nu\lambda}_{\rm HL}(k,q,r) = \frac{g^2}{2}\int_{\mathbf{p}} \frac{\partial f(\mathbf{p})}{\partial p^\beta}\, \hat{p}^\mu \hat{p}^\nu \hat{p}^\lambda \left(\frac{r^\beta}{\hat{p}\cdot q\, \hat{p}\cdot r} - \frac{k^\beta}{\hat{p}\cdot k\, \hat{p}\cdot q}\right) . \tag{6}$$

In contrast to the isotropic n-gluon HL vertices which vanish in the static limit, the anisotropic n-gluon HL vertices can be non-zero. This opens up the possibility for additional saturation mechanisms which are governed by soft non-abelian physics sensitive to these higher vertices.

Note that Arnold and Lenaghan have shown that if one considers fluctuations in the vector potential which are directed along the anisotropy direction and only depend on the longitudinal coordinate then it is possible to reduce the effective lagrangian (5) to a quadratic form which simplifies the analysis considerably [11]. They then make some additional assumptions including ignoring time-depedence in the self-energy in order to construct a toy-model which is then used to study the subsequent evolution of instabilities in the system. Their results indicate that in non-abelian gauge theories that possible non-abelian saturated states are metastable and as a result instability growth forces the system along abelian directions. However, it should be noted that by ignoring the time dependence in the potential as they have done, the instability growth rate in their model is finite at $k = 0$. This is not the case when time dependence is included in the full analysis and instead one finds that the growth rate at $k = 0$ then vanishes as shown in Fig. 1. This difference could have a significant impact on the abelianization particularly on the "abelianization length" measured in their simulations.

5. Effects on Observables

Since the unstable modes manifest themselves as poles of the propagator in the static limit, it has been argued [14] that the presence of these instabilities in general prohibits the calculation of observables in a perturbative framework, since those quantities would be plagued by unregulated divergences.

However, it turns out that at least one observable, namely the collisional energy loss of a heavy fermion, is protected from these unregulated divergencies by a mechanism dubbed "dynamical shielding" [12] (see Ref. [7] for a proof in the case of $\xi \ll 1$ and $\xi \to \infty$). As a consequence of dynamical shielding (which incidently is somewhat similar to dynamical screening of the magnetic interaction in QCD, hence the name), the collisional energy loss becomes calculable perturbatively also in the anisotropic case. Therefore, we will use the collisional energy loss as "test observable" to learn something about the effects of the instabilities on physical quantities.

Schematically, the soft contribution to the collisional energy loss is given by

$$-\left(\frac{\mathrm{d}W}{\mathrm{d}t}\right)_{\mathrm{soft}} \sim \mathrm{Im} \int \frac{d^3\mathbf{k}}{(2\pi)^3}\, k\, z\, \Delta(z,k,\theta_n)\,, \tag{7}$$

where $\Delta(z,k,\theta_n) = v^i \Delta^{ij} v^j$ is the relevant contraction of the propagator [12] and $z = \hat{\mathbf{k}} \cdot \mathbf{v}$ with $\mathbf{v}$ the velocity of the heavy fermion. The instabilities would in principle affect the integrand for $\mathrm{Re}\, z = 0$ causing singularties to be encountered along the momentum integration path. In order to see that

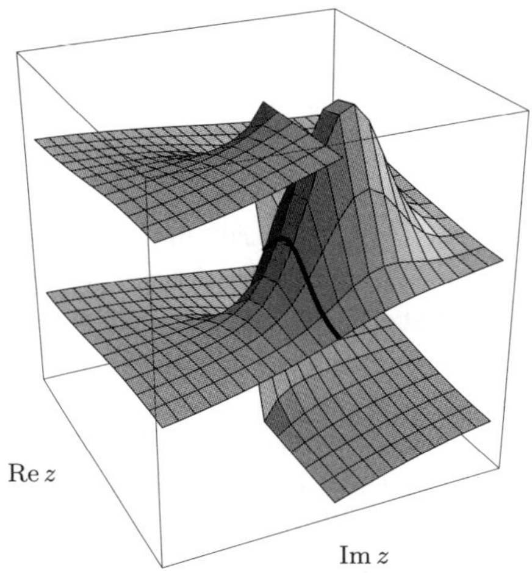

Fig. 2: Sketch of the complex z plane including the extension of the logarithm to the unphysical sheet. Also shown is how a pole in the unphysical region (mountain) has effects felt on the physical sheet. The black line indicates where the two sheets are joined together.

there are, in fact, no singularities we focus on terms of the form $(q^2+\alpha)^{-1}$. Due to the fact that in the static limit the structure function α is negative-valued we could encounter a singulartity at $q^2 = \alpha$; however, if one takes the static limit of α carefully we find

$$\lim_{z\to 0} \alpha(z) = M^2(-1+iDz) + \mathcal{O}(z^2)\ , \tag{8}$$

where M and D depend on the angle of propagation with respect to the anisotropy vector and the strength of the anisotropy. As long as D is non-vanishing the singularities are regularized because of the z in the numerator of Eq. (7) and we call the singularity "dynamically shielded". This effect can be shown to regulate all singularities which would have naively come from the presence of instabilities. Therefore, there is no catastrophic effect from the instabilities on the collisional energy loss. However, there is an effect which is associated with the presence of the instabilities; in order to understand this effect it is however necessary to extend the analysis of the collective modes of an anisotropic plasma to arbitrary Riemann sheets.

The extension to the unphysical Riemann sheets is achieved by continuing analytically the structure functions beyond their cut structures (which for finite ξ is a logarithmic cut running along the real z axis for $z^2 < 1$, in complete analogy to the isotropic case, but a square-root cut for $z^2 < \sin^2\theta_n$ for $\xi \to \infty$)[10]. More precisely, since one can continue the structure functions from either below or above the cut, the physical sheet has two neighboring unphysical Riemann sheets, which – at least in the case of finite ξ – could themselves be extended to reproduce the well-known "spiral staircase" form of the complex logarithm. Once the structure functions on the unphysical sheets are known, one may conduct the analysis of the collective modes as was done on the physical sheet, finding – among others – singularities for complex z on the neighboring unphysical sheets that extend to the region of spacelike momenta [10].

But why bother about these modes, given that they do not "live" on the physical sheet? To see that there can be an effect on physical quantities, imagine once again the spiral staircase spanned by the complex logarithm: the physical sheet would correspond to the region covered by the spiral plane from the ground floor to the first floor, while the unphysical sheet where the extra quasiparticle mode lives would correspond to the region first to second floor. Since the existence of such a mode corresponds to a pole in the propagator, there is has a mountainous dent (singularity) in the spiral plane somewhere from the first to the second floor. However, since the mountain has a finite width, its base can be felt also below the second floor, especially if its peak is near the first floor (see Fig. 2 for a sketch). Therefore, the nearer a mode on a neighboring unphysical sheet comes to the border of the physical sheet, the more will it affect the propagator on the physical sheet and consequently any physical observable sensitive to the relevant region of phase-space (in this case spacelike momentum).

This is precisely the situation encountered for the collisional energy loss: there are unstable modes on the physical sheet which, once the momentum k becomes larger than some critical value (see Fig. 1), move onto the neighboring unphysical sheets and subsequently influence the propagator on the physical sheet by the mechanism explained above. In fact, it turns out that the the contributions from these unphysical modes may drive the leading-order perturbative results for the collisional energy loss *negative* for small fermion velocities, corresponding to an energy transfer from hard to soft scales. Note, however, that this negative energy loss is *not* due to the heavy fermion in question having a sub-thermal velocity but is instead connected to the instabilities of an anisotropic system, as already noted in Ref. [15].

For further discussion of the heavy quark energy loss in an anisotropic quark-gluon plasma we refer the reader to Ref. [13]. In addition to elucidating the impact of unstable modes and associated poles on unphysical

Riemann sheets, in that paper we showed that for anisotropic systems the heavy quark collisional energy loss has an angular dependence which increases as the coupling and/or anisotropy is increased. Quantitatively, for $\alpha_s = 0.3$ and a 20 GeV bottom quark we found that the deviations from the isotropic result were on the order of 10% for $\xi = 1$ and of the order of 20% for $\xi \geq 10$. When translated into the difference between longitudinal and transverse energy loss this results in a 10% difference at $\xi = 1$, a 30% difference at $\xi = 10$, and a 50% difference at $\xi = \infty$.

6. Conclusions

In this paper we have attempted to summarize the developments which have occured in the last two years regarding the physics of a quark-gluon plasma which has a momentum-space anisotropy in the underlying parton distribution functions. We have presented expressions for the self-energy and full effective action in this case and discussed the emergence of unstable modes which cause exponential growth of the gluon field along the anisotropy direction. In addition, we discussed the expected effects such anisotropies have on observables concentrating on the heavy quark energy loss as a test observable.

As we have tried to illustrate here the physics of anisotropic plasmas is qualitatively different from isotropic ones. In the weak-coupling limit the short-time behaviour of anisotropic plasmas is dominated by the growth of unstable modes. In order to properly understand thermalization and isotropization of a quark-gluon plasma it seems necessary to take into account the effect of such unstable modes from the earliest times. We note in closing if a Vlasov description is applicable at very early times ($\tau_0 \sim Q_s^{-1}$) then instability broadening of the longitudinal momentum is much more effective than collisional broadening and it is possible that anisotropies due to expansion of the system are, in fact, never generated or reduced significantly. However, a more conservative approach calls for the Vlasov description to only be applied once the hard gluon occupation number becomes less than one which according to collisional arguments occurs at a time $\tau_1 \sim \alpha^{-3/2} Q_s^{-1}$. If this more conservative approach is applicable then the anisotropy at τ_1 can be parametrically estimated to be $\xi \sim \alpha_s^{-3/2}$ ($\xi \sim \alpha_s^{-1/2}$ if collisional broadening is taken in account) which results in a strongly anisotropic system in the weak-coupling limit. Whichever approach is used, however, it is clear that it is necessary to take into account the unstable modes.

Acknowledgements

The authors would like to thank A. Rebhan and the organizers of the SEWM conference. In addition, P.R. would like to thank the University of Helsinki for financial support and B. Heher for his support and guidance.

References

1. L. V. Gribov, E. M. Levin and M. G. Ryskin, Phys. Rept. **100**, 1 (1983); A. H. Mueller and J. w. Qiu, Nucl. Phys. B **268**, 427 (1986); J. P. Blaizot and A. H. Mueller, Nucl. Phys. B **289**, 847 (1987); L. McLerran and R. Venugopalan, Phys. Rev. D **49**, 223 (1994); **49**, 3352 (1994); **50**, 2225 (1994); I. Balitskii, Nucl. Phys. B **463**, 99 (1996); J. Jalilian-Marian, A. Kovner, L. McLerran and H. Weigert, Phys. Rev. D **55**, 5414 (1997); Yu.V. Kovchegov and A.H. Mueller, Nucl. Phys. B **529**, 451 (1998); Yu.V. Kovchegov, Phys. Rev. D **60**, 034008 (1999); **61**, 074018 (2000).
2. R. Baier, A.H. Mueller, D. Schiff, and D.T. Son, Phys. Lett. B **502**, 51 (2001).
3. E.S. Weibel, Phys. Rev. Lett. **2**, 83 (1959).
4. S. Mrówczyński, Phys. Lett. B **314**, 118 (1993); Phys. Rev. C **49**, 2191 (1994); Phys. Lett. B **393**, 26 (1997).
5. J. Randrup and S. Mrówczyński, Phys. Rev. C **68**, 034909 (2003).
6. P. Romatschke and M. Strickland, Phys. Rev. D **68**, 036004 (2003).
7. P. Romatschke, "Quasiparticle description of the hot and dense quark gluon plasma," PhD Dissertation, arXiv: hep-ph/0312152, (2003).
8. P. Arnold, J. Lenaghan, and G. D. Moore, JHEP **0308**, 002 (2003).
9. S. Mrówczyński, A. Rebhan, and M. Strickland, Phys. Rev. D **70**, 025004 (2004).
10. P. Romatschke and M. Strickland, arXiv: hep-ph/0406188, (2004).
11. P. Arnold and J. Lenaghan, arXiv: hep-ph/0408052, (2004).
12. P. Romatschke and M. Strickland, Phys. Rev. D **69**, 065005 (2004).
13. P. Romatschke and M. Strickland, arXiv: hep-ph/0408275, (2004).
14. G. Moore. QCD Kinetic theory (lecture), Quantum fields in and out of equilibrium, University of Bielefeld, (2003).
15. E.M Lifshitz and L.P. Pitaevskii, "Physical kinetics", Pergamon Press, Oxford, §30 (1981).

DECONFINEMENT AND CHIRAL SYMMETRY: COMPETING ORDERS

K. TUOMINEN

Department of Physics, P.O. Box 35,
FIN-40014 University of Jyväskylä, Finland,
Helsinki Institute of Physics, P.O. Box 64,
FIN-00014 University of Helsinki, Finland
E-mail: kimmo.tuominen@phys.jyu.fi

We investigate the interplay between chiral symmetry restoration and deconfinement both in two color QCD and adjoint QCD. In ordinary QCD we show how the behavior of the Polyakov loop near chiral phase transition is induced by the chiral order parameter, while in adjoint two color QCD one has two truly independent phase transitions. Introducing a finite baryochemical potential we find that adjoint QCD exhibits tetracritical behavior.

1. Introduction

The phase transitions in QCD and QCD-like theories have gained a lot of interest [1]. For strong interactions, with realistic quark masses, at finite temperature and zero chemical potential the chiral symmetry in not exact, but one can still follow the behavior of the condensates. Lattice calculations show that with quarks in the fundamental representation deconfinement (a rise in the Polyakov loop ℓ), coincides with chiral symmetry restoration (decrease in the chiral condensate σ) $T_{\rm d} = T_{\rm c}$ [2]. If quarks are in the adjoint representation, deconfinement and the chiral symmetry restoration become independent, $T_{\rm c} \simeq 8T_{\rm d}$ [3].

At finite chemical potential a richer phase structure emerges due to the possibility of diquark condensation which in ordinary QCD leads to super-condutive and in adjoint QCD to superfluid phenomena. We have provided a simple unified way [4] to describe the relation between deconfinement and chiral symmetry restoration in QCD and adjoint QCD, and discovered a new region of tetracritical behavior in two color adjoint QCD [5].

2. Confinement or chiral symmetry

In two color QCD with two massless quark flavors in the fundamental representation the global symmetry group $SU(4)$ breaks to $Sp(4)$. The dynamics of the chiral degrees of freedom, π^a and σ , is described by the potential [6]:

$$V_{\rm ch}[\sigma,\pi^a]=\frac{m^2}{2}{\rm Tr}\left[M^\dagger M\right]+\lambda_1{\rm Tr}\left[M^\dagger M\right]^2+\frac{\lambda_2}{4}{\rm Tr}\left[M^\dagger M M^\dagger M\right] \quad (1)$$

with $2\,M=\sigma+i\,2\sqrt{2}\pi^a\,X^a$, $a=1,\ldots,5$ and the generators X^a are provided explicitly in equation (A.5) and (A.6) of [6]. The Polyakov loop ℓ is treated as a heavy field, singlet under the chiral symmetry:

$$V_\ell[\ell]=g_0\ell+\frac{m_\ell^2}{2}\ell^2+\frac{g_3}{3}\ell^3+\frac{g_4}{4}\ell^4\,. \quad (2)$$

The interaction terms allowed by chiral symmetry are $\left(g_1\ell+g_2\ell^2\right){\rm Tr}\left[M^\dagger M\right]$, and the essential dynamics is due to the g_1 term. In the symmetry broken phase with $T<T_{c\sigma}$, the σ acquires a non-zero expectation value, which in turn induces a modification also for $\langle\ell\rangle$:

$$\langle\sigma\rangle^2\simeq-\frac{m_\sigma^2}{\lambda}\,,\quad m_\sigma^2\simeq m^2+2g_1\langle\ell\rangle\,,\quad \text{and}\quad \langle\ell\rangle\simeq-\frac{g_0}{m_\ell^2}-\frac{g_1}{m_\ell^2}\langle\sigma\rangle^2\,, \quad (3)$$

with $\lambda=\lambda_1+\lambda_2$. Near T_c the mass of the order parameter field is assumed to possess the generic behavior $m_\sigma^2\sim(T-T_c)^\nu$. For $g_1>0$ and $g_0<0$ the expectation value of ℓ behaves oppositely to that of σ : As the chiral condensate starts to decrease towards chiral symmetry restoration, the expectation value of the Polyakov loop starts to increase, signaling the onset of deconfinement. Furthermore, we find a drop in the spatial two-point correlator of the Polyakov loop near the chiral transition.

In two color QCD with two massless Dirac quark flavors in the adjoint representation, the global symmetry $SU(4)$ breaks to $O(2N_f)$. Now the chiral part of the potential is given by (1) with $2\,M=\sigma+i\,2\sqrt{2}\pi^a\,X^a$, $a=1,\ldots,9$ and the generators X^a are provided explicitly in equation (A.3) and (A.5) of [6]. The potential for the Polyakov loop is Z_2 symmetric, and the only interaction term allowed by symmetries is $g_2\ell^2\,{\rm Tr}\left[M^\dagger M\right]$ Since the interaction $g_1\ell\sigma^2$ is now forbidden, one might expect no information transfer between the fields. This is true in the case $T_c\ll T_d$, but not in the physical case in which the deconfinement happens first [3], $T_d\ll T_c$. For $T_d<T<T_c$ both symmetries are broken, and one finds that the expectation values of the two order parameter fields are again linked to each other. Furthermore, in this case the two point function of σ (ℓ) is

also infrared sensitive, $\propto \langle\sigma\rangle^2/m_\ell$ $(\propto \langle\ell\rangle^2/m_\sigma)$ near $T_{\rm d}$ $(T_{\rm c})$. Thus the two order parameter fields, a priori unrelated, do feel each other near the respective phase transitions.

While we have explicitly considered a theory with two colors and two flavors at zero chemical potential, the mechanism transferring information between the order parameter and the non order parameter field is easily carried over to describe theories with more color and flavor degrees of freedom as well as theories at finite chemical potential.

3. Tetracritical behavior in adjoint QCD

Generally in situations when two or more orders compete, the resulting phase diagram is expected to exhibit multicritical behavior when the correlation lengths associated to the two order parameters diverge simultaneously. A typical condensed matter example of multicritical behavior is the phase diagram of anisotropic antiferromagnet in a uniform magnetic field parallel to the anisotropy axis [7]. Since adjoint QCD has two well defined independent symmetries, it is reasonable to expect multicritical behavior.

Consider $O(N_1)$ and $O(N_2)$ symmetric order parameters. The effective theory, to quartic order, describing their dynamics is

$$\mathcal{L} = \frac{1}{2}(\partial_\mu \ell)^2 + \frac{1}{2}(\partial_\mu \sigma)^2 + \frac{1}{2}m_\ell^2 \ell^2 + \frac{1}{2}m_\sigma^2 \sigma^2 + \frac{\lambda}{4!}(\ell^2)^2 + \frac{g_4}{4!}(\sigma^2)^2 + \frac{g_2}{4}\ell^2\sigma^2 .$$

Here $\ell^2 = \sum_{n=1}^{N_1} \ell_n^2$ and $\sigma^2 = \sum_{m=1}^{N_2} \sigma_m^2$. This theory has three fixed points whose infrared stability depends on the value of $n = N_1 + N_2$: In the $\epsilon = 4 - D$ -expansion, to $\mathcal{O}(\epsilon^5)$ [8], for $n > 3$ the stable fixed point is the decoupled one, at which the two subsystems have independent critical behaviors. Then, for $n = 2$, the critical behavior is driven by a fixed point, which has enhanced $O(n)$ symmetry, and the system behaves as n-component Heisenberg magnet. Finally, for $n = 3$ the fixed point is the nontrivial one, so called biconical fixed point, with completely new critical exponents.

Lattice has shown [9,3] that finite temperature adjoint QCD is far from the tetracritical point: $T_{\rm dec} \ll T_{\rm ch}$. To study the effects of finite chemical potential, we study as a prototypical example the two color adjoint QCD with one Dirac flavor. This theory has $U(1)$ symmetry, i.e. the baryon number, which at finite μ is spontaneously broken by the diquark condensate.

The low energy effective theory describing this symmetry breaking pat-

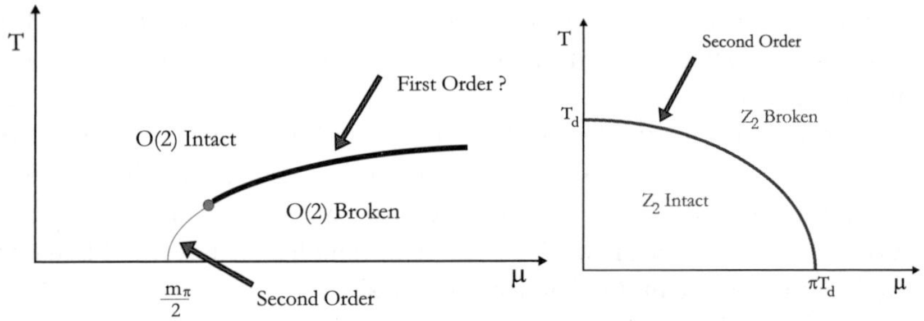

Figure 1. Schematic (T, μ) phase diagrams for diquark condensation (left panel) and deconfinement (right panel) in two color adjoint QCD with one Dirac flavor.

tern is given by

$$\mathcal{L}_{eff} = F_\pi^2 \mathrm{Tr}\left[\partial_\mu U^\dagger \partial^\mu U\right] + F_\pi^2 m_\pi^2 \mathrm{Tr}\left[U + U^\dagger\right], \quad U = e^{i\frac{\pi^a X^a}{F_\pi}} , \tag{4}$$

where $a = 1, 2$ and the chemical potential couples directly to the pions via $\partial_0 U \to D_0 U = \partial_0 - i\mu\,[U, B]$ with $B = \tau^3$. U transforms as $g\tau^1 U g^T$ for $g \in SU(2)$ and we have introduced also a Dirac mass m in the underlying theory. Such a mass appears in the effective Lagrangian as a nonzero mass for the pions, and one expects $m_\pi^2 \propto m$. For $\mu > m_\pi/2$ the $U(1) \sim O(2)$ breaks spontaneously. A schematic phase diagram is shown in the left panel of fig. 1. The existence of a second order line has been also deduced using chiral effective theory [10].

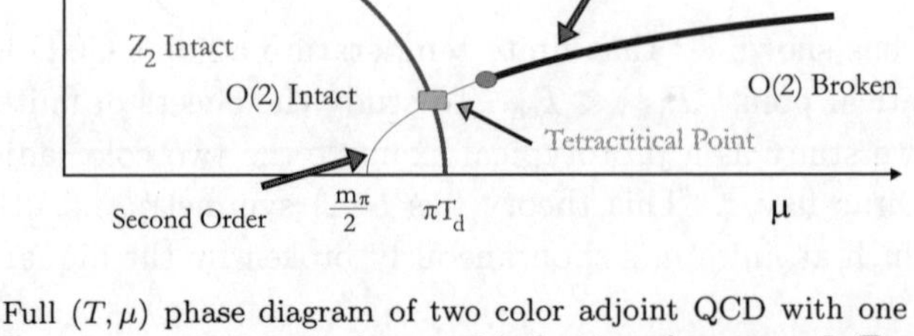

Figure 2. Full (T, μ) phase diagram of two color adjoint QCD with one Dirac flavor showing the emergence of a biconical tetracritical point when $m_\pi < 2\pi T_d$.

Recall that in the adjoint QCD also deconfinement has an exact symmetry associated to it and is an independent phase transition. A schematic deconfinement phase diagram is shown in the right panel of fig. 1, in which the critical value $\mu_c \approx \pi T_d$ was obtained by a simple bag model argument $P_g + P_q = B$. Now it is clear that taking into account both deconfinement and diquark condensation, $O(1) \oplus O(2)$ multicritcal behavior is possible by tuning m_π with respect to T_d. Such a possibility is illustrated in fig. 2. It would be interesting to see if lattice simulations at finite values of μ could see the critical lines of deconfinement and diquark condensation start to close on each other.

Acknowledgments

We thank A. Mócsy and F. Sannino for discussions and collaboration. The financial support from the Academy of Finland under project number 206024 is gratefully acknowledged.

References

1. An incomplete list: A. M. Polyakov, Phys. Lett. B **72**, 477 (1978); A. Casher, Phys. Lett. B **83**, 395 (1979); C. Adami, T. Hatsuda and I. Zahed, Phys. Rev. D **43**, 921 (1991); G. E. Brown *et al*, Nucl. Phys. A **560**, 1035 (1993); S. Digal, E. Laermann and H. Satz, Nucl. Phys. A **702**, 159 (2002); K. Rajagopal and F. Wilczek, hep-ph/0011333; Y. Hatta and K. Fukushima, Phys. Rev. D **69**, 097502 (2004) [hep-ph/0307068]; Y. Hatta, in these proceedings.
2. F. Karsch, Lect. Notes Phys. **583**, 209 (2002) [hep-lat/0106019].
3. F. Karsch and M. Lutgemeier, Nucl. Phys. B **550**, 449 (1999) [hep-lat/9812023].
4. F. Sannino, Phys. Rev. D **66**, 034013 (2002) [hep-ph/0204174]; A. Mocsy, F. Sannino and K. Tuominen, Phys. Rev. Lett. **92**, 182302 (2004) [hep-ph/0308135]; A. Mocsy, F. Sannino and K. Tuominen, Phys. Rev. Lett. **91**, 092004 (2003) [hep-ph/0301229]; A. Mocsy, F. Sannino and K. Tuominen, JHEP **0403**, 044 (2004) [hep-ph/0306069].
5. F. Sannino and K. Tuominen, Phys. Rev. D **70**, 034019 (2004) [hep-ph/0403175].
6. T. Appelquist, P. S. Rodrigues da Silva and F. Sannino, Phys. Rev. D **60**, 116007 (1999).
7. J. M. Kosterlitz, D. R. Nelson and M. E. Fisher, Phys. Rev. B **13**, 412 (1976).
8. P. Calabrese, A. Pelissetto and E. Vicari, Phys. Rev. B **67**, 054505 (2003) [cond-mat/0209580].
9. J. B. Kogut, J. Polonyi, H. W. Wyld and D. K. Sinclair, Phys. Rev. Lett. **54**, 1980 (1985); J. B. Kogut, Phys. Lett. B **187**, 347 (1987).
10. K. Splittorff, D. Toublan and J. J. M. Verbaarschot, Nucl. Phys. B **639**, 524 (2002) [hep-ph/0204076].

RELATION BETWEEN THE CHIRAL AND DECONFINEMENT PHASE TRANSITIONS

Y. HATTA

RIKEN BNL Research Center,
Brookhaven National Laboratory,
Upton, NY 11973, USA
E-mail: hatta@quark.phy.bnl.gov

Lattice QCD simulations at finite temperature have shown that the chiral phase transition in the chiral limit and the deconfinement phase transition in the quenched limit are continuously connected. I emphasize the nontriviality of this result and propose an unconventional scenario which naturally explains the existing lattice data. The continuity of the two phase transitions is a manifestation of the familiar glueball-meson mixing, which can be traced back to the properties of QCD at zero temperature.

1. Introduction and motivation

In QCD at finite temperature, there are two kinds of phase transitions in two different limits of the quark mass parameter. The chiral phase transition in the chiral limit and the deconfinement phase transition in the quenched limit. Away from these limits, the meanings of these phase transitions become less transparent. Nevertheless, the notions of deconfinement and chiral symmetry restoration are indispensable for our understanding of the quark-gluon plasma phase. In this talk we discuss the possible deep relation between the two phase transitions in the intermediate quark mass region.

In Table 1, some key properties of the two phase transitions are listed.

Table 1. QCD phase transitions: defining properties.

	chiral phase transition	deconfinement phase transition
quark mass	0	∞
symmetry	chiral symmetry	center symmetry
order parameter	quark condensate	Polyakov loop

At first sight, it does not make sense to talk about the relationship between the two phase transitions. They are defined in completely different theories in the first place. Symmetries are different, and the order parameters are different. Just by looking at the table, little do we suspect that the two phase transitions have anything to do with each other. Therefore, it is quite natural that they have long been considered as distinct phase transitions and questions like "Which phase transition occurs first?" have been addressed many times in the literature.

However, putting these theoretical speculations aside, finite temperature lattice simulations have repeatedly shown that the two phase transitions occur at the *same* critical temperature. Fig 1. is the most commonly accepted phase diagram of QCD in the temperature-quark mass plane.[1] This figure shows that, for all values of the quark mass, there is a single (crossover) phase transition which smoothly connects the chiral phase transition in the chiral limit and the deconfinement phase transition in the quenched theory.

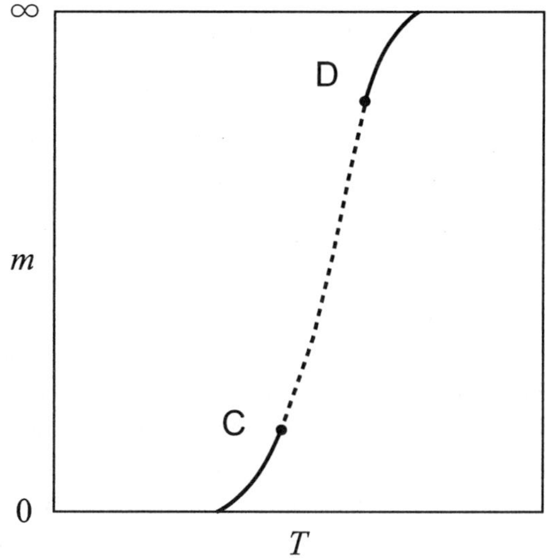

Figure 1. The QCD phase diagram in the temperature-quark mass plane. T is the temperature and m is the quark mass. Solid (dotted) lines represent first order (crossover) phase transition. C and D are second order phase transition points.

We emphasize that this is a very nontrivial result. Theorists were able to predict the order of phase transitions at $m = 0$ and $m = \infty$.[2,3] Theorists could also predict that the first order chiral (deconfinement) phase transition would turn into a crossover if the quark mass was increased from 0 (decreased from ∞). However, no one could predict the *global* structure of the phase diagram shown in Fig. 1 because *physics at the intermediate quark mass region is non-universal.* One cannot invoke the usual universality argument of phase transitions to predict anything in this region. The continuity of the two phase transitions is a consequence of the non-perturbative dynamical effect of QCD, which is far from obvious. And, of course, such a *non*-universal phenomenon is one of the most interesting aspects of a given theory. In this report, we will propose a novel scenario[4] which naturally explains the puzzling lattice data, Fig. 1.

2. The level repulsion scenario

The interplay between the two transitions is an old but interesting problem. First, Gocksch and Ogilvie observed[5] that what is responsible for the breaking of center symmetry is (the inverse of) the *constituent* quark mass rather than the current quark mass. This suggests that the Polyakov loop and the chiral dynamics are closely coupled. See, also, Ref.[6] for more recent works. So far, most of the works on this subject focused on the coupling of the Polyakov loop and the sigma field. (See, however, Ref.[7].) Here we point out that, in the presence of dynamical quarks, the deconfinement phase transition can equivalently be characterized in terms of the *glueballs.* Specifically, we have predicted[4] that the screening mass of the 0^+ electric[a] glueball goes to zero with the specified critical exponent at the second order deconfinement phase transition.[8] In the case of color SU(2), this screening state is responsible for the weak divergence of the specific heat at the Z(2) phase transition but is distinct from the true order parameter field, namely, the Polyakov loop. However, in the case of color SU(3), at the point D, nonzero expectation value of the Polyakov loop induces mixing between the glueball and the Polaykov loop. Therefore, the glueball is an equivalent critical field at D.[b] Datta and Gupta observed a significant decrease of the 0^+ glueball screening mass near the SU(2) phase transition in the quenched simulation.[9] We believe that the mass will go to zero in

[a]See, for example, Ref.[9] for the meaning of these quantum numbers. 'Electric' means that the glueball interpolating operator contains A_0's or timelike links. One can also consider 0^+ 'magnetic' glueballs (composed only from spacelike links) which, in principle, mix with the electric ones. However, near the critical temperature and above, it has been observed[10] that the mixing is very weak, could be absent.

[b]Such a mixing always takes place at generic end-points. See, Ref.[11] for an explicit example in the case of the end-point at finite density.

the infinite lattice volume limit.

Next we observe that the glueball field G must mix with the sigma field σ so that the correct massless field at D is a linear combination of the two;

$$\phi = G\cos\theta + \sigma\sin\theta, \qquad \sin\theta \approx 0. \tag{1}$$

The orthogonal linear combination with large sigma field content,

$$\phi' = -G\sin\theta + \sigma\cos\theta, \tag{2}$$

is massive. Now the key question is the behavior of the mixing angle θ as the quark mass varies. If the mixing angle remains small at small values of the quark mass, the ϕ' field, which is massive at D, would become massless at C because the critical field at C is dominantly sigma-like.[12] However, if this were the case, the coincidence of the two critical temperatures for all values of the quark mass would be a pure accident. Fig. 1 is most naturally explained by postulating that the critical field at C is again the ϕ field. Namely, the two second order phase transitions at C and D are driven by the *same* field. This is possible only if the mixing angle changes from $\theta \approx 0$ to $\approx \pi/2$.[c] Such a continuous variation of the mixing angle is typical of a *level repulsion* in quantum mechanics. Thus we have arrived at a novel scenario of the finite temperature QCD phase transition: Due to a level repulsion between the ϕ and the ϕ' fields, the ϕ field continues to be the lightest screening state for all values of the quark mass. Simultaneous divergences and peaks in various susceptibilities are simply caused by the dropping of the ϕ field screening mass.

Moreover, we conjecture that this scenario is realized at all temperatures, not only near the critical temperature. The level repulsion between the scalar glueball and the sigma meson takes place already at zero temperature as one can easily convince oneself by considering the scalar meson and glueball mass spectrum at zero temperature.[d] Therefore, our scenario can be naturally embedded in the entire phase diagram, Fig. 1. Note that this argument is made possible only when one characterizes the deconfinement phase transition in terms of the glueball. (The Polyakov loop cannot be defined at zero temperature.)

[c] The mixing angle will go to exactly $\pi/2$ at the chiral symmetry restoration point in the chiral limit.

[d] We assume that, in accordance with popular belief, the lightest glueball mass in the real world does not change appreciably from its value in the quenched theory.

3. Conclusion

We have pointed out the importance of the glueball screening states for the understanding of the QCD phase diagram for all values of the quark mass and the temperature. Compared to Polyakov loops, glueballs have been much less studied on a lattice at finite temperature with or without dynamical quarks.[9,10,13,14] We expect that further glueball measurements will provide rich information on the nature of the thermal QCD phase transition. On the other hand, it turned out to be very difficult to reproduce the lattice result, Fig. 1, in model calculations. The main difficulty is to keep track of the important coupling between the two fields in the intermediate quark mass region where there is no guiding principle (no symmetry) to construct the effective potential. (For details and discussions, see Ref.[15].) This remains to be a theoretical challenge.

Acknowledgments

This work has been done in collaboration with Kenji Fukushima. I thank the organizers of Strong and Electroweak Matter 2004 for the stimulating meeting.

References

1. See, for example, Fig. 4 of F. Karsch, E. Laermann and A. Peikert, *Nucl. Phys.* **B605**, 579 (2001).
2. L. G. Yaffe and B. Svetitsky, *Phys. Rev.* **D26**, 963 (1982).
3. R. D. Pisarski and F. Wilczek, *Phys. Rev.* **D29**, 338 (1984).
4. Y. Hatta and K. Fukushima, *Phys. Rev.* **D69**, 097502 (2004); hep-ph/0311267.
5. A. Gocksch and M. Ogilvie, *Phys. Rev.* **D31**, 877 (1985).
6. S. Digal, E. Laermann and H. Satz, *Eur. Phys. J.* **C18**, 583 (2001); K. Fukushima, *Phys. Lett.* **B553**, 38 (2003); *Phys. Rev.* **D68**, 045004 (2003); K. Tuominen, in these proceedings.
7. F. Karsch, K. Redlich and A. Tawfik, *Eur. Phys. J.* **C29**, 549 (2003).
8. See, however, A. Mocsy, F. Sannino and K. Tuominen, *JHEP* **0403**, 044 (2004).
9. S. Datta and S. Gupta, *Nucl. Phys.* **B534**, 392 (1998).
10. A. Hart, M. Laine and O. Philipsen, *Nucl. Phys.*, **B586**, 443 (2000).
11. Y. Hatta and T. Ikeda, *Phys. Rev.* **D67**, 014028 (2003).
12. S. Gavin, A. Gocksch and R. D. Pisarski, *Phys. Rev.* **D49**, 3079 (1994).
13. R. V. Gavai and S. Gupta, *Phys. Rev. Lett.* **85**, 2068 (2000).
14. N. Ishii, H. Suganuma and H. Matsufuru, *Phys. Rev.*, **D66**, 094506 (2002).
15. K. Fukushima, *Phys. Lett.* **B591**, 277 (2004); hep-ph/0403147.

RENORMALIZED POLYAKOV LOOPS, MATRIX MODELS AND THE GROSS-WITTEN POINT

ADRIAN DUMITRU

Institut für Theoretische Physik, J.W. Goethe Univ.,
Postfach 11 19 32, 60054 Frankfurt, Germany
E-mail: dumitru@th.physik.uni-frankfurt.de

JONATHAN T. LENAGHAN

Department of Physics
University of Virginia
382 McCormick Rd.
PO Box 400714
Charlottesville, VA 22904-4714, USA
E-mail: jl8bf@mail.phys.virginia.edu

The values of renormalized Polyakov loops in the three lowest representations of $SU(3)$ were measured numerically on the lattice. We find that in magnitude, condensates respect the large-N property of factorization. In several ways, the deconfining phase transition for $N = 3$ appears to be like that in the $N = \infty$ matrix model of Gross and Witten. Surprisingly, we find that the values of the renormalized triplet loop are described by an $SU(3)$ matrix model, with an effective action dominated by the triplet loop. Future numerical simulations with a larger number of colors should be able to show whether or not the deconfining phase transition is close to the Gross-Witten point.

1. Introduction

't Hooft showed that the order parameter for deconfinement in the $SU(N)$ Yang-Mills theory is a global $Z(N)$ spin [1,2] given by a thermal Wilson line which wraps around the Euclidean time direction. The thermal Wilson loop in a representation $\mathcal{R}$ is given by

$$\mathbf{L}_{\mathcal{R}}(\vec{x}) = \mathcal{P}\ \exp\left(ig\int_0^{1/T} A_0^a(\vec{x},\tau')\ \mathbf{t}^a_{\mathcal{R}}\ d\tau'\right) . \tag{1}$$

Under a gauge transformation,

$$\mathbf{L}_{\mathcal{R}}(\vec{x}) \to \Omega^\dagger_{\mathcal{R}}(\vec{x},1/T)\,\mathbf{L}_{\mathcal{R}}(\vec{x})\,\Omega_{\mathcal{R}}(\vec{x},0) . \tag{2}$$

There are important aperiodic gauge transformations:

$$\Omega_{\mathbf{N}}(\vec{x}, 1/T) = e^{i\phi}\, \Omega_{\mathbf{N}}(\vec{x}, 0) \quad , \quad e^{iN\phi} = 1 \, . \tag{3}$$

The $Z(N)$ charge $e_{\mathcal{R}}$ (defined below) determines which loops can condense above the deconfinement temperature:

$$\langle \mathrm{tr}\, \mathbf{L}_{\mathcal{R}} \rangle = 0 \quad , \quad T < T_d \quad , \quad e_{\mathcal{R}} \neq 0 \tag{4}$$

$$\langle \mathrm{tr}\, \mathbf{L}_{\mathcal{R}} \rangle \neq 0 \quad , \quad T > T_d \quad , \quad \forall\, e_{\mathcal{R}} \, . \tag{5}$$

The trace of the thermal Wilson line is the Polyakov loop [2],

$$\ell_{\mathcal{R}}(\vec{x}) = \frac{1}{d_{\mathcal{R}}} \mathrm{tr}\, \mathbf{L}_{\mathcal{R}}(\vec{x}) \, , \tag{6}$$

and is gauge invariant. We divide by the dimension of the representation, $d_{\mathcal{R}}$, so that $\ell_{\mathcal{R}} \to 1$ as $T \to \infty$.

The low-temperature, confined phase is $Z(N)$ symmetric, so the expectation value of the fundamental loop vanishes below T_d, while above T_d the $Z(N)$ symmetry is broken spontaneously:

$$\langle \ell_{\mathbf{N}} \rangle = e^{i\phi}\, |\langle \ell_{\mathbf{N}} \rangle| \neq 0 \quad , \quad e^{i\phi N} = 1 \quad , \quad T > T_d \, . \tag{7}$$

The physics around T_d is non-perturbative and it is necessary to employ lattice simulations. The expectation value of the Polyakov loop, however, is a bare quantity and contains ultraviolet divergences: the expectation value of the fundamental loop, and of loops in other representations, vanish in the continuum limit ($N_t \to \infty$), at any fixed temperature $T = 1/(N_t a)$. This is illustrated in Fig. 1.

Renormalization of the expectation value of the bare Polyakov loop gives the exponential of a divergent mass times the length of the path [3,4,5,6,7,8]:

$$|\langle \ell_{\mathbf{N}} \rangle| \sim \exp\left(-\frac{m_N^{div}}{T} \right) \quad , \quad m_N^{div} \sim \frac{1}{a} \, . \tag{8}$$

The renormalized loop, $\widetilde{\ell}_{\mathcal{R}}$, is formed by dividing the bare loop by the appropriate renormalization constant, $\mathcal{Z}_{\mathcal{R}}$:

$$\widetilde{\ell}_{\mathcal{R}} = \frac{1}{\mathcal{Z}_{\mathcal{R}}}\, \ell_{\mathcal{R}} \quad , \quad \mathcal{Z}_{\mathcal{R}} = \exp\left(-\frac{m_{\mathcal{R}}^{div}}{T} \right) \, . \tag{9}$$

We have developed a method to extract the divergent masses non-perturbatively [8]. The idea is to compute with a set of lattices, all at the same physical temperature, T, but with different values of the lattice spacing, a. The number of time steps, $N_t = 1/(aT)$, varies between these

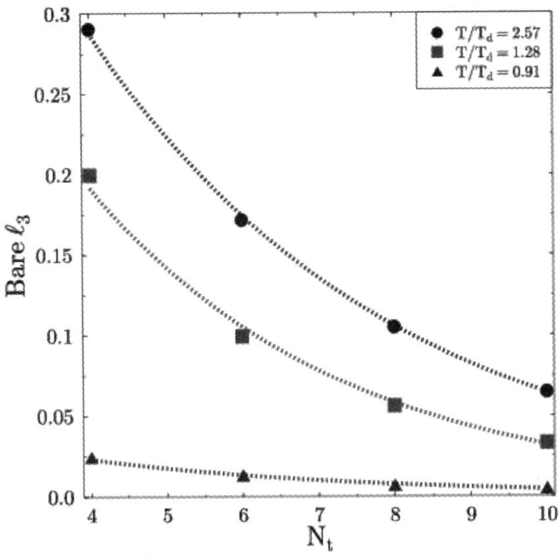

Figure 1. The bare fundamental loop as extracted from lattices with various N_t at three different temperatures.

lattices and the divergent mass, $am_{\mathcal{R}}^{div}$, can be extracted by comparing the different values of the bare Polyakov loops.

An alternate procedure was developed by Kaczmarek, Karsch, Petreczky, and Zantow [9] who obtain $\mathcal{Z}_3$ from the two point function of fundamental loops at short distances. Our numerical values for the triplet Polyakov loop agree approximately with their values. So far, they have not considered loops in other representations.

When the fundamental loop condenses it induces expectation values for all loops in higher representations, such as the octet and sextet loops. This can be precisely established in the limit of an infinite number of colors. Migdal and Makeenko observed that in $SU(N)$ gauge theories, expectation values factorize at large N [10,11,12,13,14,15,16,17]. At infinite N, factorization fixes the expectation value of any Polyakov loop to be equal to powers of those for the fundamental and anti-fundamental, $\widetilde{\ell}_{\overline{\mathbf{N}}} = (\widetilde{\ell}_{\mathbf{N}})^*$, loop [11]:

$$\left\langle \widetilde{\ell}_{\mathcal{R}} \right\rangle = \left\langle \widetilde{\ell}_{\mathbf{N}} \right\rangle^{p_+} \left\langle \widetilde{\ell}_{\overline{\mathbf{N}}} \right\rangle^{p_-} + O\left(\frac{1}{N}\right) \quad . \tag{10}$$

Hence

$$\left\langle \widetilde{\ell}_{\mathcal{R}} \right\rangle = e^{ie_{\mathcal{R}}\phi} \, \left| \left\langle \widetilde{\ell}_{\mathbf{N}} \right\rangle \right|^{p} + O\left(\frac{1}{N}\right) . \tag{11}$$

This relation defines the $Z(N)$ charge $e_{\mathcal{R}}$ of a loop in a given representation. Thus, at infinite N, any renormalized loop is an order parameter for deconfinement.

To test these large N relationships numerically, for each loop we define the difference between the measured loop and its value in the large N limit. For three colors, the expectation value of the sextet difference loop is

$$\left\langle \delta\widetilde{\ell}_6 \right\rangle = \left\langle \widetilde{\ell}_6 \right\rangle - \left\langle \widetilde{\ell}_3 \right\rangle^2 , \tag{12}$$

and that for the octet difference loop is

$$\left\langle \delta\widetilde{\ell}_8 \right\rangle = \left\langle \widetilde{\ell}_8 \right\rangle - \left| \left\langle \widetilde{\ell}_3 \right\rangle \right|^2 . \tag{13}$$

Note that the difference loops vanish both at $T \to 0$ and at $T \to \infty$. If small, they indicate that factorization is approximately satisfied.

2. Matrix Models

Assuming that Wilson lines form the degrees of freedom in an effective theory, take as the partition function

$$\mathcal{Z} = \int \Pi_i \; d\mathbf{L_N}(i) \; \exp\left(-\mathcal{S}(\ell_{\mathcal{R}}(i))\right) . \tag{14}$$

Here, i labels sites on a spatial lattice. The character expansion vastly reduces the possible couplings. Requiring the action to be $Z(N)$ invariant allows a sum over $Z(N)$ neutral loops:

$$N^2 \, \Sigma_i \Sigma_{\mathcal{R}}^{e_{\mathcal{R}}=0} \; \gamma_{\mathcal{R}} \ell_{\mathcal{R}}(i) \; . \tag{15}$$

Add nearest neighbor interactions with couplings $\beta_{\mathcal{R},\mathcal{R}'}$:

$$\mathcal{S}_{\mathcal{R}} = -\frac{N^2}{3} \Sigma_{i,\hat{n}} \Sigma_{\mathcal{R},\mathcal{R}'}^{e_{\mathcal{R}}+e_{\mathcal{R}'}=0} \; \beta_{\mathcal{R},\mathcal{R}'} \; \mathrm{Re} \; \ell_{\mathcal{R}}(i) \ell_{\mathcal{R}'}(i+\hat{n}) \, . \tag{16}$$

Kogut, Snow, and Stone showed [14] that for $N \geq 3$ in mean field approximation this model has a first order phase transition with a latent heat $\sim N^2$ and that

$$|\langle \widetilde{\ell}_{\mathbf{N}} \rangle| \approx \frac{1}{2} \text{ at } T = T_d^{+} \, . \tag{17}$$

2.1. *Three Color Example*

The simplest $N = 3$ action includes just the triplet loop:

$$\mathcal{S}_3 = -3\beta_3 \, \Sigma_{i,\hat{n}} \, \mathrm{Re} \, \ell_{\mathbf{3}}(i)\ell^*_{\mathbf{3}}(i+\hat{n}) \ . \tag{18}$$

Develop a mean field approximation by replacing all six nearest neighbors (in three space dimensions) by an average value, $\ell_0 = \langle \ell_{\mathbf{3}} \rangle$:

$$\mathcal{Z} = \int d\mathbf{L_3} \ \exp\left(+18\beta_3\ell_0 \mathrm{Re}\ell_{\mathbf{3}}\right) \equiv \exp(-9\mathcal{V}) \ . \tag{19}$$

The mean field consistency condition is

$$\ell_0 = -\frac{1}{2\beta_3}\frac{\partial}{\partial \ell_0}\mathcal{V} \ . \tag{20}$$

We have studied the condensates of the four lowest representations of $SU(3)$ as a function of the coupling β_3 [8], which are shown in Fig. 2. Our

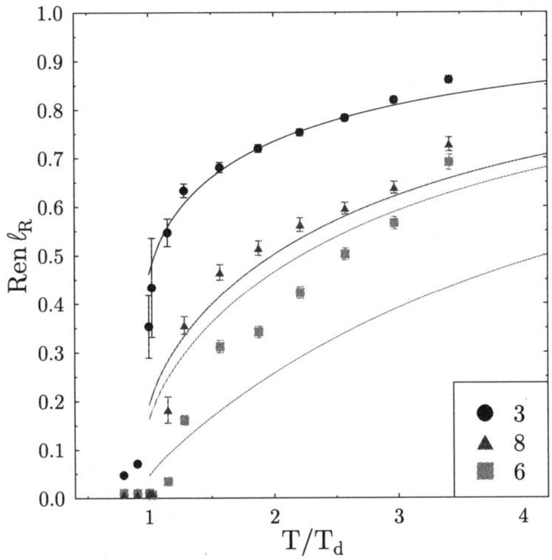

Figure 2. The condensates computed from the matrix model and compared to the lattice data[8]; for $N = 3$. The lowest line is the decuplet condensate from the $N = 3$ matrix model for which no lattice data is available.

first observation is that the ordering (fundamental, octet, sextet) agrees with that obtained from the lattice. A fit of $\langle \ell_3 \rangle(\beta_3)$ obtained within the matrix model to $\langle \ell_3 \rangle(T/T_d)$ from the lattice gives

$$\beta_3 = (0.46 \pm 0.02) + (0.33 \pm 0.02)\,\frac{T}{T_d} \ . \tag{21}$$

Remarkably, the matrix model also predicts a sizeable expectation value for the decuplet loop above T_d; so far, this has not been studied on the lattice. Also note that expectation values of $Z(3)$ neutral loops (octet and decuplet) are approximately zero in the confined phase. This is natural in a matrix model [18] but does not follow automatically in an effective theory for a scalar ℓ_3 field [19,20,21,22,23].

While β_3 is linear in the temperature, this doesn't seem to be true for all couplings. Fig. 3 illustrates this using the difference loops. Clearly, the

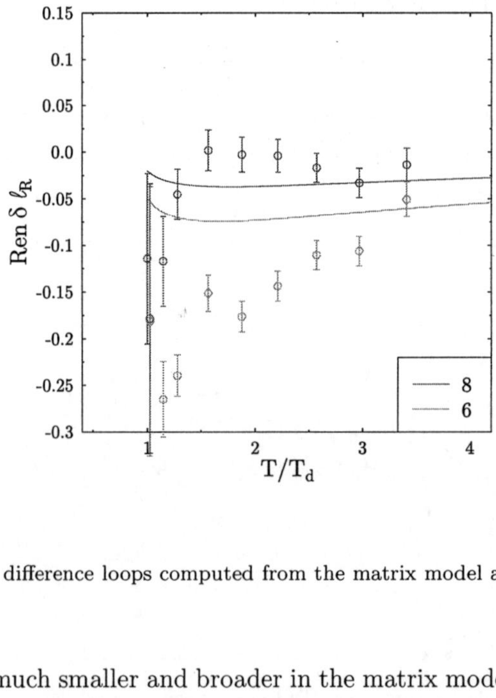

Figure 3. The difference loops computed from the matrix model and compared to the lattice data.

"spikes" are much smaller and broader in the matrix model than in the lattice data. From this we conclude that for a more quantitative interpretation of the lattice results more terms are needed in the matrix model.

3. Large–N Matrix Model and the Gross-Witten Point

At infinite N, we do not have to consider loops in higher representations as independent degrees of freedom. In mean–field approximation, we have a single–site partition function:

$$\mathcal{Z} = \int d\mathbf{L_N} \; \exp\left(N^2(2\beta\ell_0)\; \mathrm{Re}\; \ell_\mathbf{N}\right) \equiv \exp\left(-N^2\mathcal{V}_{GW}(\beta\ell_0)\right) \; , \qquad (22)$$

where $\beta \equiv \beta_N$. The mean field condition is

$$\ell_0 = -\frac{1}{2\beta}\frac{\partial}{\partial \ell_0}\mathcal{V}_{GW}(\beta\ell_0)\,, \tag{23}$$

which amounts to minimizing the mean field potential

$$\mathcal{V}_{mf}(\beta,\ell) = \beta\ell^2 + \mathcal{V}_{GW}(\beta\ell)\,. \tag{24}$$

This has been computed in the large N limit by Gross and Witten [13].

The result is nonanalytic and is given by two *different* potentials:

$$\mathcal{V}^-_{mf} = \beta(1-\beta)\ell^2 \quad , \quad \ell < \frac{1}{2\beta} \tag{25}$$

$$\mathcal{V}^+_{mf} = -2\beta\ell + \beta\ell^2 + \frac{1}{2}\log(2\beta\ell) + \frac{3}{4} \quad , \quad \ell > \frac{1}{2\beta}\,.$$

When $\beta < 1$, the theory confines and $\ell_0 = 0$. For $\beta \geq 1$ it deconfines and

$$\ell_0 = \frac{1}{2}\left(1 + \sqrt{\frac{\beta-1}{\beta}}\right).$$

At $\beta = 1$, the potential is completely *flat*, i.e. it vanishes identically for $\ell \in [0, \frac{1}{2}]$.

3.1. *Large–N and Mass of Polyakov Loop*

The connected two point function of ℓ is

$$\langle \ell^*_{\mathbf{N}}(\vec{x})\ell_{\mathbf{N}}(0)\rangle - |\langle \ell_{\mathbf{N}}\rangle|^2 \sim \frac{\exp(-m|\vec{x}|)}{|\vec{x}|} \quad , \quad |\vec{x}| \to \infty\,. \tag{26}$$

In the confined phase, $m = \sigma/T$, where σ is the string tension. In the deconfined phase, one can define $m \propto m_{Debye}$.

Computing $m^2 = \partial^2\mathcal{V}_{mf}/\partial\ell^2$ gives

$$m^2_- \approx 2(1-\beta) \quad , \quad \beta \to 1^-\,. \tag{27}$$

$$m^2_+ \approx 4\sqrt{\beta - 1} \quad , \quad \beta \to 1^+\,. \tag{28}$$

The string tension then vanishes at the transition as

$$\sigma(T) \sim (T_d - T)^{1/2} \quad , \quad T \to T_d^- \,, \tag{29}$$

and the Debye mass, as

$$m_{Debye}(T) \sim (T - T_d)^{1/4} \quad , \quad T \to T_d^+\,. \tag{30}$$

Hence, at the "Gross-Witten point" [8] there is a "critical" first order transition where the order parameter jumps at T_d, yet both m_- and m_+ vanish.

Numerical results seem to indicate that the three-color Yang-Mills theory is close to the Gross-Witten point: i) the discontinuity of the renormalized fundamental loop is approximately $1/2$; ii) the "spikes" of the difference loops are of order $1/N$ near T_d and vanish at high T; iii) the string tension and the Debye mass drop sharply near the transition [24].

4. Conclusions

It would be valuable to know from numerical simulations if the deconfining transition for more than three colors is close to the Gross–Witten point as well, or if that is unique to three colors. This is interesting, novel physics which can be obtained from lattice measurements at various N of the renormalized Polyakov loops in hot $SU(N)$ Yang-Mills theory. Further lattice simulations with higher accuracy will also constrain the couplings of the effective matrix model description of the deconfining phase transition.

Acknowledgement: This work was done in collaboration with Yoshitaka Hatta, Kostas Orginos and Robert Pisarski. We thank the organizers of SEWM 2004 for the opportunity to attend this very productive conference.

References

1. G. 't Hooft, Nucl. Phys. B **138**, 1 (1978); *ibid.* **153**, 141 (1979).
2. A. M. Polyakov, Phys. Lett. B **72**, 477 (1978); L. Susskind, Phys. Rev. D **20**, 2610 (1979).
3. J.-L. Gervais and A. Neveu, Nucl. Phys. B **163**, 189 (1980).
4. A. M. Polyakov, Nucl. Phys. B **164**, 171 (1980).
5. V. S. Dotsenko and S. N. Vergeles, Nucl. Phys. B **169**, 527 (1980).
6. I. Y. Arefeva, Phys. Lett. B **93**, 347 (1980).
7. U. M. Heller and F. Karsch, Nucl. Phys. B **251**, 254 (1985) G. Curci, P. Menotti, G. Paffuti, Z. Phys. C **26**, 549 (1985); H. Markum, M. Faber and M. Meinhart, Phys. Rev. D **36**, 632 (1987); K. Enqvist, K. Kajantie, L. Karkkainen, and K. Rummukainen, Phys. Lett. B **249**, 107 (1990); J. Kiskis, Phys. Rev. D **41**, 3204 (1990); J. Kiskis and P. Vranas, *ibid.* **49**, 528 (1994); C. P. Korthals Altes, Nucl. Phys. B **420**, 637 (1994).
8. A. Dumitru, Y. Hatta, J. Lenaghan, K. Orginos and R. D. Pisarski, arXiv:hep-th/0311223.
9. O. Kaczmarek, F. Karsch, P. Petreczky, and F. Zantow, Phys. Lett. B **543**, 41 (2002); S. Digal, S. Fortunato, and P. Petreczky, arXiv:hep-lat/0211029;

O. Kaczmarek, F. Karsch, P. Petreczky and F. Zantow, Nucl. Phys. Proc. Suppl. B **129**, 560 (2004); O. Kaczmarek, S. Ejiri, F. Karsch, E. Laermann and F. Zantow, arXiv:hep-lat/0312015; P. Petreczky and K. Petrov, arXiv: hep-lat/0405009.
10. Yu. M. Makeenko and A. A. Migdal, Phys. Lett. B **212**, 221 (1980); A. A. Migdal, Phys. Rep. **102**, 199 (1983).
11. T. Eguchi and H. Kawai, Phys. Rev. Lett. **48**, 1063 (1982); with $p_- = 0$, our (10) is their (18).
12. A. Gocksch and F. Neri, Phys. Rev. Lett. **50**, 1099 (1983); R. Narayanan and H. Neuberger, arXiv:hep-lat/0303023.
13. D. J. Gross and E. Witten, Phys. Rev. D **21**, 446 (1980).
14. J. B. Kogut, M. Snow, and M. Stone, Nucl. Phys. B **200**, 211 (1982).
15. J. Polonyi and K. Szlachanyi, Phys. Lett. B **110**, 395 (1982); J. Bartholomew, D. Hochberg, P. H. Damgaard, and M. Gross, *ibid.* **133**, 218 (1983); M. Gross and J. F. Wheater, Nucl. Phys. B **240**, 253 (1984); A. Gocksch and M. Ogilvie, Phys. Rev. D **31**, 877 (1985); P. H. Damgaard and A. Patkos, Phys. Lett. B **172**, 369 (1986); C. X. Chen and C. DeTar, Phys. Rev. D **35**, 3963 (1987); M. Ogilvie, Phys. Rev. Lett. **52**, 1369 (1987).
16. F. Green and F. Karsch, Nucl. Phys. B **238**, 297 (1984).
17. J. M. Drouffe and J. B. Zuber, Phys. Rep. **102**, 1 (1983); S. R. Das, Rev. Mod. Phys. **59**, 235 (1987); P. Di Francesco, P. H. Ginsparg and J. Zinn-Justin, Phys. Rep. **254**, 1 (1995); R. C. Brower, M. Campostrini, K. Orginos, P. Rossi, C. I. Tan and E. Vicari, Phys. Rev. D **53**, 3230 (1996); P. Rossi, M. Campostrini and E. Vicari, Phys. Rep. **302**, 143 (1998); M. Billo, M. Caselle, A. D'Adda and S. Panzeri, Intl. Jour. Mod. Phys. **A12**, 1783 (1997); Yu. Makeenko, arXiv:hep-th/0001047; M. R. Douglas, I. R. Klebanov, D. Kutasov, J. Maldacena, E. Martinec and N. Seiberg, arXiv:hep-th/0307195.
18. A. Dumitru, J.T. Lenaghan and R.D. Pisarski: in preparation.
19. B. Svetitsky and L. G. Yaffe, Nucl. Phys. B **210**, 423 (1982).
20. R. D. Pisarski, Phys. Rev. D **62**, 111501 (2000).
21. A. Dumitru and R. D. Pisarski, Phys. Lett. B **504**, 282 (2001); Phys. Lett. B **525**, 95 (2002); Phys. Rev. D **66**, 096003 (2002); A. Dumitru, O. Scavenius, and A. D. Jackson, Phys. Rev. Lett. **87**, 182302 (2001); O. Scavenius, A. Dumitru and J. T. Lenaghan, Phys. Rev. C **66**, 034903 (2002); A. Dumitru, D. Röder and J. Ruppert, arXiv:hep-ph/0311119.
22. P.N. Meisinger, T.R. Miller, and M.C. Ogilvie, Phys. Rev. D **65**, 034009 (2002); P.N. Meisinger and M.C. Ogilvie, *ibid.* **65**, 056013 (2002); K. Fukushima, *ibid.* **68**, 045004 (2003); I. I. Kogan, A. Kovner and J. G. Milhano, JHEP **0212**, 017 (2002).
23. A. Mocsy, F. Sannino and K. Tuominen, Phys. Rev. Lett. **91**, 092004 (2003); JHEP **0403**, 044 (2004); Phys. Rev. Lett. **92**, 182302 (2004); F. Sannino and K. Tuominen, Phys. Rev. D **70**, 034019 (2004).
24. O. Kaczmarek, F. Karsch, E. Laermann, and M. Lutgemeier, Phys. Rev. D **62**, 034021 (2000).

THE NATURE OF THE SOFT EXCITATION AT THE CRITICAL END POINT OF QCD *

A. JAKOVÁC

HAS [†] *and Budapest Univ. of Technology and Economics, Research Group "Theory of Condensed Matter", H-1521 Budapest, Hungary*

A. PATKÓS

Department of Atomic Physics, Eötvös University, H-1117 Budapest

ZS. SZÉP

Research Group for Statistical Physics, HAS, H-1117 Budapest
E-mail: szepzs@antonius.elte.hu

P. SZÉPFALUSY

Dept. of Physics of Complex Systems, Eötvös University, H-1117 Budapest
Research Inst. for Solid State Physics and Optics, HAS, H-1525 Budapest

Using a large flavor number expansion and a gap equation for the pion mass the chiral quark-meson model is solved at the lowest order in the fermion contributions. In the chiral limit the tricritical point (TCP) is determined analytically. The softening of the sigma particle is verified at this point and is further investigated for a physical pion mass in the neighbourhood of the critical endpoint (CEP).

1. Introduction

Recently the location of the CEP in the $\mu - T$ phase-diagram of QCD was obtained for $N_f = 2+1$ dynamical staggered quarks with physical masses [1]. One might investigate this problem using an effective low energy model also, with some nonperturbative technique. Its low energy nature might prevent us from giving the accurate location of the endpoint and the shape of the

*This work is supported by the Hungarian Research Fund (OTKA) under contract numbers F043465, T034980, T037689, and T046129.

†Hungarian Academy of Sciences

phase boundary, but one can still obtain insights into the physics near the CEP. In this contribution we focus on the nature of the soft mode.

2. The method for solving the model

In the presence of the vacuum expectation value Φ the requirement of a finite constituent quark mass $m_q = g\Phi$ as $N \to \infty$ defines the Lagrangian

$$L[\sigma, \pi^a, \psi] = -\left[\frac{\lambda}{24}\Phi^4 + \frac{1}{2}m^2\Phi^2\right]N - \left[\frac{\lambda}{6}\Phi^3 + m^2\Phi - h\right]\sigma\sqrt{N}$$
$$+\frac{1}{2}\left[(\partial\sigma)^2 + (\partial\vec{\pi})^2\right] - \frac{1}{2}m_{\sigma 0}^2\sigma^2 - \frac{1}{2}m_{\pi 0}^2\vec{\pi}^2 - \frac{\lambda}{6\sqrt{N}}\Phi\sigma\rho^2 - \frac{\lambda}{24N}\rho^4$$
$$+\bar{\psi}\left[i\partial^\mu\gamma_\mu - m_q - \frac{g}{\sqrt{N}}\left(\sigma + i\sqrt{2N_f}\gamma_5 T^a\pi^a\right)\right]\psi, \quad (1)$$

where $N = N_f^2$, $m_{\sigma 0}^2 = m^2 + \frac{\lambda}{2}\Phi^2$, $m_{\pi 0}^2 = m^2 + \frac{\lambda}{6}\Phi^2$, $\rho^2 = \sigma^2 + \vec{\pi}^2$ and h is the external field. The fermions, for which the chemical potential is introduced by the shift $\partial_t \to \partial_t - i\mu$, start contributing in the large N expansion at level $\mathcal{O}(1/\sqrt{N})$, which is between LO and NLO of the mesonic sector. To calculate their corrections to the LO, which has O(N) symmetry, we disregarded the second independent quartic coupling of the model, present for $N_f > 2$.

Due to the proliferation of graphs a resummation is needed. The best thing to do would be to use a selfconsistent pion and fermion propagator. This makes renormalisation difficult, and a consistent method for it was proposed only recently [2]. Our method consists in taking into account the N_c coloured quarks to $\mathcal{O}(1/\sqrt{N})$ perturbatively at one loop order and using for the propagator of the pions a simplified form, parameterised by a mass M determined from the gap-equation:

$$M^2 = -iG_\pi^{-1}(p=0) = m^2 + \frac{\lambda}{6}\Phi^2 + \frac{\lambda}{6}T_B(M) - 4\frac{g^2N_c}{\sqrt{N}}T_F(m_q). \quad (2)$$

The pion and fermion tadpoles ($T_{B,F}$) determine the equation of state (EoS)

$$V'_{\text{eff}} = \Phi H(\Phi^2) - h = \Phi\left[m^2 + \frac{\lambda}{6}\left(\Phi^2 + T_B(M)\right) - \frac{4g^2N_c}{\sqrt{N}}T_F(m_q)\right] - h = 0. \quad (3)$$

The renormalisation method was presented in [3]. The relation $M^2 = H(\Phi^2)$ shows that in the symmetry broken regime defined by $H(\Phi^2) = h/\Phi$ and in the chiral limit ($h = 0$) the physical pion mass M is always zero, i.e. Goldstone's theorem is fulfilled in the minimum of the effective potential. In what follows the number of flavors is set to two: $N_f = 2$.

3. Chiral case ($h = 0$): analytical determination of TCP

Since in this case $M = 0$, one obtains [4] the phase boundary by expanding the fermion tadpole in powers of Φ. The $2^{\rm nd}$ order line is given by

$$m_T^2 := m^2 + \frac{g^2\mu^2}{4\pi^2}N_c + \left(\frac{\lambda}{72} + \frac{g^2 N_c}{12}\right)T^2 = 0, \tag{4}$$

while the location of TCP comes from combining (4) with the condition

$$\frac{\bar\lambda}{6} := \frac{\lambda}{6} + \frac{g^4 N_c}{4\pi^2}\left[\frac{\partial}{\partial n}\left(\mathrm{Li}_n(-e^{-\frac{\mu}{T}}) + \mathrm{Li}_n(-e^{-\frac{\mu}{T}})\right)\Big|_{n=0} - \ln\frac{c_1 T}{M_{0B}}\right] = 0. \tag{5}$$

$\ln c_1/2 = 1 - \gamma_E + \eta$, and $\eta = \ln M_{0B}/M_{0F}$ gives a relation between M_{0B} and M_{0F} the renormalisation scales of the bosonic and fermionic tadpoles. The parameters are fixed at $T = \mu = 0$. The choice $m_{q0} = m_N/3 \approx 312.67$ MeV and $\Phi_0 = f_\pi/2$ gives $g = 6.72$. $\lambda = 400$ is fixed by requiring the agreement between the location of the complex σ-pole on the $2^{\rm nd}$ Riemann sheet and the experimentally favoured mass and width of the σ particle. Solving (4) and (5) the $2^{\rm nd}$ order phase transition, the first spinodal line and the location of TCP are found (l.h.s. of Fig. 1). On the r.h.s. we see

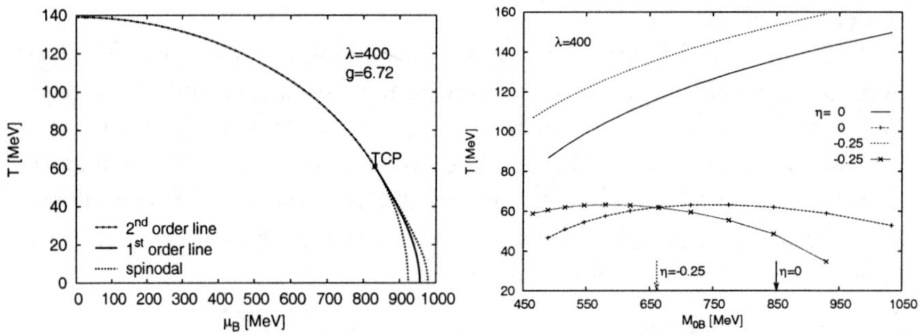

Figure 1. L.h.s.: The $T - \mu$ phase diagram for $M_{0B} = 886$ MeV and $\eta = 0$. R.h.s. the dependence of $T_c(\mu = 0)$ (upper curves) and of T_{TCP} (lower curves) on M_{0B}.

the variation of TCP and of the critical temperature with η and M_{0B}. The pole structure of the σ propagator restricts the allowed range of M_{0B} to the right from the arrow, seen on the horizontal axis, where no low scale imaginary poles appear besides the well known large scale tachyon, whose scale increases with decreasing values of λ. Although in this region one can easily obtain $T_c(\mu = 0) \in (150, 170)$ MeV, the TCP stays robustly below 70 MeV. Previous effective model studies [5] give similar low values compared with the CEP temperature obtained in [1], which could mean that the phase transition is driven by higher excited hadronic states rather than by the light d.o.f. This is supported by a resonance gas model calculation [6] used to reproduce lattice results on quark number susceptibilities and pressure.

An approximate effective potential is obtained by integrating the EoS away from equilibrium and subsequently Taylor expanding it around $\Phi = 0$. This value becomes a minimum as $T \to T_c$ and so $M \to 0$. With this, one can show the vanishing of the effective coupling in the large N limit [7] at the critical point. Here we just quote the result of [8]

$$V_{\text{approx}}(\Phi) = \frac{m_T^2}{4U^2W}\left[\frac{m_T^2}{2}\Phi^2 + \frac{\bar{\lambda}}{12}\Phi^4 + \left(\frac{W\bar{\lambda}^2}{216 m_T^2} + \frac{\kappa}{3}\right)\Phi^6\right], \tag{6}$$

where $\kappa = \frac{g^6 N_c}{(4\pi T)^2\sqrt{N}}\frac{\partial}{\partial n}\left[\text{Li}_n(-e^{\frac{\mu}{T}}) + \text{Li}_n(-e^{-\frac{\mu}{T}})\right]\Big|_{n=-2}$, $W = 1 - \frac{\lambda^2}{48\pi^2} \times \ln\frac{c_2 T}{M_0}$ and $\ln\frac{c_2}{4\pi} = \frac{1}{2} - \gamma$. The shape of $V_{\text{approx}}(\Phi)$ depends on the values of the parameters. In the temperature range of a 2nd order phase transition $W < 0$. For increasing values of μ/T, κ changes sign at 1.91. This sign change restricts the location of the TCP to the region $\mu > 2T$, where $\kappa > 0$. In the broken symmetry phase $m_T^2 < 0$ and it vanishes at T_c signalling that the coefficient of the quartic term in Φ also vanishes. Around T_c from (4) one gets $m_T^2 \sim T - T_c$, which through $m_{\text{eff}} = \left|\frac{m_T^2}{2UW}\right|$ gives the correct LO critical exponent $\nu = 1$ for the O(N) model at large N. The usual Landau type analysis applies to the square bracket in (6). The scaling exponent of the order parameter on the 2nd order line is given by the first two terms. The minimum condition gives $\Phi^2 = -\frac{3m_T^2}{\bar{\lambda}} \Rightarrow \Phi \sim (T - T_c)^\beta$, $\beta = \frac{1}{2}$. At TCP one can set $\bar{\lambda} = 0$ in (6) and keep the sixth order term. The minimum condition gives $\Phi^4 = -\frac{m_T^2}{2\kappa} \Rightarrow \Phi \sim (T - T_c)^\beta$, $\beta = \frac{1}{4}$. This mean field estimates for β were checked numerically using the exact EoS (3).

4. Case of the physical pion mass: the soft mode at CEP

In view of the gap-equation (2) and EoS (3) the form of the σ propagator can be inferred from the consistency condition $-G_\sigma^{-1}(p=0) = \frac{d^2 V_{eff}(\phi)}{d\Phi^2}$:

$$G_\sigma^{-1}(p) = p^2 - \frac{h}{\phi} - 2\Phi^2\frac{\frac{\lambda}{6} - \frac{4g^4 N_c}{\sqrt{N}} I_F(p, m_q)}{1 - \frac{\lambda}{6} I_B(p, M)}, \tag{7}$$

where $I_{B,F}$ are the pion and fermion one-loop "fish" integrals. Requiring $M = 140$ MeV at $T = \mu = 0$ one has from (2) $h/f_\pi^3 = 1.13$. The location of CEP is determined numerically by simultaneously solving (2) and (3):

$(\mu_B, T)_{\rm CEP} = (987, 12.34)$ MeV. T_{CEP} is unrealistically low, nevertheless it is of interest to study which mode becomes soft at this point. Two minima meet at CEP which flattens the effective potential. This is shown also by the diverging peak of the susceptibility $d\Phi/dh$. This static information has to show up consistently when investigating $G_\sigma(p)$ in the static limit *i.e.* first taking $p_0 \to 0$, then $|\mathbf{p}| \to 0$. The question is in which $(p_0, |\mathbf{p}|)$ region one has to look for the responsible excitations. The behaviour of the spectral function $\rho(p_0, |\mathbf{p}|) = -\frac{1}{\pi}\lim_{\varepsilon\to 0} {\rm Im}\, G_\sigma(p_0 + i\varepsilon, |\mathbf{p}|)$ for fixed $|\mathbf{p}|$ gives a hint in this respect: a peak is found in the $p_0 < |\mathbf{p}|$ region. The location of its maximum moves toward the origin with decreasing values of $|\mathbf{p}|$ and at CEP eventually the peak diverges at $p_0 = 0$ as $|\mathbf{p}| \to 0$. The nature of the analytical object on the second Riemann sheet which produces this peak will be explored further. In the NJL model a pole on the negative imaginary axis, approaching the origin as $|\mathbf{p}| \to 0$, is responsible for the peak observed in the spectral function [9].

5. Conclusions

Contrary to the chiral case, where the zero temperature σ pole continuously moves to the origin of the $2^{\rm nd}$ Riemann sheet and becomes soft at TCP, for a physical pion mass the σ pole stays massive at CEP. A space-like $(0 \le p_0 < |\mathbf{p}|)$ excitation is responsible for the flattening of the effective potential at this point. Before a more detailed investigation of the analytic structure at CEP we have to improve our method of solving the model by dynamically generating the fermion mass through a Dyson-Schwinger equation for the fermion propagator, but also one has to extend our method to the physically more appealing $SU(3)_L \times SU(3)_R$ case.

References

1. Z. Fodor, S. D. Katz, *JHEP* **0404** (2004) 050
2. J.-P. Blaizot *et al.*, *Nucl. Phys.* **A 736** (2004) 149, see also this proceedings
3. A. Jakovác, Zs. Szép, *hep-ph/0405226*, see also this proceedings
4. A. Jakovác, A. Patkós, Zs. Szép, P. Szépfalusy, *Phys. Lett.* **B582** (2004) 179
5. J. Berges and K. Rajagopal, *Nucl. Phys.* **B538** (1999) 215; O. Scavenius *et al.*, *Phys. Rev.* **C64** (2001) 045202; B.-J. Schaefer *et al.*, *nucl-th/0403039*; A. Barducci *et al.*, *Phys. Rev.* **D69** (2004) 096004
6. F. Karsch, K. Redlich, A. Tawfik, *Eur. Phys. J.* **C29** (2003) 549
7. M. Reuter, N. Tetradis, C. Wetterich, *Nucl. Phys.* **B401** (1993) 567
8. A. Jakovác, A. Patkós, Zs. Szép, P. Szépfalusy, *hep-ph/0406122*
9. H. Fujii, *Phys. Rev.* **D67** (2003) 094018

THERMODYNAMICS OF THE 1+1-DIMENSIONAL NONLINEAR SIGMA MODEL THROUGH NEXT-TO-LEADING ORDER IN $1/N$

HARMEN J. WARRINGA

Department of Physics and Astronomy, Vrije Universiteit,
De Boelelaan 1081, 1081 HV Amsterdam, The Netherlands
E-mail: harmen@nat.vu.nl

We discuss the thermodynamics of the $O(N)$ nonlinear sigma model in 1+1 dimensions. In particular we investigate the NLO $1/N$ correction to the 1PI finite temperature effective potential expressed in terms of an auxiliary field. The effective potential contains temperature-dependent divergences which cannot be renormalized properly. We argue that this problem vanishes at the minimum of the effective potential. Therefore physical quantities like the pressure are well defined and can be renormalized in a temperature-independent way. We give a general argument for the occurrence of temperature-dependent divergences outside the minimum. We present calculations of the pressure and show that $1/N$ is a good expansion. It turns out that the pressure normalized to that at infinite temperature is N-independent like the flavor independence of the same quantity in QCD.

1. Introduction

It is interesting to study the thermodynamics of the $O(N)$ nonlinear sigma model (NLSM) in $1+1$ dimensions in a $1/N$ expansion. It allows one to learn about renormalization at finite temperature (T) in a non-perturbative approach. In general one expects that temperature does not influence the ultraviolet behavior of the theory. Hence if the theory is renormalized at $T=0$, it should also be renormalized at finite T. However, this statement is not true for the effective potential as a function of the vacuum expectation value (vev) of an auxiliary field as we will see. The $1+1$-dimensional NLSM can also be used as a toy model to study QCD. This is because both theories are asymptotically free. Unlike in large N_c QCD one can calculate the pressure of the NLSM through NLO in $1/N$. By performing this calculation we were able to test the $1/N$ expansion at finite temperature.

The $O(N)$ symmetric NLSM is defined by the Lagrangian density

$$\mathcal{L} = \frac{1}{2}\partial_\mu\phi_i\partial^\mu\phi_i \,, \qquad \phi_i(x)\phi_i(x) = \frac{N}{g_b^2} \,, \qquad i = 1, \ldots, N \,, \tag{1}$$

where g_b is the bare coupling constant and ϕ_i are scalar fields which are constrained to live on a hypersphere. The NLSM is renormalizable because g_b is dimensionless. It is asymptotically free and contains infrared renormalon ambiguities [1]. The scalar fields become massive due to the interactions. The mass is non-analytic in the coupling constant.

The effective potential of the NLSM has been studied through NLO in the $1/N$ expansion at $T = 0$ [2,3,4,5,6]. The LO contribution at finite T was also investigated before [7]. We studied the NLO correction at finite T [8].

2. The effective potential

A way to implement the constraint on the ϕ fields is by introducing an auxiliary Lagrange multiplier field which we will denote by α. The action then still remains quadratic in the ϕ fields. If one integrates over the ϕ fields one finds the partition function of the NLSM which is

$$\mathcal{Z} = \int \mathcal{D}\alpha \exp\left\{ -\frac{N}{2}\mathrm{Tr}\log[-\partial^2 - i\alpha(x)] - \frac{iN}{2g_b^2}\int_0^{1/T} \mathrm{d}\tau \int \mathrm{d}x\, \alpha(x) \right\} . \tag{2}$$

The effective potential can be calculated from the partition function by expanding the α field around its vev. One can prove that the vev of the α field is purely imaginary. Therefore we write $\alpha = im^2 + \tilde{\alpha}/\sqrt{N}$. By expanding the Tr log term of Eq. (2) around im^2, one sees that $\tilde{\alpha}$ n-point vertices are suppressed with respect to the propagator in the large-N limit. Hence, the effective potential can be calculated in a $1/N$ expansion. The LO contribution stems from the classical action of the α field. The NLO correction can be found by performing the Gaussian integral over the quantum fluctuations $\tilde{\alpha}$. One finds $\mathcal{V}(m^2) = N\mathcal{V}_{\mathrm{LO}}(m^2) + \mathcal{V}_{\mathrm{NLO}}(m^2)$ where

$$\mathcal{V}_{\mathrm{LO}} = \frac{m^2}{2g_b^2} - \frac{1}{2}\sum\!\!\!\!\!\!\int_P \log(P^2 + m^2) \,, \qquad \mathcal{V}_{\mathrm{NLO}} = -\frac{1}{2}\sum\!\!\!\!\!\!\int_P \log\Pi(P, m) \,. \tag{3}$$

The Fourier transform of the inverse $\tilde{\alpha}$ propagator is given by

$$\Pi(P, m) = \frac{1}{2}\sum\!\!\!\!\!\!\int_Q \frac{1}{(P+Q)^2 + m^2}\frac{1}{Q^2 + m^2} \,. \tag{4}$$

For convenience we use the following notation, $P = (p_0, p)$ and

$$\int_P \equiv \int \frac{\mathrm{d}^2 p}{(2\pi)^2} \,, \qquad \sum\!\!\!\!\!\!\int_P \equiv T\sum_{p_0 = 2\pi n T}\int \frac{\mathrm{d}p}{2\pi} \,, \qquad \oint_P \equiv \sum\!\!\!\!\!\!\int_P - \int_P \,. \tag{5}$$

To deduce the required renormalization of the effective potential, we write $\mathcal{V} = Nm^2/2g_b^2 + ND_{\mathrm{LO}} + NF_{\mathrm{LO}} + D_{\mathrm{NLO}} + F_{\mathrm{NLO}}$, where F_{LO} and F_{NLO} are finite quantities which have to be calculated numerically. The terms D_{LO} and D_{NLO} are divergent and will be calculated analytically. The theory is infrared finite, so the divergences will arise from the high-momentum modes. To regulate the ultraviolet divergences, we introduce a momentum cutoff Λ. The quantity $\sum\!\!\!\!\!\!\int_P \log\left[f(P,m)\right]$ is finite due to Bose-Einstein suppression. Hence the divergences of the effective potential stem from $\int_P \log\left[\lim_{P\gg m,T} f(P,m)\right]$. In this way one finds $D_{\mathrm{LO}} = -m^2\log\left(\Lambda^2/m^2\right)/8\pi$. It turns out [8] that $\Pi(P,m)$ has an explicit T-dependence in the large-momentum limit. As a result we find

$$D_{\mathrm{NLO}} = -\frac{1}{8\pi}\left[\Lambda^2\log\log\left(\frac{\Lambda^2}{\bar{m}^2}\right) - \bar{m}^2\mathrm{li}\left(\frac{\Lambda^2}{\bar{m}^2}\right)\right] + \frac{m^2}{4\pi}\left[\log\left(\frac{\Lambda^2}{m^2}\right) - \log\log\left(\frac{\Lambda^2}{\bar{m}^2}\right)\right], \quad (6)$$

which is T-dependent because $\bar{m}^2 = m^2\exp\left[-J_1(\beta m)\right]$. Here $J_1(x) = 4\int_0^\infty \mathrm{d}t\, n(\omega_t)/\omega_t$ with $\omega_t = \sqrt{t^2+x^2}$ and $n(x) = \left[\exp(x)-1\right]^{-1}$. Hence the quadratic and the li-divergence have an explicit T-dependence. The logarithmic integral is defined by $\mathrm{li}(x) = \mathcal{P}\int_0^x \mathrm{d}t\, 1/\log t$.

One possibility to remove the divergences of $\mathcal{V}$ is by subtracting infinite constants independent of T and m^2. This is allowed because it does not change the shape of $\mathcal{V}$ and hence not the physics. Clearly this cannot remove the quadratic divergences, because $\Lambda^2\left[\log\log(\Lambda^2/\bar{m}^2) - \log\log(\Lambda^2/\mu^2)\right]$ is still divergent. The second possibility is to absorb T-independent divergences into g_b. This does not solve the problem of the li-divergence, since it is proportional to $\bar{m}^2$, whereas $1/g_b^2$ is proportional to m^2 in $\mathcal{V}$. The $m^2\log\log(\Lambda^2/\bar{m}^2)$ divergence can however be renormalized in this way. The two problematic divergences together are called [5] "perturbative tail" and do not arise in dimensional regularization. One can remove these divergences by subtracting the perturbative tail. This however gives rise to a T-dependent ambiguity due to a thermal infrared renormalon [8].

We therefore conclude that we cannot renormalize $\mathcal{V}$ in a T-independent way. However, in the minimum of $\mathcal{V}$, $\bar{m}^2 = \Lambda^2\exp\left(-4\pi/g_b^2\right) + \mathcal{O}(1/N)$ [8]. Furthermore one can show [2] that one only needs the LO expression for m^2 and $\bar{m}^2$ to calculate $\mathcal{V}$ in the minimum. Hence the divergences become T-independent in the minimum and they can be removed by subtracting a vacuum term. Therefore quantities which are to be calculated in the minimum are well defined. This holds for example for the pressure $\mathcal{P}$. A summary of the different ways to deal with $\mathcal{V}$ is shown in Table 1.

Table 1. Different ways to deal with the effective potential and their problems. The subscript min indicates the minimum of the effective potential.

quantity	problem
$\mathcal{V}(m^2)$	T-dependent divergences
$\mathcal{V}(m^2) - \mathcal{V}(m^2)_{T=0}$	changes shape $\mathcal{V}$ and T-dependent divergences
$\mathcal{V}(m^2) - \mathcal{V}_{\min,T=0}$	T-dependent divergences
$\mathcal{V}(m^2)$ – pert. tail	T-dependent infrared renormalon ambiguity
$\mathcal{P} = \mathcal{V}_{\min} - \mathcal{V}_{\min,T=0}$	no problem

3. Temperature-dependent divergences

In this section we will argue in a more general way why it is possible to have T-dependent divergences outside the minimum of the effective potential. These T-dependent divergences are not specific for the NLSM but also occur in the linear sigma model in $d = 3 + 1$ [9]. We add a space-time independent source J for α and define the action in the presence of J as

$$S[J] = \frac{N}{2}\mathrm{Tr}\log[-\partial^2 - i\alpha(x)] + i\left(\frac{N}{2g_b^2} + J\right)\int_0^{1/T} \mathrm{d}\tau \int \mathrm{d}x\, \alpha(x)\,. \qquad (7)$$

The quantity $Z[J]_T$ is the partition function of the theory with action $S[J]$ at temperature T. If we assume that $\log Z[0]_T$ does not contain T-dependent divergences this also holds for $\log Z[J]_T$ as long as J is T-independent. This is because any J can be incorporated in $1/g_b^2$ by a redefinition. The effective potential in the minimum is equal to $\log Z[0]_T$. Using this one can deduce from Eq. (6) that the divergences of $\log Z[0]_T$ depend on $1/g_b^2$. Therefore varying J will change the divergences of $\log Z[J]_T$.

The effective potential is given by $\mathcal{V}(m^2) = \log Z[J]_T/\beta V - Jm^2$, where J is now the current that gives the α field the vev im^2 at a certain T. If one varies T, the implicitly defined J has to change in order to keep the same vev im^2. Therefore J depends on T and as we argued before varying J changes the divergences. Outside the minimum of the effective potential J is nonzero. This implies that the effective potential outside the minimum can contain T-dependent divergences.

4. Pressure

For the arbitrary choice of $g^2(\mu = 500 \text{ a.u.}) = 10$, we calculated the pressure $\mathcal{P}$ as function of T for different values of N [8]. The results are depicted in Fig. 1. With a.u. we indicate that all results are in arbitrary units.

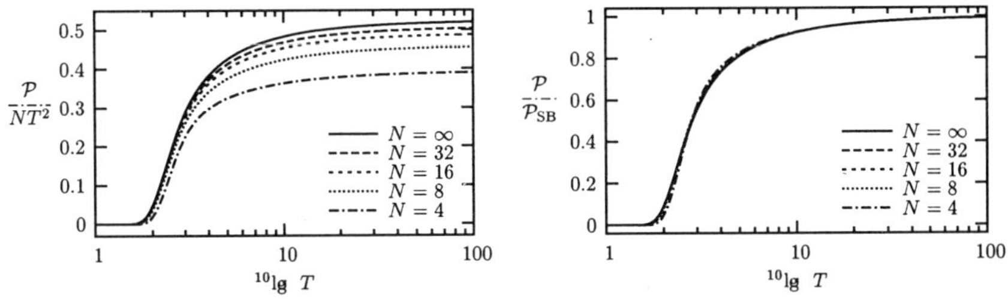

Figure 1. Pressure $\mathcal{P}$ normalized to NT^2 (left) and normalized to $\mathcal{P}(T=\infty)$ (right), as a function of T, for different values of N with $g^2(\mu = 500) = 10$.

One can see from Fig. 1 that the NLSM shows a kind of cross-over transition. This is because at low T the theory is strongly interacting and hence the degrees of freedom are effectively frozen. Due to asymptotic freedom, at high T the pressure is approaching that of a free gas of $N-1$ bosonic degrees of freedom, i.e., $\mathcal{P} = \mathcal{P}_{\mathrm{SB}} = (N-1)\pi T^2/6$. Figure 1 also shows that the $1/N$ expansion is a good expansion for the NLSM. This is because $1/N$ corrections are really of order $1/N$. The pressure divided by the pressure at $T = \infty$ is almost independent of N. Similar behavior is shown by QCD lattice calculations with and without dynamical quarks in $d = 3+1$ [10], where the same quantity is flavor independent.

Acknowledgments

This work has been carried out in collaboration with Jens O. Andersen and Daniël Boer. The author would like to thank the organizers of the strong and electroweak matter conference for the stimulating meeting and the opportunity to present this work.

References

1. F. David, Nucl. Phys. B **209**, 433 (1982).
2. R. G. Root, Phys. Rev. D **10**, 3322 (1974).
3. V. A. Novikov, M. A. Shifman, A. I. Vainshtein and V. I. Zakharov, Phys. Rept. **116**, 103 (1984).
4. C. Rim and W. I. Weisberger, Phys. Rev. Lett. **53**, 965 (1984).
5. P. Biscari, M. Campostrini and P. Rossi, Phys. Lett. B **242**, 225 (1990).
6. H. Flyvbjerg, Phys. Lett. B **245**, 533 (1990).
7. M. Dine and W. Fischler, Phys. Lett. B **105**, 207 (1981).
8. J. O. Andersen, D. Boer and H. J. Warringa, Phys. Rev. D **69** 076006 (2004).
9. J. O. Andersen, D. Boer and H. J. Warringa, arXiv:hep-ph/0408033.
10. F. Karsch, E. Laermann and A. Peikert, Phys. Lett. B **478**, 447 (2000).

LIGHT QUARK MESON CORRELATIONS AT HIGH TEMPERATURE *

E. LAERMANN[a], F. KARSCH[a], P. PETRECZKY[b], S. STICKAN[a], I. WETZORKE[c] AND S. WISSEL[a]

[a] *Fakultät für Physik, Universität Bielefeld, D-33615 Bielefeld, Germany*
[b] *Nuclear Theory Group, Physics Department, Brookhaven National Laboratory, Upton NY 11973, United States*
[c] *NIC/DESY Zeuthen, Platanenallee 6, D-15738 Zeuthen, Germany*

Lattice QCD results on finite temperature meson correlations are presented. They have been obtained from a non-perturbatively improved Wilson fermion action on quenched configurations. The observables discussed are spatial as well as temporal correlation functions for light quarks. From those, screening masses and spectral densities have been extracted, the latter through the use of the Maximum Entropy Method. Some emphasis is laid on the discussion of finite volume and finite lattice spacing effects.

1. Introduction

One of the central goals of QCD analyses at finite temperature is to clarify the existence and the properties of hadronic excitations at temperatures around the phase transition to the quark gluon plasma as well as in the plasma itself. Within lattice QCD, at zero temperature one usually computes temporal correlation functions of operators with specific hadronic quantum numbers and extracts masses and matrix elements from their behavior at large Euclidean time separations τ where the correlation function is dominated by the ground state contribution. At non-vanishing temperature this approach is hampered by the fact that the size of the system in the temporal direction is limited by the inverse temperature, $\tau < 1/T$. At tolerable computing costs one may increase the number of data points in the temporal direction, N_τ, by using anisotropic lattices in order to facilitate fits. However, also in this case one has to rely on fit ansätze where

*This work is supported by the DFG under grant FOR 339/2-1 and by the U.S. Department of Energy under contract DE-AC02-98CH10886.

on the other hand the very existence of bound states or resonances is not known a priori.

All the requested information is encoded in the spectral density $\sigma_H(p_0, \vec{p})$ of the chosen hadronic channel H. The spectral density is related to the correlation function in momentum space by

$$\Delta_H(i\omega_n, \vec{p}) = \oint \frac{dp_0}{2\pi i} \frac{\Delta_H(p_o, \vec{p})}{p_0 - i\omega_n} = \int \frac{dp_0}{2\pi} \frac{\sigma_H(p_0, \vec{p})}{p_0 - i\omega_n} + Subtr. \quad (1)$$

By a Fourier transformation in time one obtains the temporal correlator (at $\vec{p} = 0$) as

$$G_H^T(\tau) = T \sum_n e^{-i\omega_n \tau} \Delta_H(i\omega_n, 0) = \int_0^\infty dp_0 \, \sigma_H(p_0, 0) \, \frac{\cosh[p_0(\tau - 1/2T)]}{\sinh(p_0/2T)} \quad (2)$$

Of course, extracting a continuous function, σ, from data at a discrete set of lattice points, $G_H^T(\tau)$, is an ill-posed problem. However, with the rediscovery[1] of the Maximum Entropy Method (MEM) for lattice QCD it seems possible to obtain at least a most probable solution even at non-vanishing temperature (for results at $T = 0$ see also Ref. 2).

Alternatively, since long one has been studying spatial correlation functions, at $\omega_n = p_x = p_y = 0$,

$$G_H^S(z) = \int_{-\infty}^{+\infty} \frac{dp}{2\pi} \, e^{ipz} \Delta_H(0, p) = \int_{-\infty}^{+\infty} \frac{dp}{2\pi} \, e^{ipz} \int_{-\infty}^{+\infty} dp_0 \frac{\sigma_H(p_0, p)}{p_0} \quad (3)$$

These are just the Fourier transforms of Eq. 1 in z and depend on the same spectral density but the relation to σ is more involved. The spatial correlator has the advantage that the spatial separations are not limited physically and it is possible to extract screening masses from the large z behavior. However, those are not necessarily equal to physical masses.

Both correlators are subject to finite volume (L^3) and finite lattice spacing (a) corrections, parametrized as functions of $LT = N_\sigma / N_\tau$ and $aT = 1/N_\tau$, which will be addressed in this talk. The results presented are based on simulations of quenched QCD with a non-perturbatively improved fermion action.

2. Spatial correlators

At large temperatures one expects the spatial correlator to be dominated by the exchange of two free quarks with Matsubara frequencies $\pi T(2n+1)$. It is therefore instructive to compare the lattice results with the free case. The

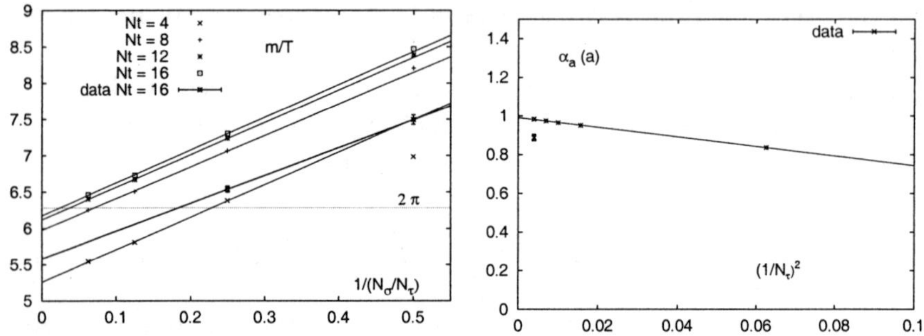

Figure 1. Pion screening mass in the free case as well as at $T = 3T_c$ on various lattices. The left part shows the volume extrapolations, the right part the residual finite lattice spacing effects in the free as well as interacting case.

free spatial correlator is known on the lattice as well as in the continuum[3] where it reads, e.g. for the pion

$$G_\pi^S(z) = \frac{N_c T}{2\pi z^2 \sinh(2\pi T z)} \left[1 + 2\pi T z \coth(2\pi T z)\right] \tag{4}$$

From the correlator one might define an effective screening mass as

$$m_{\text{screen}}^{\text{eff}}(z) = -\frac{1}{G(z)} \frac{\partial G(z)}{\partial z} \simeq 2\pi T \left\{1 + \frac{1}{2\pi T z} + \ldots\right\} \tag{5}$$

where the second equation exhibits the anticipated value of $2\pi T$ in the limit of infinite separation z as well as the leading correction. Note that at fixed z, e.g. $z = N_\sigma/4$, a leading, positive, finite volume correction proportional to N_τ/N_σ can be read off. The same behavior is seen on the lattice, Fig. 1(left) where the free lattice results are compared with the data at $3T_c$. The interacting results are slightly less but still considerably affected by the finite volume. The residual finite lattice spacing effects are shown in Fig. 1(right). In the free case the leading effect is proportional to $1/N_\tau^2 \sim a^2$, a behavior which should also be true in the interacting case since an $O(a^2)$ improved action is used. Note that the finite a corrections are working in the opposite direction to the finite L ones.

Our results on the screening masses of pion and rho as of now are summarized in Fig. 2. Below T_c, the data has been extrapolated to the chiral limit and normalized to the critical temperature. At some quark mass values where we have three different lattice volumes available we have seen finite size effects which follow the observed zero temperature behavior of masses, $m(L) - m(\infty) \sim 1/L^3$. These are not yet taken into account.

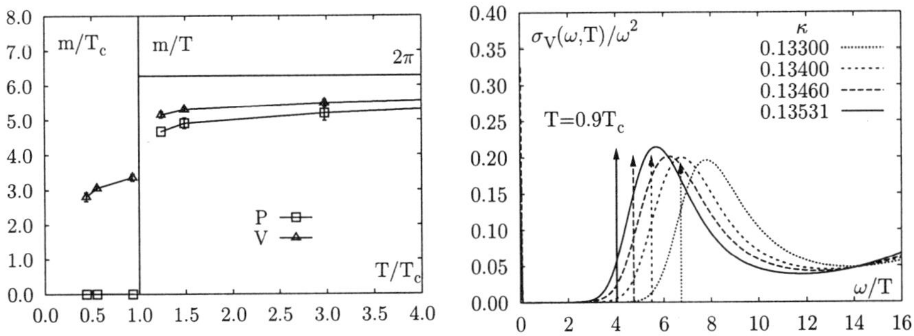

Figure 2. Left: Pion and ρ screening masses as function of the temperature. Right: The vector spectral density at $0.9T_c$ for various quark masses (κ values). The arrows denote the values of the corresponding screening masses.

Keeping this in mind, the temperature dependence seems to be rather weak. Above T_c, data obtained on the largest volumes available have been corrected for finite volume and finite a effects by normalizing with the free case finite lattice size prediction. So far, the data is some 10 to 15 % lower than the free case. Since in the free case the lattice artefacts tend to be overestimated the differences to 2π might be smaller indeed. Yet, in view of Fig. 1 it seems unlikely that in the continuum infinite volume extrapolation, yet to be done, the results match the free value almost exactly as in Ref. 4 for staggered quarks or become larger than $2\pi T$ as in an analytic calculation[5].

3. Temporal correlators

Below T_c, the MEM analysis[a] of the temporal correlation functions generally features ground state peaks which become broader and decrease in height with increasing temperature. At values of $T/T_c = 0.4$ and 0.6 the positions of the peaks coincide with the values of the screening masses, both in the pseudoscalar and the vector channel. At $T = 0.9T_c$ the ρ mass is enhanced over the screening mass, see Fig. 2, while the pion is unaffected.

At temperatures above T_c, non-perturbative properties of the spectral density have to be disentangled from the contributions of the free quark cut which will dominate at high temperatures and which has been studied at finite lattice spacing in detail in Ref. 6. On top of the free quark

[a]Tests of the MEM procedure at finite T have been presented in the talk, but are omitted here for lack of space.

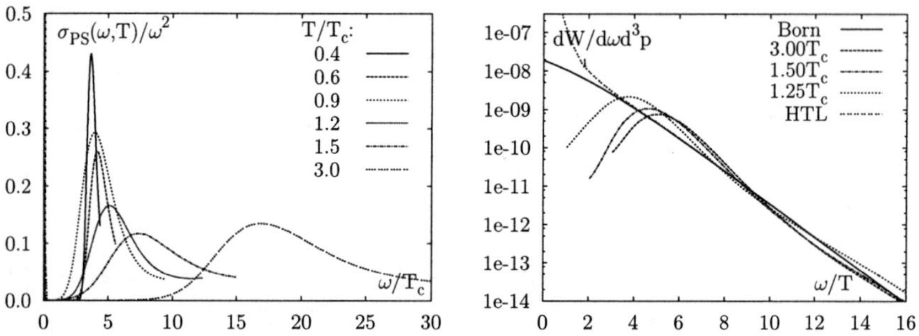

Figure 3. Left: Pseudoscalar spectral density at the temperatures investigated. Below T_c the quark mass is roughly the same for all T, above the results have been obtained roughly in the chiral limit. Right: The dilepton rate extracted from the ρ spectral density at temperatures above T_c.

contribution, the MEM results suggest the presence of resonances in the investigated temperature range up to $3T_c$. However, peaks have turned into rather broad and comparatively flat bumps whose positions move with the temperature, see Fig. 3(left). At $3T_c$ it has been checked that two different volumes ($N_\sigma/N_\tau = 2$ and 4) and two different lattice spacings ($aT = 1/12$ and $1/16$ at $N_\sigma/N_\tau = 4$) give the same results. The frequency ω of the maximum of the bump scales with the temperature, $\omega/T \simeq 6$, between $3T_c$ and $1.5T_c$ but is obtained at a lower value, $\omega/T \simeq 5$, for $1.2T_c$. These observations are corroborated by equivalent findings in the ratio of the correlation functions to their free values, in particular at the largest possible time separation of $\tau = 1/(2T)$. This ratio is (finite and) larger than 1. This fact already indicates that the spectral density at low energies is larger than in the free case and that it vanishes in the zero frequency limit. These features translate into the dilepton rate as shown in Fig. 3(right) which can be extracted immediately from the vector spectral density[7].

References

1. M. Asakawa, T. Hatsuda and Y. Nakahara, Phys. Rev. D60 (1999) 091503 and Prog. Part. Nucl. Phys. 46 (2001) 459.
2. T. Yamazaki et al., Phys.Rev. D65 (2002) 014501.
3. W. Florkowski and B. Friman, Z. Phys. A347 (1994) 271.
4. R.V. Gavai and S. Gupta, Phys. Rev. D67 (2003) 34501.
5. M. Laine and M. Vepsäläinen, JHEP 402 (2004) 4.
6. F. Karsch et al., Phys. Rev. D68 (2003) 14504.
7. F. Karsch et al., Phys. Lett. B530 (2002) 147.

CHARMONIA AT FINITE MOMENTA IN A DECONFINED PLASMA *

S. DATTA, F. KARSCH AND S. WISSEL

Fakultät für Physik, Universität Bielefeld, D-33615 Bielefeld, Germany.
E-mail: saumen,karsch,wissel@physik.uni-bielefeld.de

P. PETRECZKY †

Brookhaven National Laboratory, Upton, NY 11973, USA.
E-mail: petreczk@quark.phy.bnl.gov

I. WETZORKE

NIC/DESY Zeuthen, Platanenallee 6, D-15738 Zeuthen, Germany.
E-mail: Ines.Wetzorke@desy.de

Lattice studies of charmonium systems have indicated that in a deconfined gluonic plasma ground state charmonia survive as bound states upto temperatures $\sim 2T_c$. After surveying the methodologies used in reaching these results, we examine the behavior of these systems when the bound state is in motion with respect to the heatbath frame. We find that the finite momenta charmonia show medium modifications when the medium is deconfined; in particular, a modification of the energy-momentum dispersion relation is indicated.

The study of strongly interacting matter at high temperatures is important both for understanding QCD and for cosmological applications, and is a very active field with experimental data available from relativistic heavy ion collisions. Charmonium systems have been of great interest to the relativistic heavy ion community ever since Matsui and Satz, based on potential model studies, predicted that charmonium bound states cannot survive in a deconfined plasma, and suggested J/ψ as a probe of the confinement status of the plasma [1]. Later, more detailed potential model studies suggested that

*This research is supported by BMBF under grant no. 06BI102, and by DOE under contract DE-AC02-98CH10886.

†Goldhaber and RIKEN fellow

all charmonium bound states dissolve before or pretty soon after deconfinement [2]. However, a direct lattice investigation of the hadronic correlators have revealed that the ground state charmonia (η_c and J/ψ) survive the deconfinement transition in pure gauge theory [3,4,5], and may dissolve only at temperatures $\sim 2T_c$.

Both for understanding the mechanism of bound state dissolution in the plasma and for study of charmonium abundance in experiments, it is important to also know the fate of these bound states when they not at rest but moving with respect to the heatbath frame. The lack of covariance in finite temperature system means the pole of the inverse propagator,

$$\Delta^{-1}(p_0,p,T) = p_0^2 + p^2 + m^2 + \mathrm{Re}\Pi(p,T) + i\mathrm{Im}\Pi(p,T) \tag{1}$$

can be a generic function of spatial momentum p, leading possibly to a modification of the energy-momentum dispersion relation and a momentum dependent decay width. Such modifications have been discussed in different models for pions in hadron gas [6], and for decay width of J/ψ in hadron gas [7]. Here, we present a preliminary report of a direct lattice investigation into the fate of the ground state charmonia with a finite spatial momentum (in the heatbath frame). In this report, we will restrict ourselves to comparatively moderate temperatures $\leq 1.5T_c$.

Direct lattice investigations of the fate of the charmonium states involve analyzing the Matsubara correlators,

$$G_H(\tau,\vec{p},T) = \langle J_H(\tau,\vec{p}) J_H^\dagger(0,-\vec{p})\rangle_T \tag{2}$$

where J_H is the suitable mesonic operator, projected on the state with spatial momentum $\vec{p}$. For studying the η_c and the J/ψ, we use the point-point operators $\bar{c}\gamma_5 c$ and $\bar{c}\gamma_\mu c$, respectively. For the vector channel, here we consider the transverse component only. Through analytical continuation, the Matsubara correlator can be related to the hadronic spectral function by an integral equation:

$$G(\tau,\vec{p},T) = \int_0^\infty d\omega \sigma(\omega,\vec{p},T) \frac{\cosh(\omega(\tau - 1/2T))}{\sinh(\omega/2T)}. \tag{3}$$

While the inversion of Eq. (3) to get the spectral function from the correlators measured at a finite set of points is clearly an ill-defined problem, some progress has been made in the last few years towards tackling such problems using Bayesian tools. In the maximum entropy method [8] the positivity of the spectral function and information about the asymptotic behavior is used to find the most probable spectral function given data. Analysis of the hadronic correlators at low temperatures (below T_c)

reveal a complicated structure in the high energy regime of the spectral functions, dominated by lattice artifacts [3]. We use this structure as part of the asymptotic information in our analysis at higher temperatures [3].

Useful information about possible change of state with deconfinement can be obtained by comparing the correlators measured above T_c with the correlators reconstructed from the spectral function below T_c. We define $G_{\text{recon},T^*}(\tau,\vec{p},T)$ by substituing $\sigma(\omega,\vec{p},T)$ in Eq. (3) with $\sigma(\omega,\vec{p},T^*_)$ where T^* is usually taken as the smallest temperature below T_c available to us. This takes into account the trivial temperature dependence of the kernel and therefore, any deviation of $G(\tau,\vec{p},T)$ from $G_{\text{recon},T^*}(\tau,\vec{p},T)$ will indicate a temperature modification of the spectral function [3]. In Fig. 1 the results of such a comparison are shown for zero momentum correlators. For the pseudoscalar channel, Fig. 1 shows that the reconstructed correlator explains the measured correlators completely upto temperatures of 1.5 T_c, indicating that there is no significant change for these states upto this temperature. The spectral function extracted from the correlators, shown in the right, confirm this. For J/ψ, some small deviation of the measured correlators is seen from the reconstructed correlators, indicating a possible widening [4] or mass shift. Here we added the three (spatial) components of the vector, and the behavior is somewhat different from the four component correlator in Ref. [3] due to the change in contribution from electric conductivity below and above T_c.

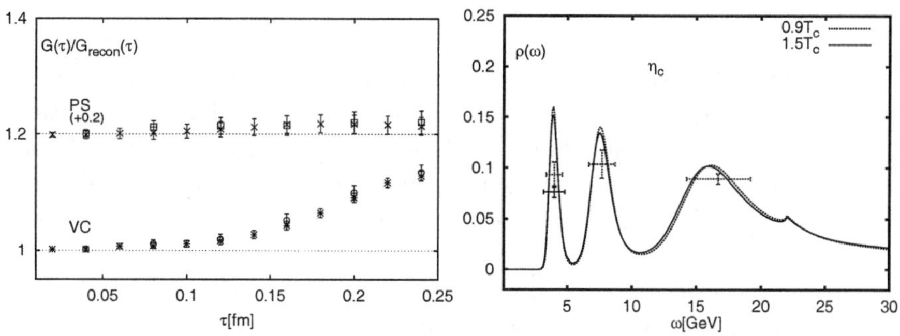

Figure 1. Comparison of zero momentum correlators at 1.5 T_c for vector and pseudoscalar charmonia with those reconstructed from spectral functions at 0.75 T_c (left). The stars and crosses correspond to a lattice with lattice spacing of 0.02 fm and the the open symbols correspond to a lattice with lattice spacing 0.02 fm. (Right) Reconstructed spectral functions at 1.5 T_c and 0.9 T_c for the pseudoscalar channel. Plotted is the rescaled spectral function $\rho(\omega) = \sigma(\omega)/\omega^2$. The default model included the peak structure at high ω, which is dominated by lattice artifacts.

For investigation of the finite momenta charmonia, we use the same analysis methods as those at zero momentum described above. Comparison of the finite momentum correlators $G(\tau, \vec{p}, T)$ at 1.1 and 1.5 T_c with the reconstructed correlators from $\sigma(\omega, \vec{p}, T)$ are presented in Fig. 2. It can be seen that with increasing spatial momentum, the correlators above T_c show larger deviations from the reconstructed correlators, signalling a medium effect already at 1.1 T_c and increasing with temperature.

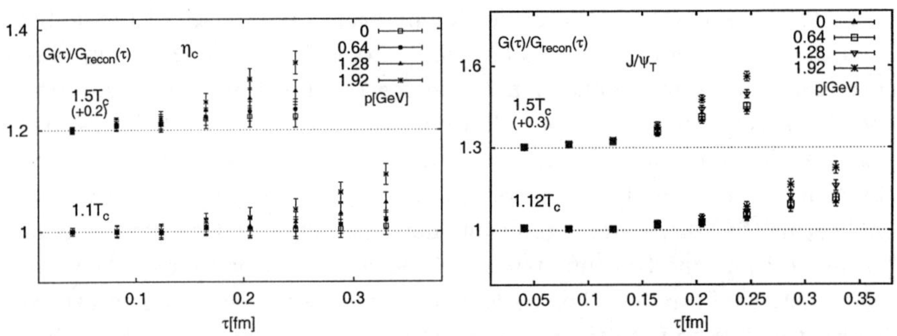

Figure 2. Comparison of correlators at 1.1 T_c and 1.5 T_c for pseudoscalar and (transverse) vector charmonia with those reconstructed from spectral function at 0.75 T_c.

In order to get further insight into the nature of the medium effect, we used maximum entropy analysis for a direct reconstruction of $\sigma(\omega, \vec{p}, T)$ above T_c. As described before, for the reconstruction we use the high energy structure of the spectral function obtained at 0.75 T_c as part of the prior information. The spectral functions at 0.75 T_c and 1.1 T_c, using such a default model, are shown in Fig. 3. Here, in order to avoid any systematics due to the small spatial extent and limited number of data points at the higher temperature, we use the same physical extent and same number of data points both below and above T_c, by omitting points at the center for the lower temperature.

Figure 3 shows two interesting features. First, a strong signal from the ground state shows up in the spectral function at 1.1 T_c even for momenta $\sim 2GeV$. Little change in peak strength is seen on crossing deconfinement, for any of the momenta. Second, there is an indication of a slower movement of the peak position above T_c, indicating a modification of the energy-momentum dispersion relation. These trends continue at 1.5 T_c, with a significant bound state showing up with a modified dispersion relation.

The J/ψ moving in the heatbath frame sees more energetic gluons than

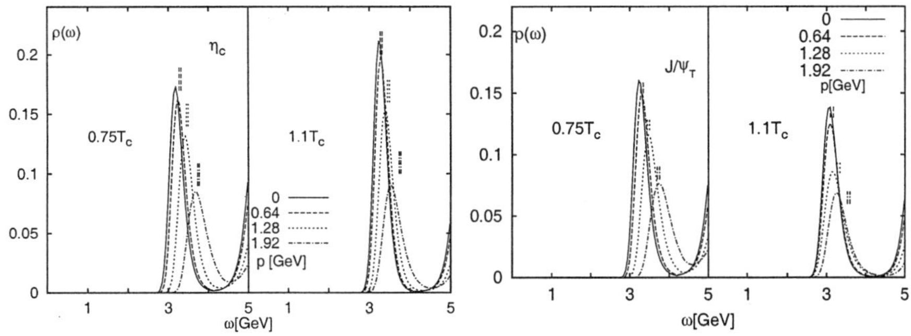

Figure 3. Spectral functions at 0.75 T_c and 1.1 T_c for different spatial momenta, for lattices with lattice spacings of 0.04 fm. The small horizontal bars indicate the expected shift of the peak using relativistic dispersion relation. The same temporal extent has been used at both the temperatures.

the J/ψ at rest. The fact that J/ψ moving at such momenta show a statistically significant peak indicates a stongly nonperturbative nature of the plasma. We would like to clarify that we are investigating here J/ψ states in an equilibriated plasma. This is very different from earlier phenomenological studies concerned with survival of J/ψ with large trasnverse momentum from a finite sized plasma, where a large transverse momentum may lead to the J/ψ spending a smaller time in the plasma, resulting in a larger survival probability [9,2]. However, such studies also need to take into account the fact that the in-medium dispersion relation for the J/ψ will be modified. A modified dispersion relation may also explain the large temperature dependence observed in the screening mass of the J/ψ and the η_c [3]; however, a quantitative statement will require a much more exact determination of the dispersion relation than is available at present.

References

1. T. Matsui and H. Satz, *Phys. Lett.* **B 178**, 416 (1986).
2. F. Karsch and H. Satz, *Z. Phys.* **C 51**, 209 (1990).
   S. Digal *et al.*, *Phys. Lett.* **B514**, 57 (2001).
3. S. Datta *et al.*, *Phys.Rev.* **D 69**, 094507 (2004).
4. T. Umeda *et al.*, hep-lat/0211003.
5. M. Asakawa and T. Hatsuda, *Phys. Rev. Lett.* **92**, 012001 (2004).
6. E. Shuryak, *Phys.Rev.* **D 42**, 1764 (1990).
7. K. Haglin and C. Gale, *Phys. Rev.* **D 63**, 065201 (2001).
8. M. Jarrel and J. E. Gubernatis, *Phys. Rept.* **269**, 133 (1996).
   M. Asakawa et al., *Prog. Part. Nucl. Phys.* **46**, 459 (2001).
9. M.-C. Chu and T. Matsui, *Phys. Rev.* **D 37**, 1851 (1988).

QCD THERMODYNAMICS: LATTICE RESULTS CONFRONT MODELS

MASSIMO D'ELIA
Dipartimento di Fisica dell'Università di Genova and INFN, I-16146, Genova, Italy
E-mail : delia@ge.infn.it

MARIA PAOLA LOMBARDO
INFN-Laboratori Nazionali di Frascati, I-00044, Frascati(RM), Italy
E-mail : lombardo@lnf.infn.it

We show that lattice results for four flavor QCD at nonzero temperature and baryon density compare well with the hadron resonance gas model up to $T \simeq 0.95T_c$, and approach a free field behaviour with a reduced effective number of flavor for $T \geq 1.5T_c$; chiral symmetry and confinement are interrelated, and we note analogies between the critical line of QCD and that of simple models with the same global symmetries.

1. QCD Thermodynamics and Imaginary μ_B

In the last four years a few lattice techniques have proven to be successful in QCD thermodynamics for $\mu_B/T < 1$.[1,2,3,4,5,6,7] While waiting for final results in the scaling limit and with physical values of the parameters, it is very useful to contrast and compare current lattice results with model calculations and perturbative studies.

The imaginary chemical potential approach[8,9,5,1,2,10] to QCD thermodynamics seems to be ideally suited for the interpretation and comparison with analytic results. In the following we review our results[1,2] from this perspective.

QCD at finite quark chemical potential μ can be simulated with ordinary methods when μ is purely imaginary. If one were able to determine thermodynamic observables with infinite accuracy, standard complex analysis arguments would guarantee that the result will be valid within the entire analytic domain, i.e. everywhere away from phase transitions. In practical numerical work one has to take into account two sources of errors: first, the analytic form of the fitting function is not known a priori; second, even if it were so, one has to deal with numerical errors on the coefficients. Cross checks among different analysis and guidance from models are thus most useful. Results from an imaginary μ have been obtained for the critical line of the two, three and two plus one flavor model,[5] as well as for four flavor.[1] Thermodynamics results – order parameter, pressure, number density – were obtained for the four flavor model.[2] The pressure was defined by introducing an integral method at fixed temperature, and the mass dependence was estimated via the Maxwell relations.[11,2,7]

2. The Critical Line, Chiral Symmetry and Confinement

We used the exact results available for the critical line of the Gross Neveu model, and for Random Matrix Models in the appropriate universality class to show that a second order polynomial in T and μ approximates well the critical line over a large μ interval.[1] The nature of the chiral transition itself can be studied with a great accuracy: the correlation between $< \bar{\psi}\psi >$ and Polyakov loop was observed at several values μ_I, either by studying the results as a function of β, and by following the Monte Carlo histories directly at $\beta_c(\mu_I)$.[1,2]

3. The Hadronic Phase

The grand canonical partition function of the Hadron Resonance Gas model[12] has a simple hyperbolic cosine behaviour. This can be framed

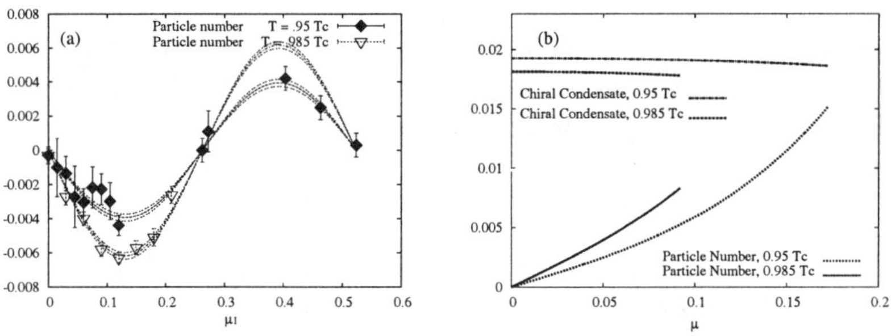

Figure 1. Hadronic Phase: (a) One Fourier coefficient fit to the particle number, showing that the Hadron Resonance Model is adequate to describe this data. (b) Compilation of the results for the chiral condensate and the particle number as a function of real chemical potential: the lines are cut in correspondence with μ_c, showing the first order character of the phase transition (inferred from the chiral condensate) and the critical density. See our extended writeups for error analysis and details.

in our discussion of the phase diagram in the temperature-imaginary chemical potential plane which suggests to use Fourier analysis in this region, as observables are periodic and continuous there.[1]

For observables which are even (O_e) or odd (O_o) under $\mu \to -\mu$ the analytic continuation to real chemical potential of the Fourier series read $O_e[o](\mu_I, N_t) = \sum_n a_F^{(n)} \cosh[\sinh](nN_tN_c\mu_I)$. In our Fourier analysis of the chiral condensate[1] and of the number density[2] - even and odd observables, respectively - we limited ourselves to $n = 0, 1, 2$ and we assessed the validity of the fits via both the value of the χ^2/d.o.f. and the stability of $a_F^{(0)}$ and $a_F^{(1)}$ given by one and two cosine [sine] fits: we found that one cosine [sine] fit describes reasonably well the data up to $T \simeq 0.985T_c$ (see Fig.1a); further terms in the expansion did not modify much the value of the first coefficients and does not particularly improve the χ^2/d.o.f.. This means that our data are well approximated by the hadron resonance gas prediction $\Delta P \propto (\cosh(\mu_B/T) - 1)$ in the broken phase up to $T \simeq 0.985T_c$. The analysis of the corrections requires better precision.

The analytic continuation (Fig. 1b) of any observable O is valid within the analyticity domain, i.e. till $\mu < \mu_c(T)$, where $\mu_c(T)$ has to be measured independently. The value of the analytic continuation of O at μ_c, $O(\mu_c)$, defines its critical value. When O is an order parameter which is zero in the quark gluon plasma phase, the calculation of $O(\mu_c)$ allows the identification of the order of the phase transition: first, when $O(\mu_c) \neq 0$, second, when $O(\mu_c) = 0$.[1,2]

4. The Hot Phase

The behaviour of the number density (Fig. 2a) approaches the lattice Stephan-Boltzmann prediction, with some residual deviation. We parametrise the deviation from a free field behavior as[13,14]

$$\Delta P(T, \mu) = f(T, \mu) P^L_{free}(T, \mu) \tag{1}$$

where $P^L_{free}(T, \mu)$ is the lattice free result for the pressure. For instance, in the discussion of Ref.[14]

$$f(T, \mu) = 2(1 - 2\alpha_s/\pi) \tag{2}$$

and the crucial point was that α_s is μ dependent.

We can search for such a non trivial prefactor $f(T, \mu)$ by taking the ratio between the numerical data and the lattice free field result $n^L_{free}(\mu_I)$

at imaginary chemical potential:

$$R(T, \mu_I) = \frac{n(T, \mu_I)}{n^L_{free}(\mu_I)} \tag{3}$$

A non-trivial (i.e. not a constant) $R(T, \mu_I)$ would indicate a non-trivial $f(T, \mu)$.

In Fig. 2b we plot $R(T, \mu_I)$ versus μ_I/T: the results for $T \geq 1.5T_c$ seem consistent with a free lattice gas, with an fixed effective number of flavors $N_f^{eff}(T)/4 = R(T)$: $N_f^{eff} = 0.92 \times 4$ for $T = 3.5T_c$, and $N_f^{eff} = 0.89 \times 4$ for $T = 1.5T_c$.

5. Lattice vs. Models: Open Issues

We give here a partial list of results together with related questions which calls for a more detailed understanding of the interrelations of analytic and numerical results.

The critical line is similar to the one of model theories with the same global symmetries: $T/T_c^2 = 1 - 0.0021(2)(\mu/T)^2$ for four flavor, and similarly for two and three flavors.[3,5,6] Can we understand the flavor dependence and the (small) coefficient of $(\mu/T)^2$ from these model?

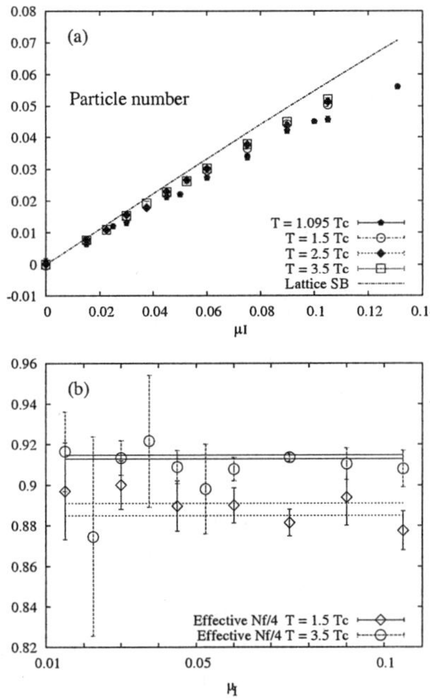

Figure 2. Hot Phase: (a) Particle number aproaching the free field behaviour (b) Ratio of the lattice results to the lattice free field: the deviation from free field can be described by an effective number of flavor $N < 4$ for $T \geq 1.5T_c$

We have observed $T_c^{chiral}(\mu) = T_c^{deconfining}(\mu)$ Can effective Lagrangians[16] account for this observation ?

In the hadronic phase $\Delta P \simeq k(1 - \cosh(\mu_B/T))$. The corrections are not completely negligible, and can be used to estimate the magnitude of further terms in the expansion. Can these corrections be modeled in some simple, intuitive way?

For $T \geq 1.5T_c$ the results are compatible with lattice Stefan Boltzmann with an active fixed number of flavor 4×0.92 for T=3.5 T_c and 4×0.89 for $T = 2.5T_c$. These finite density corrections appear to leave unchanged the free field structure, even at moderately low temperatures. How can we understand this?

Between T_c and 1.5 T_c, can we interpret the deviations from this simple behaviour by use of a rigorous perturbative analysis,[15] and / or in the framework of a strongly interacting quark gluon plasma[17]?

Acknowledgments

This work was partially supported by MIUR. MPL wishes to thank the Institute for Nuclear Theory at the University of Washington for its hospitality and the Department of Energy for partial support during the completion of this work.

References

1. M. D'Elia and M. P. Lombardo, Phys. Rev. D **67**, 014505 (2003).
2. M. D'Elia and M. P. Lombardo, arXiv:hep-lat/0406012.
3. Z. Fodor and S. D. Katz, Phys. Lett. B **534**, 87 (2002);JHEP **0404**, 50 (2004); Z. Fodor, S. D. Katz and K. K. Szabo, Phys. Lett. B **568**, 73 (2003); F. Csikor *et al.* JHEP **0405** (2004) 046; S. D. Katz, Nucl. Phys. Proc. Suppl. **129**, 60 (2004).
4. Ph. de Forcrand, S. Kim and T. Takaishi, Nucl. Phys. Proc. Suppl. **119**, 541 (2003)
5. Ph. de Forcrand and O. Philipsen, Nucl.Phys. B **642**, 290 (2002); Nucl.Phys. B **673**, 170 (2003).
6. C. R. Allton *et al.*, Phys. Rev. D **66**, 074507 (2002); Phys. Rev. D **68**, 014507 (2003).
7. R. Gavai, S. Gupta and R. Roy, Prog. Theor. Phys. Suppl.**153**, 270 (2004).
8. M. P. Lombardo, Nucl. Phys. Proc. Suppl. **83**, 375 (2000).
9. A. Hart, M. Laine and O. Philipsen, Phys. Lett. B **505**, 141 (2001).
10. P. Giudice and A. Papa, Phys. Rev. D **69**, 094509 (2004).
11. J. B. Kogut et al. Nucl. Phys. B **225**, 93 (1983).
12. F. Karsch, K. Redlich and A. Tawfik, Phys. Lett. B **571**, 67 (2003).
13. K. K. Szabo and A. I. Toth, JHEP **0306**, 008 (2003).
14. J. Letessier and J. Rafelski, Phys. Rev. C **67**, 031902 (2003).
15. A. Vuorinen, arXiv:hep-ph/0402242; A. Vuorinen, Phys. Rev. D **68**, 054017 (2003); A. Ipp, A. Rebhan and A. Vuorinen, Phys. Rev. D **69**, 077901 (2004).
16. A. Mocsy, F. Sannino and K. Tuominen, Phys. Rev. Lett. **92**, 182302 (2004)
17. See e.g.the WEB page of the RBRC Workshop *New Discoveries at RHIC: the current case for the strongly interacting QGP* `http://www.bnl.gov/riken/May14-152004workshop.htm`

SINGLET FREE ENERGIES OF A STATIC QUARK-ANTIQUARK PAIR

KONSTANTIN PETROV

Brookhaven National Laboratory, Upton, NY 11973

We study the singlet part of the free energy of a static quark anti-quark ($Q\bar{Q}$) pair at finite temperature in three flavor QCD with degenerate quark masses using $N_\tau =$ 4 and 6 lattices with Asqtad staggered fermion action. We look at thermodynamics of the system around phase transition and study its scaling with lattice spacing and quark masses.

1. Simulations and Results

Static free energy of a quark-antiquark pair is a frequently used tool to study non-perturbatively the in-medium modification of inter-quark forces. It is calculated as the difference in the free energy of the system with static quarks and the same system without them, while the temperature remains constant. Due to colour symmetry such quantity contains the singlet and the octet contributions, and the usual definition of the free energy is thus referred to as the colour-averaged. This quantity has been extensively in $SU(2)$ and $SU(3)$ gauge theories e.g. [3] and full QCD [8]. However, singlet and octet channels were considered in detail only in pure gauge theory [2, 5, 6]. In the case of full QCD the first results for singlet and octet free energy for two flavor QCD have appeared only very recently [9]. Here we present our results for 3 flavor QCD using the so-called Asqtad staggered fermion action [10] with two different lattice spacings (corresponding to $N_t = 4$ and 6) at three different quark masses.

Our analysis is to a large extent based on the gauge configurations generated by the MILC collaboration using the Asqtad action. Therefore we adopt their strategy for fixing the parameters which is described in Ref. 11. We use the most recent value of r_1 extrapolated to continuum and to the physical value of the light quark masses $r_1 = 0.317$ fm to convert the lattice units to temperature.

The free energy of a static quark anti-quark pair contains a lattice spacing dependent divergent piece, and thus needs to be renormalized. Following Ref. 5 we do so by normalizing it to the zero temperature potential at short distances where the temperature dependence of the free energy can be neglected. The static quark potential has been studied by the MILC collaboration at three different lattice spacings and various quark masses [11]. We use the following form of the zero temperature potential, which reproduces MILC data well in the whole range of masses and lattice spacings

$$r_1 V(x) = -\frac{0.44}{x} + 0.56 \cdot x + \frac{0.0125}{x^2}, \ x = r/r_1 \tag{1}$$

In Fig. 1 we show that both the potential and effective coupling constant $\alpha_s(r)$ defined as $\alpha_s(r) = 3/4r^2 dV(r)/dr$ are reproduced well. Results pre-

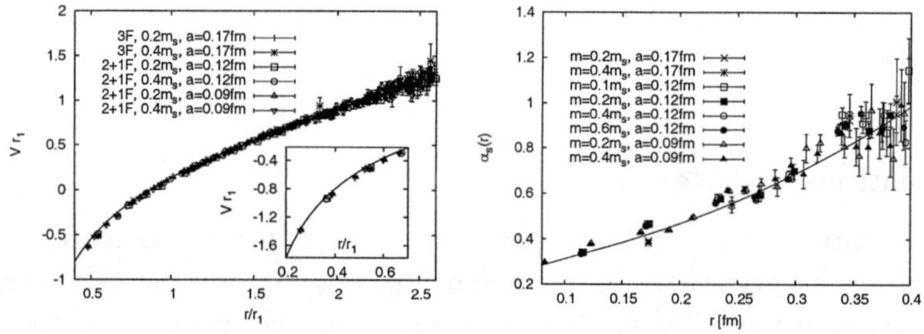

Figure 1. Zero-temperature static potential (left) and the running coupling (right) [data points from MILC collaboration]

sented here come from MILC lattices as well as our simulation for quark masses $m_{light} = 0.2m_s, 0.4m_s, 0.6m_s$ on lattices $12^3 \times 4, 8^3 \times 4$ and $12^3 \times 6$. The temperature range was $135 - 412$ MeV for the first two lattice sizes and $145 - 310$ MeV for the last one.

Following Ref. 14 the free energy of static quark-antiquark ($Q\bar{Q}$) pair in the color singlet channels is defined as

$$\exp(-F_1(r,T)/T + C) = \frac{1}{3} Tr \langle W(\vec{r}) W^\dagger(0) \rangle \tag{2}$$

Here $W(\vec{x}) = \prod_{\tau=0}^{N_\tau - 1} U_0(\tau, \vec{x})$ is the temporal Wilson line, $L(\vec{x}) = Tr W(\vec{x})$ is known as the Polyakov loop. As $W(\vec{x})$ is not gauge invariant one needs to fix a gauge. Here we will use the Coulomb gauge as advocated in [4]. This approach is exactly valid at zero temperature and is numerically true at finite temperature. One can also consider the color averaged free energy defined as

$$\exp(-F_{av}(r,T)/T + C) = \frac{1}{9} \langle L(\vec{r}) L^\dagger(0) \rangle \tag{3}$$

We start the discussion of our numerical results with the case of the quark of mass $0.4m_s$ on the $12^3 \times 4$ lattice. The corresponding numerical results for the singlet free energy are shown in Fig.2. The free energy approaches a finite value $F^i_\infty(T) = \lim_{r\to\infty} F_i(r,T)$, $i = 1, av$ at large distances which is usually interpreted as string breaking at low temperature and screening at high ones. Note that the distance where the free energy effectively flattens is temperature dependent, it becomes smaller at higher temperatures. At small distances the singlet free energy is temperature independent and coincides with the zero temperature potential, as at small distances medium effects are not important. The numerical results for other values of the quark masses are similar. The scaling with the lattice spacing is remarkably good.

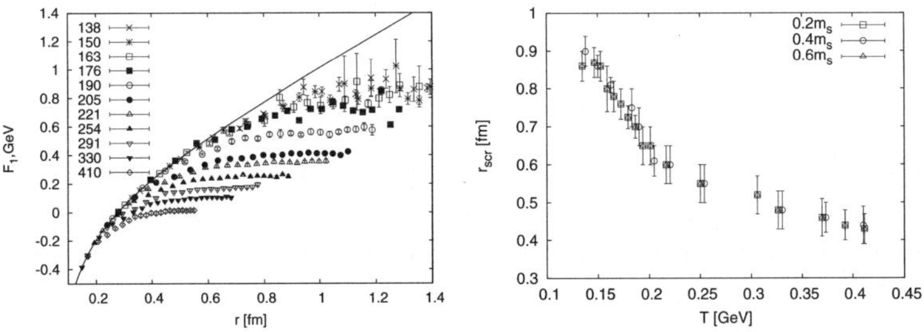

Figure 2. Singlet free energies and the effective screening radius

To characterize the range of interaction in the medium it is convenient to introduce the effective screening radius r_{scr} defined as: $F_1(r = r_{scr}, T) = 0.9F^1_\infty(T)$. Here $F^1_\infty(T)$ is the asymptotic value of the singlet free energy at infinite separation. In Fig.2. we show the values of r_{scr} for three different quark masses and $12^3 \times 4$ lattices. Certainly as $F_1(r,T)$ has statistical errors it is difficult to determine at exactly which distance r the equation $F_1(r = r_{scr}, T) = 0.9F^1_\infty(T)$ holds. We have tried to estimate this uncertainty in the values of r_{scr} and show them in Fig.2 as errors bars. At small temperatures the value of the screening radius is about $0.9fm$ and is temperature independent. As we increase the temperature r_{scr} decreases reaching the value of $0.5fm$ at the highest temperature. Note that the temperature dependence of r_{scr} is roughly the same for all quark masses. On Fig.3 (left) we plot the asymptotic value of the free energies; the quark mass dependence likely vanishes at small temperatures ($T < 150MeV$) and definitely negligible at high temperatures ($T > 250MeV$). It is however

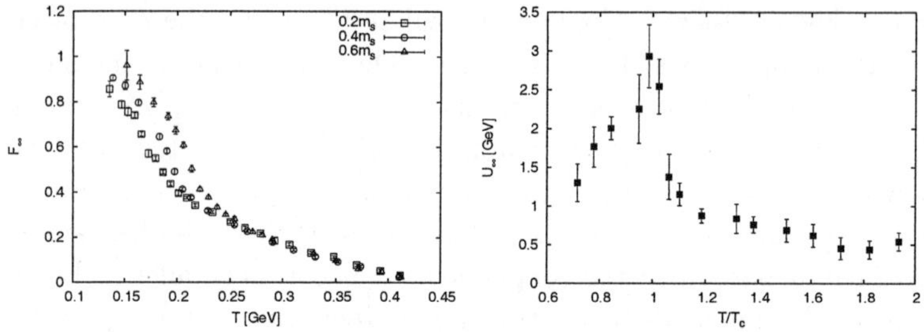

Figure 3. Infinite separation free energies for various quark masses (left) and internal energy at 0.4ms

significant in the transition region.

In the limit of very small temperatures we expect $F_\infty(T)$ to be temperature independent and related to twice the binding energy of a heavy-light ($D-$ or $B-$) meson

$$2E_{bin} = 2M_{D,B} - 2m_{c,b}. \tag{4}$$

Based on this observation in Ref. 15 it has been argued that the decrease of $F_\infty(T)$ with the temperature close to T_c implies the decrease of the $M_{D,B}$ leading to quarkonium suppression. However, $F_\infty(T)$ also contains an entropy contribution due to the presence of a static $Q\bar{Q}$ pair:

$$S_\infty(T) = -\frac{\partial F_\infty(T)}{\partial T}. \tag{5}$$

Then we can calculate the energy induced by a static quark-anti-quark pair

$$U_\infty = F_\infty + TS_\infty \tag{6}$$

Numerically the derivative with respect to the temperature in Eq. (5) was estimated using forward differences.

On the right side of Fig.3 we show the energy U_∞ as function of temperature. Both the entropy and the energy show a strong increase near T_c. This large increase in entropy and energy is probably due to many-body effects and makes the interpretation of U_∞ as the binding energy of heavy-light meson not very plausible.

2. Conclusions

The free energy gets screened beyond some distance for all temperatures as expected. For small temperature this distance, the effective screening radius, does not depend on the temperature and is about $0.9 fm$. As the temperature increases the effective screening radius decreases. Light quark mass dependence of the screening radius is negligible within our statistical accuracy. We have also identified the entropy contribution to the free energy as well as the internal energy at large distances and found that they show strong increase at T_c. We have found substantial quark mass dependence of the free energy in the vicinity of the transition.

Acknowledgment

This work has been supported by the US Department of Energy under the contract DE-AC02-98CH10886 and by the SciDAC project. It was largely based on the gauge configurations from MILC collaboration; I personally would like to thank Robert Sugar, Doug Toissant, Carleton de Tar and Urs Heller for their help.

References

1. P. Petreczky and K. Petrov, arXiv:hep-lat/0405009. Please refer to it for the full list of references.
2. N. Attig et al., Phys. Lett. B209, 65 (1988)
3. O. Kaczmarek, F. Karsch, E. Laermann, M. Lütgemeier, Phys. Rev. D62, 034021, (2000)
4. O. Philipsen, Phys. Lett. B535, 138 (2002)
5. O. Kaczmarek, F. Karsch, P. Petreczky and F. Zantow, Phys. Lett. B543, 41 (2002)
6. S. Digal, S. Fortunato and P. Petreczky, Phys. Rev. D68, 034008 (2003)
7. O. Kaczmarek, F. Karsch, P. Petreczky and F. Zantow, Nucl. Phys. B (Proc. Suppl.) 129-130, 745 (2004), hep-lat/0309121
8. F. Karsch, E. Laermann and A. Peikert, Nucl. Phys. B605, 579 (2001)
9. O. Kaczmarek et al., hep-lat/0312015
10. T. Blum et al, Phys. Rev. D55, 1133 (1997); J. F. Lagae and D.K. Sinclair, Phys. Rev. D59, 014511 (1999); K. Orginos, D. Toussaint and R.L. Sugar, Phys. Rev. D60, 054503 (1999)
11. C. Bernard *et al.* [MILC Collaboration], Nucl. Phys. B (Proc. Suppl.) 129-130, 626 (2004), hep-lat/0309118
12. C. Bernard *et al.* [MILC Collaboration], Nucl. Phys. B (Proc. Suppl.) **119**, 523 (2003), hep-lat/0209079
13. C. Bernard *et al.* [MILC Collaboration], Nucl. Phys. **A702**, 140 (2002)
14. L.D. McLerran and B. Svetitsky, Phys. Rev. D24, 450 (1981)
15. S. Digal, P. Petreczky and H.Satz, Phys. Lett. B514, 57 (2001)

CONTRIBUTIONS TO TRANSPORT THEORY FROM MULTI-PARTICLE INTERACTIONS AND PRODUCTION PROCESSES

M. E. CARRINGTON*

Department of Physics, Brandon University,
Brandon, Manitoba, R7A 6A9 Canada
and Winnipeg Institute for Theoretical Physics
Winnipeg, Manitoba, Canada

We use the Schwinger-Keldysh technique to study the collision term in the Boltzman equation for a system of quantum scalar fields. We study three three-loop diagrams in ϕ^3 model and four-loop diagrams in ϕ^4 model, and show how these diagrams correspond to multi-particle collisions and production processes.

1. Introduction

This talk is based on work done in collaboration with S. Mrowczynski [1]. In conventional formulations of transport theory one assumes that dissipation processes are governed by binary collisions. However, in systems of relativistic quantum fields at high energy density, both multi-particle interactions and production processes play important roles. The quark-gluon plasma is a prominent example of such a system, see *e.g.* [2,3,4,5,6,7].

We look at self-interacting scalar fields with cubic and quartic interaction terms and use the Schwinger-Keldysh [8,9] formulation of quantum field theory to study the collision term in the Boltzman equation beyond the binary collision approximation. Multi-particle interactions and production processes appear at three-loop level in the ϕ^3 model and at four-loops in the ϕ^4 model. We make the usual definitions for the retarded, advanced

*electronic address: carrington@brandonu.ca

and symmetric Green functions:

$$i\Delta^{+}(x,y) \stackrel{\text{def}}{=} \Theta(x_0 - y_0)\langle[\phi(x),\phi(y)]\rangle \,, \tag{1}$$

$$i\Delta^{-}(x,y) \stackrel{\text{def}}{=} -\Theta(y_0 - x_0)\langle[\phi(x),\phi(y)]\rangle \,, \tag{2}$$

$$i\Delta^{\text{sym}}(x,y) \stackrel{\text{def}}{=} \langle\{\phi(x),\phi(y)\}\rangle \,, \tag{3}$$

where the curly brackets indicate an anticommutator. The retarded, advanced and symmetric self-energies are defined similarly.

2. Transport Equation

Using standard techniques, we derive the transport equation in the limit of weak inhomogeneity and zero-width quasi-particles [1].

$$E_p\Big(\frac{\partial}{\partial t} + \mathbf{v}\cdot\nabla\Big) f(X,\mathbf{p}) + \nabla V(X,\mathbf{p})\,\nabla_p f(X,\mathbf{p}) = \Theta(p_0)C(X,p) \,, \tag{4}$$

where $V(X,p) = \Pi_\delta(X) + \text{Re}\;\Pi^{+}(Xp)$ and $\Pi_\delta(X)$ contains the tadpole contributions. The collision term is written $\Theta(p_0)C(X,p) \equiv C^{+}(X,p)$ and is given by,

$$\begin{aligned} &\Theta(p_0)A(p)\,C(p) \\ &= \frac{1}{4}\Theta(p_0)\Big[\Pi^{\text{sym}}(p)\big(\Delta^{+}(p) - \Delta^{-}(p)\big) - \big(\Pi^{+}(p) - \Pi^{-}(p)\big)\Delta^{\text{sym}}(p)\Big] \,, \end{aligned} \tag{5}$$

where we have defined the spectral function $A(x,y) \stackrel{\text{def}}{=} \langle[\phi(x),\phi(y)]\rangle$.

The calculation of a self-energy graph involves two steps: the calculation of the integrand and the computation of the integral itself. We are primarily interested in the calculation of integrands, which determine the structure of the collision term. We consider the gradient and coupling constant approximations as independent, and choose to investigate higher order terms in the coupling constant approximation, while working at lowest order in the gradient approximation. We use non-interacting Green's functions to perform the perturbative calculations of the relevant self energies, and then replace the non-interacting distribution functions by the interacting ones within the collision term. In the Keldysh representation, all contributions to the collision term have the form of 'cuts' [11]. We introduce the notion 'central cut' to refer to the cut for which all loops contain at least one cut propagator. It can be shown that contributions from non-central cuts can be included as renormalized lower loop terms [1]. Diagrams that contain tadpoles give the same contribution as the corresponding lower loop diagrams, with the addition of effective masses. This type of diagram will not be discussed. Calculations are performed using a *MATHEMATICA* program [10].

We have derived the explicit form of the contributions to the collision term of the transport equation corresponding to the processes: $2 \leftrightarrow 3$ for the ϕ^3 model, and $2 \leftrightarrow 4$ and $3 \leftrightarrow 3$ for the ϕ^4 model. Detailed results are given in reference [1]. As an example, we give below the result for processes of the form $p+q \leftrightarrow l+k+u$. There are eight self-energy graphs that need to be included in the collision term (5). They are shown in Fig. 1. The resulting contribution to the collision term is

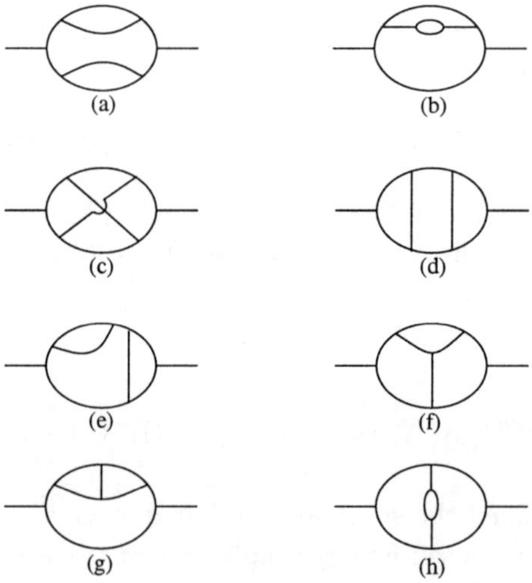

Figure 1. Three-loop contributions to the ϕ^3 self-energy.

$$
\begin{aligned}
&C^{(+)}_{3L-\phi^3}[p+q \leftrightarrow l+k+u] = \\
&\quad \frac{1}{3!}\int \frac{d^3q}{(2\pi)^3 E_q}\int \frac{d^3l}{(2\pi)^3 E_l}\int \frac{d^3k}{(2\pi)^3 E_k}\int \frac{d^3u}{(2\pi)^3 E_u} \\
&\quad \times (2\pi)^4 \delta^{(4)}(\tilde{p}+\tilde{q}-\tilde{l}-\tilde{k}-\tilde{u})\,|M|^2_{3L-\phi^3}[p+q \leftrightarrow l+k+u] \\
&\quad \times \Big[(1+f(\mathbf{p}))\,(1+f(\mathbf{q}))\,f(\mathbf{l})\,f(\mathbf{k})\,f(\mathbf{u}) \\
&\quad - f(\mathbf{p})\,f(\mathbf{q})\,(1+f(\mathbf{l}))\,(1+f(\mathbf{k}))\,(1+f(\mathbf{u}))\Big]\,, \qquad (6)
\end{aligned}
$$

where we have defined $u = p + q + k + l$ and the matrix element squared of the diagrams is shown in Fig. 2 equals

$$
\begin{aligned}
&|M|^2_{3L-\phi^3}[p+q \leftrightarrow l+k+u] = \\
&\quad \frac{g^6}{2^5}\Big| \Delta^-(\tilde{p}+\tilde{q})\big[\Delta^-(\tilde{k}+\tilde{l}) + \Delta^-(\tilde{k}+\tilde{u}) + \Delta^-(\tilde{l}+\tilde{u})\big] \\
&+ \ \Delta^-(\tilde{p}-\tilde{l})\big[\Delta^+(\tilde{q}-\tilde{k}) + \Delta^+(\tilde{q}-\tilde{u}) + \Delta^-(\tilde{k}+\tilde{u})\big] \qquad (7) \\
&+ \ \Delta^-(\tilde{p}-\tilde{k})\big[\Delta^+(\tilde{q}-\tilde{l}) + \Delta^+(\tilde{q}-\tilde{u}) + \Delta^-(\tilde{l}+\tilde{u})\big] \\
&+ \ \Delta^-(\tilde{p}-\tilde{u})\big[\Delta^+(\tilde{q}-\tilde{l}) + \Delta^+(\tilde{q}-\tilde{k}) + \Delta^-(\tilde{l}+\tilde{k})\big] \\
&+ \Delta^+(\tilde{q}-\tilde{l})\Delta^-(\tilde{k}+\tilde{u}) + \Delta^+(\tilde{q}-\tilde{k})\Delta^-(\tilde{l}+\tilde{u}) + \Delta^+(\tilde{q}-\tilde{u})\Delta^-(\tilde{l}+\tilde{k})\Big|^2 .
\end{aligned}
$$

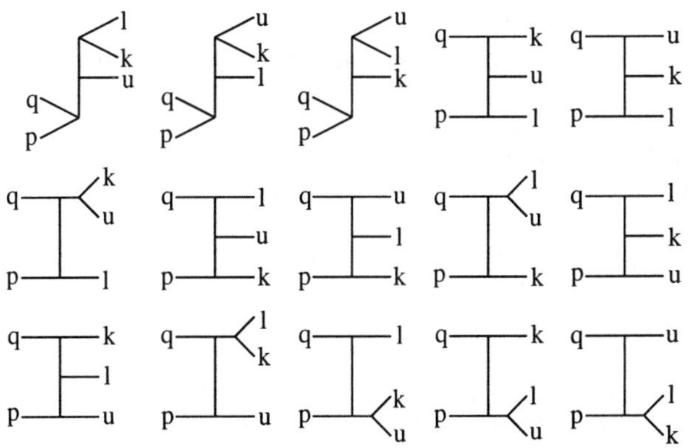

Figure 2. Scattering diagrams that contribute to the process $p + q \leftrightarrow l + k + u$ in the order they appear in Eq. (7).

3. Summary and outlook

We have performed the first systematic derivation of the transport equation of relativistic quantum fields which takes into account multi-particle and production process. Using the Schwinger-Keldysh approach, we have studied scalar fields with cubic and quartic interactions. Multi-particle processes occur at the three-loop level in ϕ^3 theory and at the four-loop level in ϕ^4 theory.

Throughout this paper, we have assumed that quasi-particles are of zero-width and that their four-momenta are on the mass-shell. This assumption is crucial since some processes which occur at one- and two-loop level, like $1 \leftrightarrow 2$ and $1 \leftrightarrow 3$ proceses, are kinematically forbidden for on mass-shell particles.

In the future we plan to study further the multi-particle collision terms derived here. Although the integrals seem to be regular for massive particles, they require very careful analysis, as the transport theory of dense gases or liquids is known to suffer serious problems at the level of multi-particle interactions, see *e.g.* [12,13].

References

1. M.E. Carrington, S. Mrowczynski, hep-ph/0406097.
2. L. Xiong and E. V. Shuryak, Phys. Rev. C **49**, 2203 (1994) [arXiv:hep-ph/9309333].
3. K. Geiger and B. Muller, Nucl. Phys. B **369**, 600 (1992).
4. D. K. Srivastava and K. Geiger, Nucl. Phys. A **647**, 136 (1999) [arXiv:nucl-th/9806050].
5. R. Baier, A. H. Mueller, D. Schiff and D. T. Son, Phys. Lett. B **502**, 51 (2001) [arXiv:hep-ph/0009237].
6. P. Arnold, G. D. Moore and L. G. Yaffe, JHEP **0301**, 030 (2003) [arXiv:hep-ph/0209353].
7. S. M. H. Wong, arXiv:hep-ph/0404222.
8. J. S. Schwinger, J. Math. Phys. **2**, 407 (1961).
9. L. V. Keldysh, Zh. Eksp. Teor. Fiz. **47**, 1515 (1964) [Sov. Phys. JETP **20**, 1018 (1965)].
10. M. E. Carrington, D. f. Hou, A. Hachkowski, D. Pickering and J. C. Sowiak, Phys. Rev. D **61**, 025011 (2000).
11. M. E. Carrington, H. Defu and R. Kobes, Phys. Rev. D **67**, 025021 (2003) [arXiv:hep-ph/0207115].
12. E.M. Lifshitz and L.P. Pitaevskii, *Physical Kinetics* (Pergamon Press, Oxford, 1981).
13. J.P. Hansen and I.R. McDonald, *Theory of Simple Liquids* (Academic Press, London, 1976).

TRANSPORT COEFFICIENTS AND THE 2PI EFFECTIVE ACTION IN THE LARGE N LIMIT*

GERT AARTS AND JOSE M. MARTINEZ RESCO

Department of Physics, The Ohio State University
174 West 18th Avenue, Columbus, OH 43210, USA
aarts@mps.ohio-state.edu, marej@mps.ohio-state.edu

We discuss the computation of transport coefficients in large N_f QCD and the $O(N)$ model for massive particles. The calculation is organized using the $1/N$ expansion of the 2PI effective action to next-to-leading order. For the gauge theory, we verify gauge fixing independence and consistency with the Ward identity. In the gauge theory, we find a nontrivial dependence on the fermion mass.

1. Motivation

In understanding the evolution of quantum fields out of equilibrium, thermalization plays an important role. For long enough times, the dynamics of a typical system can be characterized by transport coefficients, as they control the relaxation of a system towards equilibrium. For example, in heavy ion collisions, the presence of an appreciable shear viscosity would modify the ideal hydrodynamical description[1] of the aftermath of the collision.[2,3]

The nonequilibrium evolution of quantum fields and the subsequent thermalization have been studied using various approximations: mean-field dynamics (such as Hartree or leading-order large N approximations), inhomogeneous mean-field theory, the classical approximation, kinetic theory, dynamics based on truncations of the 2PI or 2PPI effective action, etc. It is clear that any scheme aimed at describing the long-time behavior should, among other things, describe correctly the approach to equilibrium of a typical system. One requirement for an approximation scheme is therefore to yield sensible results for transport coefficients. While it is clear that some approaches (mean-field theory, classical approximation) do not give the correct long-time quantum behavior, for others this might be more

*Based on a talk by G. A. and a poster by J. M. M. R.

involved.

One approach to the dynamics of quantum fields far from equilibrium that has been successful in the past few years[4−15] employs the 2PI effective action.[16] It is therefore interesting to see whether the 2PI framework gives a reliable description of transport coefficients, in the limit where a semi-analytical computation can be carried out, i.e. in a weak coupling or a large N expansion. Below we review how transport coefficients can be computed using the 2PI effective action as an organizational tool.[17] We discuss explicitly two theories in the large N limit: the shear viscosity in the $O(N)$ model[18] and the electrical conductivity and shear viscosity in large N_f QCD with massive fermions.[19]

2. 2PI effective action

The 2PI effective action offers an approach to deal with the infinite hierarchy of correlation functions in field theory, based on a variational principle for one- and two-point functions. For a bosonic field, with vanishing mean field $\langle\phi\rangle = 0$, the effective action is written as[16]

$$\Gamma[G] = \frac{i}{2}\mathrm{Tr}\ln G^{-1} + \frac{i}{2}\mathrm{Tr}\, G_0^{-1}(G - G_0) + \Gamma_2[G], \tag{1}$$

where G_0^{-1} is the free inverse propagator. $\Gamma_2[G]$ contains all two-particle irreducible diagrams without external legs. For fermionic fields the $\frac{1}{2}$'s are replaced by -1's. Although the attention is usually focused on the (one- and) two-point functions, the 2PI effective action is formulated as a generating functional with external sources, similar to the 1PI effective action, and therefore higher order n-point functions are accessible as well. In particular, there is a four-point vertex function obeying a Bethe-Salpeter equation, which in momentum space reads

$$\Gamma^{(4)}(P,K) = \Lambda(P,K) + \frac{1}{2}\int_R \Lambda(P,R)G^2(R)\Gamma^{(4)}(R,K). \tag{2}$$

This equation sums ladder diagrams with a kernel Λ obtained from the second derivative of $\Gamma_2[G]$. It plays therefore a crucial role in the computation of transport coefficients[17] in the case that these are dominated by ladder diagrams, such as in scalar[20,18] and large N_f gauge theories[19] to leading order and in weakly coupled gauge theories to leading logarithmic order.[21]

We now apply this to two theories in the large N limit: the $O(N)$ model and large N_f QCD. We consider the shear viscosity in the $O(N)$ model with massive particles both at weak coupling as well as in the large N limit.

For a weakly-coupled single-component scalar field the shear viscosity was computed some time ago by Jeon.[20] In large N_f QCD, we present the first diagrammatic calculation of the electrical conductivity and shear viscosity for massive fermions. For massless fermions, these transport coefficients were computed before by Moore[22] using kinetic theory.

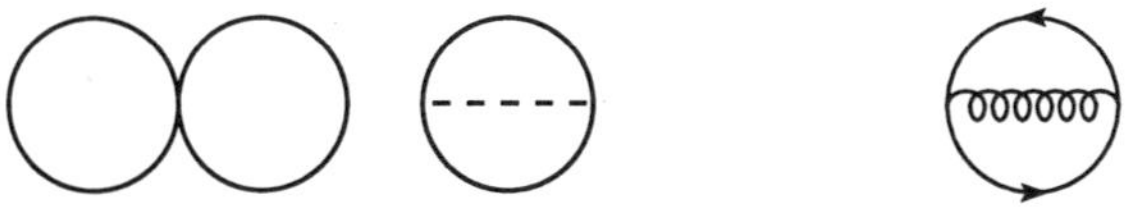

Figure 1. 2PI contribution to the effective action at LO and NLO in the large N limit in the $O(N)$ model (left) and at NLO in large N_f QCD (right). The dashed line in the $O(N)$ model sums the chain of bubbles, see Fig. 2.

The starting point is the 2PI contribution to the effective action in the $1/N$ expansion, shown in Fig. 1. In the $O(N)$ model there is a LO and a NLO contribution.[7,9] In large N_f QCD there is only a contribution at NLO. The chain of bubbles in the $O(N)$ model is indicated with the dashed line (see Fig. 2). In this formulation the scalar and gauge theory are conveniently similar. We emphasize that all results below are generated from the graphs in Fig. 1, indicating the power of the 2PI formalism.

Figure 2. Auxiliary correlator in the $O(N)$ model, summing the chain of bubbles.

Extremizing the effective action yields the Schwinger-Dyson equations for the two-point functions. For the scalar case we find

$$G^{-1} = G_0^{-1} - \Sigma, \qquad D^{-1} = D_0^{-1} - \Pi, \tag{3}$$

where G is the scalar and D the auxiliary correlator. For the gauge theory

$$S^{-1} = S_0^{-1} - \Sigma, \qquad D^{-1} = D_0^{-1} - \Pi, \tag{4}$$

with S the fermion and D the gauge field propagator. The self energies, depending on full propagators, are shown in Fig. 3. In the computation of transport coefficients, it is crucial to use dressed propagators for two reasons: to screen the so-called pinching poles,[23,20] reflecting the dependence on the finite lifetime of quasiparticles due to collisions in the plasma, and to screen the divergences due to the exchange of offshell gauge bosons[24] (this only in the gauge theory).

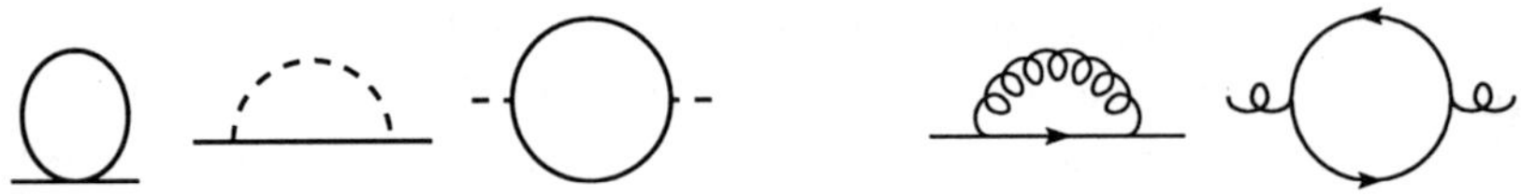

Figure 3. Self energies at LO and NLO in the large N limit in the $O(N)$ model (left) and at NLO in large N_f QCD (right).

Differentiating the self energies yields the rungs that appear in the Bethe-Salpeter equation. We find that some rungs contribute to the transport coefficients at leading order, whereas other rungs are subleading and can be neglected. When considering the Bethe-Salpeter equation for other purposes (e.g. for renormalization[25,26,27]) we expect that all rungs that follow from the effective action might have to be considered.

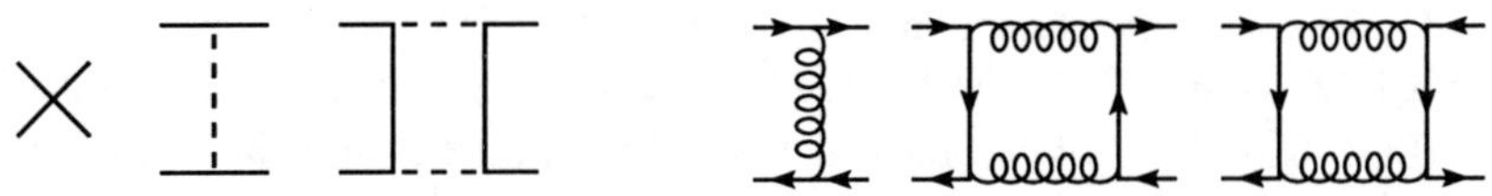

Figure 4. Kernels at LO and NLO in the large N limit in the $O(N)$ model (left) and at NLO in large N_f QCD (right).

In order to distinguish leading from subleading rungs, we need to discuss power counting. In the large N expansion, this is fairly straightforward. Positive powers of N arise from two sources: closed loops of scalar or fermionic lines and scalar or fermionic propagators suffering from pinching poles (pinching poles are screened by the scalar or fermionic thermal width $\Gamma_\mathbf{p}$ and result in contributions $\sim 1/\Gamma_\mathbf{p} \sim N$). Negative powers of N arise from the coupling constants, which are taken to scale as $1/N$ as is usually done in $1/N$ expansions. As a result we find the rungs given in Fig. 4. In the $O(N)$ model, the first point-like rung does not contribute for kinematical reasons. In large N_f QCD, the contribution from generic onshell gauge fields is subleading: the gauge field propagator appears therefore only in rungs and there is no need to consider the thermal width of onshell gauge bosons. The thermal width of an onshell fermion determined by the self energy in Fig. 3 is ill-defined.[23,28] However, the problematic part cancels against part of the contribution from the line diagram in the kernel.[23,21] Finally, we note that the two box diagrams in large N_f QCD differ only in the orientation of the fermion lines; this ensures that Furry's theorem is satisfied.

3. Summation of ladders

To sum the ladder series, we use the technique recently presented by Valle Basagoiti,[21] employing the Matsubara formalism and an effective three-point vertex $\mathcal{D}$. In terms of this vertex, the integral equations are shown in Figs. 5, 6. Note that the propagators in these diagrams are still dressed.

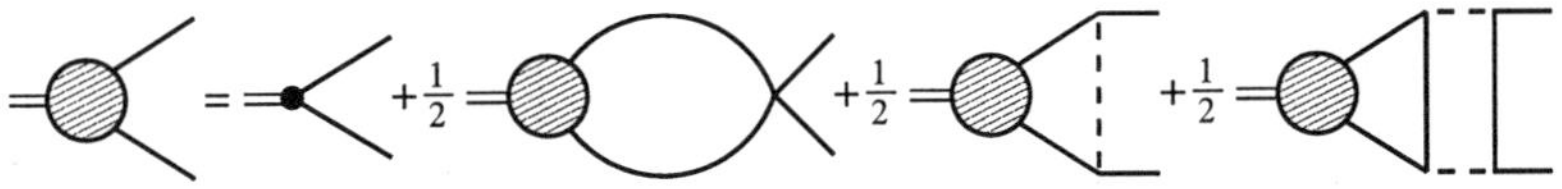

Figure 5. Integral equation for the effective vertex in the $O(N)$ model. The first closed loop on the RHS does not contribute.

We consider these equations in the kinematical configuration special for transport coefficients: the momentum entering on the left is $Q = (q^0 \to 0, \mathbf{0})$, while the momentum entering and leaving on the right is $P = (\omega_\mathbf{p}, \mathbf{p})$, where $\omega_\mathbf{p} = \sqrt{\mathbf{p}^2 + m^2}$ with m the scalar or fermion mass. Transport coefficients are then extracted from the correlator obtained by closing the lines with an insertion of the appropriate current, indicated with the small black dot.[21,17,18,19]

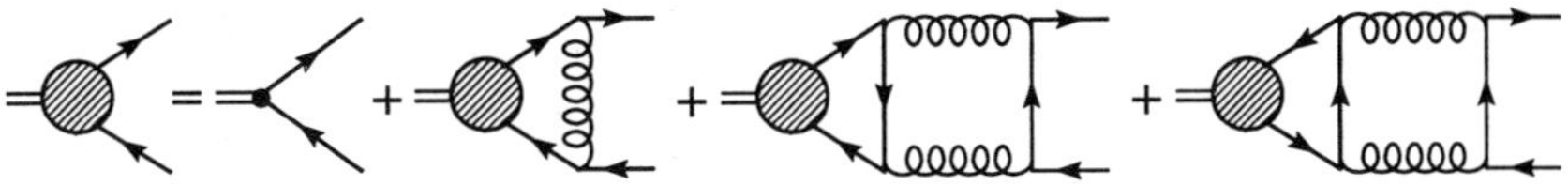

Figure 6. Integral equation for the effective vertex in large N_f QCD. In the case of the electrical conductivity, the last two contributions cancel.

The basic quantity we use to solve the integral equations is the ratio of the effective vertex $\mathcal{D}(p)$ and the thermal width $\Gamma_\mathbf{p}$. For the shear viscosity we define

$$\chi(p) = \frac{p^2}{\omega_\mathbf{p}} \frac{\mathcal{D}(p)}{\Gamma_\mathbf{p}} \times \begin{cases} 1 & O(N) \\ C_F/T_F & \text{large } N_f \text{ QCD} \end{cases} \tag{5}$$

with $C_F = (N_c^2 - 1)/2N_c$, $T_F = 1/2$. Introducing the color factors in this way removes all color factors from the integral equation. The shear viscosity in large N_f QCD then follows from a one-dimensional integral,

$$\eta = -\frac{d_F T_F}{C_F} \frac{2N_f}{15\pi^2} \int_0^\infty dp \, \frac{p^4}{\omega_\mathbf{p}} n_F'(\omega_\mathbf{p}) \chi(p) \tag{6}$$

with $d_F = N_c$. In terms of χ the integral equations read compactly as

$$\mathcal{F}(p)\chi(p) = \mathcal{S}(p) + \int_0^\infty dr\, \mathcal{H}(p,r)\chi(r), \tag{7}$$

where $\mathcal{F}(p) \propto \Gamma_{\mathbf{p}}$, $\mathcal{S}(p)$ is determined by the bare vertex, and $\mathcal{H} = \mathcal{H}_{\rm line} + \mathcal{H}_{\rm box}$ is determined by the rungs. Since $\mathcal{H}(p,r) = \mathcal{H}(r,p)$, the integral equations follow as the extremum condition from the functional

$$Q[\chi] = \int_0^\infty dp \left[\mathcal{S}(p)\chi(p) - \frac{1}{2}\mathcal{F}(p)\chi^2(p) + \frac{1}{2}\int_0^\infty dr\, \mathcal{H}(p,r)\chi(r)\chi(p)\right], \tag{8}$$

which allows for a variational treatment. The value of Q at the extremum is immediately proportional to the transport coefficient.

4. Gauge fixing and Ward identity

In applications of 2PI effective action techniques to gauge theories, gauge fixing and Ward identities have to be considered.[29,30,31] It is therefore interesting to see where gauge fixing parameters appear and why they drop out in the calculation. We use the generalized Coulomb gauge such that the gauge field propagator,

$$D^{\mu\nu} = \Delta_T P_T^{\mu\nu} + \Delta_L g^{\mu 0} g^{\nu 0} + \xi \frac{P^\mu P^\nu}{p^4}, \tag{9}$$

consists of a transverse, a longitudinal, and a gauge fixing piece. The gauge fixing parameter ξ appears in three places:

(1) thermal width. The imaginary part of the fermionic self energy is proportional to the discontinuity (spectral density) of the gauge boson propagator, i.e. to ρ_T, ρ_L. Since the gauge fixing part has no discontinuity, ξ drops out.
(2) line diagram. In the kinematical limit we consider also this contribution is proportional to the discontinuity of the gauge boson propagator and ξ drops out.
(3) box diagram. In the kinematical limit we consider *all* fermion lines are onshell and ξ drops out exactly.

One can also verify that the effective vertex in Fig. 6 and the fermion self energy are related via the Ward identity, again in the kinematical limit we consider. The analysis in this case is in fact easier than for the weakly coupled case,[32] since the contribution to the thermal width arising from soft fermions in the latter is now subleading in the large N_f expansion.[19]

When extending this approach to dynamics far from equilibrium it is necessary to include contributions that could be dropped in the leading order analysis of transport coefficients presented here. The gauge fixing dependence that will be present in that case should be suppressed for sufficiently large N_f.

5. Variational solution

The integral equations are in general too complicated to be solved analytically. In the scalar case however, we found an approximate but surprisingly accurate solution for zero mass and vanishing coupling in the limit of ultrahard momentum $p \gg T$. We find

$$\lim_{p\to\infty} \chi(p) = 6912\pi \frac{N^2}{N+2} \frac{1}{\lambda^2} \frac{p^2}{T^2}. \tag{10}$$

Using this result yields for the shear viscosity in the weakly coupled $O(N)$ model,

$$\eta_\infty = \frac{27648\zeta(5)}{\pi} \frac{N^3}{N+2} \frac{T^3}{\lambda^2} \approx 3041.9 \frac{3N^3}{N+2} \frac{T^3}{\lambda^2}. \tag{11}$$

In order to obtain the full N dependence and not just the leading order behavior $\sim N^2$ in the large N limit, we used the three-loop expansion of the 2PI effective action in the $O(N)$ model.[18] The result for the numerical prefactor is very close to the full results obtained numerically by Jeon[20] (3040) and Arnold, Moore and Yaffe[33] (3033.5) for $N = 1$.

For arbitrary values of the mass and coupling constant (limited by the presence of the Landau pole), we solve the integral equations variationally. The function χ is expanded in a set of trial functions and the remaining integrals (a one-dimensional integral for $\mathcal{S}$, three-dimensional integrals for $\mathcal{F}$ and $\mathcal{H}_{\text{line}}$, and a four-dimensional integral for $\mathcal{H}_{\text{box}}$) are performed numerically. We found that a set of four trial functions suffices.

The shear viscosity in the $O(N)$ model, normalized with the analytical result (11) in the large N limit, is shown in Fig. 7 as a function of renormalized mass at zero temperature for various values of the running coupling constant (the result is renormalization group independent). We find that the shear viscosity increases monotonically with increasing mass. This behavior was also found by Jeon[20] for $N = 1$. Interestingly, we find surprisingly little dependence (apart from that contained in η_∞) on the coupling constant, for all values allowed by the Landau pole.

For large N_f QCD the shear viscosity is shown as a function of fermion mass in Fig. 8, again for several values of the effective coupling constant

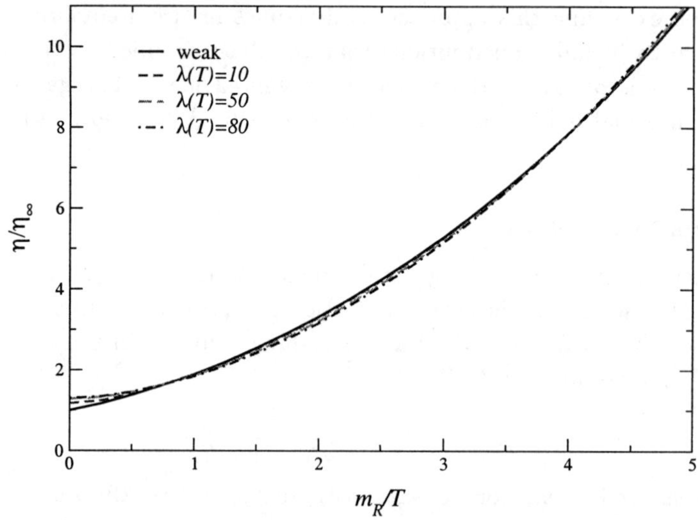

Figure 7. Shear viscosity in the $O(N)$ model vs. the renormalized mass at zero temperature for various values of the coupling constant $\lambda(\mu = T)$.

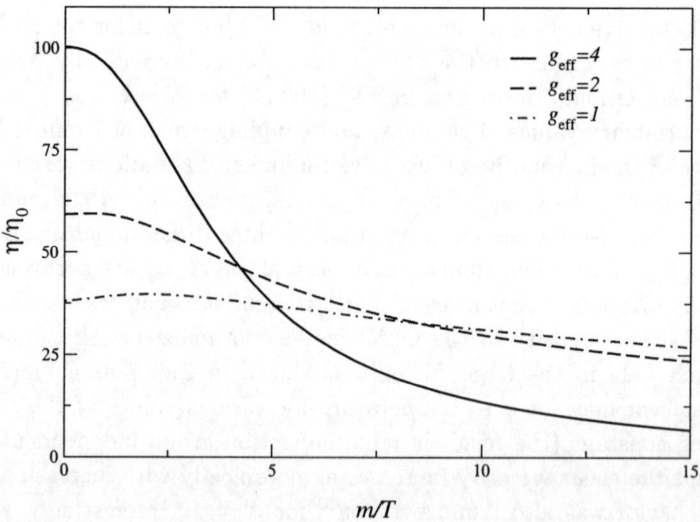

Figure 8. Shear viscosity in large N_f QCD vs. the fermion mass for various values of the effective coupling constant $g_{\rm eff}(\mu = \pi \exp^{-\gamma_E} T)$.

g_{eff} ($g_{\text{eff}}^2 = T_F g^2 N_f$). The viscosity is normalized with $\eta_0 = d_F/(T_F C_F) \times T^3/g^4$. In this case we observe a nontrivial dependence on the mass. After a slight increase, we find that the viscosity decreases with increasing mass. This behavior is caused by longitudinal gauge bosons below the lightcone. In the limit of very large mass it can be shown[34] that the viscosity goes to zero.[a] We also find a much stronger dependence on the coupling constant, compared to the scalar theory.

6. Conclusion

The diagrammatic calculation of transport coefficients in the $O(N)$ model and large N_f QCD is well organized when the 2PI effective action is used: all necessary summations are generated by the few diagrams in Fig. 1. In the scalar theory the shear viscosity increases monotonically with mass and has a dominant $1/\lambda^2$ dependence on the coupling constant. In the gauge theory, on the other hand, the dependence on the coupling constant and the fermion mass is highly nontrivial, due to longitudinal gauge bosons below the lightcone.

We emphasize that understanding transport coefficients diagrammatically provides necessary insight in the dynamics of quantum fields out of equilibrium. Our results provide further support for the applicability of truncations of the 2PI effective action to nonequilibrium QFT.

Acknowledgments

We thank Guy Moore for discussions. This work was supported by the DOE (Contract No. DE-FG02-01ER41190 and DE-FG02-91-ER4069), the Basque Government and in part by the Spanish Science Ministry (Grant FPA 2002-02037) and the University of the Basque Country (Grant UPV00172.310-14497/2002).

References

1. See e.g. P. F. Kolb and U. Heinz, in Hwa, R.C. (ed.) et al.: *Quark gluon plasma 3*, pp. 634-714 [nucl-th/0305084].
2. D. Teaney, Phys. Rev. C **68** (2003) 034913.
3. A. Muronga and D. H. Rischke, nucl-th/0407114.
4. J. Berges and J. Cox, Phys. Lett. B **517**, 369 (2001).

[a]This is not in contradiction with the conjectured lower bound on the viscosity/entropy ratio,[35] since the entropy goes to zero exponentially.

5. B. Mihaila, F. Cooper and J. F. Dawson, Phys. Rev. D **63**, 096003 (2001); *ibid.* **67** (2003) 051901; *ibid.* **67** (2003) 056003.
6. G. Aarts and J. Berges, Phys. Rev. D **64**, 105010 (2001).
7. J. Berges, Nucl. Phys. A **699**, 847 (2002).
8. G. Aarts and J. Berges, Phys. Rev. Lett. **88**, 041603 (2002).
9. G. Aarts, D. Ahrensmeier, R. Baier, J. Berges and J. Serreau, Phys. Rev. D **66**, 045008 (2002).
10. J. Berges and J. Serreau, Phys. Rev. Lett. **91** (2003) 111601.
11. J. Berges, S. Borsányi and J. Serreau, Nucl. Phys. B **660** (2003) 51.
12. B. Mihaila, Phys. Rev. D **68**, 036002 (2003).
13. S. Juchem, W. Cassing and C. Greiner, Phys. Rev. D **69** (2004) 025006; nucl-th/0401046.
14. D. J. Bedingham, Phys. Rev. D **69**, 105013 (2004).
15. T. Ikeda, Phys. Rev. D **69**, 105018 (2004).
16. J. M. Cornwall, R. Jackiw and E. Tomboulis, Phys. Rev. D **10** (1974) 2428.
17. G. Aarts and J. M. Martínez Resco, Phys. Rev. D **68** (2003) 085009.
18. G. Aarts and J. M. Martínez Resco, JHEP **0402**, 061 (2004).
19. G. Aarts and J. M. Martínez Resco, in preparation.
20. S. Jeon, Phys. Rev. D **52**, 3591 (1995).
21. M. A. Valle Basagoiti, Phys. Rev. D **66**, 045005 (2002).
22. G. D. Moore, JHEP **0105**, 039 (2001).
23. V. V. Lebedev and A. V. Smilga, Physica A **181**, 187 (1992).
24. G. Baym, H. Monien, C. J. Pethick and D. G. Ravenhall, Phys. Rev. Lett. **64** (1990) 1867.
25. H. van Hees and J. Knoll, Phys. Rev. D **65**, 025010 (2002); *ibid.* 105005 (2002).
26. J. P. Blaizot, E. Iancu and U. Reinosa, Phys. Lett. B **568** (2003) 160; Nucl. Phys. A **736**, 149 (2004).
27. F. Cooper, B. Mihaila and J. F. Dawson, hep-ph/0407119.
28. J. P. Blaizot and E. Iancu, Phys. Rev. D **55**, 973 (1997).
29. A. Arrizabalaga and J. Smit, Phys. Rev. D **66**, 065014 (2002).
30. E. Mottola, in *Proceedings of SEWM2002*, Heidelberg, Germany, 2-5 Oct 2002 [hep-ph/0304279].
31. M. E. Carrington, G. Kunstatter and H. Zaraket, hep-ph/0309084.
32. G. Aarts and J. M. Martínez Resco, JHEP **0211**, 022 (2002).
33. P. Arnold, G. D. Moore and L. G. Yaffe, JHEP **0011**, 001 (2000).
34. G. D. Moore, private communication.
35. P. Kovtun, D. T. Son and A. O. Starinets, JHEP **0310** (2003) 064.

THERMAL FEATURES FAR FROM EQUILIBRIUM: PRETHERMALIZATION

SZABOLCS BORSÁNYI

Institute of Theoretical Physics, University of Heidelberg, Philosophenweg 16, Heidelberg 69120, Germany

The phenomenon of prethermalization and the subsequent steps of thermalization are analyzed in the framework of the chiral quark model. We solve the quantum equations of motion of the field theory derived from the 2PI effective action and study the time scales of equilibration. We find that already after a 0.6 fm/c long period of time some equilibrium features appear, even though the system is still far from equilibrium. This might be an ingredient for understanding the success of ideal hydrodynamic description.

1. Introduction

One of the crucial issues in heavy ion physics is the early thermalization of the excited quark and gluon fields: The extreme success of hydrodynamic calculations started earlier than 1 fm/c require the theoretical establishment of the early existence of an equation of state.[1]

The proper description of equilibration is a longstanding challenge in particle physics. Kinetic theories are popular tools for studying the series of incoherent collisions. By neglecting the off-shell processes the easiest way is to follow the elastic collisions only, but these deliver thermalization times by orders of magnitudes larger than desired. Equilibration may be enhanced by inelastic scatterings so that it completes before chemical freeze-out[2]. This way, thermalization times reduce to 2-3 fm/c, which is still longer than the early use of flow equations would require[3].

Recent field theoretical developments enable the nonequilibrium treatment of quantum fields by performing a systematic ladder resummation[4]. The available numerical techniques and resources enable us to solve the time evolution of 3+1 dimensional quantum fields in a self-consistent approximation scheme. These methods are not limited to scalar fields any more, nowadays the 3+1 dimensional chiral quark model (with two quark flavors and 4 scalars) may also be routinely solved[5].

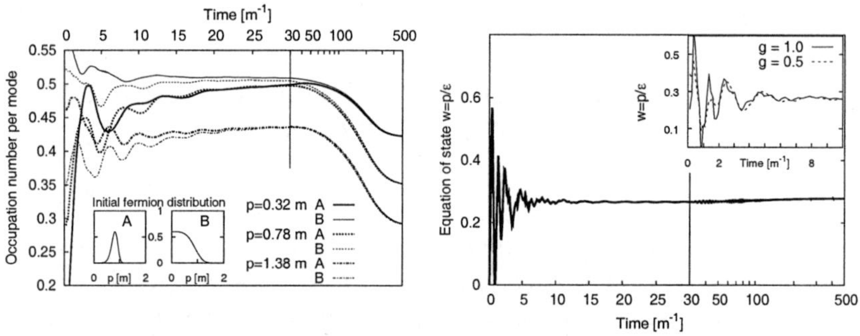

Figure 1. *Left:* The fermionic occupation number $n^{(f)}(t;p)$ for three different momentum modes as a function of time. The evolution is shown for two different initial conditions with *same* energy density. The dynamics becomes rather quickly insensitive to the initial distributions (A) and (B) displayed in the insets – much before the modes settle to their final values. *Right:* The ratio of pressure and energy density as a function of time. One observes that an approximately time-independent equilibrium form of the equation of state builds up very early. The inset demonstrates for two different couplings that this prethermalization time is independent of the interaction details. The insensitivity of bulk quantities to the late-time thermalization scale is in sharp contrast to mode-by-mode quantities. The long-time behavior is shown on a logarithmic scale for $t \geq 30\,m^{-1}$.

Thermalization time is not a single number, of course. Different degrees of thermalization (LTE, kinetic and chemical equilibration) can have very different time scales ranging from the rapid convergence of bulk quantities to the slow equilibration of energy spectra[6].

The model we consider is built up from two massless fermionic degrees of freedom and four scalars interacting according to the following Lagrangian:

$$L = \left\{ \bar{\psi} i \partial\!\!\!/ \psi + \tfrac{1}{2}\left[(\partial\sigma)^2 + (\partial\pi)^2\right] + \tfrac{g}{N_f}\bar{\psi}\left[\sigma + i\gamma_5\tau^a\pi^a\right]\psi - V(\sigma^2+\pi^2)\right\},$$

where τ^a denote the standard Pauli matrices. We consider a quartic scalar self-interaction $V(\sigma^2+\pi^2) = m^2\left(\sigma^2+\pi^2\right)/2 + \lambda\left(\sigma^2+\pi^2\right)^2/(4!N_f^2)$. The employed couplings are taken to be $g = 1$ and $\lambda = 1$. Though these couplings are smaller than heavy ion applications would require, we will be able to draw conclusions for the physical values since our main result will turn out to be coupling independent. The smaller-than-natural coupling leads to a spectacular separation of the various time scales.

We use the two-particle irreducible (2PI) effective action to two-loop order, which includes direct scattering as well as off-shell and memory effects[5]. We numerically solve the nonequilibrium gap equation and obtain the evolution of the propagator $G(t_1, t_2, |\vec{x}_1 - \vec{x}_2|)$ for each degree of freedom. To simplify our equations we assume homogeneity and isotropy in space.

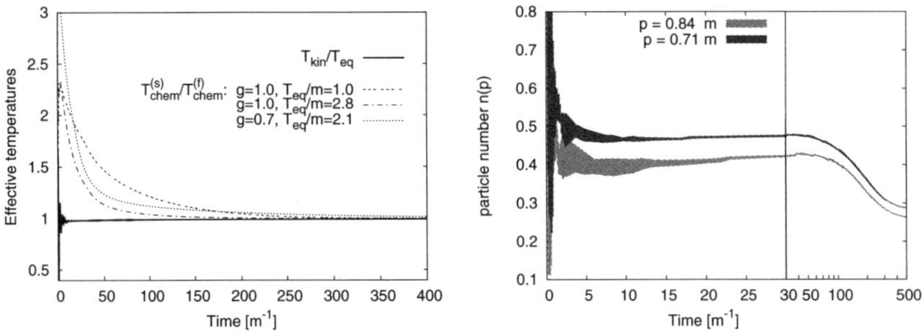

Figure 2. *Left:* The solid line shows the kinetic temperature $T_{\text{kin}}(t)$ as a function of time. In contrast to the mode-by-mode quantities, the kinetic temperature, which includes all momentum modes, exhibits the phenomenon of prethermalization. Also shown is the ratio of the scalar and the fermion chemical temperatures, $T^{(s)}_{\text{chem}}(t)/T^{(f)}_{\text{chem}}(t)$, for different values of the coupling g and T_{eq}. One observes hat chemical equilibration does not happen on the prethermalization time scale. *Right:* On the damping time scale a non-equilibrium variant of the fluctuation–dissipation relation is dynamically realized. This gives a particle number definition even for strongly coupled systems. The grey bands demonstrate the particle numbers and the consistency of their determination.

2. Time scales of thermalization

Thermalization is loss of initial information. Our dynamics is conservative, still, virtually, we loose information if we do not read out all the internal variables but only the time-local propagator $G(t_1 \approx t_2, |\vec{x}_1 - \vec{x}_2|)$. This time-local propagator stores the information on the particle number distribution[5]. The evolution of these occupation numbers for three different momenta is shown in Fig. 1left. We also show the convergence for two different initial spectra with equal energy density. Although the details of the initial distribution are mainly washed out in the rather early evolution $(15|m|^{-1})$, the spectrum is still far from equilibrium. It takes by an order of magnitude longer to approach the thermal spectrum (for small coupling).

We may split the observed time scales to three epoches. The separation of these scales we find for perturbative couplings only, in a strong coupling application the first two might coincide. (The time values correspond to the parameters used in the presented plots.)

(1) *Prethermalization* — loss of coherence $(t < 5|m|^{-1})$

Prethermalization is a universal far-from-equilibrium phenomenon, its time scale is given by the relevant scales of the dynamics (temperature, mass) but it is independent of the coupling, as shown in Fig. 1right.

Contrary to the mode-by-mode occupation numbers (Fig. 1left) the bulk, averaged quantities are often insensitive to the details of the spectrum as it is also observed in classical field theory [7].

We find that this short time is enough for very rapid establishment of an almost constant ratio of pressure over energy density (Fig. 1right), as well as a kinetic temperature based on average kinetic energy (Fig. 2left, for definitions see Ref.[6]). These are important ingredients of the early equation of state. This level of equilibration is one of the important preconditions of the use of hydrodynamic equations.

(2) *Damping* — decay of individual modes ($t < 15|m|^{-1}$)

The individual excitations are washed out so that a smooth but still far-from-continuum spectrum is formed. We find that this relaxation time agrees with the inverse width ($\gamma(\vec{p};t)$) of the nonequilibrium spectral function[5].

We know that the damping time (obtained from the imaginary part of the self energy or from the width of the spectral function) is the decay rate of an individual signal excited over the ensemble[8]. This relaxation time may be much shorter than thermalization time[9]. In the leading order perturbation theory relaxation time is proportional to the coupling-squared. We also see that this relaxation time scale is not a subject of spectacular evolution.

Within this time scale the generically independent symmetric propagator (F) and spectral function (ρ) evolve to fulfill a relation (in Wigner coordinates): $F(\omega,\vec{p};t) = \pm(n(\omega;t)+\frac{1}{2})\rho(\omega,\vec{p};t)$, which is known as fluctuation–dissipation relation in equilibrium. Here the nontrivial issue is the momentum ($\vec{p}$) independence of $n(\omega;t)$. This enables us to translate it to $n(\vec{p};t)$ using the dispersion relation defined by the peak of the spectral function. The existence of the relation above provides a particle number definition $n(\vec{p};t)$ not tainted by quasi-particle picture.[10] Fig 2right shows by shaded bands how consistently the particle number $n(\vec{p};t)$ is determined from the evolution of different $\vec{k}$ modes with $\omega_k \approx \omega_p \pm \gamma_p$.

(3) *Thermalization* — loss of spectral information ($t \lesssim 500|m|^{-1}$)

Kinetic equilibration is a result of the collective evolution of the particle spectra. At the same time the particle abundances in the various species are equilibrated as well. Whether this longer epoch fits into the short time range before chemical freeze-out is still a matter of discussion.

3. Discussion

Direct heavy ion applications require higher couplings than in the example above. Tuning the coupling to $g = 2.5 - 3$ one observes that both relaxation and thermalization time shrinks considerably, approaching the prethermalization time. The elementary oscillators are then over-damped, and the estimated prethermalization time will also give a hint for the damping time scale. With these couplings one can easily obtain thermal spectra in the fermionic sector within $1 - 3$ fm/c.

We find that prethermalization time can be estimated by the inverse average kinetic temperature[6]. In particular, we obtained the following quantitative relation: $T\,t_{pt} \approx 2 - 2.5$. If one attempts to use the concept of prethermalization for the case of heavy ion collisions one has to replace this temperature by the relevant scale of the dynamics. This we assume to be the saturation scale: $Q_s \approx 1$ GeV at RHIC energies. Estimating the prethermalization time by 3 times $1/Q_s$ one has: $t_{\text{prethermalization}} \approx 0.6\text{fm}/c$ This is the very close to the instant that is usually taken as initial time for the hydrodynamic evolution[11,12].

If the complete local equilibration cannot be established, it is worth thinking if this partial equilibration can already establish the use of flow equations. We do believe that the study of partial equilibration can help us overcome the discrepancy between the long thermalization time predicted by kinetic theories and the short times used by hydrodynamic models.

The author acknowledges the fruitful discussions and collaborations on this work with J. Berges, J. Serreau and C. Wetterich.

References

1. P. F. Kolb, arXiv:nucl-th/0407066, contribution to this meeting.
2. R. Baier, A. H. Mueller, D. Schiff and D. T. Son, Phys. Lett. **B502** (2001) 51; Phys. Lett. **539** (2002) 46.
3. J. Serreau, Nucl. Phys. **A715**, 805c (2003)
4. For a recent review see J. Berges and J. Serreau, arXiv:hep-ph/0302210.
5. J. Berges, S. Borsanyi and J. Serreau, Nucl. Phys. B **660** (2003) 51
6. J. Berges, S. Borsanyi and C. Wetterich, "Prethermalization," arXiv:hep-ph/0403234, to appear in Phys. Rev. Lett.
7. S. Borsanyi, A. Patkos and D. Sexty, Phys. Rev. D **68** (2003) 063512
8. S. Borsanyi and Z. Szep, Phys. Lett. B **508** (2001) 109
9. G. Aarts, G. F. Bonini and C. Wetterich, Nucl. Phys. B **587** (2000) 403
10. J. Berges, Sz. Borsányi C. Wetterich, in preparation
11. P. F. Kolb, J. Sollfrank and U. W. Heinz, Phys. Rev. C **62** (2000) 054909
12. T. Hirano and Y. Nara, arXiv:nucl-th/0404039;

QCD PHASE DIAGRAM AT SMALL BARYON DENSITIES FROM IMAGINARY μ: STATUS REPORT

O. PHILIPSEN*

Dept. of Physics and Astronomy, University of Sussex,
Brighton BN1 9QH, UK
E-mail: o.philipsen@sussex.ac.uk

PH. DE FORCRAND

ETH, CH-8093 Zürich, Switzerland, and
CERN, Physics Dept. , TH Unit, CH-1211 Geneva 23, Switzerland
E-mail: forcrand@phys.ethz.ch

We summarize our recent results for the (T, μ)-phase diagram of $N_f = 3$ QCD. Finding a strong variation of the critical endpoint μ_c with the quark mass m, we point out that an endpoint within reach of experiment requires fine-tuned quark masses. We further discuss the strategy and first data towards extending our results to the physical $N_f = 2 + 1$ theory.

1. Introduction

Currently, several approaches are used to side-step the sign problem of lattice QCD in attempts to compute the low density QCD phase diagram. These are based on multi-parameter reweighting [1], Taylor expansions [2] and analytic continuation from simulations at imaginary μ [3], for which there is no sign problem [4]. While reweighting is exponentially difficult on large volumes, calculation of Taylor coefficients becomes increasingly complicated with growing order. By contrast, a simulation at imaginary $\mu = i\mu_I$ is no more difficult than one at $\mu = 0$. Moreover, it is the only approach for which two simulation parameters are varied, resulting in uncorrelated statistical errors for different values of μ_I. Monte Carlo results for imaginary μ thus involve no approximations. Within the circle of analyticity around $\mu = 0$, they can be fitted by truncated Taylor series in $(\mu_I/T)^2$ and

*Talk given by O.P., PPARC Advanced Fellow

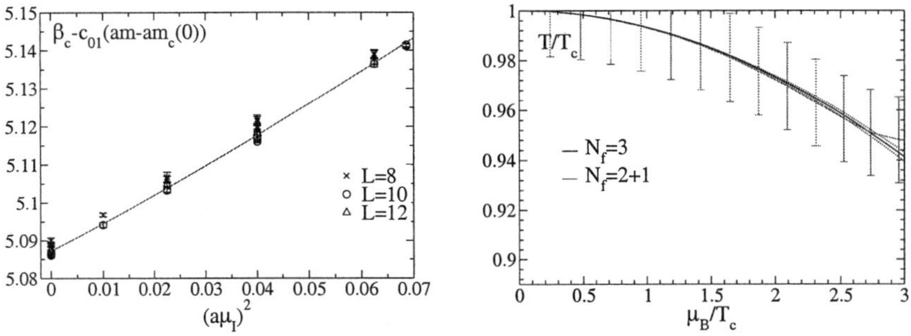

Figure 1. Left: The pseudo-critical gauge coumpling for various quark masses and lattice sizes as a function of imaginary $\mu = i\mu_I$. Right: $N_f = 3$ [5] and $N_f = 2+1$ [1] results.

analytically continued to real μ. While truncating the series introduces an approximation, it is well controlled because it can be compared to the full result at imaginary μ.

The task for lattice simulations is to first map out the phase diagram in the $(m_{u,d}, m_s)$ quark mass plane at $\mu = 0$, i.e. to find the critical line $m_s^c(m_{u,d})$ of second order transitions, separating the first order region from the crossover region. In a second step, we wish to know the surface emanating from this line in the $\mu \neq 0$ direction [6]. So far, even at $\mu = 0$ we know only one point of the critical line, corresponding to degenerate quark masses, $m_c(\mu = 0), N_f = 3$, and only for one coarse lattice spacing, $N_t = 4$. In a recent paper we have computed the $\mu \neq 0$ phase diagram for the $N_f = 3$ theory and determined the quark mass dependence of the critical endpoint [5]. Here we summarize these results and extend them to the physical case of $N_f = 2 + 1$. The general properties of QCD with complex chemical potential as well as analyticity properties of observables are discussed in the literature [3,5].

2. The pseudo-critical line for $N_f = 3$

For $N_f = 3$, we computed the pseudo-critical gauge couplings at the phase boundary, and their dependence on imaginary chemical potential as well as the quark mass. The data are shown in Figure 1 for different lattice sizes. They are well fitted by a polynomial of degree one in the (am) and two in $(a\mu_I)^2$. We have explicitly checked that additional polynomial terms are statistically insignificant. The figure shows that, to good approximation, the leading μ_I^2-term suffices over the interval considered, but our accuracy is good enough to also determine the next-to-leading coefficient. Analytic

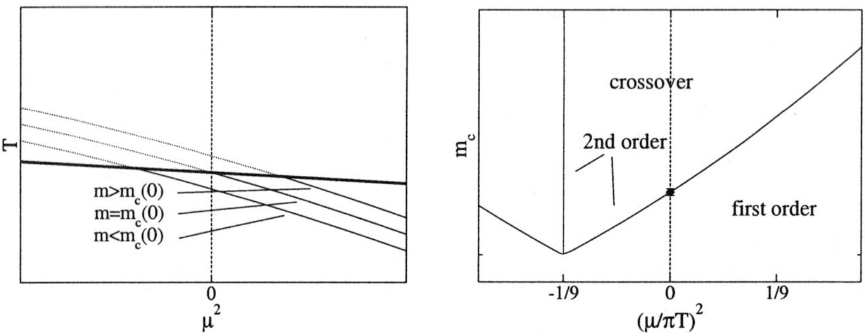

Figure 2. Left: Phase diagram for different quark masses, dotted lines denote crossover. The bold curve $T^*(\mu^2)$ represents the line of endpoints. Right: Lines of critical quark mass separating first order and crossover regions.

continuation and conversion to physical units yields

$$\frac{T_c(\mu,m)}{T_c(\mu=0,m_c(0))} = 1 + 1.94(2)\left(\frac{m-m_c(0)}{\pi T_c(0,m_c)}\right)$$
$$+0.602(9)\left(\frac{\mu}{\pi T_c(0,m_c)}\right)^2 + 0.23(9)\left(\frac{\mu}{\pi T_c(0,m_c)}\right)^4 . \quad (1)$$

Note that the natural expansion parameters are $m/(\pi T), \mu^2/(\pi T)^2$, leading to Taylor coefficients of order one. This is understood in terms of the finite temperature Matsubara modes $\sim \pi T$, which set the energy scale. Since the strange quark mass and its difference with $m_c(0)$ are small compared to πT_c, the critical temperature is fairly insensitive to quark masses in the light quark regime. It is then sensible to compare our $N_f = 3$ results with the $N_f = 2+1$ curve of Fodor and Katz [1] (Figure 1 right), and excellent agreement is found. Note that $T_c(\mu)$ decreases only at the percent level over the whole range of baryon chemical potentials considered here, which thus has a very weak effect. This is a natural consequence of the smallness of the effective expansion parameter with Taylor coefficients of order one.

3. The critical endpoint and its quark mass dependence

In the three flavor theory, we have a 3d parameter space, (T, μ, m). The pseudo-critical temperature, $T_c(\mu, m)$, represents a surface in this space. On this surface there is a line of critical points, $T^*(\mu^2) = T_c(\mu^2, m_c(\mu))$, separating first order transitions from crossover. Projections of this situation to the $(T, \mu^2), (m, \mu^2)$ planes are shown in Figure 2. As the quark mass is increased from $m_c(0)$, the critical endpoint moves to a finite μ_c,

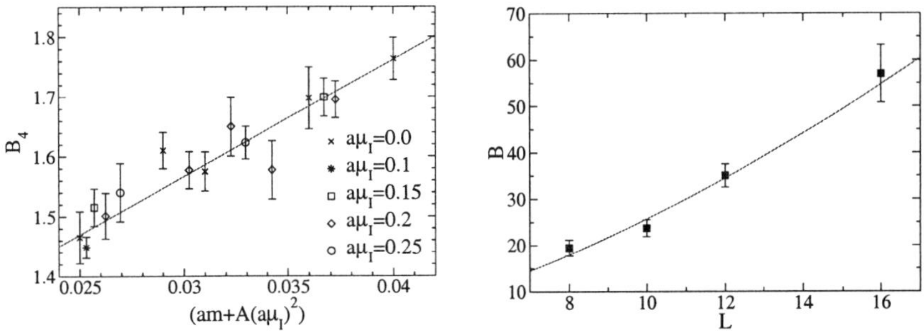

Figure 3. Left: The Binder cumulant on an L=8 lattice. Right: Finite size scaling of the fit parameter B. The line corresponds to 3d Ising scaling.

while a decrease moves it to imaginary μ_c. The change of m_c with μ, or vice versa, is best seen in the (m, μ^2)-plane, where it forms a critical line separating regions of first order transitions and crossover. We determine this line around $\mu = 0$ by measuring the Binder cumulant of $\bar{\psi}\psi$. In the infinite volume limit it assumes a universal value at a 3d Ising critical point,

$$B_4(m_c, \mu_c) = \frac{\langle(\delta\bar{\psi}\psi)^4\rangle}{\langle(\delta\bar{\psi}\psi)^2\rangle} \to 1.604, \quad V \to \infty, \tag{2}$$

while it is smaller or larger in first order and crossover regions, respectively. Data for various quark masses, as well as its universal finite size scaling, are shown in Figure 3. Fitting the data to a Taylor expansion about the critical point, $B_4(am, a\mu) = 1.604 + B\left(am - am_c(0) + A(a\mu)^2\right) + \ldots$, the parameter A represents the first coefficient of the series

$$\frac{m_c(\mu)}{m_c(\mu = 0)} = 1 + 0.84(36)\left(\frac{\mu}{\pi T}\right)^2 + \ldots \tag{3}$$

Like for $T_c(\mu)$, the leading Taylor coefficient is of order one. This implies that the change in $m_c(\mu)$ is very weak up to chemical potentials of $\mu_B = 3\mu \sim 500$ MeV. Conversely, $\mu_c(m)$ is extremely sensitive to the precise value of the quark mass. We believe that this statement holds for all discretizations, because the μ-independent multiplicative mass renormalization drops out of the ratio in (3), which should have only additive $O(a^2)$ corrections. This means that for growing quark masses $m > m_c(0)$, the critical point $\mu_c(m)$ will rapidly shoot to very large values. Hence a small μ_c requires a fine-tuned quark mass.

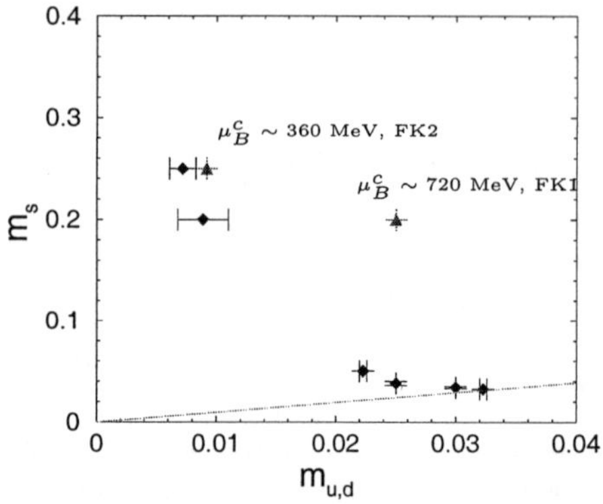

Figure 4. Measured critical line separating 1.O. and crossover for $N_f = 2+1$ at $\mu = 0$ (diamonds), and finite μ-endpoints of Fodor and Katz [1] (triangles).

4. Towards the $N_f = 2+1$ phase diagram

In the case of physical interest, we have a 4d parameter space $(T, \mu, m_{u,d}, m_s)$. The critical surface can be determined in a two-step procedure. First, we need to complete the $(m_{u,d}, m_s)$ phase diagram at $\mu = 0$, i.e. find the line $m_s^c(m_{u,d}, \mu = 0)$. The results of this step are shown in Figure 4. As the light quark mass is decreased, strongly non-linear behavior is encountered, and linear extrapolations from the $N_f = 3$ result to the physical point are clearly ruled out. In a second step in progress, we compute the change of this line with finite μ_I, leading to the critical surface $m_s^c(m_{u,d}, \mu)$. It is intriguing to consider the simulation point of Fodor and Katz [1] for physical quark mass values (FK2) in this diagram. This point is only slightly displaced from our critical line $m_s^c(m_{u,d}, \mu = 0)$, but corresponds to $\mu_c \sim 120$ MeV. This is consistent with our observation of a rapid change of μ_c for small variations of the quark masses. Since quark masses are very cut-off sensitive, one thus has to expect large shifts of the critical point in physical units as the continuum limit is approached.

References

1. Z. Fodor and S. D. Katz, JHEP **0404**, 050 (2004); JHEP **0203**, 014 (2002).
2. C. R. Allton *et al.*, Phys. Rev. D **66**, 074507 (2002).
3. P. de Forcrand and O. Philipsen, Nucl. Phys. B **642**, 290 (2002); M. D'Elia and M. P. Lombardo, Phys. Rev. D **67**, 014505 (2003).
4. For a comparative review, see E. Laermann and O. Philipsen, Ann. Rev. Nucl. Part. Sci. **53**, 163 (2003).
5. P. de Forcrand and O. Philipsen, Nucl. Phys. B **673**, 170 (2003).
6. See the contribution of C. Schmidt, these proceedings.
7. F. Karsch, E. Laermann and C. Schmidt, Phys. Lett. B **520**, 41 (2001).

TWO LOOP RENORMALISATION OF THE MAGNETIC COUPLING IN HOT QCD AND SPATIAL WILSON LOOP

P. GIOVANNANGELI

Centre de Physique Théorique
163, avenue de Luminy,
13288, Marseille cedex 09, France
E-mail: giovanna@cpt.univ-mrs

With the known two loop renormalisation of the magnetic coupling, the 4D result of the spatial Wilson loop are compared to the prediction from the magnetostatic sector.

1. Effective theories for hot QCD

Perturbation theory in hot QCD suffer from infrared (IR) divergences. These divergences can be best confronted by constructing effective field theories for the low-energy dynamics. For the partition function and other static observables, the relevant theory is the Dimensionally Reduced (DR) effective theory [1,2]. The main idea of DR is to separate systematically the physical scales of the quark-gluon plasma and to construct, for each of these scales, an effective action. The procedure is as follows:

- At distances of order $1/T$, physics is described by the non-static modes. But, if one wants to go to larger distances, one meets with IR divergences, due to massless static modes.
- At distances of order $1/gT$, however, electrostatic modes get screened by the Debye mechanism, like in a classical electromagnetic plasma. This thermal mass is going to regulate the electric IR divergences. At these scale, QCD is described by a 3D electrostatic Lagrangian L_E that reads:

$$L_E = Tr(\vec{D}(A)A_0)^2 + m_E^2 Tr A_0^2 + \lambda_E (Tr(A_0^2))^2 + \\ + \bar{\lambda}_E \big(Tr(A_0)^4 - \frac{1}{2}(Tr A_0^2)^2\big) + \frac{1}{2} Tr F_{ij}^2 + \delta L_E. \qquad (1)$$

with as parameters $g_E(g,T)$, $m_E(g,T)$, ... and δL_E represent higher orders operators of relative order g^4.

- Furthermore, at distances of order $1/g^2T$, magnetostatic modes are the dominant ones. At these distances QCD is described by a confining three-dimensional (3D) Yang-Mills theory L_M that is non-perturbative.

$$L_M = \frac{1}{2}TrF_{ij}^2 + \delta L_M. \tag{2}$$

with a gauge coupling $g_M(g_E, m_E, ...)$. δL_M is of relative order g^3.

If one wants to compute the thermal average of an observable, then each of these three scales will have its own contribution to the final result.

From the relative order of the truncation in (1) and (2) we see that g_M is needed to relative order g^2. In the next section, I will briefly present the result for the coupling g_M of the 3D magnetostatic Yang-Mills theory L_M at two loop order.

2. Two-loop determination of the magnetostatic action

The basic idea behind the effective actions eqns (1) and (2) is that one can compute with both in the region of momenta $p \sim g^2T$. To know what the parameters of the latter are in terms of those of the former requires computing two-point functions, three point functions etc. in both theories and match them. In the matching the diagrams of the pure 3d Yang-Mills theory drop out.

Here we will follow a well-known shortcut [3] by introducing a background field B_i in L_E:

$$\begin{aligned} \vec{A} &= \frac{1}{g_E}\vec{B} + \vec{Q} \\ A_0 &= g_E Q_0. \end{aligned} \tag{3}$$

We calculate the fluctuations around the background in a path integral:

$$\exp -\frac{1}{g_M^2}S_M(B) = \int DQ_0 DQ_i \exp\big(- S_E - \frac{1}{\xi}Tr(D_iQ_i)^2\big). \tag{4}$$

We added a general background gauge term. The resulting action $S_M(B)$ is gauge invariant to all loop orders and the renormalization of the coupling is identified from the background field two point function at a momentum $p = O(g^2T)$.

$$\exp\left(-\frac{1}{g_M^2}S_M(B)\right) = \exp\left(-\frac{1}{g_E^2}S_M(B)\right)\left(1+(F_1^{tr}+F_2^{tr}+...)S_M(B)\right). \quad (5)$$

F_i is the sum of all Feynman diagrams for the two point function of the background field with i loops and reads:

$$F_i = F_i^{tr}\ (\delta_{lm}p^2 - p_l p_m). \quad (6)$$

This leaves us with the relation, using eq.(5):

$$\frac{1}{g_M^2} = \frac{1}{g_E^2} - F_1^{tr} - F_2^{tr}, \quad (7)$$

with

$$g_E^2 F_1^{tr} = -\frac{1}{48}\frac{g_E^2 N}{\pi m_E} \quad (8)$$

$$g_E^2 F_2^{tr} = -\frac{19}{4608}\left(\frac{g_E^2 N}{\pi m_E}\right)^2. \quad (9)$$

This is the main result [4].

3. Spatial Wilson loop

We want now to test the applicability of 3D physics at medium high T. Let us consider an observable which has, unlike the pressure whose dominant contribution is the Stefan-Boltzmann term due to hard modes, its dominant contribution from the 3D Yang-Mills sector. Such an observable can be the spatial Wilson loop in the fundamental representation:

$$W(L) = TrP\exp\left(i\oint_L g\vec{A}\cdot d\vec{l}\right). \quad (10)$$

As L is purely spatial, it measures the magnetic flux in the plasma. The case where $\vec{A}$ is in an irreducible representation made of k quarks is described in this volume by C.P. Korthals Altes. The thermal average of this spatial Wilson loop shows area behaviour with a surface tension $\sigma(T)$. As it is a purely magnetic quantity, we expect from dimensional arguments that $\sqrt{\sigma} = cg_M^2$ where c is a nonperturbative proportionality constant. Indeed, as the average of the loop is due to long distance correlation, hard modes will not have any effects on the thermal average. In the same way, for soft modes, we can integrate out the A_0 field, which is what we have done in the previous section while constructing L_M, as the loop does not depend on A_0 ! Finally, we have:

$$\langle W(L)\rangle = \exp -\sigma A(L) = \frac{\int D\vec{A} W(L) \exp -S_M(A)}{\int D\vec{A} \exp -S_M(A)}, \tag{11}$$

and it gives, as δL_M is of relative order g^3,

$$\sigma(T) = c^2 g_M^4 (1 + O(g^3)). \tag{12}$$

Now the aim is to fit with our formula for g_M the proportionality constant c, and to study its extension for finite T. For that, we have to go to the lattice. We are going to fit $\frac{T}{\sqrt{\sigma}}$ has a function of $\frac{T}{T_c}$. Indeed, g_M is a function of g_E and m_E which are functions of g and T. So, for $N = 3$:

$$g_M^{-2} = g^{-2} T^{-1} (1 + \frac{g}{16\pi} + \frac{19 g^2}{512\pi^2}) \tag{13}$$

with

$$g^{-2} = \frac{11}{8} (Log(\frac{T}{T_c}) + Log(\frac{T_c}{\Lambda_{\bar{M}S}}) + 1.09835...). \tag{14}$$

So thanks to these formulas, one can fit $\frac{T}{\sqrt{\sigma}}$ versus $\frac{T}{T_c}$ in order to determin c and $\frac{T_c}{\Lambda_{\bar{M}S}}$.

We have taken data points for the $SU(3)$ spatial string tension from [6,9] for $T > 2T_c$. So we have 10 points. The fit is shown in fig 1.

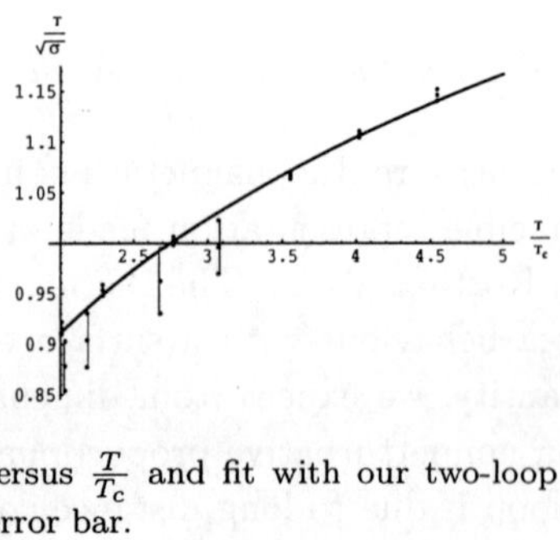

Figure 1. Plot of $\frac{T}{\sqrt{\sigma}}$ versus $\frac{T}{T_c}$ and fit with our two-loop formula. Each of the 10 points is shown with its error bar.

We have found the following results:

$$\frac{T_c}{\Lambda_{\overline{MS}}} = 1.78(12) \tag{15}$$

$$c = 0.505(11). \tag{16}$$

Then we have also fitted data with $T > 2.5T_c$. This leaves us with 6 points. We obtained

$$\frac{T_c}{\Lambda_{\overline{MS}}} = 1.57(20) \tag{17}$$

$$c = 0.488(18) \tag{18}$$

Let us draw our conclusions.

Despite the fact that the region of confidence is larger, the values of c are still incompatible with refs. [8,7,5]. To see compatibility one obviously needs string data for higher T. It might also be recommendable to use other Monte Carlo updates, like the one described in ref. [11], which is a version of the Lüscher-Weisz algorithm [10].

References

1. P. H. Ginsparg, Nucl. Phys. B **170** (1980) 388.
2. T. Appelquist and R. D. Pisarski, Phys. Rev. D **23** (1981) 2305.
3. L. Abbott, Nucl. Phys.B**185**, 189, 1981.
4. P. Giovannangeli, Phys. Lett. B **585**(2004) 144 [arXiv:hep-ph/0312307].
5. F. Karsch, E. Laermann and M. Lütgemeier, Phys. Lett. B **346** (1995) 94 [arXiv:hep-lat/9411020].
6. G. Boyd, J. Engels, F. Karsch, E. Laermann, C. Legeland, M. Lütgemeier and B. Petersson, Nucl. Phys. B **469** (1996) 419 [arXiv:hep-lat/9602007].
7. B. Lucini and M. Teper,Phys. Rev. D **66** (2002) 097502, arXiv:hep-lat/0206027
8. M. J. Teper,Phys. Rev. D **59**, 014512 (1999), hep-lat/9804008
9. M. Lütgemeier, Nucl. Phys. Proc. Suppl. **42** (1995) 523.
10. M. Lüscher and P. Weisz, JHEP **0109** (2001) 010 [arXiv:hep-lat/0108014].
11. H. B. Meyer, JHEP **0401** (2004) 030 [arXiv:hep-lat/0312034].

THERMODYNAMICS OF DECONFINED QCD AT SMALL AND LARGE CHEMICAL POTENTIAL

ANDREAS IPP *

Institut für Theoretische Physik, Technische Universität Wien
Wiedner Hauptstr. 8-10/136, A-1040 Vienna, Austria
E-mail: ipp@hep.itp.tuwien.ac.at

We present large N_f QCD/QED as a test bed for improved pressure calculations, show how to apply the hints obtained on optimized renormalization scales at large N_f to finite $N_f = 2$, and compare the results to recent lattice data.

1. Introduction

In the deconfined phase of QCD, strict perturbative calculations of thermodynamic potentials show poor convergence when approaching the phase transition. There have been a number of attempts to overcome this problem by reorganization or partial resummation of the perturbative expansions, like HTL perturbation theory[1] or Φ-derivable approximations for 2PI skeletons[2], but so far independent verification of these models was only possible through lattice simulations[3]. It is therefore instructive to consider the exactly solvable special case of large number of flavors (large N_f), in which the improvements above can be tested[4]. The large N_f limit can furthermore be easily extended to finite chemical potential[5]. Also, effects relevant to full QCD, like the anomalous specific heat at low temperatures, can be readily studied in the large-N_f limit[6,7].

2. Large N_f

The limit of large number of flavors is formed by sending N_f to infinity, while keeping the combination $g^2 N_f$ as well as N_c of the order of $O(1)$. The diagrams contributing to the free energy are at leading order (LO) only the fermion loop, while at next-to-leading order (NLO) an infinite

*Work supported by the Austrian Science Foundation FWF, project no. 16387-N08.

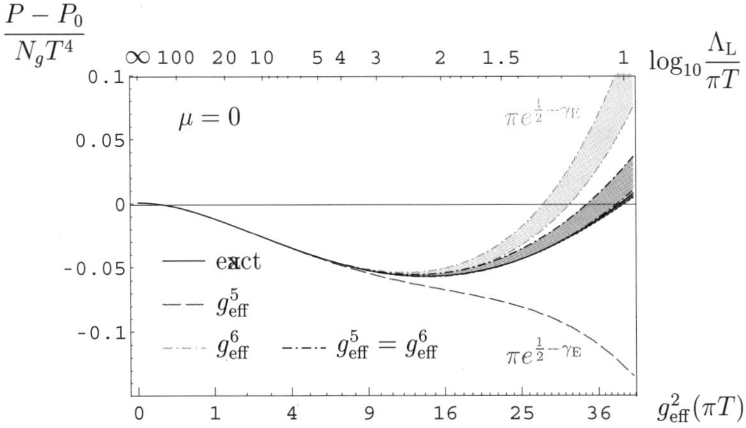

Figure 1. Exact result for the interaction pressure at zero chemical potential as a function of $g_{\rm eff}^2(\bar\mu_{\rm MS} = \pi T)$. The dashed line is the perturbative result, evaluated with renormalization scale $\bar\mu_{\rm MS} = \bar\mu_{\rm FAC}$; the light band includes the numerically determined coefficient to order $g_{\rm eff}^6$ (with its estimated error) also at $\bar\mu_{\rm FAC}$. The result marked "$g_{\rm eff}^5 = g_{\rm eff}^6$" corresponds to choosing $\bar\mu_{\rm MS}$ such that the order-$g_{\rm eff}^6$ coefficient vanishes and retaining all higher-order terms contained in the plasmon term $\propto m_E^3$.

number of ring diagrams consisting of a boson loop with any number of fermion loop insertions have to be summed up[4]. We can treat massless QCD and ultrarelativistic QED at the same time by defining an effective coupling as $g_{\rm eff}^2 \equiv g^2 N_f/2$ for QCD and $g_{\rm eff}^2 \equiv e^2 N_f$ for QED. Large N_f contains a Landau pole of the order of $\Lambda_{\rm L} \sim \mu \exp(6\pi^2/g_{\rm eff}^2)$, but the resulting ambiguity for the thermal pressure at NLO is suppressed by a factor $(\max({\rm T},\mu)/\Lambda_{\rm L})^4$.

After subtracting off the vacuum part of the ring diagrams and applying Schwinger-Dyson resummation[4], the NLO thermal pressure is given by

$$\frac{P_{\rm NLO}}{N_g} = -\int \frac{d^3q}{(2\pi)^3} \int_0^\infty \frac{dq_0}{\pi} \Big[2\left([n_b + \tfrac{1}{2}] {\rm Im} \ln (q^2 - q_0^2 + \Pi_T + \Pi_{\rm vac}) \right. \\ \left. - \tfrac{1}{2} {\rm Im} \ln (q^2 - q_0^2 + \Pi_{\rm vac}) \right) \\ + \left([n_b + \tfrac{1}{2}] {\rm Im} \ln (\tfrac{q^2 - q_0^2 + \Pi_L + \Pi_{\rm vac}}{q^2 - q_0^2}) - \tfrac{1}{2} {\rm Im} \ln (\tfrac{q^2 - q_0^2 + \Pi_{\rm vac}}{q^2 - q_0^2}) \right) \Big] \quad (1)$$

with the bosonic distribution function $n_b(\omega) = 1/(e^{\omega/T} - 1)$ and the gauge-boson self energy functions Π_T and Π_L. These cannot be given in closed form except for their imaginary parts[8], but are represented by one-dimensional integrals. We therefore have to evaluate the integrals numerically. Parts proportional to n_b can be safely integrated in Minkowski space, but terms without n_b are potentially logarithmically divergent. We compute them by introducing a Euclidean invariant cutoff[4].

3. Numerical results

Figure 1 shows the numerical result for $\mu = 0$ as a function of $g^2_{\rm eff}(\bar{\mu}_{\rm MS} = \pi T)$. For small coupling the coefficients to order $g^6_{\rm eff}$ which are not yet known analytically can be extracted numerically[5]. The large renormalization scale dependences of successive perturbative approximations to order $g^5_{\rm eff}$ beyond $g^2_{\rm eff} \sim 4$ can be fixed by applying "fastest apparent convergence" (FAC) in the m_E^2 parameter of dimensional reduction. Using $\bar{\mu}_{\rm MS} = \bar{\mu}_{\rm FAC} \equiv \pi e^{1/2-\gamma} T$ we obtain good agreement up to $g^2_{\rm eff} \sim 9$. The result can be further improved by the procedures explained below Fig. 1.

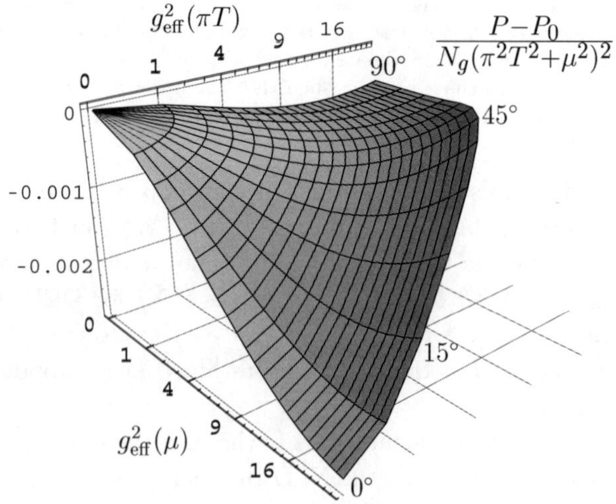

Figure 2. Exact result for the large-N_f interaction pressure $P - P_0$ normalized to $N_g(\pi^2T^2 + \mu^2)^2$ as a function of $g^2_{\rm eff}(\bar{\mu}_{\rm MS})$ with $\bar{\mu}^2_{\rm MS} = \pi^2T^2 + \mu^2$ and $\phi = \arctan \frac{\pi T}{\mu}$.

For non-vanishing chemical potential μ we use the fermionic distribution function

$$n_f(k, T, \mu) = \frac{1}{2}\left(\frac{1}{e^{(k-\mu)/T} + 1} + \frac{1}{e^{(k+\mu)/T} + 1}\right) \tag{2}$$

which enters via the gauge boson self-energy expressions Π_T and Π_L. In Fig. 2 we display our exact results for the interaction pressure $P - P_0 \propto N_f^0$

for the entire μ-T plane (but reasonably below the scale Landau pole). The figure shows a kink at $\phi = 45°$ corresponding to $\mu = \pi T$ indicating that a simple scaling behavior for the pressure at small chemical potentials would break down at larger chemical potentials[5].

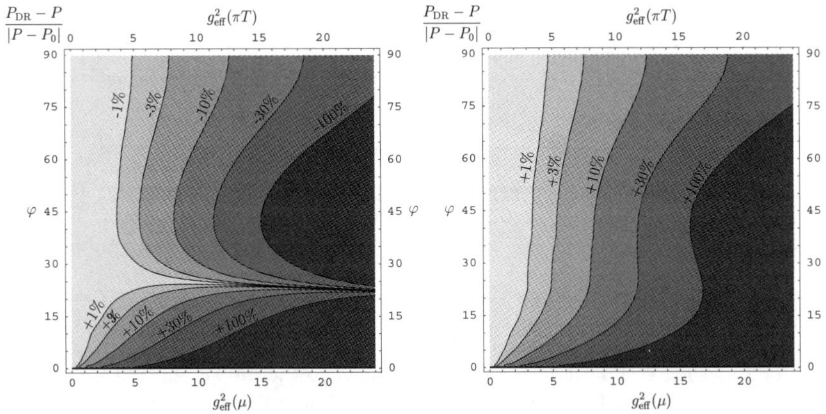

Figure 3. Percentage errors of the perturbative result for the interaction part of the pressure to order $g^5_{\rm eff}$ in the large-N_f limit as a function of $\varphi = \arctan(\pi T/\mu)$ and $g^2_{\rm eff}(\bar\mu_{\rm MS})$ at $\bar\mu_{\rm MS} = \sqrt{\pi^2 T^2 + \mu^2}$ for two choices of $\bar\mu_{\rm MS}$: Fastest apparent convergence of P as well as m_E^2 (FAC-m, left panel), and of g_E^2 (FAC-g, right panel). The brightest area corresponds to an error of less than 1%, the darkest ones to an error of over 100%.

Figure 3 shows a comparison between the perturbative result obtained by dimensional reduction[9] through order g^5 with complete analytic dependence on arbitrary T and μ to the exact result in the large N_f limit. In the left panel the renormalization scale is $\bar\mu_{\rm MS} = \bar\mu_{\rm MS}^{\rm FAC-m}$. An alternative choice for the renormalization scale is to set $\bar\mu_{\rm MS}$ such that the $g^4_{\rm eff}$ correction in the dimensional reduction coupling parameter g_E^2 is put to zero (FAC-g, right panel). The accuracy of the results is comparable in both cases and decreases slowly with increasing chemical potential, apart from an accidental zero of the error; for $\varphi \lesssim 18°$, i.e. $T \lesssim 0.1\mu$, the errors eventually start to grow rapidly, marking the breakdown of dimensional reduction. This is precisely the region where non-Fermi-liquid effects lead to anomalous $T \ln T^{-1}$ terms in the entropy and specific heat[6,7].

In Fig. 4 we display the FAC optimized results[10] at finite $N_f = 2$ for the difference $\Delta P = P(T,\mu) - P(T,0)$ for various μ/T corresponding to recent lattice results[11] assuming $T_0 \equiv T_c^{\mu=0} = 0.49\,\Lambda_{\rm QCD}$ [12]. At $T/T_0 = 2$ our FAC-g and FAC-m results exceed the not-yet-continuum-extrapolated

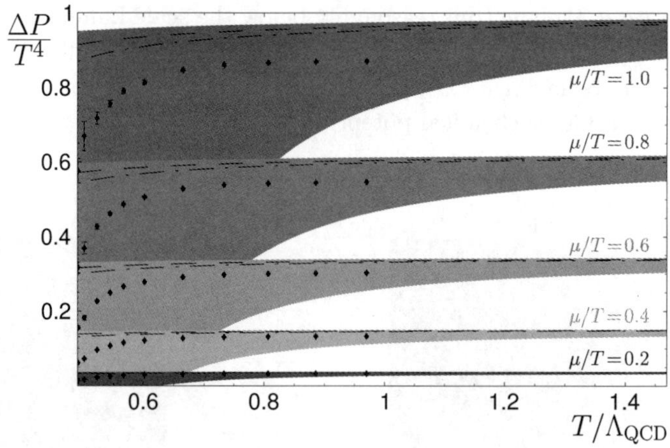

Figure 4. The difference $\Delta P = P(T,\mu) - P(T,0)$ divided by T^4 for $N_f = 2$ using the unexpanded three-loop result from dimensional reduction for $\mu/T = 0.2, \ldots, 1.0$ (bottom to top). Shaded areas correspond to a variation of $\bar{\mu}_{\rm MS}$ around the FAC-m choice by a factor of 2; dashed and dash-dotted lines correspond to the FAC-g and FAC-m results, respectively. Also included are recent lattice data (not yet continuum-extrapolated!) assuming $T_c^{\mu=0} = 0.49\,\Lambda_{\rm QCD}$.

lattice data consistently by 10.5% and 9%, respectively, which is roughly the expected discretization error[13].

References

1. J. O. Andersen, E. Braaten, and M. Strickland, Phys. Rev. Lett. **83** (1999) 2139–2142; R. Baier and K. Redlich, Phys. Rev. Lett. **84** (2000) 2100.
2. J. P. Blaizot, E. Iancu, and A. Rebhan, Phys. Rev. Lett. **83** (1999) 2906–2909; Phys. Rev. **D63** (2001) 065003; A. Peshier, Phys. Rev. **D63** (2001) 105004.
3. G. Boyd *et. al.*, Nucl. Phys. **B469** (1996) 419–444; CP-PACS Collaboration, M. Okamoto *et. al.*, Phys. Rev. **D60** (1999) 094510.
4. G. D. Moore, JHEP **0210**, 055 (2002); A. Ipp, G. D. Moore, and A. Rebhan, JHEP **0301**, 037 (2003).
5. A. Ipp and A. Rebhan, JHEP **0306**, 032 (2003).
6. A. Ipp, A. Gerhold, and A. Rebhan, Phys. Rev. **D69**, 011901 (2004).
7. A. Gerhold, A. Ipp, and A. Rebhan, hep-ph/0406087; see also contribution in these proceedings.
8. A. Ipp, PhD thesis TU Vienna 2003, hep-ph/0405123.
9. A. Vuorinen, Phys. Rev. **D68**, 054017 (2003).
10. A. Ipp, A. Rebhan and A. Vuorinen, Phys. Rev. D **69**, 077901 (2004)
11. C. R. Allton *et al.*, Phys. Rev. **D68**, 014507 (2003).
12. S. Gupta, Phys. Rev. **D64**, 034507 (2001).
13. F. Karsch, E. Laermann, and A. Peikert, Phys. Lett. **B478**, 447 (2000).

EVADING THE INFRARED PROBLEM OF THERMAL QCD

Y. SCHRODER

Center for Theoretical Physics, MIT, Cambridge, MA 02139, USA
and
Faculty of Physics, University of Bielefeld, D-33501 Bielefeld, Germany

Due to asymptotic freedom, QCD is guaranteed to be accessible to perturbative methods at asymptotically high temperatures. However, in 1979 Linde has pointed out the existence of an "infrared wall", beyond which an infinite number of Feynman diagrams contribute. Following a proposal by Braaten and Nieto, it is shown explicitly how the limits to computability that this infrared problem poses can be overcome in the framework of dimensionally reduced effective theories.

1. Introduction

The theory of strong interactions, Quantum Chromodynamics (QCD), is guaranteed to be accessible to perturbative methods once one of its parameters, the temperature T, is increased to asymptotically high values. This statement relies solely on the well-known property of asymptotic freedom[1].

In practice, however, calculations of corrections to the behavior of an ideal gas of quarks and gluons, the limit that is formally realized at infinite T, are obstructed by infrared divergences[2]: For every observable one sets out to compute, there exists an order of the perturbative expansion to which an infinite number of Feynman diagrams contribute ("infrared wall").

No method is known how to re-sum these infinite classes of diagrams, a fact that seriously obstructs progress in the field of thermal QCD, a field that presently receives attention particularly due to its relevance to the ongoing program of heavy-ion collisions at RHIC, where one of the main focuses is to explore the phase diagram of QCD.

For the QCD pressure – as an example of a static thermodynamic observable – it has been shown explicitly how the limits to computability that the infrared problem poses can be overcome in the framework of dimensionally reduced effective theories[3,4,5]. The key idea is to measure the effect of ultra-soft ("magnetic") gluons by lattice Monte–Carlo simulations in 3–dimensional (3d) pure gauge theory (MQCD), and to match this theory up to full thermal QCD in perturbation theory, via 3d gauge + adjoint Higgs theory (EQCD).

The expansion of the QCD pressure in the effective theory framework, up to the order where infrared contributions are relevant, is now known analytically[5]. The $\mathcal{O}(g^6)$ coefficient is non–perturbative, but *computable.* All other effects (N_f- or μ_f-dependence; orders g^7, ...) are perturbative.

2. Status of the QCD Pressure

Below, we specify the contributions to the pressure $p_{\rm QCD} = p_G + p_M + p_E$ from each physical scale individually, for the slightly more general case of gauge group SU(N_c) and N_f quark flavors, mainly following Ref. [5]. We will work at zero quark masses $m_{q_i} = 0$ and vanishing chemical potential $\mu_f = 0$, and display all dependence on the $\overline{\rm MS}$ scale by $L \equiv \ln \frac{\bar\mu}{4\pi T}$.

• Contributions from the ultra-soft scale g^2T, i.e. from MQCD:

$$\frac{p_G(T)}{\mu^{-2\epsilon}} = d_A 16\pi^2 T^4 \hat{g}_M^6 \left[\alpha_G \left(\frac{1}{\epsilon} + 8L - 8\ln(8\pi\hat{g}_M^2)\right) + \beta_G + \mathcal{O}(\epsilon)\right], \quad (1)$$

where $d_A = N_c^2 - 1$, $\alpha_G = \frac{43}{96} - \frac{157}{6144}\pi^2$ is a perturbative 4-loop coefficient, and β_G is non-perturbative, requiring two ingredients: a lattice-measurement in MQCD, and a perturbative computation which allows to match between the lattice and the continuum regularization schemes. For the latter, a 4-loop lattice-regularized computation is needed, which could possibly be accomplished with methods used in Refs. [6,7]. The matching condition reads

$$\hat{g}_M^2 \equiv \frac{N_c g_M^2}{16\pi^2 T} = \hat{g}_E^2 + \mathcal{O}(\hat{g}_E^4 \hat{m}_E^{-1}) \,. \quad (2)$$

• Contributions from the soft scale gT, i.e. from EQCD:

$$\begin{aligned}\frac{p_M(T)}{\mu^{-2\epsilon}} = d_A 16\pi^2 T^4 \Big\{ & \hat{m}_E^3 \left[\frac{1}{3} + \mathcal{O}(\epsilon)\right] \\ & + \hat{g}_E^2 \hat{m}_E^2 \left[-\frac{1}{4\epsilon} + \left(-L + \frac{1}{2}\ln \hat{m}_E^2 + \ln 2 - \frac{3}{4}\right) + \mathcal{O}(\epsilon)\right] \\ & + \hat{g}_E^4 \hat{m}_E \left[\left(-\frac{89}{24} - \frac{\pi^2}{6} + \frac{11}{6}\ln 2\right) + \mathcal{O}(\epsilon)\right] \\ & + \hat{g}_E^6 \left[\alpha_M \left(\frac{1}{\epsilon} + 8L - 4\ln \hat{m}_E^2 - 8\ln 2\right) + \beta_M + \mathcal{O}(\epsilon)\right] \\ & + \hat{\lambda}_E^{(1)} \hat{m}_E^2 \left[\frac{\hat{n} - 2}{4} + \mathcal{O}(\epsilon)\right] + \hat{\lambda}_E^{(2)} \hat{m}_E^2 \left[\frac{1 - 3\hat{n}}{4} + \mathcal{O}(\epsilon)\right] \\ & + \mathcal{O}(\hat{g}_E^8 \hat{m}_E^{-1}, \hat{\lambda}_E^2 \hat{m}_E) \Big\} \,, \end{aligned} \quad (3)$$

with $\hat{n} \equiv \frac{N_c^2-1}{N_c^2}$, the 4-loop coefficients $\alpha_M = \frac{43}{32} - \frac{491}{6144}\pi^2$, $\beta_M \approx -1.391512$ and the matching parameters

$$\hat{m}_E^2 \equiv \left(\frac{m_E}{4\pi T}\right)^2 = \hat{g}^2 \left[\tilde{\alpha}_{\rm E4} + (2\tilde{\alpha}_{\rm E4}L + \tilde{\alpha}_{\rm E5})\,\epsilon + \mathcal{O}(\epsilon^2)\right] + \hat{g}^4 \left[\left(2\hat{\beta}_0\tilde{\alpha}_{\rm E4}L + \tilde{\alpha}_{\rm E6}\right) + \bar{\beta}_{\rm E2}\epsilon + \mathcal{O}(\epsilon^2)\right] + \mathcal{O}(\hat{g}^6)\,, \quad (4)$$

$$\hat{g}_E^2 \equiv \frac{N_c g_E^2}{16\pi^2 T} = \hat{g}^2 + \hat{g}^4 \left[\left(2\hat{\beta}_0 L + \tilde{\alpha}_{\rm E7}\right) + \bar{\beta}_{\rm E3}\epsilon + \mathcal{O}(\epsilon^2)\right] + \mathcal{O}(\hat{g}^6)\,, \quad (5)$$

$$\hat{\lambda}_E^{(1)} \equiv \frac{N_c^2 \lambda_E^{(1)}}{16\pi^2 T} = \hat{g}^4 \left[4 + \mathcal{O}(\epsilon)\right] + \mathcal{O}(\hat{g}^6)\,, \quad (6)$$

$$\hat{\lambda}_E^{(2)} \equiv \frac{N_c \lambda_E^{(2)}}{16\pi^2 T} = \hat{g}^4 \left[\frac{4}{3}(1-z) + \mathcal{O}(\epsilon)\right] + \mathcal{O}(\hat{g}^6)\,, \quad (7)$$

where, for brevity, $z \equiv N_f/N_c$ and

$$\tilde{\alpha}_{\rm E4} = \frac{2+z}{6}\,, \quad \tilde{\alpha}_{\rm E6} = \frac{1}{3}\tilde{\alpha}_{\rm E4}(6\hat{\beta}_0\gamma + 5 + 2z - 8z\ln 2) - \frac{z}{2}\hat{n}\,, \quad (8)$$

$$\tilde{\alpha}_{\rm E5} = 2\tilde{\alpha}_{\rm E4}Z_1 + \frac{z}{6}(1 - 2\ln 2)\,, \quad \tilde{\alpha}_{\rm E7} = 2\hat{\beta}_0\gamma + \frac{1}{3} - \frac{8}{3}z\ln 2\,, \quad (9)$$

and $\bar{\beta}_{\rm E2}$ (see Sec. 3) and $\bar{\beta}_{\rm E3}$ remain to be computed by perturbatively matching suitable correlators computed in thermal QCD and in EQCD.

• Contributions from the hard scale $2\pi T$, i.e. from thermal QCD:

$$\frac{p_E(T)}{\mu^{-2\epsilon}} = d_A 16\pi^2 T^4 \frac{1}{16}\frac{1}{45}\left\{\left[1 + \frac{7}{4}\frac{z}{\hat{n}}\right] + \hat{g}^2\left[\tilde{\alpha}_{\rm E2} + \mathcal{O}(\epsilon)\right] + \hat{g}^4\left[\tilde{\alpha}_{\rm E4}\frac{180}{\epsilon} + (180\cdot 6\tilde{\alpha}_{\rm E4} + 2\hat{\beta}_0\tilde{\alpha}_{\rm E2})L + \tilde{\alpha}_{\rm E3} + \mathcal{O}(\epsilon)\right] + \hat{g}^6\left[\frac{\tilde{\beta}_{\rm E1}^{(\rm div)}}{\epsilon} + \tilde{\beta}_{\rm E1}^{(L^2)}L^2 + \tilde{\beta}_{\rm E1}^{(L)}L + \tilde{\beta}_{\rm E1} + \mathcal{O}(\epsilon)\right] + \mathcal{O}(\hat{g}^8)\right\}, \quad (10)$$

with $\tilde{\alpha}_{\rm E2} = -\frac{5}{4}(4+5z)$ and, writing $Z_1 \equiv \frac{\zeta'(-1)}{\zeta(-1)}$ and $Z_3 \equiv \frac{\zeta'(-3)}{\zeta(-3)}$,

$$\tilde{\alpha}_{\rm E3} = 180(\tilde{\alpha}_{\rm E4})^2\gamma + 5\left[\left(\frac{116}{5} + \frac{220}{3}Z_1 - \frac{38}{3}Z_3\right) + \frac{z}{2}\left(\frac{1121}{60} - \frac{157}{5}\ln 2 + \frac{146}{3}Z_1 - \frac{1}{3}Z_3\right) + \frac{z^2}{4}\left(\frac{1}{3} - \frac{88}{5}\ln 2 + \frac{16}{3}Z_1 - \frac{8}{3}Z_3\right) + \frac{z}{4}\hat{n}\left(\frac{105}{4} - 24\ln 2\right)\right], \quad (11)$$

and unknown coefficients $\beta_{\rm E1}$, which can be determined e.g. by a 4-loop computation of vacuum diagrams in thermal QCD. Since $p_{\rm QCD}$ is physical, the divergent and scale-dependent parts of $\beta_{\rm E1}$ are related to the other coefficients introduced in the above, serving as a valuable check on this open computation. Specifically, from 2-loop running of the 4d gauge coupling

$$\hat{g}^2 \equiv \frac{N_c g^2(\bar{\mu})}{16\pi^2} = \hat{g}^2(\bar{\mu}_0) + \hat{g}^4(\bar{\mu}_0)(-2\hat{\beta}_0\ell) + \hat{g}^6(\bar{\mu}_0)(4\hat{\beta}_0^2\ell^2 - 2\hat{\beta}_1\ell) \quad (12)$$

with beta-function coefficients $\hat{\beta}_0 = \frac{11-2z}{3}$, $\hat{\beta}_1 = \frac{34}{3} - \frac{10}{3}z - z\hat{n}$, $\overline{\rm MS}$ scale parameter $\bar{\mu}^2 = 4\pi e^{-\gamma}\mu^2$ and $\ell \equiv \ln\frac{\bar{\mu}}{\bar{\mu}_0} = L - \ln\frac{\bar{\mu}_0}{4\pi T}$, one can fix

$$\tilde{\beta}_{\rm E1}^{\rm (div)} = 180\left[4\hat{\beta}_0\tilde{\alpha}_{\rm E4}L + \tilde{\alpha}_{\rm E6} + \tilde{\alpha}_{\rm E4}\tilde{\alpha}_{\rm E7} - 4(\alpha_G + \alpha_M)\right], \quad (13)$$

$$\tilde{\beta}_{\rm E1}^{(L^2)} = 180\left[20\hat{\beta}_0\tilde{\alpha}_{\rm E4} + \tilde{\beta}_{\rm E2}^{(L^2)} + \tilde{\alpha}_{\rm E4}\tilde{\beta}_{\rm E3}^{(L^2)}\right] + 4\hat{\beta}_0^2\tilde{\alpha}_{\rm E2}\,, \quad (14)$$

$$\tilde{\beta}_{\rm E1}^{(L)} = 180\left[4\tilde{\alpha}_{\rm E6} + 6\tilde{\alpha}_{\rm E4}\tilde{\alpha}_{\rm E7} - 2\hat{\beta}_0\tilde{\alpha}_{\rm E5} - 32(\alpha_G + \alpha_M) + \tilde{\beta}_{\rm E2}^{(L)} + \tilde{\alpha}_{\rm E4}\tilde{\beta}_{\rm E3}^{(L)}\right] + 2\hat{\beta}_1\tilde{\alpha}_{\rm E2} + 4\hat{\beta}_0\tilde{\alpha}_{\rm E3}\,. \quad (15)$$

3. Determination of β_{E2}

To determine $\bar{\beta}_{E2}$ in Eq. (4), one can e.g. match the pole masses of the A_0 propagator. In thermal QCD, writing $\Pi_{00}^{ab}(k_0 = 0, \vec{k}) = \delta^{ab}\Pi(k^2)$, one needs to solve $k^2 + \Pi(k^2) = 0$ at $k^2 = -m_{\rm pole}^2$. Inserting a loop expansion for the self-energy Π and noting that the leading-order solution gives a perturbatively small $m_{\rm pole}^2 \sim g^2$, one can Taylor-expand to get

$$\hat{m}_{\rm pole}^2 \equiv \left(\frac{m_{\rm pole}}{4\pi T}\right)^2 = \hat{\Pi}_1(0) + \hat{\Pi}_2(0) - \hat{\Pi}_1(0)\Pi_1'(0) + \mathcal{O}(\hat{g}^6)\,. \quad (16)$$

The bare gluon self-energies can be deduced from the literature[3] as

$$\hat{\Pi}_1(0) \equiv \frac{\Pi_1(0)}{16\pi^2T^2} = \hat{g}_B^2\left[(d-2)^2\hat{I}_b(1) - 2z(d-2)\hat{I}_f(1)\right], \quad (17)$$

$$\Pi_1'(0) = \hat{g}_B^2\left[\frac{1}{6}(-22+7d-d^2)\hat{I}_b(2) + \frac{z}{3}(d-2)\hat{I}_f(2)\right], \quad (18)$$

$$\hat{\Pi}_2(0) \equiv \frac{\Pi_2(0)}{16\pi^2T^2} = \hat{g}_B^4\Big[\left(2z\hat{I}_f(1) - (d-2)\hat{I}_b(1)\right)(12-8d+d^2)\hat{I}_b(2) + \hat{n}z(d-4)(d-2)\left(\hat{I}_b(1) - \hat{I}_f(1)\right)\hat{I}_f(2)\Big], \quad (19)$$

where we have used scaled bosonic and fermionic 1-loop tadpole integrals $\hat{I}(x) = 16\pi^2(4\pi T)^{2x-4}T\sum_{n=-\infty}^{\infty}\mu^{2\epsilon}\int\frac{d^{d-1}p}{(2\pi)^{d-1}}\frac{1}{(\omega_n^2+\vec{p}^2)^x}$, with $\omega_n = 2n\pi T$ for bosons and $(2n+1)\pi T$ for fermions,

$$\hat{I}_b(x) = \mu^{2\epsilon}\frac{2^{2x-3}}{\sqrt{\pi}}(\sqrt{\pi}T)^{d-4}\zeta(1+2x-d)\frac{\Gamma(x+\frac{1-d}{2})}{\Gamma(x)} \quad (20)$$

and correspondingly $\hat{I}_f(x) = (2^{2x+1-d} - 1)\hat{I}_b(x)$.

In EQCD, the solution of $k^2 + m_E^2 + \Pi_E(k^2) = 0$ at $k^2 = -m_{\rm pole}^2$ is simply $m_{\rm pole}^2 = m_E^2$, since again treating $m_{\rm pole}^2$ as perturbatively small and Taylor-expanding, there is no scale left in Π_E, such that it vanishes in $\overline{\rm MS}$. Renormalizing Eq. (16) via $\hat{g}_B^2 = Z_{\hat{g}}^2 \hat{g}^2$ with $Z_{\hat{g}}^2 = 1 - \hat{g}^2\hat{\beta}_0/\epsilon + \mathcal{O}(\hat{g}^4)$ and comparing with Eq. (4), one reproduces Eqs. (8–9) and finally obtains the coefficients in $\bar{\beta}_{\rm E2} = 6\hat{\beta}_0\tilde{\alpha}_{\rm E4}L^2 + \tilde{\beta}_{\rm E2}^{(L)}L + \tilde{\beta}_{\rm E2}$, which are

$$\begin{aligned}
\tilde{\beta}_{\rm E2}^{(L)} &= 4\hat{\beta}_0\tilde{\alpha}_{\rm E4}(2\gamma + Z_1) + \frac{1}{9}(20 + 29z + 2z^2) \\
&\quad - 2z\,(\hat{n} + 3\ln 2) - \frac{4}{3}z^2 \ln 2\;, \qquad (21) \\
\tilde{\beta}_{\rm E2} &= \frac{1}{4}\hat{\beta}_0\tilde{\alpha}_{\rm E4}\left(16\zeta'(1) + \pi^2\right) + \frac{2}{3}\tilde{\alpha}_{\rm E4}Z_1(6\hat{\beta}_0\gamma + 5 + 2z - 8z\ln 2) \\
&\quad + \frac{2}{9}\gamma(5 + 10z - (19 + 2z)z\ln 2) + \frac{2}{9} + \frac{z}{18}(7 + 6\ln 2 - 16\ln^2 2) \\
&\quad + \frac{z^2}{9}(1 - 2\ln 2 + 4\ln^2 2) - \frac{z}{6}\hat{n}(3 + 6\gamma + 6Z_1 + 10\ln 2)\;. \qquad (22)
\end{aligned}$$

4. Outlook

The future should see a completion of the above setup, thereby establishing a first example of successfully computing an observable beyond the infrared wall. Once this is achieved, thermal QCD in its high-temperature phase will again be amenable to perturbative calculations, opening up numerous opportunities to precisely compute observables that might become relevant to the RHIC program, to future accelerators, and to cosmology. Indeed, the next term in the series, formally of order $\mathcal{O}(g^7)$, requires the corrections of order $\mathcal{O}(\hat{g}_E^4/\hat{m}_E)$ to Eq. (2) (known), of order $\mathcal{O}(\hat{g}_E^8/\hat{m}_E)$ to Eq. (3) (5-loop vacuum diagrams in EQCD), and of order $\mathcal{O}(\hat{g}^6)$ to Eq. (4) (3-loop 2-point functions in thermal QCD), and will then certainly be within reach.

References

1. http://nobelprize.org/physics/laureates/2004
2. A. D. Linde, Phys. Lett. B **96** (1980) 289; D. J. Gross, R. D. Pisarski and L. G. Yaffe, Rev. Mod. Phys. **53** (1981) 43.
3. E. Braaten and A. Nieto, Phys. Rev. D **53** (1996) 3421.
4. K. Kajantie et. al., Phys. Rev. Lett. **86** (2001) 10.
5. K. Kajantie et.al., Phys. Rev. D **67** (2003) 105008.
6. F. Di Renzo, A. Mantovi, V. Miccio and Y. Schroder, JHEP **0405** (2004) 006.
7. B. Alles, M. Campostrini, A. Feo and H. Panagopoulos, Phys. Lett. B **324** (1994) 433.

CHIRAL MESONS IN HOT MATTER

A. GÓMEZ NICOLA, F.J. LLANES-ESTRADA AND J.R.PELÁEZ
Departamentos de Física Teórica I y II,
Facultad de Ciencias Físicas, Universidad Complutense, 28040 Madrid, SPAIN

We review our recent work on thermal meson properties within the Chiral Perturbation Theory framework. We will focus on the pion electromagnetic form factor, stressing its importance for Relativistic Heavy Ion Collisions. We obtain model-independent predictions (based only on chiral symmetry) such as the temperature dependence of the pion electromagnetic charge radius. Imposing unitarity allows to describe the thermal effects of resonances such as the ρ meson.

1. Introduction

In order to describe the meson gas formed after a Relativistic Heavy Ion Collision (RHIC) for temperatures below Chiral Symmetry Restoration, Chiral Perturbation Theory (ChPT) is the most general theoretical scheme [1] and its predictions are based only on Chiral Symmetry. When applied to the low-T meson gas, one can analyze static quantities such as the pressure [2,3], quark condensate [1,2,3] and pion decay constants [1,4]. All these quantities point towards Chiral Symmetry Restoration. In addition, the analysis of the pion dispersion relation $p^2 = m_\pi^2 + g(p_0, |\vec{p}|; T)$ shows that to leading order $O(T^2)$, g is real and depends only on temperature but not on momenta [1] while corrections to the pion width start to $O(T^4)$ [4,5].

Recently, we have shown that ChPT is equally useful to describe dynamical quantities such as scattering amplitudes [6] or form factors [7]. Furthermore, combining Chiral Symmetry Breaking and Unitarity provides a fruitful approach to describe the thermal behaviour of resonances [8] such as the ρ, whose medium properties are crucial to explain the excess observed in the dilepton spectrum data [9], which has motivated a huge amount of theoretical work devoted to explain in-medium resonances [10].

2. Pion form factors at finite temperature

The main contribution to the dilepton rate comes from the pion annihilation channel via a virtual photon exchange $\pi^+\pi^- \to \gamma^* \to e^+e^-$. The dilepton rate is then proportional to the pion electromagnetic form factor [11] which for $T \neq 0$ has the following general form:

$$\begin{aligned} \langle \pi^+(p)\pi^-(p')|V_0(0)|0\rangle &= q_0 F_t(S_0, |\vec{S}|, q_0; T) \\ \langle \pi^+(p)\pi^-(p')|V_k(0)|0\rangle &= q_k F_s(S_0, |\vec{S}|, q_0; T) + S_k q_0 G_s(S_0, |\vec{S}|, q_0; T) \end{aligned} \quad (1)$$

with V_μ the electromagnetic current, $S = p + p'$, $q = p - p'$ and F_t, F_s, G_s even functions in q_0 by charge conjugation invariance. These functions are related by the gauge invariance Ward identity $\langle \pi\pi|\partial_\mu V^\mu|0\rangle = 0$. We have

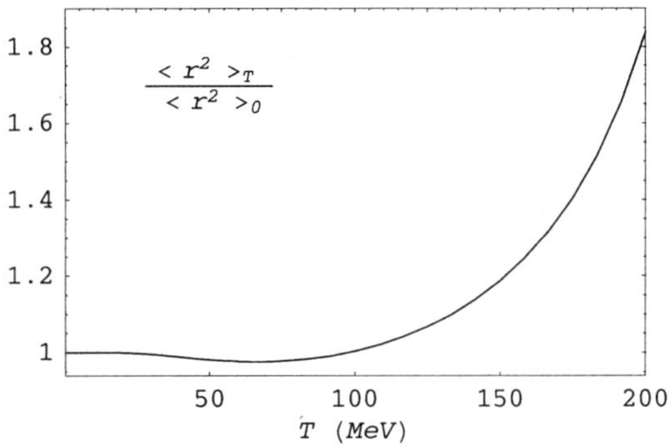

Figure 1. The pion electromagnetic charge radius at finite temperature

calculated the one-loop ChPT form factors, whose detailed results can be found in [7]. This allows for model-independent predictions. For instance, defining the pion electromagnetic charge and radius as:

$$\begin{aligned} Q_T &= \lim_{|\vec{S}|\to 0^+} F_t(0, |\vec{S}|; T) \\ \langle r^2\rangle_T &= -\frac{6}{Q_T} \left.\frac{dF_t(0, |\vec{S}|; T)}{d|\vec{S}|^2}\right|_{|\vec{S}|=0} \end{aligned} \quad (2)$$

which takes into account Lorentz covariance breaking at $T \neq 0$, one finds the result shown in Figure 1. The radius remains almost constant for low T, which is consistent with the Vector Meson Dominance (VMD) idea that $\langle r^2\rangle \sim M_\rho^{-2}$ [13], provided M_ρ changes little at low temperatures. On the other hand, the radius increases significantly for higher temperatures, dominated by a Q_T reduction (charge screening). Estimating the critical temperature of deconfinement by requiring that the electromagnetic pion volume $V(T) = (4\pi/3)\langle r^2\rangle_T^{3/2}$ equals the inverse pion density gives $T_c \simeq 200$ MeV, very close to the prediction of ChPT for chiral symmetry restoration [1,2,3]. This gives support to the idea that both transitions take place nearly at the same point. Our results for the pion charge radius confirm those in [14] based on QCD sum rules.

3. Thermal unitarity and resonances

In the center of mass frame ($\vec{p} = -\vec{p'}$, corresponding to back to back dilepton pairs) the ChPT pion form factor satisfies the following perturbative unitarity relation for energies above the two-pion threshold $S_0 > 2m_\pi$:

$$\mathrm{Im}F^{(1)}(S_0 + i\epsilon; T) = \sigma_T(S_0) a_{11}^{(0)}(S_0^2) \left[F^{(0)}(S_0^2)\right]^* \tag{3}$$

with:

$$\sigma_T(S_0) = \sqrt{1 - \frac{4m_\pi^2}{S_0^2}} \left[1 + \frac{2}{e^{S_0/2T} - 1}\right] \tag{4}$$

The superscripts (0) and (1) indicate the LO and NLO ChPT contributions respectively and a_{IJ} denote pion scattering partial waves with definite isospin I and angular momentum J. Note that $F^{(0)} = 1$ but we have displayed it in the above equation for consistency with what follows. Here, σ_T is the thermal two-particle phase space, which can be understood by writing it as $\sigma_T = \sigma_0 \left[(1 + n_B)(1 + n_B) - n_B^2\right]$ where $n_B = [\exp(S_0/2T) - 1]^{-1}$ is the Bose-Einstein distribution function for each pion. Hence, the phase space is enhanced at finite temperature due to the difference between induced emission and absorption processes in the thermal bath [15]. This effect contributes to the resonance thermal width enhancement [8]. The partial waves a_{IJ} also satisfy a thermal perturbative unitarity relation [6] which is the $T = 0$ one with σ_0 replaced by σ_T, as in (3).

Thermal perturbative unitarity is the key point to describe resonances within ChPT. In fact, inspired by the success of the so called Inverse Amplitude Method (IAM) at $T = 0$ [16] we have constructed unitarized partial waves and form factors demanding *exact* thermal unitarity. For instance, $\mathrm{Im}F^{IAM} = \sigma_T a_{11}^{IAM} \left[F^{IAM}\right]^*$. The remaining freedom is fixed by matching the ChPT series at low energies. This approximation is valid for small pion densities n_B, which is very reasonable for the temperatures and energies we are interested in. The IAM generates the σ ($I = J = 0$) and ρ

$(I = J = 1)$ poles in the energy complex plane and their thermal behaviour is the expected one. In Figure 2 we have plotted the results for the poles and for the unitarized form factor. The ρ widens in the thermal bath, effect also visible in the form factor, peaked at the ρ mass. Such widening is in perfect agreement with the expectations from dilepton experiments and previous analysis [10,12]. Our resulting M_ρ remains almost constant for $T < 100$ MeV (consistently with VMD) and decreases slightly for higher T. This result points in the same direction as recent measurements by the STAR collaboration [17] but we have ignored other medium effects such as baryon density. The in-medium ρ mass remains certainly an open issue [10]. As for the σ pole, the results in Figure 2 show a behaviour compatible with Chiral Symmetry Restoration: m_σ decreases sharply, hence approaching m_π which increases softly with T [1]. The σ width increases at first by phase space enhancement but as $m_\sigma \to 2m_\pi$ the decay $\sigma \to 2\pi$ is reduced and so does the width. A similar behaviour has been found in [18].

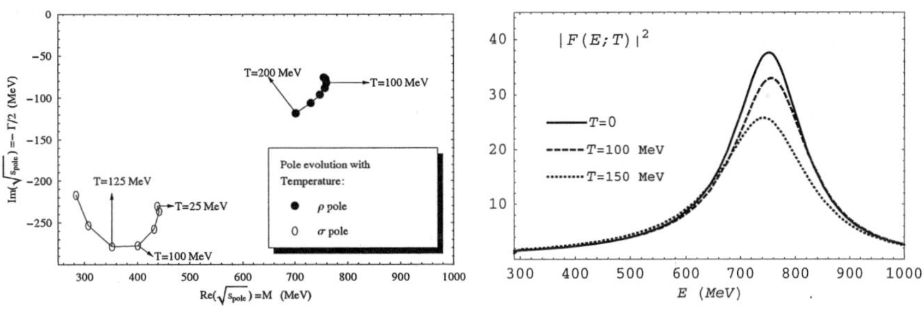

Figure 2. Thermal poles (left pannel) and unitarized form factor (right pannel).

Summarizing, ChPT and its unitarization programme can be extended successfully to finite temperature and provide a promising framework for future applications in the context of Relativistic Heavy Ion Collisions.

Acknowledgments

This work is supported by the Spanish research projects FPA2000-0956,PB98-0782, BFM2000-1326 and FPA2004-02602.

References

1. J. Gasser and H. Leutwyler, *Phys.Lett.* **B184**, 83 (1987).
2. P.Gerber and H.Leutwyler, *Nucl.Phys.***B321**, 387 (1989).
3. A.Dobado, and J.R.Peláez, *Phys.Rev* **D59**, 034004 (1999); J.R.Peláez, *Phys.Rev* **D66**, 096007 (2002).
4. R.D.Pisarski and M.Tytgat, *Phys.Rev.* **D54**, 2989 (1996). D.Toublan, *Phys.Rev.* **D56**, 5629 (1997). J.M.Martinez Resco and M.A.Valle Basagoiti, *Phys.Rev.* **D58**, 097901 (1998).
5. J.L. Goity and H.Leutwyler *Phys.Lett.* **B228**, 517 (1989).
6. A. Gómez Nicola, F. J. Llanes-Estrada and J. R. Peláez, *Phys.Lett.* **B550**, 55 (2002).
7. A. Gómez Nicola, F. J. Llanes-Estrada and J. R. Peláez, hep-ph/0405273.
8. A. Dobado, A. Gómez Nicola, F. Llanes-Estrada and J. R. Peláez, *Phys.Rev.* **C66**, 055201 (2002).
9. G.Agakichiev *et al* [CERES Collaboration] *Phys. Rev. Lett.* **75**, 1272 (1995); *Phys.Lett.* **B442**, 405 (1998); *Nucl.Phys.* **A 661**, 23 (1999); D.Adamova *et al* [CERES/NA45 Collaboration] , *Phys. Rev. Lett.*, **91** 042301 (2003).
10. M.Dey, V.L.Eletsky and B.L.Ioffe, *Phys.Lett.* **B252**, 620 (1990); R.D.Pisarski, *Phys.Rev.* **D52**, 3773 (1995); G. Q. Li, C. M. Ko and G. E. Brown, *Phys.Rev.Lett.* **75**, 4007 (1995). V. Koch and C. Song, *Phys. Rev.* **C54**, 1903 (1996). R.Rapp and J.Wambach, *Adv.Nucl.Phys* **25**:1 (2000); V. L. Eletsky, M. Belkacem, P. J. Ellis and J. I. Kapusta, *Phys. Rev.* **C64**, 035202 (2001). H.-J. Schulze and D.Blaschke, *Phys.Lett.* **B386**, 429 (1996); *Part.Nucl.Lett.* **119**, 27 (2004).
11. K.Kajantie, J.Kapusta, L.McLerran and A.Mekjian, *Phys.Rev.* **D34**, 2746 (1986). C.Gale and J.I.Kapusta, *Phys.Rev.* **C35**, 2107 (1987)
12. C.Song and V.Koch, *Phys.Rev.* **C54** 3218 (1996).
13. J.F.Donoghue, C.Ramirez and G.Valencia, *Phys.Rev.* **D39**, 1947 (1989); G. Ecker, J. Gasser, A. Pich and E. de Rafael, *Nucl.Phys.* **B321**, 311 (1989).
14. C. A. Dominguez, M. Loewe and J. S. Rozowsky, *Phys.Lett.* **B335**, 506 (1994).
15. H. A. Weldon, *Ann. Phys.* **214**, 152 (1992).
16. T. N. Truong, *Phys.Rev.Lett.***61**, 2526 (1988); A. Dobado, M.J.Herrero and T.N. Truong, *Phys.Lett.* **B235**, 134 (1990); A. Dobado and J.R. Peláez, *Phys.Rev.***D47**, 4883 (1993); *Phys.Rev.***D56**, 3057 (1997); A.Dobado, M.J.Herrero, J.R.Peláez and E.Ruiz Morales, *Phys.Rev.* **D62**, 055011 (2000); A.Gómez Nicola and J.R.Peláez, *Phys.Rev.***D65**:054009 (2002).
17. J. Adams *et al.* [STAR Collaboration], *Phys.Rev. Lett.* **92**:092301 (2004).
18. S.Chiku and T.Hatsuda, *Phys.Rev.***D57**, R6 (1998); *Phys.Rev.***D58**, 076001 (1998). K.Yokokawa, T.Hatsuda, A.Hayashigaki and T.Kunihiro, *Phys.Rev.* **C66**, 022201 (2002).

THERMAL PRODUCTION OF AXINOS IN THE EARLY UNIVERSE

A. BRANDENBURG AND F. D. STEFFEN *

DESY Theory Group, Notkestrasse 85, D-22603 Hamburg, Germany
E-mail: Arnd.Brandenburg@desy.de, Frank.Daniel.Steffen@desy.de

We compute the thermal axino production rate in supersymmetric QCD to leading order in the gauge coupling. Using hard thermal loop resummation and the Braaten–Yuan prescription, we obtain a finite result in a gauge-invariant way, which takes into account Debye screening in the hot quark–gluon–squark–gluino plasma. The relic axino density from thermal reactions in the early Universe is evaluated assuming the axino is the lightest supersymmetric particle and stable due to R-parity conservation. From the comparison with the WMAP results, we find that axinos could provide the dominant part of cold dark matter, for example, for an axino mass of 100 keV and a reheating temperature of 10^6 GeV.

1. Introduction

In supersymmetric extensions of the standard model in which the strong CP problem is solved by the Peccei–Quinn (PQ) mechanism, the axino $\tilde{a}$ arises naturally as the fermionic superpartner of the axion.[1] Thus, the axino is electrically and color neutral and its interactions with the MSSM particles are suppressed by the PQ scale $f_a/N \gtrsim 5 \times 10^9\,$GeV. If the axino is the lightest supersymmetric particle (LSP) and if R-parity is conserved, axinos are stable and could provide the dominant part of cold dark matter.

In this talk we present the computation of the thermal axino production rate at high temperatures and discuss its cosmological implications. With inflation governing the earliest moments of the Universe, any initial population of axinos was diluted away and the thermal production of axinos set in at the reheating temperature T_R. We restrict our investigation to $f_a/N > T_R \gtrsim 10^4\,$GeV, where the $\mathrm{U}(1)_{\mathrm{PQ}}$ symmetry is broken and axino production from decays of particles out of equilibrium is negligible.[2] The results presented are extracted from Ref. 3 where more details can be found.

*Speaker at the Conference.

2. Thermal Axino Production in Supersymmetric QCD

We concentrate on the axino–gluino–gluon interactions given by the dimension-5 interaction term

$$\mathcal{L}_{\tilde{a}\tilde{g}g} = i\,\frac{g^2}{64\pi^2(f_a/N)}\,\bar{\tilde{a}}\,\gamma_5\,[\gamma^\mu,\gamma^\nu]\,\tilde{g}^a\,G^a_{\mu\nu} \tag{1}$$

with the strong gauge coupling g, the gluon field strength tensor $G^a_{\mu\nu}$, the gluino $\tilde{g}$, and N being the number of quarks with PQ charge. The resulting $2 \to 2$ scattering processes in supersymmetric QCD and the corresponding squared matrix elements are listed in Table 1, where $s = (P_1 + P_2)^2$ and $t = (P_1 - P_3)^2$ with P_1, P_2, P_3, and P referring to the particles in the given order. Working in the limit, $T \gg m_i$, the masses of all particles involved have been neglected. Sums over initial and final spins have been performed. For quarks and squarks, the contribution of a single chirality is given.

The processes B, F, G, and H lead to a logarithmic collinear singularity due to t-channel and u-channel exchange of soft (massless) gluons. Here screening effects in the hot quark–gluon–squark–gluino plasma (QGSGP) have to be taken into account. With hard thermal loop (HTL) resummation,[4] this can be done systematically in a gauge-invariant way. Following the Braaten–Yuan prescription,[5] we introduce a momentum scale $k_{\rm cut}$ such that $gT \ll k_{\rm cut} \ll T$ in the weak coupling limit, $g \ll 1$. This separates soft gluons with momentum transfer of order gT from hard gluons with momentum transfer of order T. In the region of soft momentum trans-

Table 1. Squared matrix elements for axino ($\tilde{a}$) production in two-body processes involving quarks (q_i), squarks ($\tilde{q}_i$), gluons (g^a), and gluinos ($\tilde{g}^a$) in the high-temperature limit, $T \gg m_i$, with the SU(3) color matrices f^{abc} and T^a_{ji}

	process i	$\lvert\mathcal{M}_i\rvert^2 / \frac{g^6}{128\pi^4(f_a/N)^2}$
A	$g^a + g^b \to \tilde{g}^c + \tilde{a}$	$4(s + 2t + 2\frac{t^2}{s})\lvert f^{abc}\rvert^2$
B	$g^a + \tilde{g}^b \to g^c + \tilde{a}$	$-4(t + 2s + 2\frac{s^2}{t})\lvert f^{abc}\rvert^2$
C	$\tilde{q}_i + g^a \to q_j + \tilde{a}$	$2s\lvert T^a_{ji}\rvert^2$
D	$g^a + q_i \to \tilde{q}_j + \tilde{a}$	$-2t\lvert T^a_{ji}\rvert^2$
E	$\bar{\tilde{q}}_i + q_j \to g^a + \tilde{a}$	$-2t\lvert T^a_{ji}\rvert^2$
F	$\tilde{g}^a + \tilde{g}^b \to \tilde{g}^c + \tilde{a}$	$-8\frac{(s^2+st+t^2)^2}{st(s+t)}\lvert f^{abc}\rvert^2$
G	$q_i + \tilde{g}^a \to q_j + \tilde{a}$	$-4(s + \frac{s^2}{t})\lvert T^a_{ji}\rvert^2$
H	$\tilde{q}_i + \tilde{g}^a \to \tilde{q}_j + \tilde{a}$	$-2(\frac{t}{2} + 2s + 2\frac{s^2}{t})\lvert T^a_{ji}\rvert^2$
I	$q_i + \bar{q}_j \to \tilde{g}^a + \tilde{a}$	$-4(t + \frac{t^2}{s})\lvert T^a_{ji}\rvert^2$
J	$\tilde{q}_i + \bar{\tilde{q}}_j \to \tilde{g}^a + \tilde{a}$	$2(\frac{s}{2} + 2t + 2\frac{t^2}{s})\lvert T^a_{ji}\rvert^2$

fer, $k < k_{\rm cut}$, we use the HTL-resummed gluon propagator, which takes into account Debye screening in QGSGP in terms of the supersymmetric thermal gluon mass $m_g = gT\sqrt{(N_c + n_f)/6}$, where $N_c = 3$ and $n_f = 6$ are respectively the number of colors and color triplet and anti-triplet chiral multiplets. Computing the imaginary part of the thermal axino self-energy with the ultraviolet (UV) momentum cutoff $k_{\rm cut}$, the soft contribution to the production rate of axinos with energies $E \gtrsim T$ is obtained

$$\left.\frac{d\Gamma^{\tilde{a}}}{d^3p}\right|_{\rm soft} = f_F(E)\frac{3g^4(N_c^2-1)m_g^2T}{4096\pi^8(f_a/N)^2}\left[\ln\left(\frac{k_{\rm cut}^2}{m_g^2}\right) - 1.379\right], \tag{2}$$

where $f_F(E) = [\exp(E/T+1)]^{-1}$. In the region of hard momentum transfer, $k > k_{\rm cut}$, the bare gluon propagator can be used. From the squared matrix elements of the $2 \to 2$ processes (cf. Table 1) weighted with appropriate multiplicities, statistical factors, and phase space densities, the corresponding hard contribution can then be obtained conveniently

$$\left.\frac{d\Gamma_{\tilde{a}}}{d^3p}\right|_{\rm hard} = \frac{g^6\,(N_c^2-1)}{32\,\pi^4\,(f_a/N)^2}\left[(N_c+n_f)\frac{T^3\,f_F(E)}{128\pi^4}\,\ln\left(\frac{2^{1/3}T}{k_{\rm cut}}\right) + ...\right] \tag{3}$$

with $k_{\rm cut}$ as the infrared (IR) cutoff for gluon momentum transfers. The expression represented by the ellipses can be found in Ref. 3. It is independent of $k_{\rm cut}$. Thus, by summing the soft and hard contributions, the artificial $k_{\rm cut}$ dependence cancels and the finite leading order rate is obtained

$$\frac{d\Gamma_{\tilde{a}}}{d^3p} = \left.\frac{d\Gamma_{\tilde{a}}}{d^3p}\right|_{\rm soft} + \left.\frac{d\Gamma_{\tilde{a}}}{d^3p}\right|_{\rm hard}. \tag{4}$$

In Fig. 1a, we show the normalized thermal axino production rate $1/\Gamma_{\tilde{a}}\, d\Gamma_{\tilde{a}}/d(E/T)$ as a function of E/T for temperatures of $T = 10^6$ GeV, 10^7 GeV, 10^8 GeV, and 10^9 GeV. For $E \lesssim T$, we expect modifications as our computation is restricted to the thermal production of hard ($E \gtrsim T$) axinos.[3] Here and below the 1-loop running of the strong coupling in the MSSM is taken into account by replacing g with $g(T) = [g^{-2}(M_Z)+3\ln(T/M_Z)/(8\pi^2)]^{-1/2}$, where the value of the strong coupling at the Z-boson mass M_Z, $g^2(M_Z)/(4\pi) = 0.118$, is used as input. For $E/T \lesssim 2$, the rate turns negative except for very high temperatures. With $g(T = 10^6\,\text{GeV}) = 0.986$ and $g(T = 10^9\,\text{GeV}) = 0.880$, this unphysical behavior follows from the extrapolation of our result from $g \ll 1$ to $g \approx 1$. As higher-order corrections might become sizable for $T \lesssim 10^6$ GeV, new techniques are needed to compute the thermal production rate reliably also for $g \gtrsim 1$. At present, however, our result derived in a gauge-invariant way supersedes previous gauge-dependent and cutoff-dependent estimates.[2]

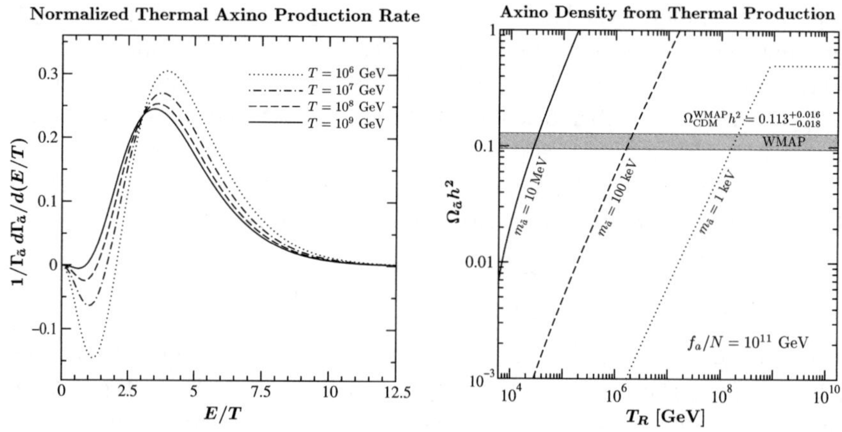

Figure 1. (a) The normalized thermal axino production rate $1/\Gamma_{\tilde{a}}\, d\Gamma_{\tilde{a}}/d(E/T)$ as a function of E/T for $T = 10^6$ GeV (dotted line), 10^7 GeV (dash-dotted line), 10^8 GeV (dashed line), and 10^9 GeV (solid line). The curves result from (4) derived for $E \gtrsim T$. (b) The axino density parameter $\Omega_{\tilde{a}} h^2$ as a function of T_R for $f_a/N = 10^{11}$ GeV and $m_{\tilde{a}} = 1$ keV (dotted line), 100 keV (dashed line), and 10 MeV (solid line). The grey band indicates the WMAP result on the relic density of cold dark matter (2σ error).

3. Stable Axinos as Dark Matter in the Universe

By comparing the axino interaction rate for $f_a/N = 10^{11}$ GeV with the Hubble parameter H for the early radiation-dominated epoch, an axino decoupling temperature of $T_D \approx 10^9$ GeV is estimated.[6] For $T_R < T_D$, axinos have not been in thermal equilibrium after inflation. Here the evolution of the axino number density $n_{\tilde{a}}$ with cosmic time t is described by the Boltzmann equation with a collision term $C_{\tilde{a}}$ accounting for both the axino production and disappearance processes in the primordial plasma,

$$\frac{dn_{\tilde{a}}}{dt} + 3Hn_{\tilde{a}} = C_{\tilde{a}} \ . \tag{5}$$

The disappearance processes can be neglected for T_R sufficiently below T_D. Then, by integrating the production rate (4), we obtain the collision term

$$C_{\tilde{a}} \approx \frac{(N_c^2-1)}{(f_a/N)^2}\frac{3\zeta(3)g^6T^6}{4096\pi^7}\left[\ln\left(\frac{1.380\,T^2}{m_g^2}\right)(N_c+n_f)+0.4336\,n_f\right] , \tag{6}$$

where $\zeta(3) \approx 1.2021$. Assuming conservation of entropy per comoving volume, the Boltzmann equation can be solved analytically. The resulting present (t_0) axino density parameter $\Omega_{\tilde{a}} h^2 = m_{\tilde{a}} n_{\tilde{a}}(t_0) h^2/\rho_c$ with $\rho_c/h^2 = 3.6 \times 10^{-9}$ GeV depends on the axino mass $m_{\tilde{a}}$, the PQ scale f_a/N, and the reheating temperature T_R in the following way

$$\Omega_{\tilde{a}} h^2 = 5.5\, g^6 \ln\left(\frac{1.108}{g}\right)\left(\frac{m_{\tilde{a}}}{0.1\ \mathrm{GeV}}\right)\left(\frac{10^{11}\,\mathrm{GeV}}{f_a/N}\right)^2\left(\frac{T_R}{10^4\,\mathrm{GeV}}\right), \quad (7)$$

where $g = g(T_R)$. In Fig. 1b, this result is illustrated as a function of T_R for $f_a/N = 10^{11}\,\mathrm{GeV}$ and $m_{\tilde{a}} = 1$ keV, 100 keV, and 10 MeV. For T_R above T_D, axinos are in thermal equilibrium so that $\Omega_{\tilde{a}} h^2$ is independent of T_R as shown for $m_{\tilde{a}} = 1$ keV by the horizontal line. There will be a smooth transition instead of a kink once the axino disappearance processes are taken into account. The grey band indicates the WMAP result[7] on the cold dark matter density (2σ error) $\Omega_{\mathrm{CDM}}^{\mathrm{WMAP}} h^2 = 0.113^{+0.016}_{-0.018}$.

4. Conclusion

For $m_{\tilde{a}} = 100$ keV and $T_R \approx 10^6\,\mathrm{GeV}$, the relic axino density (obtained with $f_a/N = 10^{11}\,\mathrm{GeV}$) agrees with the WMAP result on the cold dark matter density. Although relatively light for being cold dark matter, axinos with $m_{\tilde{a}} = 100$ keV could still explain large-scale-structure formation, the corresponding power spectrum, and the early reionization observed by WMAP. Thus, as far as abundance and structure formation are concerned, axinos from thermal production in the early Universe are indeed a viable solution of the cold dark matter problem.[8] This conclusion has already been drawn in Ref. 2 based on a gauge-dependent and cutoff-dependent estimate of the relic axino abundance from thermal production. Since our result for $\Omega_{\tilde{a}} h^2$ is smaller by a factor of ten, the upper limit on T_R for a given value of $m_{\tilde{a}}$ is relaxed by one order of magnitude. This can be important for models of inflation and for the understanding of the baryon asymmetry in the Universe. Already $T_R \approx 10^6\,\mathrm{GeV}$ is relatively small and excludes, for example, thermal leptogenesis.

References

1. H. P. Nilles and S. Raby, *Nucl. Phys.* **B198**, 102 (1982);
   J. E. Kim and H. P. Nilles, *Phys. Lett.* **B138**, 150 (1984).
2. L. Covi *et al.*, *J. High Energy Phys.* **05**, 033 (2001).
3. A. Brandenburg and F. D. Steffen, arXiv: hep-ph/0405158.
4. E. Braaten and R. D. Pisarski, *Nucl. Phys.* **B337**, 569 (1990).
5. E. Braaten and T. C. Yuan, *Phys. Rev. Lett.* **66**, 2183 (1991).
6. K. Rajagopal, M. S. Turner, and F. Wilczek, *Nucl. Phys.* **B358**, 447 (1991).
7. D. N. Spergel *et al.*, *Astrophys. J. Suppl.* **148**, 175 (2003).
8. A. Brandenburg and F. D. Steffen, arXiv:hep-ph/0406021.

THE 2-PI-1/N APPROXIMATION APPLIED TO TACHYONIC PREHEATING

ANDERS TRANBERG,
ALEJANDRO ARRIZABALAGA AND JAN SMIT
Institute of Theoretical Physics, University of Amsterdam,
Valckenierstraat 65, 1018XE, Amsterdam, The Netherlands.

We study the process of tachyonic preheating after inflation using the 2PI-1/N approximation to the quantum dynamics, and compare with the classical approximation.

1. Tachyonic preheating

We consider an $O(N)$ model of scalar (Higgs) fields ϕ_a, $a = 1, \cdots, N$, with a time dependent quadratic term

$$S = -\int d^4x \left(\frac{\partial^\mu \phi_a \partial_\mu \phi_a}{2} + \frac{\mu_{\text{eff}}^2(t)}{2}\phi_a\phi_a + \frac{\lambda}{24N}(\phi_a\phi_a)^2 \right). \quad (1)$$

In hybrid inflation models, the inflaton (σ) may be coupled to the scalar fields in such a way that it triggers symmetry breaking as it rolls off its potential,

$$\mu_{\text{eff}}^2(t) = \tilde{\mu}^2 - \lambda_{\sigma\phi}\sigma^2(t). \quad (2)$$

We consider the limiting case when μ_{eff}^2 flips its sign instantaneously,

$$\mu_{\text{eff}}^2(t) = \mu^2,\ t < 0, \quad \mu_{\text{eff}}^2(t) = -\mu^2,\ t > 0. \quad (3)$$

After the flip, the amplitude of momentum modes with $|\mathbf{k}| < \mu$ grows exponentially,

$$\phi_{\mathbf{k}} \propto e^{\pm i\sqrt{\mu_{\text{eff}}^2 + k^2}\, t} = e^{\mp\sqrt{\mu^2+k^2}\, t}. \quad (4)$$

This process is known as tachyonic preheating[1]. As the modes grow, eventually the back-reaction from the quartic (interaction) term will become important. Through scattering, the system will thermalize to some temperature, T_{reh}. Because of the high occupation numbers, it is believed that the classical approximation to the dynamics is reliable, and all numerical simulations of tachyonic preheating[1,2,3,4] and baryogenesis during tachyonic preheating[5,6] have used classical dynamics.

2. The 2PI-1/N approximation

Out-of-equilibrium dynamics of quantum fields has been studied recently using so-called Φ-derived equations of motion, in terms of the 2PI effective action[7,8]. These equations of motion concern the full, dressed propagator $G(x,y) = -i\langle\phi(x)\phi(y)\rangle$,

$$\delta(x,y) = \left(G_0^{-1}(x,z) - \Sigma(x,z)\right) G(z,y), \tag{5}$$

The self-energy is determined by

$$\Sigma(x,y) = 2\frac{\delta\Phi[G]}{\delta G(x,y)} \tag{6}$$

from (some truncation of) the functional Φ, the sum of all 2PI skeleton diagrams with full propagator lines and bare vertices. We use an expansion in $1/N$ to NLO, which is known to be able to cope with the large occupation numbers expected to be generated during the tachyonic instability[9]. The initial condition is $O(N)$ symmetric and so the mean field $\bar{\phi}$ is zero and $G_{ab}(x,y) = \delta_{ab}G(x,y)$. Also, we assume the system to be homogeneous, so that $G(x,y) = G(t,t',\mathbf{x}-\mathbf{y})$. The full equations of motion and additional details can be found elsewhere[9,10].

3. Classical approximation

In the classical approximation, a sample of initial configurations is drawn from an ensemble reproducing the (real part of the) initial quantum two-point correlators. We take the initial condition to be a zero temperature (vacuum) free-field state, with the correlators

$$\frac{1}{2}\langle\{\hat{\phi}^a_{\mathbf{k}},\hat{\phi}^b_{-\mathbf{k}}\}\rangle = \frac{1}{2\,\omega_k}\delta_{ab}, \quad \frac{1}{2}\langle\{\hat{\pi}^a_{\mathbf{k}},\hat{\pi}^b_{-\mathbf{k}}\}\rangle = \frac{\omega_k}{2}\delta_{ab}, \tag{7}$$

$$\frac{1}{2}\langle\{\hat{\pi}^a_{\mathbf{k}},\hat{\phi}^b_{-\mathbf{k}}\}\rangle = 0, \quad \omega_k^2 = k^2 + \mu^2. \tag{8}$$

This determines the initial conditions for the classical as well as the quantum evolution (for quantum, adding $\langle[\hat{\phi}^a_{\mathbf{k}},\hat{\pi}^b_{-\mathbf{k}}]\rangle = i\delta_{ab}$). For the classical case we only initialize the unstable modes ($|\mathbf{k}| < \mu$). The classical sample is evolved using classical equations of motion derived from eq. (1). Observables are averaged over the ensemble. In the limit of $\lambda = 0$ (quadratic approximation), the quantum and classical evolution are identical, and we can solve for the correlators in time,

$$\langle\hat{\phi}^a_{\mathbf{k}}(t)\hat{\phi}^b_{-\mathbf{k}}(t)\rangle = \frac{1}{2\omega_k^+}\left[1+\left(\frac{\omega_k^{+2}}{\omega_k^{-2}}-1\right)\sin^2(\omega_k^- t)\right]\delta_{ab}, \tag{9}$$

$$\langle\hat{\pi}^a_{\mathbf{k}}(t)\hat{\pi}^b_{-\mathbf{k}}(t)\rangle = \frac{\omega_k^{-2}}{2\omega_k^+}\left[1+\left(\frac{\omega_k^{+2}}{\omega_k^{-2}}-1\right)\cos^2(\omega_k^- t)\right]\delta_{ab}, \tag{10}$$

$$\langle\hat{\phi}^a_{\mathbf{k}}(t)\hat{\pi}^b_{-\mathbf{k}}(t)\rangle = \left[\frac{\omega_k^-}{4\omega_k^+}\left(\frac{\omega_k^{+2}}{\omega_k^{-2}}-1\right)\sin(2\omega_k^- t)+\frac{i}{2}\right]\delta_{ab}. \tag{11}$$

with $\omega_k^\pm = \sqrt{\pm\mu^2+k^2}$. As observables we look at "occupation numbers" n_k and "frequencies" ω_k defined as

$$n_k + 1/2 = \sqrt{\langle\hat{\phi}^a_{\mathbf{k}}(t)\hat{\phi}^a_{-\mathbf{k}}(t)\rangle\langle\hat{\pi}^b_{\mathbf{k}}(t)\hat{\pi}^b_{-\mathbf{k}}(t)\rangle/N}, \tag{12}$$

$$\omega_k = \sqrt{\langle\hat{\pi}^a_{\mathbf{k}}(t)\hat{\pi}^a_{-\mathbf{k}}(t)\rangle/\langle\hat{\phi}^b_{\mathbf{k}}(t)\hat{\phi}^b_{-\mathbf{k}}(t)\rangle}. \tag{13}$$

We apply this definition to the quadratic, quantum and classical (with $n_k + 1/2 \to n_k$) approximations alike.

4. Results

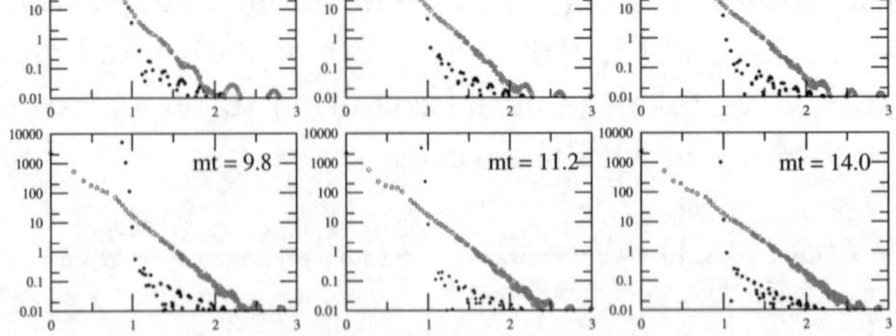

Figure 1. The occupation numbers (n_k vs. k/μ) during the initial instability for the NLO dynamics (grey) compared to the quadratic approximation (black). Initially the agreement is perfect, but as back-reactions set in, the growth is stopped in the NLO and scattering moves power from the low to the higher momentum modes. (m in the plot is the μ of eq. (3).)

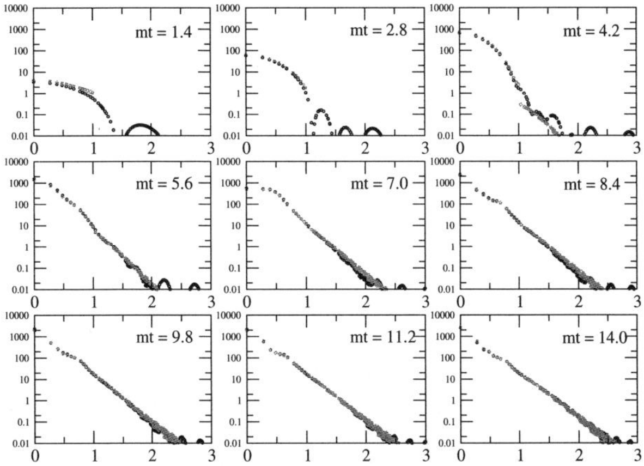

Figure 2. The NLO dynamics (black) compared to the classical approximation (grey). Only the unstable modes are initialized classically. The classical approximation approximates the initial growth well, and as back-reactions set in, the un-initialized modes become populated and reproduce the NLO distribution.

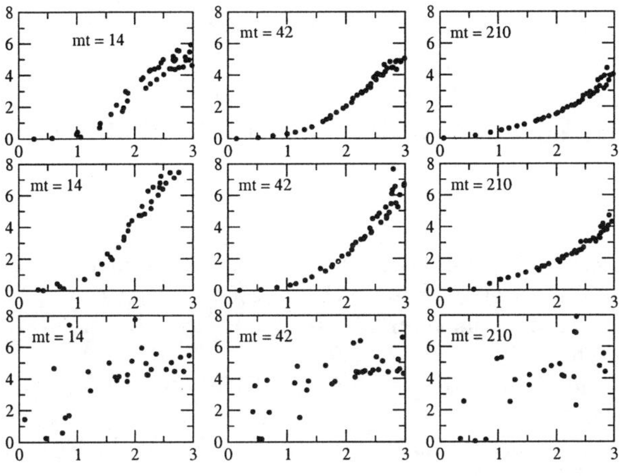

Figure 3. At later times, equilibration begins to take place for the NLO (top) and classical (middle) approximations (shown is $\ln(1 + 1/n_k)$ vs. ω_k/μ). We include the Hartree approximation (bottom) for comparison, which in known to have no equilibration. At the latest time a straight line is emerging in the low momentum range, suggesting that the spectrum is approaching a Bose-Einstein distribution[10] .

5. Conclusion

We have compared the 2PI-1/N-NLO and classical approximations when applied to tachyonic preheating. The two agree well at early times ($\mu t = 14$)

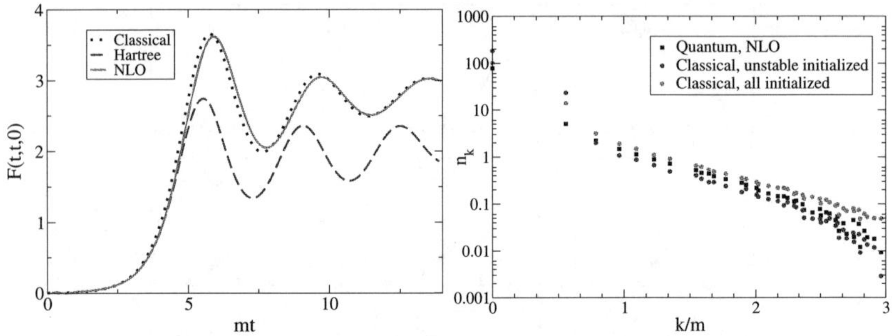

Figure 4. Left: $\langle\phi(x)\phi(x)\rangle$ correlator for different approximations during and just after the initial roll-off. Classical and NLO agree nicely. Right: The lowest momentum modes at the latest time ($\mu t = 210$), comparing NLO to classical. The agreement is reasonable, although the very lowest modes are surprisingly off.

validating both approaches for simulations of preheating. The NLO truncation includes scattering and equilibration, and for longer times we find kinetic equilibration on a time scale of a few hundred in μ-units. The classical approximation gives qualitatively similar results, although the very lowest modes are off at late times. Additional comparisons including Hartree dynamics and different classical initialization schemes can be found elsewhere[10].

Acknowledgments: This work is supported by FOM/NWO.

References

1. G. N. Felder, J. Garcia-Bellido, P. B. Greene, L. Kofman, A. D. Linde and I. Tkachev, Phys. Rev. Lett. **87**, 011601 (2001) [arXiv:hep-ph/0012142].
2. A. Rajantie, P. M. Saffin and E. J. Copeland, Phys. Rev. D **63**, 123512 (2001) [arXiv:hep-ph/0012097].
3. J. I. Skullerud, J. Smit and A. Tranberg, JHEP **0308**, 045 (2003) [arXiv:hep-ph/0307094].
4. S. Borsanyi, A. Patkos and D. Sexty, Phys. Rev. D **68**, 063512 (2003) [arXiv:hep-ph/0303147].
5. J. Garcia-Bellido, M. Garcia-Perez and A. Gonzalez-Arroyo, Phys. Rev. D **69**, 023504 (2004) [arXiv:hep-ph/0304285].
6. A. Tranberg and J. Smit, JHEP **0311**, 016 (2003) [arXiv:hep-ph/0310342].
7. J. M. Cornwall, R. Jackiw and E. Tomboulis, Phys. Rev. D **10**, 2428 (1974).
8. J. Berges and J. Cox, Phys. Lett. B **517**, 369 (2001) [arXiv:hep-ph/0006160].
9. J. Berges, Nucl. Phys. A **699**, 847 (2002) [arXiv:hep-ph/0105311].
10. A. Arrizabalaga, J. Smit and A. Tranberg, In preparation.

NONEQUILIBRIUM DYNAMICS IN SCALAR HYBRID MODELS

JÜRGEN BAACKE and ANDREAS HEINEN

Institut für Physik, Universität Dortmund, D-44221 Dortmund , Germany

We study by numerical simulations the transition from the metastable "false vacuum" to the broken symmetry phase in the preheating stage after cosmic inflation in a scalar hybrid model. We take quantum fluctuations and their back reaction into account by applying a one-loop bubble-resummation.[1]

1. Introduction

At present there is a wide class of inflationary models, having in common the aim of explaining various kinds of cosmological puzzles[2]. One of the consecutive questions raised is how inflation has ended. In a scenario called "Hybrid inflation"[3] the inflationary expansion ends by a phase transition from a metastable "false vacuum" to the true, i.e. broken symmetry vacuum. The subsequent phase of (p)reheating[4] after inflation is characterized by a phase of abundant particle production due to spinodal amplification of quantum modes and due to parametric resonance. We will study some aspects of this phase in the following.

2. Effective action and renormalized equations of motion

The Lagrangian for the Hybrid model is given by

$$\mathcal{L} = \frac{1}{2}\partial_\mu \Phi \partial^\mu \Phi + \frac{1}{2}\partial_\mu X \partial^\mu X - \frac{1}{2}m^2\Phi^2 - \frac{1}{2}g^2\Phi^2 X^2 - \frac{\lambda}{4}(X^2 - v^2)^2 \, . \quad (1)$$

The field X has an effective mass square $m_X^2(\Phi) = -\lambda v^2 + g^2\Phi^2$ and becomes unstable if the absolute value of Φ is lower than $\Phi_c = \frac{v\sqrt{\lambda}}{g}$ (spinodal region). Both quantum fields Φ and X are assumed to have classical expectation values, i.e. $\langle\Phi\rangle = \phi$ (inflaton field) and $\langle X\rangle = \chi$ (symmetry breaking field).

We will use here the so-called *two-particle point-irreducible* (2PPI) effective action (EA) formalism. [5,6] The renormalized equations of motion

(EOM) that describe the nonequilibrium dynamics of classical mean fields and their quantum fluctuations, all coupled to each other, are derived from the EA functional

$$\Gamma[\phi,\chi,\Delta_{\{\phi\phi,\phi\chi,\chi\chi\}}] = S[\phi,\chi] + \frac{g^2}{2}\int d^Dx\,(\Delta_{\phi\phi}\Delta_{\chi\chi} + 2\Delta_{\phi\chi}^2) \tag{2}$$
$$+\frac{3\lambda}{4}\int d^Dx\Delta_{\chi\chi}^2 + \Gamma^{2\mathrm{PPI}}[\phi,\chi,\mathcal{M}^2_{\{\phi\phi,\phi\chi,\chi\chi\}}]\,,$$

$\Gamma^{2\mathrm{PPI}}[\phi,\chi,\mathcal{M}^2_{\{\phi\phi,\phi\chi,\chi\chi\}}] =$ 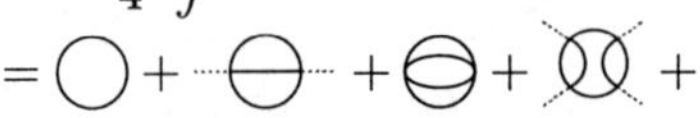 $+\ldots\,.$ (3)

The classical EOM follow from $\frac{\delta\Gamma}{\delta\phi}=0$ and $\frac{\delta\Gamma}{\delta\chi}=0$, while the gap equations are derived from $\frac{\delta\Gamma}{\mathcal{M}^2_{ij}}=0$ with $ij=\phi\phi,\chi\chi$ or $\phi\chi$. The propagator G fulfills a constraint equation

$$(G^{-1})_{ij}(x,x') = i\left(\Box\delta_{ij} + \mathcal{M}^2_{ij}(x)\right)\delta^{(D)}(x-x')\,. \tag{4}$$

It is dressed by a resummation of local self-energy insertions. Assuming spatial homogeneous fields, $G_{ij}(t,t';\mathbf{p})$ can be decomposed in momentum space into mode functions via $G_{ij}(t,t';\mathbf{p}) = \sum_{\alpha=1}^{2}\frac{1}{2\omega_\alpha}f_i^\alpha(t;p)f_j^{*\,\alpha}(t';p)$. The quantities ω_α ($\alpha=1,2$) have to be determined self consistently at the initial time ($t=0$). For practical calculations one has to truncate the infinite series in Eq.(3). Here we will only keep the one-loop term which leads to a summation of bubble diagrams.[a] It is given by $\Gamma^{(1)}[\mathcal{M}^2_{\phi\phi},\mathcal{M}^2_{\chi\chi},\mathcal{M}^2_{\phi\chi}] = \frac{i}{2}\mathrm{Tr}\ln[G^{-1}]$. The tadpole insertions derived from $\Gamma^{(1)}$ denote

$$\Delta_{ij}^{(1)}(t) = -2\frac{\delta\Gamma^{(1)}}{\delta\mathcal{M}^2_{ij}(t)} = \sum_{\alpha=1}^{2}\int\frac{d^{D-1}p}{(2\pi)^{D-1}}\frac{1}{2\omega_\alpha}\mathrm{Re}\left[f_i^\alpha(t;p)f_j^{\alpha\,*}(t;p)\right]\,. \tag{5}$$

These quantities have quadratic and logarithmic divergencies and have to be renormalized. It turns out[1] that the EA is renormalized by a simple vacuum counter term $\delta E_{\mathrm{vac}} = -\delta\xi[(\mathcal{M}^2_{\phi\phi})^2 + 2(\mathcal{M}^2_{\phi\chi})^2 + (\mathcal{M}^2_{\chi\chi})^2]$, where in dimensional regularization $\delta\xi = -\frac{1}{64\pi^2}(\frac{2}{\epsilon}-\gamma+1+\ln 4\pi)$. The vacuum counter term δE_{vac} indeed reflects a resummation of the standard counterterms derived from a counterterm Lagrangian.[5] The renormalized gap equations for the variational parameters $\mathcal{M}^2_{ij}$ take the form

$$\mathcal{M}^2_{\mathrm{R},\,\phi\phi}(t) = m^2 + g^2\left[\chi^2(t) + \Delta_{\chi\chi}(t)\right] - 4g^2\,\delta\xi\mathcal{M}^2_{\mathrm{R},\chi\chi}(t)\,, \tag{6}$$
$$\mathcal{M}^2_{\mathrm{R},\,\chi\chi}(t) = -\lambda v^2 + g^2\left[\phi^2(t) + \Delta_{\phi\phi}(t)\right] + 3\lambda\left[\chi^2(t) + \Delta_{\chi\chi}(t)\right]$$
$$-4g^2\,\delta\xi\mathcal{M}^2_{\mathrm{R},\,\phi\phi}(t) - 12\lambda\,\delta\xi\mathcal{M}^2_{\mathrm{R},\,\chi\chi}(t)\,, \tag{7}$$
$$\mathcal{M}^2_{\mathrm{R},\,\phi\chi}(t) = 2g^2\left[\phi(t)\chi(t) + \Delta_{\phi\chi}(t)\right] - 8g^2\,\delta\xi\mathcal{M}^2_{\mathrm{R},\,\phi\chi}(t)\,. \tag{8}$$

---

[a]Equivalently one can take the *local part* of a two-loop-2PI approximation.

The final set of gap equations, with all finite parts from dimensional regularization, is rather lengthy[1] and we don't write it down here. However, one can see from Eqs. (6)-(8) that the gap equations form a (3×3) system of linear equations, whose coefficient matrix has to be diagonalized by a time-independent rotation matrix. This fixes the various mass and coupling constant counterterms, obtained from a standard counterterm Lagrangian, in a non-perturbative way.

After all the classical EOM are given by

$$\ddot{\phi}(t) + \mathcal{M}^2_{\mathrm{R},\,\phi\phi}(t)\phi(t) + \mathcal{M}^2_{\mathrm{R},\,\phi\chi}(t)\chi(t) - 2g^2\chi^2(t)\phi(t) = 0\,, \quad (9)$$

$$\ddot{\chi}(t) + \mathcal{M}^2_{\mathrm{R},\,\chi\chi}(t)\chi(t) + \mathcal{M}^2_{\mathrm{R},\,\phi\chi}(t)\phi(t) - 2\lambda\chi^3(t) - 2g^2\phi^2(t)\chi(t) = 0 \quad (10)$$

Inserting the decomposition of $G_{ij}(t,t;\mathbf{p})$ into mode functions $f_i^\alpha(t,p)$ in Eq. (4) the system of EOM for the $f_i^\alpha(t,p)$ denotes explicitly as

$$\ddot{f}^\alpha_\phi(t;p) + \mathbf{p}^2 f^\alpha_\phi(t;p) + \mathcal{M}^2_{\mathrm{R},\,\phi\phi}(t) f^\alpha_\phi(t;p) + \mathcal{M}^2_{\mathrm{R},\,\phi\chi}(t) f^\alpha_\chi(t;p) = 0\,, \quad (11)$$

$$\ddot{f}^\alpha_\chi(t;p) + \mathbf{p}^2 f^\alpha_\chi(t;p) + \mathcal{M}^2_{\mathrm{R},\,\chi\chi}(t) f^\alpha_\chi(t;p) + \mathcal{M}^2_{\mathrm{R},\,\phi\chi}(t) f^\alpha_\phi(t;p) = 0\,. \quad (12)$$

3. Results

The EOM (9)-(12) are solved numerically. After renormalization the momentum integrals are finite and are carried out using a non-equidistant momentum discretization. In order to simulate the phase transition at the end of inflation we choose the initial amplitudes $\phi(0) > \phi_c$ and $\chi(0) \approx 0$. In particular we take $\phi(0) = 1.2$ and $\chi(0) = 10^{-7}$. The other parameters are here $m^2 = 0$, $v = 1$, $g = 2\lambda$, $\lambda = 1$.

The time evolution of the classical fields ϕ and χ and of the variational masses is displayed in Fig. 1. One observes three time regimes: (i) the first regime is characterized by a slow-roll of the field ϕ, where quantum fluctuations are almost negligible; (ii) in the spinodal region the mass $\mathcal{M}^2_{\chi\chi}$ becomes negative (Fig. 1b). Gaussian fluctuations build up (left column in Fig. 2) and drive $\mathcal{M}^2_{\chi\chi}$ back to positive values. Along the way the amplitude of χ growths exponentially; (iii) at later times the momentum spectra display several spikes (see e.g. the right column in Fig. 2) and the symmetry breaking field oscillates around a nonzero value with a relatively constant amplitude. The details in the dynamics depend on the parameters in the model while the characteristic features are very similar.[1]

In Fig. 3 we display the time and space evolution of correlations between the fluctuations. One can see that spatial correlations build up in the spinodal region and propagate by twice the speed of light. The correlations

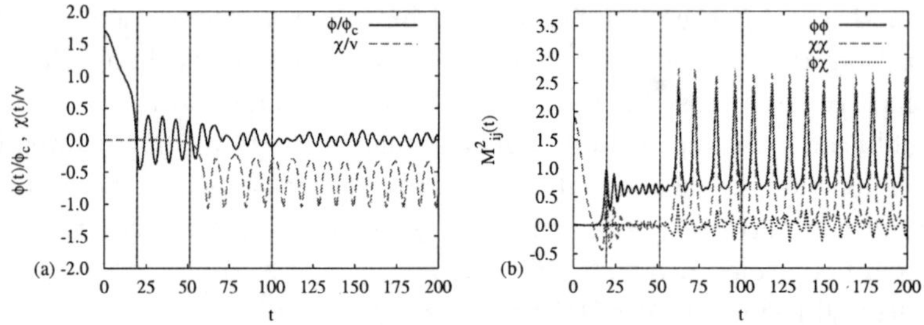

Figure 1. Time evolution of (a) the classical fields $\phi(t)/\phi_c$ (solid line) and $\chi(t)/v$ (dashed line), (b) the effective masses $\mathcal{M}^2_{ij}(t)$ with $ij = \phi\phi$ (solid line), $ij = \chi\chi$ (dashed line) and $ij = \phi\chi$ (dotted line); the vertical dotted lines indicate the times 19.5, 51.3 and 100.2; parameters: $g^2 = 2\lambda$, $\phi(0) = 1.2$, $\chi(0) = 1.0 \times 10^{-7}$, $m^2 = 0$, $\lambda = 1$ and $v^2 = 1$.

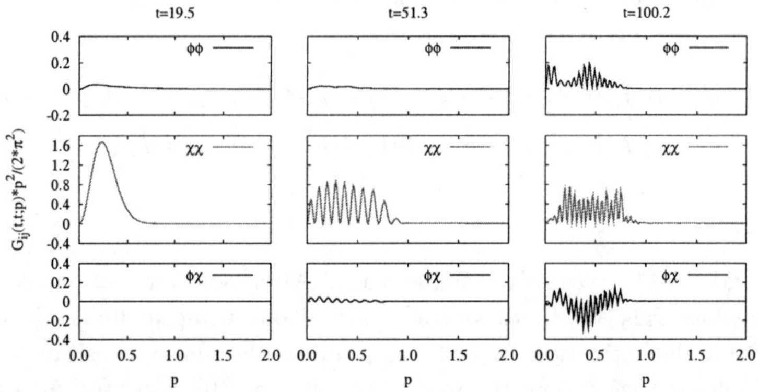

Figure 2. Momentum spectra $G_{ij}(t, t; \mathbf{p})p^2/(2\pi^2)$ for the simulation in Fig. 1 at the times $t = 19.5$ (left column), $t = 51.3$ (middle column) and $t = 100.2$ (right column).

decrease once the field χ starts to oscillate around a non-zero expectation value.

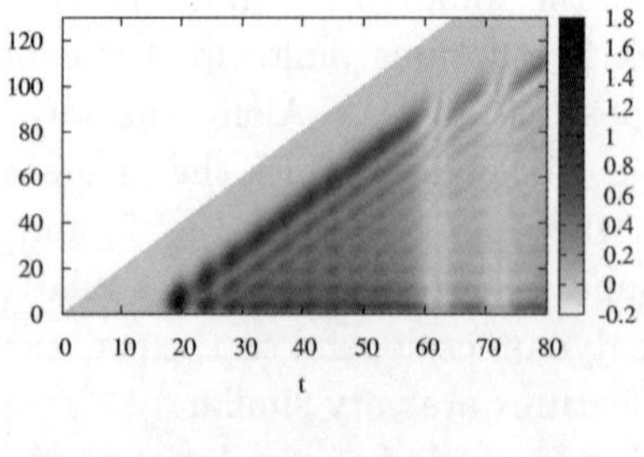

Figure 3. Correlation function $rC_{\chi\chi}(r,t) = r\int \frac{d^3p}{(2\pi)^3} e^{i\mathbf{p}\cdot\mathbf{r}} \sum_\alpha \frac{1}{2\omega_\alpha} \mathrm{Re}[f^\alpha_\chi f^{*\,\alpha}_\chi]$ for the simulation in Fig. 1. The propagation velocity is $r/t = 2$ (twice the speed of light).

The decoherence time $t_{\rm dec}(p)$ for which a given mode becomes classical is defined by $|F_{ij}(t_{\rm dec}, \mathbf{p})| = 1$ with $F_{ij}(t, \mathbf{p}) = \mathrm{Im}[\sum_\alpha \frac{f_i^{\alpha *} \dot{f}_j^\alpha}{2\omega_\alpha}]$. The modes right to the curves in Fig. 4 never become classical. The functional dependence is not proportional to p^2, as found in models with a slow quench where $\mathcal{M}_{\chi\chi}^2(t) \propto (t_0 - t)$.

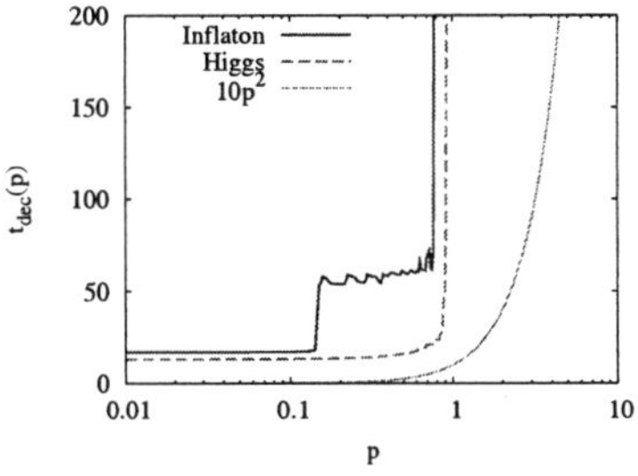

Figure 4. Decoherence time $t_{\rm dec}(p)$ for which a given mode p becomes "classical" for the simulation in Fig. 1; the solid line represents the inflaton and the dashed line the Higgs modes; the dotted line corresponds to $t_{\rm dec} \propto p^2$.

4. Conclusions

We have analyzed by numerical simulations the transition from the metastable phase to the broken symmetry phase in the Hybrid model. We observed that the back reaction of quantum fluctuations plays an important role and leads, e.g., to a dynamical stabilization in the spinodal region. The approximation used here provides reliable information on the nonequilibrium evolution at earlier times. However, one has to improve further in order to cure, e.g. the lack of dissipation at later times. While there are various improvements, once scattering is taken into account via non-local two-loop diagrams, even in the 2PPI resummation scheme [6], one ultimately wants to achieve a universal late time behavior, i.e. reaching quantum thermalization. The full treatment of the reheating phase remains a desirable goal for future investigations.

AH thanks the GK 841 for partial financial support.

References

1. J. Baacke and A. Heinen, Phys. Rev. D **69**, 083523 (2004)
2. see e.g., D. H. Lyth and A. Riotto, Phys. Rept. **314**, 1 (1999).
3. A. D. Linde, Phys. Lett. **B249**, 18 (1990).
4. see e.g., G. N. Felder *et al.*, Phys. Rev. Lett. **87**, 011601 (2001); E. J. Copeland, D. Lyth, A. Rajantie and M. Trodden, Phys. Rev. **D64**, 043506 (2001); J. Garcia-Bellido, M. Garcia Perez and A. Gonzalez-Arroyo, Phys. Rev. **D67**, 103501 (2003); S. Borsanyi, A. Patkos and D. Sexty, Phys. Rev. **D68**, 063512 (2003).
5. H. Verschelde and M. Coppens, Phys. Lett. **B287**, 133 (1992).
6. J. Baacke and A. Heinen, Phys. Rev. D **67**, 105020 (2003); *ibid.* **68**, 127702 (2003)

PHOTON MASS IN INFLATION AND NEARLY MINIMAL MAGNETOGENESIS

TOMISLAV PROKOPEC

Institute for Theoretical Physics & Spinoza Institute,
Leuvenlaan 4, 3584 CE Utrecht, The Netherlands;
Email: T.Prokopec@phys.uu.nl

We argue that the dynamics of photons canonically coupled to scalars in de Sitter inflation gets modified by a photon mass term. This might have lead to generation of potentially observable magnetic field correlated over cosmological scales today.

1. Introduction

One of the important unresolved puzzles of modern astrophysics and cosmology is the origin of large scale magnetic fields of the Universe. It is conceivable that galactic magnetic fields are explained by the dynamo mechanism, by which an initial small seed field at the time of galaxy formation was amplified to the microgauss strength observed today. Seed fields of magnitudes $B_{\rm seed} \sim 10^{-20}$ Gauss and correlated over distances larger than about 100 parsecs could have been created by the Biermann battery mechanism induced by the large scale structure dynamics at the time of galaxy formation. On the other hand, it seems that no currently available astrophysical scenario can explain recent cluster fields [1] observed in several galaxy clusters.

Based on a study of scalar electrodynamics (ΦED) during inflation [2,3,4,5,6,7,8,9] and subsequent radiation and matter epochs, in this talk I will argue that the observed large scale magnetic fields could have originated in cosmological inflation.

2. The Model

Cosmological inflation is an epoch of a rapid (accelerated) expansion of the Universe. For simplicity, here we model inflation by de Sitter space-time. In conformal coordinates it is characterized by the metric, $g_{\mu\nu} = a^2\eta_{\mu\nu}$, where $\eta_{\mu\nu} = {\rm diag}(-1, 1, 1, 1)$ and the scale factor $a = -1/(H\eta)$, H is the Hubble parameter and η conformal time. In conformal coordinates the Lagrangian of scalar electrodynamics reduces to

$$\mathbf{L}_{\Phi\mathrm{ED}} = -\frac{1}{4}\eta^{\mu\nu}\eta^{\rho\sigma}F_{\mu\rho}F_{\nu\sigma} - a^2\eta^{\mu\nu}(D_\mu\Phi)^*(D_\nu\Phi) - a^4(m_\phi^2 + \xi\mathrm{R})\Phi^*\Phi\,, \quad (1)$$

where $F_{\mu\rho} = \partial_\mu A_\rho - \partial_\rho A_\mu$ is the photon field strength, $D_\mu = \partial_\mu - ieA_\mu$ is the covariant derivative, m_ϕ the scalar field mass, R the curvature scalar, which in $D = 4$ and de Sitter inflation reads $\mathrm{R} = 12H^2$, and A_μ is the gauge (photon) field. When $m_\phi^2 + \xi\mathrm{R} \neq 0$, conformal invariance of Φ is broken during inflation, resulting in a growth of scalar fluctuations primarily on superhorizon scales. This is formally encoded by the Chernikov-Tagirov [11] (Bunch-Davies) Feynman propagator, which in $D = 4$ reads

$$iG_F(y) = \frac{H^2}{4\pi^2}\Gamma\Big(\frac{3}{2} - \mu\Big)\Gamma\Big(\frac{3}{2} + \mu\Big)\,{}_2F_1\Big(\frac{3}{2} - \mu, \frac{3}{2} + \mu, 2; 1 - \frac{y}{4}\Big)\,, \quad (2)$$

where $\mu = [(3/2)^2 - (m_\phi^2 + \xi\mathrm{R})/H^2]^{1/2}$, and $y = a(\eta)a(\eta')\Delta x^2$ is the de Sitter invariant length, $\Delta x^2 = -(|\eta - \eta'| - i\epsilon)^2 + \|\vec{x} - \vec{x}'\|^2$. When expanded in $s \equiv (3/2) - \mu \approx (m_\phi^2/3H^2) + 4\xi$ ($|s| \ll 1$), the propagator simplifies to [3]

$$iG_F(y) \simeq \frac{H^2}{4\pi^2}\Big\{\frac{1}{y} - \frac{1}{2}\ln(y) + \frac{1}{2s} - 1 + \ln(2)\Big\}\,. \quad (3)$$

Note that this propagator is still fully de Sitter invariant. In the limit when $m_\phi \to 0$ and $\xi \to 0$ however, the propagator diverges (2), and a natural choice is an $SO(3,1)$ invariant propagator [12,3],

$$iG_F|_{m_\phi=0} \simeq \frac{H^2}{4\pi^2}\Big\{\frac{1}{y} - \frac{1}{2}\ln(y) + \frac{1}{2}\ln\big(a(\eta)a(\eta')\big) - \frac{1}{4} + \ln(2)\Big\}\,. \quad (4)$$

The de Sitter invariance $SO(4,1)$ is broken by the term $\propto \ln\big(a(\eta)a(\eta')\big)$. We shall use both propagators (3) and (4) to study how scalars influence the photon dynamics during inflation and its consequences today.

During inflation the photon dynamics in ΦED (1) is governed by the modified Maxwell equation,

$$\eta^{\mu\nu}\partial_\mu F_{\nu\rho} + \eta_{\rho\mu}\int d^4x'[{}^\mu\Pi^\nu_{\mathrm{ret}}](x,x')A_\nu(x') = 0\,, \quad (5)$$

where $[{}^\mu\Pi^\nu_{\mathrm{ret}}]$ denotes the retarded vacuum polarization tensor of the theory. We have calculated [3,5] the one-loop photon vacuum polarization tensor by making use of the Schwinger-Keldysh formalism, suitable for study of time dependent problems. Our polarization tensor is renormalized by the technique of dimensional regularization. Further, it is manifestly transverse, and hence gauge invariant. We have performed both calculations, for photons coupled to a massive light scalar (3) [3], as well as to a massless minimally coupled scalar (4) [5]. In both cases the photon acquires a mass through the coupling to the infrared scalar modes. The result for the case of coupling to massive excitations can be neatly summarized by the effective Lagrangian [3]

$$\mathbf{L}_{\Phi\mathrm{ED}}^{\mathrm{eff}} = \mathbf{L}_{\Phi\mathrm{ED}} - \frac{1}{2}a^2\eta^{\mu\nu}A_\mu A_\nu \,, \tag{6}$$

such that the photon obeys a Proca equation of motion, with the mass

$$m_\gamma^2 \simeq \frac{3\alpha H^4}{\pi(m_\phi^2 + \xi \mathrm{R})} \,, \tag{7}$$

where $\alpha = e^2/(4\pi)$ denotes the fine structure constant. This result is obtained in a generalized Loren(t)z gauge, $\eta^{\mu\nu}\partial_\mu(a^2 A_\nu) = 0$. Both transverse and longitudinal vector excitations are endowed by the mass (7). This represents an alternative, gravity induced, mechanism for gauge field mass generation, in which the longitudinal photon degree of freedom is provided by the infrared modes of the (complex) scalar field. As a consequence of mass generation, scaling of the electric and magnetic fields during inflation changes from the vacuum scaling, $\vec{E}_{\mathrm{vac}}, \vec{B}_{\mathrm{vac}} \propto a^{-2}$ to $\vec{E} \propto a^{-\frac{3}{2}-\frac{1}{2}[1-4(m_\gamma/H)^2]^{1/2}}$ and $\vec{B} \propto a^{-\frac{5}{2}+\frac{1}{2}[1-4(m_\gamma/H)^2]^{1/2}}$. If the photon couples to a massless minimally coupled scalar, then the magnetic field remains massless, while the electric field acquires a time dependent mass, $m_\gamma^2(a) = (2\alpha H^2/\pi)[\ln(a) - 1/4]$, [4] such that the fields scale differently, $\vec{E} \propto a^{-3/2}$ and $\vec{B} \sim \vec{B}_{\mathrm{vac}} \propto a^{-2}$.

3. Nearly minimal magnetogenesis

We shall now argue that, provided the scalar field evolution remains de Sitter invariant (3), a certain choice of the couplings leads to magnetic field generation during inflation.

A general solution of the photon mode equation associated with the Proca effective theory (6) can be written in terms of the Hankel function,

$$A_\mu = \frac{1}{2}\sqrt{-\pi\eta}H_\nu^{(1)}(-k\eta)\varepsilon_\mu^{(1)} \,, \qquad \nu = \frac{1}{2}\left(1 - 4m_\gamma^2/H^2\right)^{1/2} \tag{8}$$

where $\varepsilon_\mu^{(1)}$ denotes the photon polarization vector. This needs to be matched to the plane wave solutions of the radiation era, $A_\mu^{(\mathrm{rad})} = (2k)^{-1/2}\mathrm{e}^{\mp ik\eta}$. These modes describe correctly the photon dynamics in radiation era, provided conductivity is small and the photon mass decays nonadiabatically during radiation era. After performing the mode matching, we arrive at the following ensemble averaged magnetic field in radiation era,

$$\langle \vec{B}^2(\vec{x},\eta,\ell)\rangle = \int \frac{dk}{k}\mathbf{P}_B(k,\eta,\ell)|w(k,\ell)|^2\,, \tag{9}$$

where w denotes a window function, *e.g.* $w = \exp(-k^2\ell^2/2) \equiv \int d^3x W(\vec{x})\exp(i\vec{k}\cdot\vec{x})$, $W = (2\pi\ell^2)^{-3/2}\exp\big(-\|\vec{x}\|/(2\ell^2)\big)$, and $\mathbf{P}_B(k,\eta,\ell)$ is the field spectrum (defined in analogous manner as the spectrum of cosmological perturbations),

$$\mathbf{P}_B = \frac{1}{a^4}\frac{\Gamma^2(\nu)}{8\pi^3}\big(\nu-\frac{1}{2}\big)^2(2H)^{2\nu+1}k^{3-2\nu}\sin^2\big((k/H)(a-1)\big)\,. \tag{10}$$

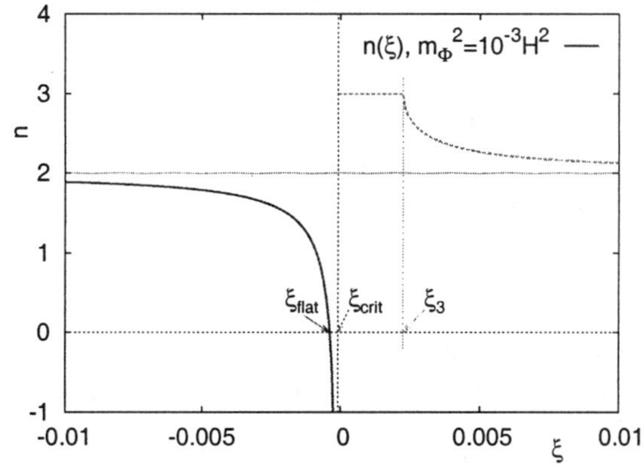

Figure 1. Spectral index n as a function of the scalar field coupling to curvature ξ and the scalar mass m_ϕ^2.

This spectrum is characterized by the spectral index $n = 3 - 2\nu$ on subhorizon scales, and by $n = 5 - 2\nu$ on superhorizon scales. Figure 1 shows n on subhorizon scales as a function of ξ and m_ϕ^2. Note that for $\xi_{\rm flat} = -(\alpha/8\pi) - m_\phi^2/12H^2$, one gets a flat spectrum, $n = 0$. In this case the field strength is typically $B \sim 10^{-10}$ Gauss on all scales, such that it is potentially observable by the next generation of CMBR measurements. The region of couplings $\xi \in (\xi_{\rm flat},\xi_{\rm crit})$ $(\xi_{\rm crit} = -m_\phi^2/12H^2)$ leads to a growth in magnetic energy during inflation, such that its backreaction on the background space-time may become important. This may change the course of the Universe's evolution, and thus $n < 0$ cannot be trusted. The values corresponding to $\xi > \xi_3 = (\alpha/\pi) - m_\phi^2/12H^2$ have already been proposed earlier [9,8,7] as a model for magnetic field generation, powerful enough to seed the galactic dynamo.

Finally, we recall that radiation era may be endowed by a large conductivity. In this case the photon dynamics is governed by the suitably modified [2] Bödeker-Langevin theory [10],

$$(a\sigma\partial_t + \vec{k}^2)A^T = a^3\zeta^T,\ \langle\zeta^T(\vec{k},\eta)\zeta^{T'}(\vec{k}',\eta)\rangle = \frac{2\sigma T}{a^4}\delta^{TT'}(2\pi)^3\delta(\eta-\eta')\delta(\vec{k}-\vec{k}') \tag{11}$$

and the spectrum (10) is modified to

$$\mathbf{P}_B = \frac{1}{a^4}\frac{\Gamma^2(\nu)}{2\pi^3}(2H)^{2\nu-1}k^{5-2\nu}\exp\Big(-\frac{2k^2}{H\sigma}(1-1/a)\Big) + \mathbf{P}_B^{\rm th}\,. \tag{12}$$

This suggests the following definition of the conductivity momentum, $k_\sigma = \sigma H(t)/2$, above which the plasma conductivity destroys any primordial spectrum. Neglecting the thermal contribution, $\mathbf{P}_B^{\rm th} \propto k^3$ ($k \gg k_\sigma$), it is worth noting that the spectrum (12) gives a magnetic field that is smaller than (10) on all scales. The spectrum (10) may be understood as an upper bound and the spectrum (12) as a lower bound on the magnetic field spectrum obtained by a more realistic treatment of the photon dynamics during radiation and matter eras.

References

1. C. Vogt and T. A. Ensslin, arXiv:astro-ph/0309441.
2. T. Prokopec and E. Puchwein, Phys. Rev. D **70** (2004) 043004 [arXiv:astro-ph/0403335].
3. T. Prokopec and E. Puchwein, JCAP **0404** (2004) 007 [arXiv:astro-ph/0312274].
4. T. Prokopec and R. P. Woodard, Annals Phys. **312** (2004) 1 [arXiv:gr-qc/0310056].
5. T. Prokopec, O. Törnkvist and R. P. Woodard, Annals Phys. **303** (2003) 251-274 [arXiv:gr-qc/0205130].
6. T. Prokopec, O. Törnkvist and R. P. Woodard, Phys. Rev. Lett. **89**, 101301 (2002) [arXiv:astro-ph/0205331].
7. T. Prokopec and R. P. Woodard, Am. J. Phys. **72** (2004) 60 [arXiv:astro-ph/0303358].
8. K. Dimopoulos, T. Prokopec, O. Törnkvist and A.-Ch. Davis, Phys. Rev. D **65** (2002) 063505 [arXiv:astro-ph/0108093].
9. A.-Ch. Davis, K. Dimopoulos, T. Prokopec and O. Törnkvist, Phys. Lett. B **501** (2001) 165 [arXiv:astro-ph/0007214].
10. D. Bödeker, Nucl. Phys. B **559** (1999) 502 [arXiv:hep-ph/9905239].
11. N. A. Chernikov and E. A. Tagirov, Annales Poincare Phys. Theor. A **9** (1968) 109.
12. B. Allen and A. Folacci, Phys. Rev. D **35** (1987) 3771.

TRANSPORT EQUATIONS FOR CHIRAL FERMIONS TO ORDER $\hbar$ AND ELECTROWEAK BARYOGENESIS

S. WEINSTOCK*

Nuclear Theory Group, Brookhaven National Laboratory,
Upton, NY 11951-5000, USA
E-mail: weinstock@bnl.gov

M. G. SCHMIDT, T. PROKOPEC

Institut f"ur Theoretische Physik, Universit"at Heidelberg
Philosophenweg 16,
69120 Heidelberg, Germany
E-mail: m.g.schmidt,t.prokopec@thphys.uni-heidelberg.de

Based on the Closed Time Path formalism for problems in non-equilibrium field theory, we derive the relevant kinetic equations for electroweak baryogenesis. These are semiclassiclal Boltzmann equations which are accurate to order $\hbar$.

1. Introduction

The idea of producing the observed baryon asymmetry in the universe during a strong first order electroweak phase transition is appealing since it is based on physics that in principle will become experimentally accessible in the near future. In a first order electroweak phase transition bubbles of the new phase with a non-vanishing Higgs expectation value nucleate and grow. The wall of such bubbles interacts in a CP-violating way with the particles in the plasma and leads to non-equilibrium conditions. The created CP-asymmetry is then converted into a baryon asymmetry by the sphaleron interaction in front of the wall. The B-asymmetry is transported inside the bubble, where the sphaleron is turned off due to the Higgs expectation value. It is well known that this scenario does not work in the Standard Model, which has no electroweak phase transition, but is feasible

*Work supported by the Alexander von Humboldt Foundation and by the U.S. Department of Energy under Grant No. DE-AC02-98CH10886

in supersymmetric extensions like the MSSM or NMSSM, or for example in the two Higgs doublet model.

In order to compute the actual amount of generated baryons, it is essential to understand the transport phenomena in presence of the bubble wall. We have to find kinetic equations for particles, whose masses vary due to the interaction which the changing background field. We have to go beyond the classical order, since CP-violation, essential for the baryon generation, appears at the level of quantum corrections. Starting with the equation of motion for the nonequilibrium Green functions in the Schwinger-Keldysh formalism, we apply a gradient expansion with respect to space-time variations of background fields. This leads to the establishment of a quasiparticle picture and to Boltzmann-type transport equations for the dynamics of the quasiparticles. For details of the work and more references see[1].

2. Flow terms

We work in a model with chiral fermions $\psi_{L,R}$ that couple to scalars ϕ *via* a Yukawa interaction $L_{\rm int}$:

$$L = i\bar{\psi}\,\partial\!\!\!/\,\psi - \bar{\psi}_L m \psi_R - \bar{\psi}_R m^* \psi_L + (\partial_\mu \phi)^\dagger\,(\partial^\mu \phi) - \phi^\dagger M^2 \phi + L_{\rm int}\,. \quad (1)$$

Due to the interaction with the background Higgs field the masses m and M^2 are space-time dependent and they may contain complex elements which provide CP-violation. The Schwinger-Dyson equations for this system lead to the Kadanoff-Baym equations for the scalar and fermionic Wightman functions $\Delta^{<,>}$ and $S^{<,>}$. In the Wigner representation the latter is

$$\Big(k\!\!\!/ + \frac{i}{2}\,\partial\!\!\!/ - m_h e^{-\frac{i}{2}\overleftarrow{\partial}\cdot\partial_k} - i\gamma^5 m_a e^{-\frac{i}{2}\overleftarrow{\partial}\cdot\partial_k}\Big) S^{<,>} = C_\psi\,, \quad (2)$$

with the fermionic collision term

$$C_\psi = \frac{1}{2}e^{-i\diamond}\left(\{\Sigma^>\}\{S^<\} - \{\Sigma^<\}\{S^>\}\right), \quad (3)$$

where $\diamond\{a\}\{b\} \equiv (1/2)(\partial_x a\cdot\partial_k b - \partial_k a\cdot\partial_x b)$, and $\Sigma^{<,>}$ denotes the fermionic self-energy. We already omitted terms on the left hand side of the equations which represent collisional broadening due to multiparticle effects, because we found arguments which indicate that, in a weakly coupled regime, they represent a small correction to the quasiparticle picture.

Assuming that the relevant mass parameter depends only on one spatial coordinate, which models reasonably well the limit of large, almost planar bubble walls, the discussion of the fermionic kinetic equation begins with the observation that the spin operator

$$S_z(k) = (1/\tilde{k}_0)\left(k_0\gamma^0 - \vec{k}_{\parallel}\cdot\vec{\gamma}\right)\gamma^3\gamma^5 \tag{4}$$

is a conserved quantity. Here $\tilde{k}_0 = \mathrm{sign}(k_0)(k_0^2 - k_x^2 - k_y^2)^{1/2}$. With the corresponding spin projector the fermionic Wigner function can accordingly be cast into a spin diagonal form:

$$S_s = iP_s\left[s\gamma^3\gamma^5 g_0^s - s\gamma^3 g_3^s + \mathbb{1}g_1^s - i\gamma^5 g_2^s\right]. \tag{5}$$

In stationary situations the densities g_a^s are real functions of z and momentum. Therefore the Kadanoff-Baym equations result in two sets of equations for these functions, constraint and kinetic equations. At first order in gradients (or equivalently, to linear order in $\hbar$) the constraint equations allow to express $g_{1,2,3}^s$ in terms of g_0^s. When the collisional broadening is neglected, a quasiparticle picture remains valid to order $\hbar$, such that the constraint equation for the density g_0^s is solved by the on-shell form

$$g_0^{<s}(k,x) = 2\pi\sum_{\pm}\frac{\delta(k_0 \mp \omega_{\pm s})}{2\omega_{\pm s}Z_{\pm s}}\,|\tilde{k}_0|\,n_s\,, \tag{6}$$

where $n_s(k,x)$ is a distribution function, to be determined by solving the kinetic equation, and Z_s is a normalization factor. The dispersion relation acquires a CP violating, spin dependent shift due to the interaction with the background field:

$$\omega_s(k,z) = \omega_0 - \frac{s}{2\tilde{\omega}_0\omega_0}|m|^2\partial_z\theta\,. \tag{7}$$

Here $\omega_0 = (\vec{k}^2 + |m|^2)^{\frac{1}{2}}$, $\tilde{\omega}_0 = (k_z^2 + |m|^2)^{\frac{1}{2}}$, and θ is the phase of the complex mass. The densities of particles and anti-particles with spin s are the on-shell projections

$$f_{s+}(\vec{k},x) = n_s(\omega_s,\vec{k},x) \tag{8}$$
$$f_{s-}(\vec{k},x) = n_s^{cp}(\omega_s,\vec{k},x) = 1 - n_{-s}(-\omega_s,-\vec{k},x)\,. \tag{9}$$

Besides an equilibrium part and a CP-even deviation from the equilibrium density, these particle densities contain spin dependent contributions $\delta f_{s\pm}$. From these we form two combinations, which are related to vector and axial density:

$$\delta f_s^v \equiv \delta f_{s+} - \delta f_{s-}\quad,\quad \delta f_s^a \equiv \delta f_{s+} - \delta f_{-s-}\,. \tag{10}$$

From the other set of equations we obtain a kinetic equation for g_0^s. This leads to a Boltzmann equation for the vector part, which does not have a semiclassical source,

$$(\partial_t + \vec{v}_0 \cdot \nabla_{\vec{x}} + F_0 \partial_{k_z}) \delta f_s^v = \int_0^\infty \frac{dk_0}{\pi} \left[K_0^s(k) - K_0^{cps}(k) \right], \qquad (11)$$

but there is a CP-violating semiclassical source $sS^{\rm flow}$ in the equation for the distribution function related to the axial density,

$$(\partial_t + \vec{v}_0 \cdot \nabla_{\vec{x}} + F_0 \partial_{k_z}) \delta f_s^a + sS^{\rm flow} = \int_0^\infty \frac{dk_0}{\pi} \left[K_0^s(k) - K_0^{cp-s}(k) \right], \qquad (12)$$

where $\vec{v}_0 = \vec{k}/\omega_0$ and $F_0 = -(\partial_z |m|^2)/(2\omega_0)$. The collisional contributions on the right hand side are essentially the traces of the spin projected fermionic collision term. The relevant CP-violating sources are

$$S^{\rm flow} = -\frac{1}{2} v_w f_0 (1 - f_0) \beta \Big\{ \delta F + F_0 \frac{\delta\omega}{\omega_0} + F_0 \delta\omega \beta (1 - 2 f_0) + F_0 \delta Z \Big\} \qquad (13)$$

$$+ \frac{1}{2} \Big\{ \cdots \Big\} \delta f^{\rm even}, \qquad (14)$$

where the definitions

$$\delta F \equiv \frac{\partial_z (|m|^2 \partial_z \theta)}{\omega_0 \tilde{\omega}_0}, \quad \delta\omega \equiv \frac{|m|^2 \partial_z \theta}{\omega_0 \tilde{\omega}_0}, \quad \delta Z \equiv \frac{\omega_0}{\tilde{\omega}_0^2} \delta\omega \qquad (15)$$

have been made. The first line contains sources that come directly from the shift in the energy, and some of them are unaccounted for in the existing literature. New is also the second line, which indicates a CP-violating source that is generated by a potentially existing CP-even deviation of the complete distribution function from the equilibrium distribution.

This solution applies in a similar form also to to a situation where we have several fermionic particles mixing via a mass matrix. In this case we first diagonalize the mass matrix and then obtain corresponding results for the diagonal components of the Wigner function in the mass eigenbasis, provided no eigenvalues $|m|$ are nearly degenerate, that is $\hbar k \cdot \partial \ll \delta(|m|)$.

The treatment for scalar particles goes along the same lines, although without the complication caused by the spin structure in the fermionic case. We find quasiparticles living on the mass shell $\omega_\phi = (\vec{k}^2 + M^2)^{1/2}$, which obey a Boltzmann equation without semiclassical sources in the flow term.

3. Collision term

Here we study the effects of the Yukawa interaction term

$$L_{\rm int} = -y \phi \bar{\psi}_L \psi_R - y^* \phi \bar{\psi}_R \psi_L . \qquad (16)$$

For the computation of the collision term (3) we expand the self energies in the coupling constant and truncate at the one-loop order. For the fermionic self energy this leads to

$$\Sigma^{<>}(k) = iy^2 \int \frac{d^4k'd^4k''}{(2\pi)^8}(2\pi)^4\delta^4(k-k'+k'') \tag{17}$$

$$\left[\Delta^{><}(k'')P_L G^{<>}(k')P_R + \Delta^{<>}(k'')P_R G^{<>}(k')P_L\right] . \tag{18}$$

For the Wigner functions we use the expressions found above. We decompose the fermionic function $g_0 = g_{0,eq} + \delta g$, where the first part contains an equilibrium Fermi-Dirac distribution, and the second is a first order deviation from equilibrium. After linearizing with respect to gradients we find two types of contributions. The terms containing δg constitute ordinary relaxation terms. They are exactly what one would expect in the collision term of a Boltzmann equation. The terms without δg do not vanish because of the modified structure of the fermionic Wigner function due to the shift in the energy. These terms constitute collisional sources, which are of first order in gradients and suppressed by the wall velocity. Their parametric dependence is qualitatively different from the semiclassical force source.

The collision term for the scalar equation is evaluated in the same way. Besides the expected relaxation terms, it contains a collisional source, which is of the same parametric form as the fermionic one.

Numeric and analytical estimates show, however, that the collisional sources are typically suppressed compared to the flow term sources, so that they seem not to be important for baryogenesis.

4. Conclusions

This is the first derivation of semiclassical kinetic equations for electroweak baryogenesis that provides a unified treatment of flow and collision terms, and the first ever calculation of CP-violating sources arising from nonlocal (loop) effects. However, these collisional sources apparently are not relevant for baryogenesis. We also compute additions to the sources that already have been found in papers using the WKB-method. Numerical estimates show that these are capable of changing the resulting baryon asymmetry by a factor two to three. For a precise result a more detailed evaluation of the kinetic equations in a specific model has to be made.

References

1. M. G. Schmidt, T. Prokopec, S. Weinstock *hep-ph/0312110 and hep-ph/0406140*, to appear in Annals of Physics.

THE GAPLESS 2SC PHASE *

MEI HUANG, IGOR A. SHOVKOVY

Institut für Theoretische Physik,
J.W. Goethe-Universität,
D-60054 Frankurt/Main, Germany

Recent results on the gapless 2SC phase are reviewed. These include: the thermal stability under the constraint of the local charge neutrality condition, the properties at zero and nonzero temperatures, and the color screening properties.

1. Introduction

Because the interaction between two quarks in the color anti-triplet channel is attractive, sufficiently cold dense quark matter is a color superconductor[1]. It is very likely that the color superconducting phase may exist in the core of compact stars, where bulk matter should satisfy the charge neutrality condition as well as β-equilibrium. For a three-flavor quark system, when the strange quark mass is small, the color-flavor-locked (CFL) phase [2] is favorable [3]. For the two-flavor quark system, the charge neutrality condition plays a nontrivial role. In the ideal two-flavor color superconducting (2SC) phase, the paired u and d quarks have the same Fermi momenta. Because u quark carries electrical charge 2/3, and d quark carries electrical charge $-1/3$, it is easy to check that quark matter in the ideal 2SC phase is positively charged. To satisfy the electrical charge neutrality condition, roughly speaking, twice as many d quarks as u quarks are needed. This induces a large difference between the Fermi surfaces of the two pairing quarks, i.e., $\mu_d - \mu_u = \mu_e \approx \mu/4$, where μ, μ_e are chemical potentials for quark and electron, respectively. Naively, one would expect that the requirement of the charge neutrality condition will destroy the ud Cooper pairing in the 2SC phase.

However, it was found in Ref. 4 that a charge neutral two-flavor color superconducting (N2SC) phase does exist. Comparing with the ideal 2SC

*Talk given by Mei Huang.

phase, the N2SC phase found in Ref. 4 has a largely reduced diquark gap parameter, and the pairing quarks have different number densities. The latter contradicts the paring ansatz [5]. It is natural to think that the N2SC phase found in Ref. 4 is an unstable Sarma state [6]. In Ref. 7, it was shown that the N2SC phase is a thermal stable state under the restriction of the local charge neutrality condition. As a by-product, which comes out as a very important feature, it was found that the quasi-particle has zero-energy excitation spectrum. Thus this phase was named the "gapless 2SC (g2SC)" phase. In the following, we first show the thermal stability of the g2SC phase, then we discuss its properties at zero as well as at nonzero temperatures, at last we present our most recent results about the chromomagnetic instability in the g2SC phase.

2. The gapless 2SC phase and its thermal stability

Bulk quark matter inside the neutron star should be neutral with respect to the color charge as well as the electrical charge. The color superconducting phase in full QCD is automatically color neutral [8]. In the Nambu–Jona-Lasinio (NJL) type model, the color neutrality can be satisfied by tuning the chemical potential μ_8 for the color charge. The value of μ_8 for a charge neutral 2SC phase is very small [4,12]. Correspondingly, the electrical charge neutrality can be satisfied by tuning the chemical potential μ_e for the electrical charge. The value of μ_e is determined by the electrical charge neutrality condition, a rough estimate gives $\mu_e \approx \mu/4$.

The ground state is determined by solving the gap equation together with the charge neutrality condition. It is found that the ground state of charge neutral two-flavor quark matter is very sensitive to the diquark coupling constant G_D [7]:

$$\begin{aligned} G_D/G_S &\gtrsim 0.8, && \Delta > \delta\mu, \ \ \text{2SC}, \\ 0.7 \lesssim G_D/G_S &\lesssim 0.8, && \Delta < \delta\mu, \ \ \text{g2SC}, \\ G_D/G_S &\lesssim 0.7, && \Delta = 0, \ \ \text{NQM}. \end{aligned} \tag{1}$$

Where $\delta\mu \equiv \mu_e/2$, G_S is the quark-antiquark coupling constant, and "NQM" indicates the normal phase of quark matter. The most interesting case is the g2SC phase, which exists in a diquark coupling regime $0.7 \lesssim G_D/G_S \lesssim 0.8$. Even though this regime is narrow, it is worth to mention that, either from the Fierz transformation ($G_D/G_S = 0.75$) or by fitting the vacuum baryon mass ($G_D/G_S \simeq 2.26/3$) [9], the value of the ratio G_D/G_S sits inside this regime.

The g2SC phase indicated by the order parameter $\Delta < \delta\mu$, resembles the unstable Sarma state [6]. For the flavor asymmetric ud quark system, i.e., μ_e is a free parameter and there is no constraint from the charge neutrality condition, the solution $\Delta < \delta\mu$ of the gap equation indeed corresponds to the maximum of the thermodynamical potential $\Omega_{u,d,e}$.

However, bulk quark matter inside the neutron star should be charge neutral. A nonzero net electrical charge density n_Q will cause an extra energy $\Omega_{Coulomb} \sim n_Q^2 V^{2/3}$ (V is the volume of the system) by the repulsive Coulomb interaction. The total thermodynamical potential of the whole system is given by $\Omega = \Omega_{Coulomb} + \Omega_{u,d,e}$. The energy density grows with increasing the volume of the system, as a result, it is almost impossible for matter inside stars to remain charged over macroscopic distances. So, the correct way to find the ground state of the homogeneous neutral u, d quark matter is to minimize the thermodynamical potential along the neutrality line $\Omega|_{n_Q=0} = \Omega_{u,d,e}|_{n_Q=0}$ with $\Omega_{Coulomb}|_{n_Q=0} = 0$. The g2SC phase corresponds to the global minimum of the thermodynamical potential along the charge neutrality line, thus it is a stable state under the restriction of the charge neutrality condition.

3. The g2SC phase at zero and nonzero temperatures

As we already mentioned, at zero temperature, in the g2SC phase, the pairing quarks have different number densities [4,7]. This is different from the 2SC phase when $\delta\mu < \Delta$.

It is the quasi-particle spectrum that makes the g2SC phase different from the 2SC phase. The excitation spectrum for the ideal 2SC phase ($\delta\mu = 0$) include: two free blue quarks, which do not participate in the Cooper pairing, and four quarsi-particle excitations (linear superpositions of $u_{r,g}$ and $d_{r,g}$) with an energy gap Δ. If there is a small mismatch ($\delta\mu < \Delta$) between the Fermi surfaces of the pairing u and d quarks, $\delta\mu$ induces two different branches of quasi-particle excitations. One branch moves up with a larger energy gap $\Delta + \mu$, another branch moves down with a smaller energy gap $\Delta - \mu$. If the mismatch $\delta\mu$ is larger than the gap parameter Δ, the lower dispersion relation for the quasi-particle crosses the zero-energy axis. Thus we call the phase with $\Delta < \delta\mu$ the gapless 2SC (g2SC) phase. In the g2SC phase, there are only two gapped fermionic quasiparticles, and the other four quasiparticles are gapless.

In a superconducting system, when one increases the temperature at a given chemical potential, thermal motion will eventually break up the quark

Cooper pairs. In the weakly interacting Bardeen-Copper-Schrieffer (BCS) theory, the transition between the superconducting and normal phases is usually of second order. The ratio of the critical temperature $T_c^{\rm BCS}$ to the zero temperature value of the gap $\Delta_0^{\rm BCS}$ is a universal value[10] $r_{\rm BCS} = T_c^{\rm BCS}/\Delta_0^{\rm BCS} \approx 0.567$. In the ideal 2SC phase, the ratio of the critical temperature to the zero temperature value of the gap is also the same as in the BCS theory[11]. However, the g2SC phase has very different properties at nonzero temperatures[12]. The ratio T_c/Δ_0 is not a universal value. It is infinity when $G_D/G_S \to 0.68$ and approaches $r_{\rm BCS}$ when G_D/G_S increases. The temperature dependence of the gap is a nonmonotonic behavior. In some cases, the diquark gap could have sizable values at finite temperature even if it is exactly zero at zero temperature.

4. Chromomagnetic instability in the g2SC phase

Because the g2SC phase has four gapless modes and two gapped modes, one may think that the low energy (large distance scale) properties of the g2SC phase should interpolate between those in the normal phase and those in the 2SC phase. However, its color screening properties do not fit this picture.

One of the most striking properties of an ordinary superconductor is its Meissner effect, i.e., the superconductor expels the magnetic field. Using the linear response theory, the induced magnetic current j_i^{ind} responses to the external magnetic field A_j is related by the response magnetic part of the photon polarization tensor Π_{ij}, i.e., $j_i^{ind} = \Pi_{ij} A^j$. The response function has two components, diamagnetic and paramagnetic part. In the static and long-wavelength limit, for the normal metal, the paramagnetic component cancels exactly the diamagnetic component. In the superconducting phase, the paramagnetic component is quenched by the energy gap and produces a net diamagnetic response. Thus the ordinary superconductor is a prefect diamagnet. In cold dense quark matter, the gauge bosons connected with the broken generators obtain masses in the ideal 2SC phase [13] as well as in the CFL phase [14], which indicate the Meissner screening effect in these phases.

However, in the g2SC phase, it is found, the Meissner screening masses of the five gluons, corresponding to the five broken generators of the $SU(3)_c$ group, are *imaginary* [15]. This is because, in the static and long-wavelength limit, the paramagnetic contribution to the magnetic part of these five gluon polarization tensors becomes dominant. In condensed matter, this phenomenon is called the paramagnetic Meissner effect (PME) [16]. The imaginary Meissner screening mass indicates a chromomagnetic instability of the g2SC phase. There are, several possibilities to resolve the instability. One is through the gluon condensate, which may not change the structure the g2SC phase. It is also possible that the instability drives the homogeneous system to an inhomogeneous phase, like the crystalline phase or the vortex lattice phase. This problem remains to be clarified in the future. It is also very interesting to know whether the chromomagnetic instability develops in the gapless CFL phase[17].

Acknowledgements The work of M. H. was supported by the Alexander von Humboldt-Foundation, and by the NSFC under Grant Nos. 10105005, 10135030. The work of I.A.S. was supported by Gesellschaft für Schwerionenforschung (GSI) and by Bundesministerium für Bildung und Forschung (BMBF).

References

1. R. Rapp, T. Schäfer, E. V. Shuryak and M. Velkovsky, Phys. Rev. Lett. **81**, 53 (1998); M. Alford, K. Rajagopal, and F. Wilczek, Phys. Lett. B **422**, 247 (1998).
2. M. G. Alford, K. Rajagopal and F. Wilczek, Nucl. Phys. **B537**, 443 (1999).
3. M. Alford and K. Rajagopal, JHEP **0206**, 031 (2002). A.W. Steiner, S. Reddy and M. Prakash, Phys. Rev. D **66**, 094007 (2002).
4. M. Huang, P.F. Zhuang and W.Q. Chao, Phys. Rev. D **67**, 065015 (2003).
5. K. Rajagopal and F. Wilczek, Phys. Rev. Lett. **86**, 3492 (2001).
6. G. Sarma, J. Phys. Chem. Solids **24**, 1029 (1963).
7. I. Shovkovy and M. Huang, Phys. Lett. B **564**, 205 (2003).
8. A. Gerhold and A. Rebhan, Phys. Rev. D **68**, 011502 (2003); D. D. Dietrich and D. H. Rischke, Prog. Part. Nucl. Phys. **53**, 305 (2004).
9. D. Ebert, L. Kaschluhn and G. Kastelewicz, Phys. Lett. **B264**,420 (1991).
10. J. R. Schrieffer, *Theory of Superconductivity* (Benjamin, New York, 1964).
11. R.D. Pisarski and D.H. Rischke, Phys. Rev. D **61**, 051501 (2000).
12. M. Huang and I. Shovkovy, Nucl. Phys. A **729**, 835 (2003).
13. D. H. Rischke, Phys. Rev. D **62**, 034007 (2000); D. H. Rischke and I. A. Shovkovy, Phys. Rev. D **66**, 054019 (2002).
14. D. H. Rischke, Phys. Rev. D **62**, 054017 (2000).
15. M. Huang and I. A. Shovkovy, hep-ph/0407049, to appear in Phys. Rev. D; M. Huang and I. A. Shovkovy, hep-ph/0408268.
16. A. K. Geim, S. V. Dubonos, J. G. S. Lok, M. Henini, and J. C. Maan, Nature **396**, 144 (1998).
17. M. Alford, C. Kouvaris and K. Rajagopal, Phys. Rev. Lett. **92**, 222001 (2004); M. Alford, C. Kouvaris and K. Rajagopal, hep-ph/0406137.

GAPLESS CFL AND ITS COMPETITION WITH MIXED PHASES*

MARK ALFORD

Physics Department, Washington University, St. Louis, MO 63130, USA

CHRIS KOUVARIS AND KRISHNA RAJAGOPAL

Center for Theoretical Physics, Massachusetts Institute of Technology, Cambridge, MA 02139, USA

We recently argued that as the density of quark matter decreases, there is a continuous transition from the color-flavor-locked (CFL) phase to a gapless "gCFL" phase. The reason is the growing importance of the strange quark mass, and the constraint of electric/color neutrality. In this paper we discuss mixed phases that achieve neutrality by charge separation, and might offer an alternative to the gCFL phase. We find that none of the obvious mixtures is favored relative to gCFL.

1. Introduction

There is a strong possibility that quark matter may occur in the core of compact ("neutron") stars, and at sufficiently high density it is expected to be color superconducting: the attractive QCD interaction between the quarks at the Fermi surface causes them to condense in Cooper pairs. At asymptotic densities, where the up, down and strange quarks can be treated on an equal footing and the potentially disruptive strange quark mass can can be neglected, quark matter is in the color-flavor locked (CFL) phase [1], in which quarks of all three colors and all three flavors form Cooper pairs [2].

However a very interesting question is what happens at lower densities. In dense matter with quark chemical potential μ and pairing gap Δ, the strange quark mass has negligible impact on CFL pairing when $M_s^2 \ll \mu\Delta$. At compact-star densities $\mu \sim 350$ to 500 MeV, $M_s \sim 80$ to 500 MeV, and $\Delta \sim 10$ to 100 MeV, so it is quite possible that the strange quark has a strong influence on the pairing.

*Talk given by C. Kouvaris.

Starting at asymptotic density and reducing the density by reducing μ, we expect CFL pairing to be disrupted, at first by kaon condensation[3] (which we do not discuss), and then by the transition to gapless CFL (gCFL)[4,5] with the appearance of gapless modes. In this paper we assume that this happens before the transition from quark matter to hadronic matter. We first review some of the features of gCFL, and then compare it with various charge-separated mixed phases, finding that they are not energetically favored even before we include their extra energy costs arising from Coulomb (color-)electric fields and the surface tension.

2. The gapless CFL phase

Stable bulk matter has to be neutral under all the gauged symmetries (color and electromagnetic), and equilibrated under all interactions including the weak interaction. These conditions are imposed by introducing chemical potentials coupled to the gauged charges: μ_e couples to negative electric charge, and μ_3, μ_8 couple to the color generators T_3 and T_8, generators of the Cartan subalgebra of the color group. μ_e, μ_3, μ_8 are determined by requiring that the corresponding charge densities vanish. The pairing ansatz we use is

$$\langle \psi_a^\alpha C\gamma_5 \psi_b^\beta \rangle \sim \Delta_1 \epsilon^{\alpha\beta 1}\epsilon_{ab1} + \Delta_2 \epsilon^{\alpha\beta 2}\epsilon_{ab2} + \Delta_3 \epsilon^{\alpha\beta 3}\epsilon_{ab3} \tag{1}$$

The gap parameters Δ_1, Δ_2 and Δ_3 describe down-strange, up-strange and up-down Cooper pairs, respectively. Above a critical $M_s^2/\mu = 2\Delta$, the CFL phase is replaced by a new gapless CFL (gCFL) phase[4]. The defining properties of the gapless CFL phase arise in its dispersion relations, not in its pattern of gap parameters. However, it is useful for orientation to list the patterns of gap parameters for all the phases we shall discuss:

$$\begin{array}{ll} \text{CFL}: \Delta_3 \simeq \Delta_2 = \Delta_1 = \Delta_{CFL} & \text{gCFL}: \Delta_3 > \Delta_2 > \Delta_1 > 0 \\ \text{g2SC}: \Delta_3 > 0, \quad \Delta_1 = \Delta_2 = 0 & \text{2SCus}: \Delta_2 > 0, \quad \Delta_1 = \Delta_3 = 0 \end{array} \tag{2}$$

The 2SCus phase has the same free energy as 2SC at $M_s = 0$, and to leading order in M_s if their respective nonzero gap parameters have the same value[6]. However, an NJL model calculation indicates that Δ_2 in 2SCus is always less than Δ_3 in 2SC, so 2SCus is never favored[5].

3. Mixed phases

The gCFL phase can be viewed as a distortion of the CFL phase, induced by the stress of a non-zero strange quark mass, combined with the constraints of color and electromagnetic neutrality. However, it is possible for

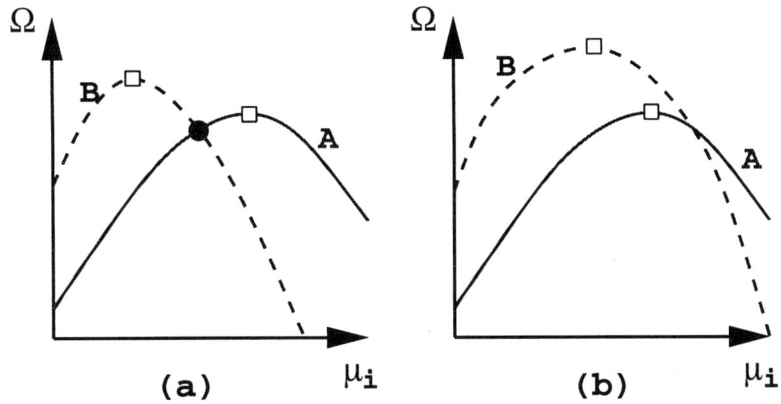

Figure 1. Schematic illustration of the dependence of the free energy Ω on a gauged chemical potential μ_i, showing conditions for the occurrence of mixed phases. Charge $Q_i = -\partial\Omega/\partial\mu_i$ is given by the slope. Squares mark the neutral points. Panel (a): at the neutral value of μ_i for each phase, the other phase has lower free energy, so there is a coexistence point (black dot) for oppositely-charged phases with lower free energy than either neutral phase. Depending on Coulomb and surface energy costs, a mixed phase may exist there. Panel (b): phase B has higher free energy than phase A at the value of μ_i where A is neutral. The two phases never coexist with opposite charge, so no mixed phase is possible.

neutrality to be achieved by a mixture of oppositely charged phases at the same pressure. This happens in the two-flavor case, where a gapless 2SC phase exists[7], but a mixed phase is free-energetically favored[8]. We now ask whether such a mixed phase might be preferred over the gCFL phase.

The two possible situations are schematically illustrated in Fig. 1, which shows generic free energy curves $\Omega(\mu_i)$ for two phases A and B. In Fig. 1b there is no coexistence point and hence no mixed phase is possible. In Fig. 1a there is a coexistence point of oppositely-charged phases, and its free energy is lower than that of either neutral phase, so if the energy of the electric fields induced by the charge separation and the surface tension of the interface is small enough, a neutral mixed phase will be free-energetically preferred over either homogeneous neutral phase.

In quark matter we have gauge charges associated with color as well as electromagnetism. However, it seems unlikely that mixed phases involving color-charged components could form. The color gauge coupling is strong, so separating color charges into different domains will produce color-electric fields with very high energy cost. It seems likely, therefore, that mixed phases will consist of components that are individually color-neutral, with different values of the color chemical potentials μ_3 and μ_8, but which are electrically charged, with a common electric chemical potential μ_e. In this case there are still color-electric energy costs, associated with the color electric field induced at the phase boundary by the gradient in μ_3 and μ_8. We will not try to calculate these interface costs [9,8], since our argument is that most of the mixed phases are excluded even before we include the electrostatic and interface energy costs.

In the region $(M_s^2/\mu) > 2\Delta$ where gCFL is the free-energetically favored homogeneous neutral phase, the other less-favored quark matter phases are 2SC, 2SCus, and unpaired quark matter (2). Note that there is no CFL solution, charged or neutral, in this region. We have compared the mixed phases with neutral gCFL. We performed calculations in an NJL model at $\mu = 500$ MeV, with coupling chosen so that $\Delta_{CFL} = 25$ MeV at $M_s = 0$[4,5]. When we vary M_s at fixed μ we find the gCFL region is 47 MeV $< M_s^2/\mu <$ 130 MeV. At the upper limit there is a convergence of the free energies of g2SC, gCFL, and unpaired quark matter, so there is a first-order transition from gCFL to unpaired, with a possible window of g2SC[10]. We have performed calculations of the various possible mixtures. We now discuss the results.

(1) <u>unpaired+gCFL</u>. For color neutral unpaired and gCFL phases, the situation is typically that of Fig. 1(b) so mixed phases are ruled out. This is true for all values of M_s^2/μ except for a range of a few MeV just below $M_s^2/\mu = 130$ MeV, where the neutral gCFL and neutral unpaired free energies cross. There, a mixed phase may arise, although it may be superseded by other more favorable possibilities such as the crystalline phase[11].

(2) <u>2SC+2SCus</u>. We find the situation of Fig 1a: a neutral mixed phase exists. However, its free energy is higher than that of gCFL even before electrostatic and surface energy costs are included. At $M_s^2/\mu = 80$ MeV, it has $\Omega = -14.93 \times 10^6$ MeV^4, vs. $\Omega_{\text{gCFL}} = -18.01 \times 10^6$ MeV^4. (These free energies are both measured relative to that of neutral unpaired quark matter.) We have checked that a neutral 2SC+2SCus mixed phase is also free-energetically unfavored relative to homogeneous gCFL at $M_s^2/\mu = 51.2$ MeV, which is just above the CFL$\rightarrow$gCFL transition.

(3) <u>gCFL+2SCus</u>. The μ_e dependence of the free energies of these two phases is as in Fig 1b, so a mixed phase is not possible. We have verified this at $M_s^2/\mu = 51.2$ MeV and 80 MeV.

(4) <u>gCFL+2SC</u>. At $M_s^2/\mu = 51.2$ MeV we find the situation of Fig. 1b, so no mixed phase is possible. At $M_s^2/\mu = 80$ MeV, the free-energy dependence is of the type shown in Fig. 1a, so a mixed phase is possible, and since

gCFL itself is one of the components its free energy is necessarily lower than that of neutral gCFL. However, $\Omega_{gCFL}(\mu_e)$ depends *very* weakly on μ_e: the $\Omega_{gCFL}(\mu)$ parabola is very shallow. This suppresses the mixed phase in two ways. (i) the free energy of the mixed phase is only lower than that of neutral gCFL by a very small margin (0.0012×10^6 MeV^4 at $M_s^2/\mu = 80$ MeV), so there is very little chance that it will survive once electrostatic and surface costs are included; (ii) The charge density of gCFL, proportional to $\partial\Omega/\partial\mu_e$, is very small so the mixed phase must be dominantly gCFL, with a tiny admixture of 2SC, to achieve neutrality. It is known that such highly asymmetric mixed phases have the highest electrostatic energy costs[9].

We conclude that the gCFL phase is a very strong candidate for the "next phase down in density" after CFL. None of the mixed phases that we investigated was able to compete with it. It remains to be seen what role kaon condensation (in CFL and gCFL) plays, and whether a completely different phase such as the crystalline (LOFF) phase[11] can compete with gCFL at sufficiently low density.

Acknowledgements: We thank J. Bowers for helpful conversations. This research was supported in part by DOE grants DE-FG02-91ER40628 and DF-FC02-94ER40818.

References

1. M. G. Alford, K. Rajagopal and F. Wilczek, Nucl. Phys. B **537**, 443 (1999).
2. For reviews, see K. Rajagopal and F. Wilczek, hep-ph/0011333; M. G. Alford, Ann. Rev. Nucl. Part. Sci. **51**, 131 (2001); G. Nardulli, Riv. Nuovo Cim. **25N3**, 1 (2002); S. Reddy, Acta Phys. Polon. B **33**, 4101 (2002); T. Schäfer, hep-ph/0304281.
3. P. F. Bedaque and T. Schafer, Nucl. Phys. A **697**, 802 (2002); D. B. Kaplan and S. Reddy, Phys. Rev. D **65**, 054042 (2002).
4. M. Alford, C. Kouvaris and K. Rajagopal, Phys. Rev. Lett. **92**, 222001 (2004).
5. M. Alford, C. Kouvaris and K. Rajagopal, hep-ph/0406137.
6. M. Alford and K. Rajagopal, JHEP **0206**, 031 (2002).
7. I. Shovkovy and M. Huang, Phys. Lett. B **564**, 205 (2003); M. Huang and I. Shovkovy, Nucl. Phys. A **729**, 835 (2003).
8. S. Reddy and G. Rupak, nucl-th/0405054.
9. M. G. Alford, K. Rajagopal, S. Reddy and F. Wilczek, Phys. Rev. D **64**, 074017 (2001).
10. S. B. Ruster, I. A. Shovkovy and D. H. Rischke, hep-ph/0405170.
11. M. G. Alford, J. A. Bowers and K. Rajagopal, Phys. Rev. D **63**, 074016 (2001); J. A. Bowers and K. Rajagopal, Phys. Rev. D **66**, 065002 (2002); R. Casalbuoni et. al. hep-ph/0404090.

TRANSPORT COEFFICIENTS IN COLOR SUPERCONDUCTING QUARK MATTER

CRISTINA MANUEL*

Instituto de Física Corpuscular
Universitat de València -C.S.I.C.
Edificio Institutos de Paterna,
Apt. 2085, 46071 València. Spain
E-mail: Cristina.Manuel@ific.uv.es

Transport coefficients of dense quark matter are needed to study properties of compact stars. They can tell us about the cooling, vibrational and rotational properties of the star. We report below a computation of the shear viscosity in the CFL phase at low temperature T. CFL quark matter behaves as a superfluid, and at low T its transport properties are dominated by the collisions of superfluid phonons.

1. Introduction

The microscopic understanding of electromagnetic superconductivity is based on the seminal works of Bardeen, Cooper and Schrieffer. The BCS theory explains the main phenomenological properties of the superconductors, such as the Meissner effect, and the appearance of an energy gap Δ in the quasiparticle spectrum. It also explains the fact that superconductors are both good heat conductors and superfluids, with exponentially suppressed transport coefficients $\sim \exp[-\Delta/T]$ for low temperatures $T \ll \Delta$.

At very high density QCD is in a color superconducting phase[1,2]. Our present microscopic knowledge of this phenomena relies on the BCS techniques adapted to dense quark matter. From first principles one can evaluate the values of the fermionic gaps and gluon masses. There are some macroscopic properties of dense quark matter that have not been yet extensively studied. Should we expect the transport coefficients in a color superconductor to behave as in an electromagnetic superconductor? Naively, one would say so, but the answer is slightly more elaborated. Quarks have

*Work supported by MCYT (Spain) under grant FPA 2001-3031.

different quantum numbers (spin, color, flavor), and the way they condense depends on several parameters, such as the quark masses and chemical potentials. Then one talks about different superconducting phases (CFL, 2SC, crystalline phases, gapless phases, etc), which differ basically on the way the quarks pair, and on the local or global symmetries that are spontaneously broken. Because the hydrodynamic regime of a system depends on the low energy spectrum of the theory, we could simply expect that transport coefficients of the different superconducting phases to behave very differently.

Let us insist on the relevance of transport phenomena in astrophysics[3]. While with the equation of state one can determine the mass and radius of a star, with the transport coefficients one can study its cooling, and its vibrational and rotational properties. Thus, they are essential for detecting signatures of quark matter in compact stars in any of its possible phases.

It has been established that the viscosities put stringent tests to astrophysical models for very rapidly rotating stars, such as for millisecond pulsars. This is based on the existence of r(otational)-mode instabilities in all relativistic rotating stars[4], which are only suppressed by sufficiently large viscosities. Madsen actually ruled out[5] quark matter in both the CFL and 2SC phases in millisecond pulsars, because of their low viscosities. This last conclusion, however, was based on a wrong estimate, as it was assumed a behavior of the viscosities as in an electromagnetic superconductor $\sim \exp[-\Delta/T]$, for values of the temperature T of the order of 0.1 MeV and below. These estimates are not correct. In collaboration with Antonio Dobado and Felipe Llanes-Estrada we have computed the shear viscosity in the low T regime of the CFL phase[6] that I report below. A previous effort of studying transport properties in the 2SC phase was done in [7].

2. Transport in the CFL phase

Transport properties in the CFL phase of QCD at low T are not dominated by the quarks, which certainly give a contribution of the sort $\sim \exp[-\Delta/T]$. In this phase the diquark condensate breaks spontaneously the baryon symmetry $U_B(1)$, and CFL quark matter is a superfluid. Chiral symmetry is also spontaneously broken. Associated to those breakings, there are Goldstone bosons, which are light degrees of freedom. In addition, there is an unbroken $U(1)$ subgroup whose gauge boson, a combination of the photon and one gluon, is massless at zero temperature and can be viewed as the in-medium photon. A CFL quark star is electrically neutral, both for the real and in-medium electromagnetism, but at finite temperature one may also expect to find electrons. All the above mentioned particles are light and their contribution to the transport coefficients in this phase is bigger than that of the gapped quarks. Let us mention that we are considering a CFL quark star after the deleptonization era, so that all the neutrinos have escaped from the star.

Chiral symmetry is not an exact symmetry of QCD. Therefore, the associated (pseudo) Goldstone bosons are massive. Their masses are estimated to be in the range of the tens of MeV. At sufficiently low temperatures, their contribution to the viscosities is Boltzmann suppressed, and transport is dominated by the lightest particles. In particular, the highest contribution[8] is given by the Goldstone boson associated to baryon symmetry breaking, the superfluid phonon, which remains massless. There is also a contribution of the in-medium electromagnetism, but this turns out to be negligible.

The low energy effective field theory for the superfluid phonon has been constructed by Son[9]. He realized that the different coefficients of the corresponding effective Lagrangian can be fixed with the knowledge of the equation of state of dense quark matter. At high density, the effective Lagrangian reads

$$L_{\rm eff} = \frac{3}{4\pi^2}\left[(\partial_0\varphi - \mu)^2 - (\partial_i\varphi)^2\right]^2 \ . \tag{1}$$

The equations of motion associated to φ can be interpreted as hydrodynamic equations for a relativistic superfluid[9], where the velocity is given by the gradient of the phase of the condensate, as in Landau's two fluid model of superfluidity. At finite T they need modifications, as the different collective modes are thermally excited, and they conform a different - and dissipative - fluid. At very low T, dissipative effects are dominated by the collisions of the superfluid phonons.

Superfluidity in He^4 is understood as a consequence of Bose-Einstein condensation. There is also a spontaneous breaking of a global $U(1)$ symmetry, with the corresponding appearance of a massless collective mode (or Goldstone boson, in the high energy language). The phonon moves at the speed of sound v of the system. While in principle He^4 and the CFL matter have little in common, the physics of the collective mode or phonon in the two cases exhibits many similarities, and one can then talk about universality. In particular, we found the same momentum and T dependence for their damping rates. Also the mean free path has the same T dependence.

Universality aspects of these two systems are under investigation.

The physics associated to the superfluid phonon is peculiar. Even if at finite T almost all particles attain a thermal mass, this is not so for the phonon, as thermal effects do not represent a violation of the $U_B(1)$ symmetry. The superfluid phonon suffers Landau damping, which for $p_0, p \ll T$ reads

$$\mathrm{Im}\,\Pi(p_0, \mathbf{p}) \approx \frac{8\pi^5}{1215} \frac{T^4}{v^8\mu^4} \frac{p_0^3}{p}\, \Theta(v^2p^2 - p_0^2) \ . \tag{2}$$

Evaluated on-shell, it gives account of its finite lifetime in the thermal bath.

Shear viscosity describes the relaxation of the momentum components perpendicular to the direction of transport, and it is usually dominated by large-angle collisions. In the CFL fluid it is not the case. The reason is that the differential cross section of binary collisions mediated by phonon exchange is divergent for small-angle collisions. It is only regulated by the finite width of the phonon, which effectively amounts to a resummation of a certain class of diagrams in the computation. Ultimately, it turns out that it is more efficient to achieve a large-angle collision by the addition of many small-angle scatterings.

With Son's effective field theory, it is possible to compute the mean free path associated to both large-angle and small-angle binary collisions. The first one behaves[8] as $\sim \mu^8/T^9$ and for the values corresponding to a cold compact star, it exceeds the typical radius of a star $R \sim 10$ km, thus being irrelevant for transport. We have computed numerically the mean free path for the second ones, which parametrically behave as $\sim \mu^4/T^5$. It also exceeds the radius of a star for $T \ll 0.06$ MeV, the value at which the CFL fluid in the star would behave as a perfect superfluid, showing then no dissipation.

In Ref.[6] we have computed numerically the shear viscosity in the CFL phase within kinetic theory, taking into account the dominant processes in the collision term. The result is consistent with a parametric behavior of the sort $\eta \sim \mu^4/T$. This means that the number of collisions necessary to make a large-angle collision from many small-angle ones do not depend parametrically on T, as one could naively expect. This is also what happens for superfluid He^4, see Ref.[10].

Our result could be improved by taking into account two different corrections. Namely, a better determination of the phonon interactions (determined with a more accurate equation of state of CFL quark matter), and the inclusion of the LPM effect in the collision term. We do not expect, however, that neither of these two corrections would change drastically the final result of the shear viscosity.

3. Conclusions

Viscosities of the different color superconducting phases are needed for an analysis of r-mode instabillities of compact stars, which tell us about the maximum stable frequency of the star. We have focused our efforts to study the shear viscosity of the CFL phase, while the bulk viscosities are also needed and will be the subject of a different project. Our results indicate that for $T \ll 0.06$ MeV the viscosity in the phonon fluid is absent and it cannot damp the r-modes of millisecond pulsars. However, other sources of dissipation should be considered as well. In particular, in a rotating superfluid there are vortices. Particles scattering off vortices might contribute substantially to the viscosity of the rotating star. An analysis of the friction of the superfluid and the surrounding crust might also be relevant for the study of r-modes. So it is premature to discard the CFL quark matter in a millisecond pulsar, and more work is needed to draw firm conclusions about this point. Obviously, it would be equally important to evaluate the viscosities of other color superconducting phases.

References

1. K. Rajagopal and F. Wilczek,"The condensed matter physics of QCD," hep-ph/0011333.
2. M. Alford, K. Rajagopal and F. Wilczek, *Nucl. Phys.* **B537**, 443 (1999).
3. F. Weber, "Strange quark matter and compact stars," arXiv:astro-ph/0407155.
4. N. Andersson, *Astrophys. J.* **502**, 708 (1998).
5. J. Madsen, *Phys. Rev. Lett.* **85**, 10 (2000).
6. C. Manuel, A. Dobado and F. J. Llanes-Estrada, "Shear viscosity in a CFL quark star," arXiv:hep-ph/0406058.
7. D. F. Litim and C. Manuel, *Phys. Rev. Lett.* **87**, 052002 (2001).
8. I. A. Shovkovy and P. J. Ellis, *Phys. Rev.* **C66**, 015802 (2002).
9. D. T. Son, "Low-energy quantum effective action for relativistic superfluids," arXiv:hep-ph/0204199.
10. H. J. Maris, Phys. Rev. A **8**, 1980 (1973).

RENORMALIZATION AND RESUMMATION IN FINITE TEMPERATURE FIELD THEORIES

A. JAKOVÁC*

HAS Research Group "Theory of Condensed Matter" and TU Budapest, H-1521 Budapest, Hungary

ZS. SZÉP†

Research Group for Statistical Physics of the Hungarian Academy of Sciences, H-1117 Budapest, Hungary

Using resummation in perturbation theories at finite temperature or in non-equilibrium is unavoidable to obtain consistent results. Resummation, however, is often in conflict with renormalization. In this talk we give two possible solutions to problems of this type in static resummation. One is the use of auxiliary finite temperature UV divergent counterterms. The other is to use a renormalization scheme fitted to the environment. We also suggest a way how one could proceed in case of dynamic resummations.

1. Introduction

In perturbation theories at finite temperature and non-equilibrium environment we are often faced with infrared (IR) divergent diagrams. Although this divergency can be of physical origin, but more often it is just an artefact of a not properly organized perturbation series, and it disappears after summing up a certain subset of diagrams.

A zero temperature example is the on-shell singularity of the perturbation series of a propagator. In this case one uses 1PI resummation (Schwinger-Dyson equations) that makes possible to shift the inverse propagator. At finite temperature a similar phenomenon occurs since the environment modifies the properties of the propagating particles. As a consequence we have to count with temperature dependent mass terms[1], or self-energies[2]. In these cases the Schwinger-Dyson resummation is again

*e-mail: jakovac@planck.phy.bme.hu

†e-mail: szepzs@achilles.elte.hu

necessary to avoid divergent perturbation series. Coupling constant resummation is needed near a second order phase transition point[3] or in case of non-Abelian gauge theories[2].

In many cases it happens, however, that new, unexpected, temperature (or, more generally, environment) dependent ultraviolet (UV) divergences appear. An illustrative example is the Φ^4 model. Here the one-loop correction to the self-energy at finite temperature and in dimensional regularization reads as

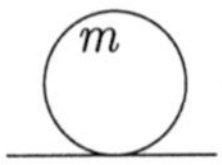

$$= \frac{m^2}{16\pi^2}\left[-\frac{1}{\varepsilon} + \gamma_E - 1 + \ln\frac{m^2}{4\pi\mu^2}\right] + \frac{1}{2\pi^2}\int_m^\infty d\omega\,\sqrt{\omega^2 - m^2}\,n(\omega). \quad (1)$$

This diagram is UV divergent, so we need a mass counterterm $\delta m^2 = \frac{m^2}{16\pi^2}\,\frac{1}{\varepsilon}$. If we resum the mass we effectively substitute $m^2 \to M_T^2$, a temperature-dependent mass term, and the same diagram now requires $\delta m^2 = \frac{M_T^2}{16\pi^2}\,\frac{1}{\varepsilon}$. This is T and ε dependent in the same time, in contradiction with the physical picture that short distance physics cannot be affected by the physics in the IR.

In order to understand the problem deeper let us try to understand, why, in general, resummation and renormalization can be in conflict. The theory that defines the system is, in fact, the bare theory. The renormalized theory is already a resummation in the bare theory: the bare parameters are divided into sum of infinite parts (counterterms) and these parts are taken into account at different levels of perturbation theory. In this subdivision the organizing principle is the UV relevance of the diagrams. There is a subtle balance between the diagrams and counterterms, and a series of theorems ensure[4] that this method finally works consistently for a certain set of theories (the perturbatively renormalizable ones).

If we encounter other type of divergencies at finite energy scales we are forced to do another resummation. In the course of this IR resummation we order the diagrams according to their IR relevance. "Normal" and counterterm diagrams, however, have different IR relevance in general, and so IR resummation will segregate them. So UV divergences do not drop out, the resummed theory will be UV inconsistent.

The above analysis suggest immediately a way out of this problem: we should resum the appropriate counterterm diagrams together with the most IR relevant "normal" diagrams[5]. Here we describe how to do that.

2. Counterterm resummation by hand

We demonstrate the method for the mass resummation, but it will be applicable for other cases (coupling constant resummations) as well. We start from the generic procedure of resummation: we add to and subtract from the Lagrangian the same term and treat the two parts in different order in perturbation theory:

$$-\mathcal{L}_{\text{mass}} = \underbrace{\frac{m^2 + \Delta M^2}{2}\varphi^2}_{\text{tree level}} + \underbrace{\left(\frac{\delta m^2}{2}\varphi^2 - \frac{\Delta M^2}{2}\varphi^2\right)}_{\text{one loop}}, \tag{2}$$

where the subscripts under the brace denote the order where the given term starts to contribute. If $m^2 = m^2_{\text{bare}}$, $\delta m^2_{\text{bare}} = 0$ and $m^2_{\text{bare}} + \Delta M^2 = m^2_R$, then we describe renormalization. If ΔM^2 is temperature dependent, then we have the "thermal counterterm" method[6]. In this case, however, as we have seen, δm^2 must be temperature dependent to make the theory finite at one loop level.

To resum counterterms we can follow the same strategy, *but now at one loop level.* We write

$$-\mathcal{L}_{\text{mass}} = \underbrace{\frac{m_R^2 + \Delta M^2}{2}\varphi^2}_{\text{tree level}} + \underbrace{\left(\frac{\delta m^2 + \delta m_T^2}{2}\varphi^2 - \frac{\Delta M^2}{2}\varphi^2\right)}_{\text{one loop}} + \underbrace{\left(-\frac{\delta m_T^2}{2}\varphi^2\right)}_{\text{two loop}}. \tag{3}$$

δm_T^2, referred as "compensating counterterm" should be temperature dependent *and* divergent (in Φ^4 model $\delta m_T^2 = \frac{\Delta M^2}{16\pi^2}\frac{1}{\varepsilon}$); but its value is irrelevant from the point of view of the complete theory, being subtracted one loop later. Otherwise it is a normal counterterm, its value should be determined order by order. In mass-independent schemes (where $\delta m^2 = zm^2$, where z is mass-independent), however, we expect that the shift of the tree level mass $m^2 \to m^2 + \Delta M^2$ will imply the mass counterterm shift $\delta m^2 = zm^2 \to zm^2 + z\Delta M^2$, so we get $\delta m_T^2 = z\Delta M^2$. This is not the most general case (any finite term could be added), but an especially comfortable one. In this case, namely, we arrive at

$$\Pi^{\text{resum},\,\overline{\text{MS}}}(m^2 + \Delta M^2) = \Pi^{\overline{\text{MS}}}(m^2)\Big|_{m^2 \to m^2 + \Delta M^2}, \tag{4}$$

ie. we do the *renormalization first,* and then substitute the resummed mass into the finite expression. It can be proven[5] that this choice is in fact consistent to all orders in perturbation theory.

3. Resummation scheme

We can make the procedure of the previous subsection more systematic by noting that the value of the tree level mass in renormalized perturbation theory is not fixed, it is in fact the consequence of the scheme chosen. The previous subsection was, from this point of view, the description of a scheme where the tree level mass is $m^2_{\overline{\rm MS}}+\Delta M^2$. We will call this scheme *resummation scheme.* In a more generic framework resummation scheme is a scheme that suits well the perturbation theory in an external environment[7]. We define it that the infinite pieces of the counterterms cancel the UV divergences, while the finite pieces should diminish IR sensitivity of the theory.

As this definition shows, counterterms in a resummation scheme will depend on the environment. Therefore changing the environment will imply a change in the scheme; but results in different schemes are not comparable[a]. We should project the results onto a fixed reference scheme (we chose $\overline{\rm MS}$ for concreteness). Since we change from one scheme to another, the projections can be done by adapting the renormalized parameters of the Lagrangian[4]. In case of Φ^4 theory there exist functions

$$m = m(m_{\overline{\rm MS}}, \lambda_{\overline{\rm MS}}, T), \quad \lambda = \lambda(m_{\overline{\rm MS}}, \lambda_{\overline{\rm MS}}, T), \quad \zeta = \zeta(m_{\overline{\rm MS}}, \lambda_{\overline{\rm MS}}, T), \quad (5)$$

that for any n-point function

$$G^{(n)}_{\overline{\rm MS}}(p_i; m_{\overline{\rm MS}}, \lambda_{\overline{\rm MS}}) = \zeta^n G^{(n)}_{\rm resum}(p_i; m, \lambda). \quad (6)$$

In perturbation theory this equation holds up to a given order: the difference is the effect of the resummation performed by the resummation scheme.

Technically we determine (5) from the condition that the bare parameters should be the same[4]: $m^2_{\overline{\rm MS}} + \delta m^2_{\overline{\rm MS}} = m^2 + \delta m^2$ in case of mass resummation. In the on-shell scheme we obtain from this condition

$$m^2 = m^2_{\overline{\rm MS}} + \Pi^{\overline{\rm MS}}(m^2). \quad (7)$$

This matching equation is just the gap equation in the language of the resummation.

At one loop level an equivalent form could have been reached if we had wrote $m^2_{\overline{\rm MS}}$ as the argument of $\Pi^{\overline{\rm MS}}$. The two cases correspond to different resummations: one with $\Pi^{\overline{\rm MS}}(m^2)$ yields the super-daisy, while the other with $\Pi^{\overline{\rm MS}}(m^2_{\overline{\rm MS}})$ the daisy resummation.

[a]For example the mass in the finite temperature on shell scheme is always m^2: there is no explicit dependence on the temperature!

Resummation scheme can be used to resum any parameters of the Lagrangian in a momentum-independent way (or at a fixed momentum). For example we can resum the coupling constant in Φ^4 theory. The details are in Ref.[5]. The results for the matching equations (5) read in high temperature expansion

$$\frac{1}{\lambda} - \frac{1}{\lambda_{\overline{\rm MS}}} = \frac{T}{16\pi M} - \frac{3}{32\pi^2}\ln\frac{T^2}{e\tilde{\mu}^2}$$

$$m^2 - m^2_{\overline{\rm MS}} = \frac{\lambda_{\overline{\rm MS}}T^2}{24} - \frac{\lambda_{\overline{\rm MS}}Tm}{8\pi} + \frac{\lambda_{\overline{\rm MS}}m^2}{32\pi^2}\ln\frac{T^2}{\tilde{\mu}^2}. \qquad (8)$$

At high temperature $m^2 = m^2_{\overline{\rm MS}}(T) + \frac{1}{24}\lambda_{\overline{\rm MS}}(T)T^2$, where the arguments refer to the renormalization scale in $\overline{\rm MS}$ scheme. Near the critical point $m \sim T - T_c$, signaling a second order phase transition. The coupling constant at high temperatures behaves as $\lambda \sim \sqrt{\lambda_{\overline{\rm MS}}(T)}$, characteristic to a 3D, dimensionally reduced theory. Near the critical point $\lambda \sim m \to 0$.

4. Outlook

So far we have spoken about momentum-independent resummations. The whole procedure goes through, however, if momentum dependence is concentrated only to the IR regime: for example when we have $\lambda(p \to \infty) = \lambda_{\rm resum}$. In fact the UV behavior is much more under control, and it is not affected by the IR problems of the theory. So a possibility for momentum dependent cases would be to make a partial resummation of the IR momentum dependence, and leave the UV part for perturbation theory. This work is in progress.

References

1. L. Dolan and R. Jackiw, Phys.Rev. D 9 (1974) 3320.
2. E. Braaten and R. Pisarski, Nucl.Phys. B 337 (1990) 569.; B 339 (1990) 662.
3. M. D'Attanasio and M. Pietroni, Nucl.Phys. B472 (1996) 711; D. Boyanovsky, H.J. de Vega and M. Simionato, Phys.Rev. D63 (2001) 045007
4. J. Collins, *Renormalization* (Cambridge University Press, 1984.)
5. A. Jakovác and Zs. Szép, hep-ph/0405226
6. J. Frenkel, A. Saa and J. Taylor, Phys.Rev. D 46 (1992) 3670; R. Parwani, Phys.Rev. D 45 (1994) 4965.
7. D. O'Connor and C.R. Stephens, Int.J.Mod.Phys. A9 (1994) 2805; Erratum-ibid. A9 (1994) 5851.

RENORMALIZATION AND GAUGE SYMMETRY FOR 2PI EFFECTIVE ACTIONS*

U. REINOSA

Institut für Theoretische Physik, Technische Universität Wien, Wiedner Hauptstrasse 8-10/136, A-1040 Wien, Austria

We explore a method to recover symmetry identities in the 2PI formalism. It is based on non-perturbative approximations to the 1PI effective action. We discuss renormalization questions raised by this technique.

1. Introduction

In the 2PI resummation scheme, the quantum fluctuations encoded in the effective action are expressed in terms of 1- and 2- point fields ϕ and G via the so-called CJT loop expansion[1]: $\Gamma[\phi, G] = S_0[\phi] + (i/2)\text{Tr } \ln G^{-1} + (i/2)\text{Tr } G\, G_0^{-1} - i\Gamma_{\text{int}}[\phi, G]$, where $S_0[\phi]$ denotes the free, classical action and $\Gamma_{\text{int}}[\phi, G]$ is made of two-particle irreducible diagrams (2PI) with vertices arising from the shifted theory $S[\phi + \varphi] - S[\phi] - \varphi \cdot \delta S/\delta\varphi|_{\varphi=\phi}$ and propagator G. The physical point of the system corresponds to the stationnarity point of Γ. Such a formulation of the quantum theory is useful in many situations, from describing the thermodynamics of the equilibrated quark-gluon plasma[2] to describing far from equilibrium dynamics and equilibration of quantum fields[3]. However this formalism presents important difficulties in the case of gauge theories where G does not fulfill usual Ward-Takahashi identities. We explore here a way to define n-point functions satisfying the usual Ward-Takahashi identities, in particular a transverse propagator. This approach has already been followed in the case of the $O(N)$ model[4] and allows to recover the Goldstone theorem in the case of a spontaneously broken symmetry. We apply it here to abelian gauge theories. We furthermore complete the scalar analysis[4] by explaining how to renormalize the scheme.

*Work in collaboration with J. Berges, S. Borsányi & J. Serreau from the Institut for Theoretical Physics, University of Heidelberg, Germany.

2. Symmetries

The idea is to construct n-point functions $\Gamma^{(n)}$ from the derivatives of $\Gamma[\phi] \equiv \Gamma[\phi, \bar{G}[\phi]]$ that is the 2PI effective action evaluated at the stationnary propagator in presence of a mean field value ϕ. If both definitions $\bar{G}$ and $\Gamma^{(2)}$ for the propagator agree in the exact theory, this is not the case when one considers approximations. $\bar{G}$ is then interpreted as a mere tool to build up resummations whereas $\Gamma^{(2)}$ is a more reliable object as it satisfies usual symmetry identities, at least in the $O(N)$ model[4] or in QED.

It is worth noting here that the symmetry is still present at the level of the 2PI effective action. In the case of QED, we thus have a 2PI Ward-Takahashi identity[5], even within a truncation. From this one can infer a symmetry property for the effective action $\Gamma[\mathcal{A}_\mu, \bar{\Psi}, \Psi]$ constructed as explained in the previous paragraph. More precisely, one can show that the effective action:

$$\Gamma^*[\mathcal{A}_\mu, \bar{\Psi}, \Psi] = \Gamma[\mathcal{A}_\mu, \bar{\Psi}, \Psi, \bar{G}_{\mu\nu}, \bar{S}] - \frac{1}{2}\mathcal{A}^\nu \left\{ \mu^2 + \frac{1}{\lambda}\partial^2 \right\} \mathcal{A}_\nu \,, \tag{1}$$

is invariant under a usual gauge transformation[a] (μ and λ denote respectively an infrared regulator for the photon and the gauge fixing parameter in the covariant gauge; $\bar{G}_{\mu\nu}$ and $\bar{S}$ denote the stationnary propagators in presence of mean fields). From this effective action it is then easy to construct transverse n-point functions, in particular a trustworthy propagator.

3. Renormalization

The aim of this section is to show how to renormalize the functional $\Gamma[\phi]$ constructed as explained in the previous section. We will concentrate here in the case of a 4–dimensional Z_2–symmetric scalar theory with φ^4 interaction and leave the inclusion of fermions and gauge symmetry for subsequent publications[5]. The functional $\Gamma[\phi]$ is nothing but the 1PI effective action which is known to be renormalizable in a 1PI loop expansion[6]. However, in the present case, the functional is constructed from approximations at the level of the 2PI effective action, and thus renormalization has to be revisited. We will not give here a complete proof of renormalizability[5] but rather insist on a few points which deserve special attention.

Part of the discussion can be followed directly on the CJT[1] formula for the 2PI effective action. The analysis of divergences is based on standard

[a] This can be directly checked on the CJT formula using the transformation of the auxiliary propagators: $\delta\bar{G} = 0\,, \quad \delta\bar{S}(x,y) = i\Big\{\Lambda(x) - \Lambda(y)\Big\}\bar{S}(x,y)$.

BPH[7] techniques applied to diagrams with resummed propagators. For such an analysis, it is convenient to use a field expansion of the effective action: $\Gamma_{\rm int}[\phi, G] = \sum_{n=0}^{\infty} \Gamma_{2n}[G]\,\phi^{2n}$. For a given n, the approximation for $\Gamma_{2n}[G]$ consists of a certain number of diagrams with $2n$ external legs. The strategy is then to draw boxes in order to track divergences: 2-point boxes represent mass and field strength singularities whereas 4-point boxes represent coupling singularities. Such divergences are absorbed in diagrams with lower number of loops, namely the topologies obtained by shrinking BPH boxes into points [b]. As far as 2-point singularities are concerned, the result is quite simple: The diagrams in the effective action being 2PI, the only 2-point boxes one can draw are either boxes on 0 legs diagrams, with one of the lines of the diagram lying outside the box, or boxes drawn on two legs diagrams, encapsulating the whole diagram. These two situations correspond to the two cross diagrams in Fig. 1 which are the only ones carrying mass and field strength counterterms in the CJT formula.

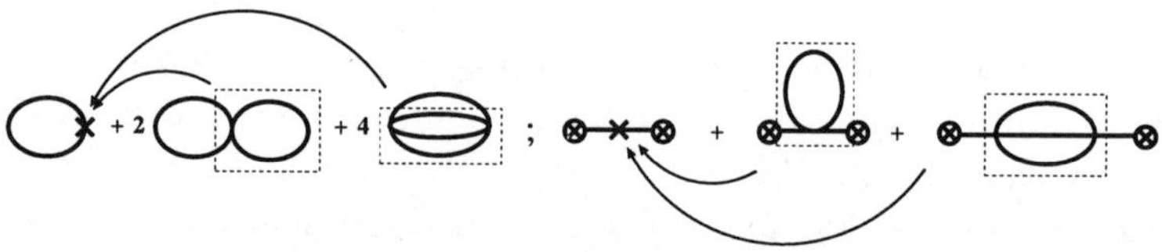

Figure 1. Renormalization of 2-point singularities using standard BPH techniques on resummed diagrams. Only two diagrams absorb these singularities.

The previous analysis misses part of the 4-point (coupling) singularities related to the non-perturbative nature of the propagator. These divergences emerge from an overlap between the logarithmic behaviour of the propagator and that of the 2PI diagram under study and call for analysing the diagrammatic content of the resummed propagator $\bar{G}$.[8,9] In the present case, to easily disentangle such divergences, it is much more convenient to work with proper vertices, that is, derivatives of the effective action. We work here in the symmetric phase but the very same counterterms can be used in the broken phase[5]. The structure of proper vertices is easily obtained by taking derivatives of the equation (which is itself derived from

[b]Thus not all the approximations are renormalizable. For an approximation to be renormalizable, it has to be such that whenever one includes a diagram, one has also to include all the diagrams related to it by the BPH procedure. In general we will use a 2PI loop expansion which is renormalizable.

the CJT formula and the stationarity condition):

$$i\frac{\delta^2\Gamma}{\delta\phi_1\delta\phi_2} = -G_{0,12}^{-1} + \frac{\partial^2\Gamma_{\rm int}}{\partial\phi_1\partial\phi_2} + \frac{\partial^2\Gamma_{\rm int}}{\partial\phi_1\partial G_{ab}}\bar{G}_{ac}\bar{G}_{bd}\frac{\delta\bar{\Sigma}_{bd}}{\delta\phi_2}\,, \tag{2}$$

where ∂ and δ denote partial and total derivatives respectively. Thus proper vertices will consist of loop diagrams involving partial derivatives of $\Gamma_{\rm int}$ with respect to ϕ or G (which we call 2PI kernels) and total derivatives of $\bar{\Sigma}$, connected via the resummed propagator $\bar{G}$. For example in the symmetric phase, the 2-point function consists of the first two terms on the r.h.s. of Eq. (2) whereas the 4-point proper vertex reads:

$$i\Gamma^{(4)}_{1234} = \frac{\partial^4\Gamma_{\rm int}}{\partial\phi_1\cdots\partial\phi_4} + \left(\frac{\partial^3\Gamma_{\rm int}}{\partial\phi_1\partial\phi_2\partial G_{ab}}\,\bar{G}_{ac}\bar{G}_{bd}\,\frac{\delta^2\bar{\Sigma}_{bd}}{\delta\phi_3\delta\phi_4} + \text{perm.}\right)\,, \tag{3}$$

where 'perm.' denotes the two permutations $(12,34) \to (13,24)$ and $(12,34) \to (14,23)$ of the term between brackets. The total derivative appearing on the r.h.s. satisfies the following integral equation:

$$\frac{\delta^2\bar{\Sigma}_{ab}}{\delta\phi_1\delta\phi_2} = \left(2\frac{\partial^3\Gamma_{\rm int}}{\partial G_{ab}\partial\phi_1\partial\phi_2}\right) + \left(2\frac{\partial^2\Gamma_{\rm int}}{\partial G_{ab}\partial G_{cd}}\right)\bar{G}_{ce}\bar{G}_{df}\,\frac{\delta^2\bar{\Sigma}_{ef}}{\delta\phi_1\delta\phi_2}\,. \tag{4}$$

Renormalization of $\Gamma[\phi]$ then amounts to first renormalizing the 2PI kernels, next total derivatives of $\bar{\Sigma}$ and finally the proper vertices. The BPH approach described previously renormalizes almost all the kernels. Only two kernels deserve much more attention due to the non-perturbative character of the propagator. These kernels are the 2-point kernels $\partial\Gamma_{\rm int}/\partial G$ and $\partial^2\Gamma_{\rm int}/\partial\phi^2$. $\partial\Gamma_{\rm int}/\partial G$ has been extensively discussed in Refs.[8,9]. There it was shown that the coupling singularities hidden in the kernel are exactly accounted for by a Bethe-Salpeter equation for a 4-point auxiliary function $\bar{V} = \bar{\Lambda} + \bar{\Lambda}\,\bar{G}^2\,\bar{V} = \bar{\Lambda} + \bar{V}\,\bar{G}^2\,\bar{\Lambda}$, where $\bar{\Lambda} = 2\partial^2\Gamma_{\rm int}/\partial G^2$. Renormalization of $\bar{V}$ is performed via a shift of the tree level (local) contribution to the kernel $\bar{\Lambda}$. At the level of the effective action, this contribution corresponds to the "eight" diagram depicted in Fig. 2. The same results apply to the kernel $\partial^2\Gamma_{\rm int}/\partial\phi^2$. Coupling singularities are enterily described by a second Bethe-Salpeter equation for V (this equation can be solved in terms of $\bar{V}$): $V = \Lambda + V\,\bar{G}^2\,\bar{\Lambda} = \Lambda + \Lambda\,\bar{G}^2\,\bar{V}$, with $\Lambda = 2\partial^3\Gamma_{\rm int}/\partial G\partial\phi^2$. Renormalization amounts to a shift of the tree level (local) contribution to the kernel Λ. At the level of the effective action, this contribution corresponds to the "tadpole" diagram depicted in Fig. 2 (let us note here that self-consistency and the 2PI character of the resummation are crucial for the renormalization program to work at all[8,5]). Indeed, this last result allows, in the symmetric phase, to also renormalize the 2-point function which is

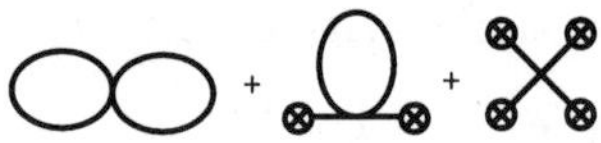

Figure 2. Diagrams related to the renormalization of divergences arising from the non-perturbative structure of the propagator. They are all local.

nothing but the kernel $\partial^2\Gamma_{\rm int}/\partial\phi^2$. Renormalization of the 4-point function is not so straightforward as it involves the quantity $\delta^2\bar{\Sigma}/\delta\phi^2$. However the latter is nothing but V as it satisfies the same self-consistent equation (see Eq. (4)). Thus $\delta^2\bar{\Sigma}/\delta\phi^2$ is finite. We can now analyze divergences in the 4-point function. To this purpose, it is convenient to express $\delta^2\bar{\Sigma}/\delta\phi^2 = V$ in terms of $\bar{V}$: $i\Gamma^{(4)} = \partial^4\Gamma_{\rm int}/\partial\phi^4 + 2\Big(\Lambda\,\bar{G}^2\,\boxed{\Lambda} + \Lambda\,\bar{G}^2\,\boxed{\bar{V}\,\bar{G}^2\,\Lambda} + \text{perm.}\Big)$. The boxes drawn in the previous equation show that all the subdivergences are indeed already taken into account by the renormalization of V. There only remain overall divergences which are absorbed in a shift of the tree level contribution (local) to $\partial^4\Gamma_{\rm int}/\partial\phi^4$. At the level of the effective action, this contribution arises from the third diagram depicted in Fig. 2.

Once all the kernels have been renormalized together with $\bar{V}$ and V, renormalization of total derivatives of $\bar{\Sigma}$, $\delta^p\bar{\Sigma}/\delta\phi^p$ (with $p > 2$) and proper vertices $\Gamma^{(n)}$ (with $n > 4$) is straightforward. Indeed one can show[5] that it is possible to express the latter in terms of diagrams made of finite kernels, or Bethe-Salpeter functions $\bar{V}$ and V, and only involving finite loops.

To extend the results on renormalization to QED, one first needs to incorporate fermions[5]. Next one needs to discuss spurious UV singularities in the auxiliary photon propagator $\bar{G}_{\mu\nu}$[5].

References

1. J. M. Cornwall, R. Jackiw and E. Tomboulis, Phys. Rev. D **10** 2428 (1974) .
2. J.-P. Blaizot, E. Iancu and A. Rebhan, Phys. Rev. D **63** 065003 (2001).
3. J. Berges, Nucl. Phys A **699** 351 (2002). J. Berges, S. Borsányi and J. Serreau, Nucl. Phys. B **660** 51 (2003).
4. H. van Hees and J. Knoll, Phys. Rev. D **66** 025028 (2002).
5. J. Berges, S. Borsányi, U. Reinosa and J. Serreau, work in progress.
6. J. Zinn-Justin, Quantum Field Theory and Critical Phenomena, Oxford Science Publications.
7. N. N. Bogoliubov, O. S. Parasiuk, Acta Math. **97** 227 (1957). K. Hepp, Commun. Math. Phys. **2** 301 (1966).
8. J.-P. Blaizot, E. Iancu and U. Reinosa, Phys. Lett. B **568** 160 (2003), Nucl. Phys. A **736** 149 (2004).
9. H. van Hees and J. Knoll, Phys. Rev. D **65** 025010 and 105005 (2002).

OUT-OF-EQUILIBRIUM MASSLESS SCHWINGER MODEL*

RAMON F. ALVAREZ-ESTRADA

Departamento de Fisica Teorica I, Facultad de Ciencias Fisicas, Universidad Complutense, 28040 Madrid, Spain. E-mail: rfa@fis.ucm.es

An isolated statistical many-particle system, with microscopic dynamics described by massless Quantum Electrodynamics (QED) in one spatial dimension (massless Schwinger Model) is studied in out-of-equilibrium cases. This constitutes a model of Non-Equilibrium Relativistic Quantum Field Theory. The initial condition is chosen to correspond to local equilibrium but, on the other hand, to a global out of equilibrium situation, with a spatially-inhomogeneous distribution of temperatures on suitably large length scales. The out-of-equilibrium generating functionals are discussed and some partial results are outlined.

1. A Possible Non-Equilibrium Initial State

The study of multiparticle systems formed by relativistic microscopic entities, with dynamics described by Relativistic Quantum Field Theory, in non-equilibrium situations has become an important and active, albeit rather difficult, research subject. See [1,2] for introductions to statistical relativistic quantum (field-theoretic) systems at thermal equilibrium and see [3] for non-equilibrium studies. It may be interesting to analyze some models in which: i) certain features, specifically genuine of non-equilibrium field-theoretic situations, are treated in detail, at the expense of ii) simplifying and omitting other features (ultraviolet divergences, non-abelian gauge invariance,..) as much as possible. Accordingly, we shall consider an isolated (closed) and large statistical system, formed by very many photons and massless electrons and positrons, described by (superrenormalizable) QED in one spatial dimension (the massless Schwinger model). The present study treats several aspects of that model in a somewhat more precise way than before [4] and outlines some partial results.

*This work is supported by Ministerio de Ciencia y Tecnologia(Proyecto FPA2000-0956) and by Ministerio de Educacion y Ciencia (Proyecto FPA2004-02602), Spain

The quantum hamiltonian operator is $H_Q = \int dx h_Q(x)$, the operator $h_Q(x)$ being the standard unrenormalized quantum hamiltonian density for QED. By assumption, at some initial time t_0, the system is in a non-equilibrium situation, characterized by the initial non-equilibrium (spatially inhomogeneous) density operator: $\rho_{Q,in} = \exp[-\beta_{eq}\int dx\lambda(x)_1 h_Q(x)]$. Previously [4], various arguments were given to justify this proposal and its possible physical interest. β_{eq} is a constant, with $0 < \beta_{eq} < +\infty$. $\lambda_1(x) = \lambda_1$ is some given function, to be regarded as known. $\beta_{eq}\lambda_1(x)$ is determined, in turn, by some given "observable" $u(x)$, characterizing the average energy density in the initial non-equilibrium state. Thus, $u(x)$ is given (up to a suitable subtraction of the vacuum energy density) by $[Tr\rho_{Q,in}]^{-1}.[Tr(h_Q(x)\rho_{Q,in})]$. Tr denotes the usual trace operation. λ_1 is assumed to be finite and continuous for any x, to vary slowly on suitably large length scales and to approach unity as $\mid x \mid \to +\infty$. Thus, β_{eq}^{-1} can be regarded as the equilibrium temperature at large distances ($\mid x \mid \to +\infty$) and λ_1 describes a spatial inhomogeneity in the system for finite x, at the initial time. $\lambda_1 \equiv 1$ for any x corresponds to exact thermal equilibrium at the initial time. Notice that the initial distribution describes a situation with vanishing chemical potential.

We shall recall some functional integrals for matrix elements of $\rho_{Q,in}$ [4]. Let $\int_{\beta_{eq}} \equiv \int dx \int_0^{\beta_{eq}} d\tau$, τ being an "euclidean time". Let $\mid A_{\mu,1};\psi_1\rangle$ be a generic simultaneous eigenstate of the quantized electromagnetic ($A_{Q;\mu}(t_0,x)$) and of the electron-positron ($\psi_Q(t_0,x)$) field operators, at time t_0, with eigenvalues $A_{\mu,1}(x)$ and $\psi_1(x)$. And so on for the eigenbra $\langle A_{\mu,2};\psi_2 \mid$ for the operators $A_{Q;\mu}(T,x)$ and $\bar{\psi}_Q(T,x)$), at time T, with eigenvalues $A_{\mu,2}(x)$ and $\bar{\psi}_2(x)$, respectively. We shall write a functional integral representation for the matrix element $\langle A_{\mu,2};\psi_2 \mid \rho_{Q,in} \mid A_{\mu,1};\psi_1\rangle$:

$$\langle A_{\mu,2};\psi_2 \mid \rho_{Q,in} \mid A_{\mu,1};\psi_1\rangle = \int [DA_{\mu,in}] \int [D\bar{\psi}_{in} D\psi_{in}] \times f_2^{(+)} \exp\left[-\int_{\beta_{eq}} s\right] \tag{1}$$

$$s = \frac{1}{2\lambda_1}E_{in}^2 + \frac{F^2}{2\alpha} + s_{ep} + s_{QED} \tag{2}$$

$$s_{QED}[A_{\mu,in};\bar{\psi}_{in},\psi_{in}] = -\lambda_1 ie[\bar{\psi}_{in}\gamma_\mu A_{\mu,in}\psi_{in}] \tag{3}$$

$$s_{ep} = \bar{\psi}_{in}\gamma_0\frac{\partial\psi_{in}}{\partial\tau} + \lambda_1\bar{\psi}_{in}\gamma_1(\partial/\partial x)\psi_{in} \tag{4}$$

with $f_2^{(+)} = \exp[+\int dx\bar{\psi}_{in}(\beta_{eq},x)\gamma_0\psi_{in}(\beta_{eq},x)]$. γ_0 and γ_1 are standard two-dimensional Dirac matrices. e is the (unrenormalized) electric charge. The above fermionic functional integral is carried out over the Grassmann fields $\bar{\psi}_{in}(\tau,x)$ and $\psi_{in}(\tau,x)$, with the boundary conditions $\psi_{in}(\tau=0,x)=\psi_1(x)$ and $\bar{\psi}_{in}(\tau=\beta_{eq},x)=\bar{\psi}_2(x)$. In turn, the above bosonic functional integral is carried out over all classical fields $A_{\mu,in}=A_{\mu,in}(\tau,x)$, $\mu=0,1$, with the boundary conditions $A_{\mu,in}(\tau=0,x)=A_{\mu,1}(x)$ and $A_{\mu,in}(\tau=\beta_{eq},x)=A_{\mu,2}(x)$. $F(=\partial_\mu A_{\mu,in})$ is the gauge-fixing function. We have factored out the corresponding Faddev-Popov determinant. E_{in} is the real classical electric field (appearing in the euclidean lagrangian) [2]. We recall that, at complete thermal equilibrium (with $\lambda_1=1$ for any x), there exists the so-called trivial sector (namely, that without zero modes), in which the analysis of the model simplifies somewhat [5,6]. In our out-of-equilibrium analysis with $\lambda_1\neq 1$, we restrict ourselves to work in a sector which would simplify to the trivial one if $\lambda_1=1$. Thus, we consider what could well be called the out-of-equilibrium trivial sector.

The above "action" s has a specific dependence on $\lambda_1(\neq 1)$, which characterizes the actual non equilibrium state. The vanishing of the variations of s with respect to $\bar{\psi}_{in}$ and ψ_{in} (for given $A_{\mu,in}$) yields the following Dirac-like equations of motion, respectively:

$$\gamma_0\frac{\partial\psi_{in}}{\partial\tau}+\lambda_1\gamma_1\frac{\partial\psi_{in}}{\partial x}-\lambda_1 ie\gamma_\mu A_{\mu,in}\psi_{in}=0 \tag{5}$$

$$\frac{\partial\bar{\psi}_{in}}{\partial\tau}\gamma_0+\frac{\partial(\lambda_1\bar{\psi}_{in})}{\partial x}\gamma_1+\lambda_1 ie\bar{\psi}_{in}\gamma_\mu A_{\mu,in}=0 \tag{6}$$

We now introduce the "current" J_μ, $\mu=0,1$, through: $J_0\equiv\bar{\psi}_{in}\gamma_0\psi_{in}$, $J_1\equiv\lambda_1\bar{\psi}_{in}\gamma_1\psi_{in}$. A remark (or partial result) is that Eqs. (5) and Eq. (6) yield "current" conservation: $\partial J_0/\partial\tau+\partial J_1/\partial x=0$. The specific dependence of J_1 on λ_1 is responsible for that "current" conservation.

2. The Generating Functional for the Initial State

The analogy with Equilibrium Thermal Field Theory [1,2] indicates the convenience of studying the properties of the generating functional $Z_{in}=Z_{in}[j_{\mu,in};\bar{\xi}_{in},\xi_{in}]$ determined by $\rho_{Q,in}$ and depending on "external" sources $j_{\mu,in}=j_{\mu,in}(\tau,\mathbf{x})$ (for the electromagnetic field) and $\bar{\xi}_{in}=\bar{\xi}_{in}(\tau,x)$, $\xi_{in}=\xi_{in}(\tau,x)$ (for the fermions). It reads::

$$Z_{in}=\int[DA_{\mu,in}]\int[D\bar{\psi}_{in}D\psi_{in}]$$

$$\times\exp\left[\int_{\beta_{eq}}(-s+j_{\mu,in}A_{\mu,in}+\bar{\xi}_{in}\psi_{in}+\bar{\psi}_{in}\xi_{in})\right] \tag{7}$$

This functional integral is carried out with standard boundary conditions for bosons and fermions. Z_{in} would yield, through standard functional derivatives with respect to its sources, correlation functions for the initial non equilibrium state. The functional integral over the electron-positron fields in Z_{in} for the non-equilibrium situation (say, $\lambda_1 \neq 1$) boil down, through standard techniques, to determine the two-point fermionic correlation function $G(\tau, x; \tau', x')$, for given $A_{\mu,in}$. A partial result is that, in the trivial sector, the crucial dependence of $G(\tau, x; \tau', x')$ on $A_{\mu,in}$ can be displayed in a compact and specific way, up to the additional complication of certain λ_1-dependent ordinary linear differential equations, as we shall now outline. In fact, $G(\tau, x; \tau', x')$ fulfills:

$$\left[\gamma_0 \frac{\partial}{\partial \tau} + \lambda_1 \gamma_1 \frac{\partial}{\partial x} - \lambda_1 i e \gamma_\mu A_{\mu,in}\right] G(\tau, x; \tau', x') = \delta(\tau - \tau'; \beta_{eq}).\delta(x - x') \tag{8}$$

$\delta(\tau; \beta_{eq}) \equiv (\beta_{eq})^{-1} \sum_{n=-\infty}^{+\infty} \exp i\omega_n \tau$, $\omega_n = \pi(2n+1)/\beta_{eq}$, $n = 0, \pm 1, \pm 2, \ldots$, and $\delta(x - x')$ denotes the Dirac delta function. By extending methods employed previously for the equilibrium case [6], the solution of Eq. (8) can be expressed as:

$$G(\tau, x; \tau', x') = \exp ie[\Phi(\tau, x) - \Phi(\tau', x')] G_0(\tau, x; \tau', x') \tag{9}$$

$$\left[\gamma_0 \frac{\partial}{\partial \tau} + \lambda_1 \gamma_1 \frac{\partial}{\partial x}\right] G_0(\tau, x; \tau', x') = \delta(\tau - \tau'; \beta_{eq}).\delta(x - x') \tag{10}$$

$$\left[\frac{\partial^2}{\partial \tau^2} + \lambda_1 \frac{\partial}{\partial x}(\lambda_1 \frac{\partial}{\partial x})\right] \Phi(\tau, x) = \left[\gamma_0 \frac{\partial}{\partial \tau} + \lambda_1 \gamma_1 \frac{\partial}{\partial x}\right] \lambda_1 \gamma_\mu A_{\mu,in} \tag{11}$$

Notice that the τ dependences are separable in both Eqs. (10) and Eq. (11). Eq. (11) is to be solved with periodic boundary conditions regarding the τ dependence. Thus, $G(\tau, x; \tau', x')$ depends on the solutions of the resulting $\lambda_1(x)$-dependent linear ordinary differential equations. The above $G(\tau, x; \tau', x')$ is a necessary ingredient in order to find the functional determinant associated to the fermionic functional integral. In turn, the latter functional determinant is required for the remaining functional integral over $A_{\mu,in}$.

3. Generating Functional for Time Evolution: Some Comments

The density operator describing the actual system at any time $t > t_0$ and determined, at $t = t_0$, by the above initial density operator $\rho_{Q,in}$

is: $\rho_Q(t,t_0) = U_Q(t,t_0)\rho_{Q,in}U_Q(t,t_0)^+$. $U_Q(t,t_0) = \exp[-i(t-t_0)H_Q]$ is the evolution operator and the superscript + denotes the adjoint. Let T be a fixed, but very large, time : $T \gg t_0$. Let $\mid A_{\mu,T}; \psi_T\rangle$ be a generic simultaneous eigenstate of the quantized field operators $A_{Q;\mu}(T,x)$, and $\psi_Q(T,x)$ at time T and so on for the corresponding bra. We consider the matrix element $\langle A_{\mu,T}; \psi_T \mid \rho_Q(T,t_0) \mid A_{\mu,T}; \psi_T\rangle$. At this stage, we introduce external sources: i) $J_{\mu,+}(t_+,x)$ (bosonic) and $\bar{\xi}_+(t_+,x)$, $\xi_+(t_+,x)$ (fermionic), ii) $J_{\mu,-}(t_-,x)$ (bosonic) and $\bar{\xi}_-(t_-,x)$, $\xi_-(t_-,x)$ (fermionic). Those sources are, respectively, associated to the functional integrals corresponding to the matrix elements of : i') $U_Q(T,t_0)$, ii') $U_Q(T,t_0)^+$. No sources are introduced for $\rho_{Q,in}$. After the inclusion of all those sources, we take the trace over all $\mid A_{\mu,T}; \psi_T\rangle$. All that gives rise to the generating functional $Z[J_{\mu,+}, \bar{\xi}_+, \xi_+; J_{\mu,-}, \bar{\xi}_-, \xi_-]$. The latter is the restriction of a more general generating functional $Z[J_{\mu,+}, \bar{\xi}_+, \xi_+; j_{\mu,in}, \bar{\xi}_{in}, \xi_{in}; J_{\mu,-}, \bar{\xi}_-, \xi_-]$, given through Eqs. (7), (10-12) of our previous report [4], when one omits all sources for $\rho_{Q,in}$ ($j_{\mu,in} = \bar{\xi}_{in} = \xi_{in} = 0$). It is convenient to employ the same Dirac matrices (for instance, the euclidean ones) in all integrands, namely, in the Minkowski and in the euclidean ones. In the case $\lambda_1(x) \equiv 1$ (the trivial sector), a computation of the real-time Green's functions in Thermo-Field Dynamics, equivalent to finding $Z[J_{\mu,+}, \bar{\xi}_+, \xi_+; J_{\mu,-}, \bar{\xi}_-, \xi_-]$, has been reported [7]. The process of performing all functional integrations yielding $Z[J_{\mu,+}, \bar{\xi}_+, \xi_+; J_{\mu,-}, \bar{\xi}_-, \xi_-]$ (with $\lambda_1(x) \neq 1$ and also in the trivial sector) is far more complicated than that for Z_{in}.

References

1. J.I. Kapusta *Finite-Temperature Field Theory* (Cambridge University Press, Cambridge, 1993).
2. M. Le Bellac *Thermal Field Theory* (Cambridge University Press, Cambridge, 1996).
3. K.C. Chou, Z. B. Su, B. L. Hao and L. Yu, *Phys. Rep.* **118**, 1 (1985).
4. R.F. Alvarez-Estrada, "Initial States and Time Evolution for Out-of-Equilibrium Matter" in *Strong and Electroweak Matter 2002*, Proc. of the SEWM2002 Meeting, Heidelberg, 2-5 October 2002, ed. Michael G. Schmidt (World Scientific, Singapore, 2002).
5. I. Sachs and A. Wipf, Helv. Phys. Acta, *Helv. Phys. Acta* **65**, 652 (1992).
6. R.F. Alvarez-Estrada and A. Gomez Nicola *Phys. Rev. D* **57**, 3618 (1998).
7. F. Ruiz Ruiz and R. F. Alvarez-Estrada *Phys. Lett. B* **180**, 153 (1986).

SELFCONSISTENT CALCULATIONS OF HADRONS AT FINITE TEMPERATURE

C. BECKMANN

Institut für Theoretische Physik,
J. W. Goethe-Universität,
D-60325 Frankfurt am Main, Germany

A set of selfconsistent Schwinger-Dyson equations for the Walecka-Model is derived within the framework of the Cornwall-Jackiw-Tomboulis formalism. In the Hartree approximation these equations are solved by propagators with T- and μ-dependent masses. The results are compared to tree-level calculations. The difference is most pronounced for the mass of the scalar meson.

1. Introduction

Understanding the properties of hot and dense nuclear matter is one of the most important goals in the investigation of heavy ion collisions as well as astrophysical objects like neutron stars. A common method in these investigations is the use of effective hadronic models which describe nuclear matter in terms of hadronic degrees of freedom instead of the fundamental quarks. A very popular model is the Walecka-model, also known as QHD I, which was in this form introduced in 1974[1] based on an idea of Johnson and Teller[2]. It describes nuclear properties as a balance of an attractive scalar interaction, the σ meson, and a repulsive vector interaction, the ω meson. This model will be used in our work. Normally, calculations in these hadronic models are performed in the tree-level approximation, i.e. propagators do not include any self-interactions in the form of momentum loops. At low temperature, this approach might be appropriate, but at higher temperature the question arises, whether higher-order loop contributions lead to significant changes in the observables. Especially the terms corresponding to the creation of light mesons are expected to make a noticeable effect already at moderate temperatures.

A common way to include higher order diagrams is the loop expansion where the effective action is expanded around its classical value by including terms with increasing numbers of momentum loops[3]. However, up to two loop order this approximation is non convergent in the Walecka model[4]. A more advanced approach for the consideration of higher-order loop diagrams is the expansion scheme developed by Cornwall, Jackiw, and Tomboulis[6] (called CJT-formalism in the following). The CJT formalism is a resummation scheme to derive a selfconsistent set of Schwinger-Dyson equations. It is equivalent to the Φ-derivable approximation introduced by Baym and Grinstein[7]. In the CJT approximation an effective action is constructed that leads to selfconsistent Schwinger-Dyson equations which contain the fully dressed propagators. By solving these equations certain classes of diagrams with a given number of momentum loops are resummed.

2. Walecka model in Hartree approximation

In this section the CJT formalism is applied to the Walecka-model. A similar investigation in an equivalent formalism has been done by Mishra et al.[8]. The model Lagrangian is

$$\mathcal{L} = \overline{\psi}\left(\mathrm{i}\gamma_\mu\partial_\mu - m_\psi - g_\sigma\sigma - g_\omega\gamma^\mu\omega_\mu\right)\psi + \frac{1}{2}\partial^\mu\sigma\partial_\mu\sigma - \frac{1}{2}m_\sigma^2\sigma^2$$
$$-\lambda\sigma^4 + \frac{1}{2}m_\omega^2\omega_\mu\omega^\mu - \frac{1}{4}(\partial_\mu\omega_\nu - \partial_\nu\omega_\mu)(\partial^\mu\omega^\nu - \partial^\nu\omega^\mu) \quad . \qquad (1)$$

Note the quartic self-interaction which was introduced by Boguta and Bodmer in 1977[5] to correct the large incompressibility of the original model. After the substitution $\phi \longrightarrow \phi + \hat{\phi}$ for all fields ϕ, where ϕ is splitted into its vacuum expectation value and the fluctuations, the interaction term of the Lagrangian becomes

$$\mathcal{L}_{\text{int}} = \overline{\psi}g_\sigma\sigma\psi + \overline{\psi}g_\omega\gamma^0\omega_0\psi + \lambda(\sigma^4 + 4\hat{\sigma}\sigma^3) \quad . \qquad (2)$$

Up to two loops this corresponds to the Feynman diagrams shown in Fig. 1.

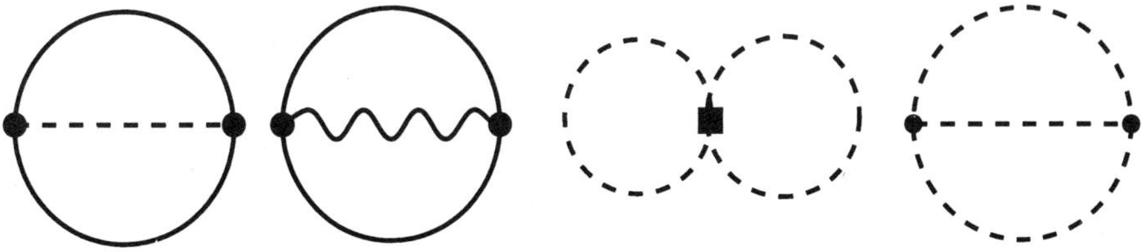

Figure 1. The two particle irreducible Feynman-diagrams which form the V_2-part of the effective potential.

For now only the third diagram is considered. In the following we refer to this approximation as the Hartree approximation. For the ω meson the mean-field approximation is used, i.e. $\langle\omega^\mu\rangle = \delta_{\mu 0}\omega_0$. The parameters of the model were adapted to the properties of infinite nuclear matter. The values are $g_\sigma = -7.197$, $g_\omega = 7.557$ and $\lambda = 35.557$ for the calculations presented here.

Applying the CJT-formalism leads to the following equations for the fields and the effective masses, denoted by capital M with the index indicating the field:

$$M_\sigma^2 = m_\sigma^2 + 12\lambda\left(\hat{\sigma}^2 + \int\frac{\mathrm{d}^3\mathbf{q}}{(2\pi)^3}\frac{1}{E(M_\sigma)}\left\{\frac{1}{\exp\left(\frac{E(M_\sigma)}{T}\right)-1}+\frac{1}{2}\right\}\right) \quad ,$$

$$M_\omega^2 = m_\omega^2 \quad ,$$

$$M_\psi = m_\psi + g_\sigma\hat{\sigma} \quad ,$$

$$0 = m_\sigma^2\hat{\sigma} + 4\lambda\hat{\sigma}^3 + 12\lambda\hat{\sigma}\int\frac{\mathrm{d}^3\mathbf{q}}{(2\pi)^3}\frac{1}{E(M_\sigma)}\left\{\frac{1}{\exp\left(\frac{E(M_\sigma)}{T}\right)-1}+\frac{1}{2}\right\}$$

$$+4g_\sigma\int\frac{\mathrm{d}^3\mathbf{q}}{(2\pi)^3}\frac{M_\psi}{E(M_\psi)}(n+\overline{n}-1) \quad ,$$

$$0 = -m_\omega^2\hat{\omega}_0 + 4g_\omega\int\frac{\mathrm{d}^3\mathbf{q}}{(2\pi)^3}(n-\overline{n}) \quad .$$

Here $\hat{\sigma}$ and $\hat{\omega}$ stand for the vacuum expectation values of the fields, n and $\bar{n}$ are the particle and anti-particle distribution functions, respectively. The energy is denoted by E.

The most pronounced effect due to the Hartree term is visible for the mass of the σ field which is shown in Fig.2. The effect of the two-particle irreducible diagrams gets larger for increasing temperature. As the chemical potential is increased, the difference between tree-level and the Hartree-approximation is suppressed. The maximum of the difference is about 20 %. We find that the inclusion of just the double bubble-diagram allready gives a noticable difference compared to the calcualation with one-particle-irreducible contributions only.

3. Outlook

The calculations in Hartree-approximation show that higher order contributions are in fact relevant for calculations at high temperatures. The next

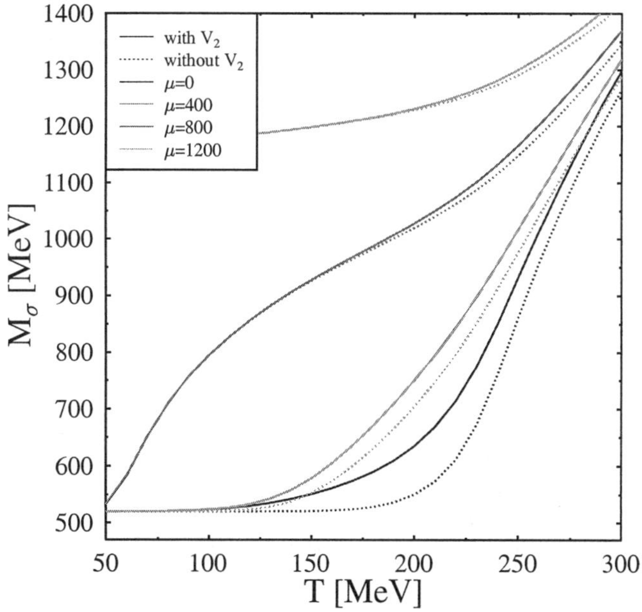

Figure 2. The mass of the σ-field as a function of temperature for different chemical potentials μ. The full lines are the calculations in Hartree-approximation, the dotted lines are the calculations on tree-level.

step is thus to inlcude the other two-particle-irreducible diagrams, too. For the diagrams with three-line-vertices the simple quasi-particle ansatz presented above is not longer applicable. Now the dressed propagator can depend on the internal momentum as well as the external momentum, so one has to solve set of coupled integral equations. The propagators have to be expressed by spectral functions to account for a finte decay-width. Furthermore the vector mesons have to be decomposed into projectors and one has to deal with the dirac structure of the fermions.

4. Conclusion

Calculations at finite temperature have been done selfconsistently in the framework of a Walecka-model. The Cornwall-Jackiw-Tomboulis-formalism has been applied. In the Hartree-approximation the mass of the σ-meson is enhanced up to 20 %, so an investigation of more two-particle irreducible contributions to the effective potential is promising.

Acknowledgments

The author is grateful to Dirk H. Rischke for fruitful discussions.

References

1. J. D. Walecka, *Annals Phys.* **83** (1974) 491
2. M. Johnson, E. Teller, *Phys. Rev.* **98** (1955) 783
3. R. Jackiw, *Phys. Rev.* D **9** (1974) 1686
4. R. J. Furnstahl, R. J. Perry, B. D. Serot, *Phys. Rev.* **40** (1989) 321
5. J. Boguta, A. R. Bodmer, *Nucl. Phys.* A **292** (1977) 413
6. J. M. Cornwall, R. Jackiw, E. Tomboulis, *Phys. Rev.* D **10** (1974) 2428
7. G. Baym, G. Grinstein, *Phys. Rev.* D bf 15 (1977) 2897
8. A. Mishra, P. K. Panda, W. Greiner *J. Phys. G.* **27** (2001) 1561

FERMION PRODUCTION IN CLASSICAL FIELDS

D. D. DIETRICH

Laboratoire de Physique Théorique, Université Paris XI, Orsay, France

Here, the formalism for the production of fermion-antifermion pairs by vacuum polarisation due to classical fields is investigated according to and extending Refs.[1,2]. The issue of gauge invariance is discussed for the calculation of the expectation value for the number of produced pairs from the exact retarded and/or advanced propagator. These correlators are given as solutions of the equation of motion for the Dirac Green's functions in the classical field. The general characteristics of the two point functions in field configurations expected for ultrarelativistic heavy-ion collisions are lined out. Calculations for a model for the respective radiation field are presented in a covariant-derivative expansion scheme.

Following the derivation in[3], the expectation value for the number of produced pairs can be determined by calculating the overlap of a negative energy fermion – corresponding to a positive energy antifermion – with a positive energy fermion, after propagation through the field:

$$M(p,q) = \int d^3x d^3y \psi_p^\dagger(x) G_R(x,y)\gamma^0\psi_q(y). \tag{1}$$

Use has been made of the fact that solutions of the Dirac equation at x and y are connected by the retarded propagator which in turn can be expressed with the retarded single-particle scattering-operator $\mathcal{T}_R$:

$$G_R(x,y) = G_R^0(x-y) + \int d^4\xi d^4\eta G_R^0(x-\xi)\mathcal{T}_R(\xi,\eta)G_R^0(\eta-y).$$

Free solutions are linked by the free retarded propagator G_R^0 and positive and negative energy solutions are orthogonal to each other, whence integrating the square of the overlap over the phase space(s) yields:

$$\langle n\rangle = \int \frac{d^3q}{2(2\pi)^3\omega_q}\frac{d^3p}{2(2\pi)^3\omega_p}\left|\bar{u}(q)\mathcal{T}_R(q,-p)v(p)\right|^2, \tag{2}$$

with the unit spinors $\bar{u}(q)$ and $v(p)$ and where $\omega_p = \sqrt{|\vec{p}|^2+m^2}$ and $\omega_q = \sqrt{|\vec{q}|^2+m^2}$. Gauge invariance can be discussed better in an alternative expression[4], where the derivatives act only on the two-point function:

$$\langle n\rangle = \int \frac{d^3q}{2(2\pi)^3\omega_q} d^4x d^4y e^{iq\cdot(x-y)} \mathrm{tr}\left\{(\not{q}+m)[i\not{\partial}_x - m]G_{+-}(x,y)[m - i\not{\partial}_y]\right\}.$$

Even the trace over the correlator $G_{+-}(x,y) = \langle 0_{in}|\,\bar{\psi}(y)\psi(x)\,|0_{in}\rangle$ is still gauge dependent: $\mathrm{tr}\left\{\Omega(x)G(x,y)\Omega^\dagger(y)\right\} \neq \mathrm{tr}\left\{G(x,y)\right\}$.

I. One way to generate a gauge-invariant result is to project onto gauge-dependent solutions in a vacuum configuration of the gauge field [instead of free states as in Eq.(1)]. However, the purely partonic interpretation of the result is lost in the process: In all gauges other then the one in which the gauge-dependent coincides with the free solution, the state contains a superposition of an arbitrary number of virtual gluons. Further, the interpretation and the result depend on the choice of that special gauge. This method is equivalent to multiplying with adequate gauge (Wilson) links. There the choice of the path is the supplementary degree of freedom that has to be fixed. Additionally, in the course of the derivation of Eq.(2), q and p were interpreted as the momenta of particle and antiparticle, respectively. Here, this stays evident only in the one gauge of reference.

II. If purely fermionic states are to be preserved in all gauges, gauge invariance can be achieved by averaging the expectation value over the entire gauge group. After a coordinate transformation from d^4xd^4y to $d^4[(x+y)/2]d^4(x-y)$ one investigates $d\langle n\rangle/d^4(x-y)$ and finds that its average over the gauge group $d\langle\bar{n}\rangle/d^4(x-y)$ is equal to zero but for $x=y$. This is similar to lattice calculations, where the final result contains the four-volume $\mathcal{V}$ of the lattice as an additional factor. Thus one would obtain $\langle\bar{n}\rangle/\mathcal{V}$, there. As $x-y$ is the conjugate coordinate to the fermion momentum q this implies that the entire dependence on that momentum is only due to the geometry of the phase space and the factor $\not{q}+m$ in the trace:

$$\left.\frac{d\langle\bar{n}\rangle}{d^4(x-y)}\right|_{x=y} = \int \frac{d^4q}{(2\pi)^3}\theta(+q_0)\delta(q^2-m^2)\mathrm{tr}\{(\not{q}+m)L\}$$

$$L = \int \frac{d^4l}{(2\pi)^4}\frac{d^4k}{(2\pi)^3}\theta(-k_0)\delta(k^2-m^2)\mathcal{T}_R(l,k)(\not{k}-m)\mathcal{T}_A(k,l) \qquad (3)$$

The needed retarded and advanced propagators can be written as

$$iG_{\substack{R\\A}}(x,y,\vec{k})\gamma^0 = \pm\int \frac{d^3k}{(2\pi)^3} e^{i\vec{k}\cdot(\vec{x}-\vec{y})}\theta[\pm(x_0-y_0)]G_H(x,y,\vec{k})$$

with the homogeneous solution

$$G_H(x,y,\vec{k}) = \mathcal{P}\exp\left\{i\int_{y_0}^{x_0} d\xi_0\gamma^0[i\gamma^j\partial_j(x) + \gamma^j k_j + \not{A}(\xi_0,\vec{x}) - m]\right\}$$

Now consider the special case of ultrarelativistic heavy-ion collisions. They are frequently described by means of colour charges moving along the light-cone[5]. This picture provides the adequate framework for the description of particles with low longitudinal momentum[4,1]. The classical Yang-Mills equations have to be solved for that current. In Lorenz gauge the field consists of Weizsäcker-Williams sheets and the radiation field inside the forward light-cone. The former represent the Coulomb fields of the charges boosted into the closure of the Lorentz group, whence they are δ-distributions in the light-cone coordinates:

$$A^{WW}_{+,-}(x) = \sum_{n_{L,R}=1}^{N_{L,R}} \delta\left[x_{-,+} - (b^{L,R}_{-,+})_{n_{L,R}}\right] a^{L,R}_{n_{L,R}}.$$

with $2\pi a^{L,R}_{n_{L,R}} = -gt_a(t^{L,R}_a)_{n_{L,R}} \ln \lambda|\vec{x}_T - (\vec{b}^{L,R}_T)_{n_{L,R}}|$ and where λ regularises the logarithm but does not appear in the field tensor. t_a are the generators of $SU(3)_c$. $(t^{L,R}_a)_{n_{L,R}}$ represent the colour of the charges. It should be mentioned that the colour neutrality of each nucleus has to be ensured and that the charges of the first nucleus precess in the field of the second and vice versa, causing further sheet-like contributions that can be incorporated by generalising the charges[5,1]. However, the general form of the field – continuous inside the forward light-cone and singular but integrable on the forward light-cone – allows to express the retarded propagator by a finite number of addends[a]:

$$\begin{aligned} G_H(x,y) &= G^{rad}_H(x,y) + \\ &+ \int d^4\xi G^{rad}_H(x,\xi)[T^L_\pm(\xi)\delta(\xi_-) + T^R_\pm(\xi)\delta(\xi_+)]G^{rad}_H(\xi,y) + \\ &+ \int d^4\xi d^4\eta G^{rad}_H(x,\xi)T^L_\pm(\xi)\delta(\xi_-)G^{rad}_H(\xi,\eta)T^R_\pm(\eta)\delta(\eta_+)G^{rad}_H(\eta,y) + \\ &+ \int d^4\xi d^4\eta G^{rad}_H(x,\xi)T^R_\pm(\xi)\delta(\xi_+)G^{rad}_H(\xi,\eta)T^L_\pm(\eta)\delta(\eta_-)G^{rad}_H(\eta,y), \end{aligned}$$

because higher orders in the T cannot contribute, as lines of constant x_- or x_+ can only be crossed once[4,1]. There $G^{rad}_H(\eta,y)$ represents the homogeneous solution in the radiation field only and:

$$T^{L,R}_\pm(x) + \rho^{+,-} = \rho^{+,-}(\pm 1)^{N_{L,R}}\mathcal{P}\prod_{n_{L,R}=1}^{N_{L,R}} \exp[\pm i a^{L,R}_{n_{L,R}}(\vec{x}_T)].$$

[a]The dependence on $\vec{k}$ is not shown below.

$\mathcal{P}$ indicates that the above factors are to be ordered in such a way that, if $m_{L,R} > n_{L,R}$ then $(b^{L,R}_{-,+})_{m_{L,R}} > (b^{L,R}_{-,+})_{n_{L,R}}$. The subscript $\pm$ stand for the sign of the time difference $x_0 - y_0$ and $2\rho^{\pm} = 1 \pm \gamma^0\gamma^3$. Here, the ultra-relativistic limit has already been taken: In other words, in very energetic nuclear collisions the nuclei are highly Lorentz-contracted. For infinitely high energies $\gamma \to \infty$, the $(b^{L,R}_{\pm})_{n_{R,L}}$ go to zero like γ^{-1}. Even now, the longitudinal structure is still manifest in the path-ordering of the products. However, care has to be taken, as potentially important, because qualitatively different contributions to observables could be neglected, although the expression for the propagator in this limit would still be correct. If finite longitudinal distances were kept, every charge would have to be treated on its own and the previous equation would contain contributions with up to $N_L + N_R$ insertions.

In order to proceed from here, the exact solution for the radiation field would be needed which requires transverse lattice calculations beyond the scope of this paper. Therefore, the propagator in a model for the radiation field A^{rad} is to be investigated. By virtue of longitudinally boost-invariant boundary conditions on the light cone, it is often taken as being invariant under longitudinal boosts, too. Although, this is not necessarily the case just such a field is chosen here: $\tau_0 A_{\pm} = \pm x_{\pm} a \exp\{-x_+ x_- / \tau_0^2\}$, which satisfies the Lorenz and Fock-Schwinger gauge-conditions.

Evaluating Eq.(2) for this field can be interpreted according to **I** with the above gauge of reference. The single-particle scattering-operators to the lowest order in an expansion in the transverse components of the covariant derivatives read $\mathcal{T}_R = \gamma_+ \mathcal{T}_R^+ + \gamma_- \mathcal{T}_R^-$[b] where

$$\mathcal{T}_R^{\pm}(p,-q) = \mp a\tau_0 \frac{f(r_+ r_- \tau_0{}^2)}{(r_+ r_- \tau_0{}^2) r_\pm} - a^2 \sum_{\mu\nu} \frac{(\pm i a \tau_0)^\mu}{\mu!} \frac{(\mp i a \tau_0)^\nu}{\nu!} \times$$
$$\times \frac{f[r_+ r_- \tau_0{}^2/(\mu+\nu+2)] - f[p_\pm r_\mp \tau_0{}^2/(\mu+\nu+2)]}{r_\pm{}^2[(\mu+1)q_\mp - (\nu+1)p_\mp]}$$

which could be partially resummed into hypergeometric functions. $f(\sigma) = \sigma^2 e^{-\sigma}\mathrm{Ei}(\sigma) - \sigma$ and $r_\pm = p_\pm + q_\pm$. Note that an expansion in covariant derivatives and/or the mass preserves the properties of the propagators under gauge transformations. Evaluating the integrand yields here:

$$|\bar{u}(q)\mathcal{T}_R(q,-p)v(p)|^2 = \sum_n [8(q\cdot\mathcal{T}_R)(p\cdot\mathcal{T}_R{}^\dagger) - 4(p\cdot q - m^2)(\mathcal{T}_R\cdot\mathcal{T}_R{}^\dagger)]$$

[b]Here $(2\pi)^2\delta^{(2)}(\vec{p}_T - \vec{q}_T)/\sqrt{2}$ has been factored out.

The sum runs over the eigenvalues λ_n of the colour matrix $a = \sum_n \lambda_n |n\rangle \langle n|$, where all occurences of a in $\mathcal{T}$ are replaced by λ_n. If a light-cone gauge, *e.g.* $A_- = 0$ was chosen as reference, then $\mathcal{T}_R^- = 0$ and

$$\mathcal{T}_R^+(p,-q) = -2a\tau_0 \frac{f(r_+ r_- \tau_0{}^2)}{(r_+ r_- \tau_0{}^2) r_+} - 4a^2 \sum_{\mu\nu} \frac{(+2ia\tau_0)^\mu}{\mu!} \frac{(-2ia\tau_0)^\nu}{\nu!} \times$$
$$\times \frac{f[r_+ r_- \tau_0{}^2/(\mu+\nu+2)] - f[p_+ r_- \tau_0{}^2/(\mu+\nu+2)]}{r_+{}^2[(\mu+1)q_- - (\nu+1)p_-]}$$

would have to be used.

The expression for the advanced single-particle scattering-operator needed in the case **II** can be obtained from $\mathcal{T}_A(p,q) = -\mathcal{T}_R^*(q,p)$. For the present model, the integration over d^4l in Eq.(3) can be carried out analytically. However, the result – moreso after carrying out the trace – is to lengthy in order to be spelled out here, but will be presented elsewhere[6]. Further, it is not yet clear how the momentum variable k in Eq.(3) can be related to the antiparticle momentum p in Eq.(2). More reflection on that point is also required.

Acknowledgements

The author would like to thank A. Mosteghanemi for his help. Useful and informative discussions with F. Guerin, G. Korchemsky, A. Mishra, A. Mueller, J. Reinhardt, D. Rischke, D. Schiff, and S. Schramm are acknowledged gratefully. This project was supported financially by the German Academic Exchange Service (DAAD).

References

1. D. D. Dietrich, arXiv:hep-th/0402026.
2. D. D. Dietrich, Phys. Rev. D **68** (2003) 105005 [arXiv:hep-th/0302229];
   D. D. Dietrich, G. C. Nayak, and W. Greiner, Phys. Rev. D **64** (2001) 074006 [arXiv:hep-th/0007139]; J. Phys. G **28** (2002) 2001 [arXiv:hep-ph/0202144]; arXiv:hep-ph/0009178;
   G. C. Nayak, D. D. Dietrich, and W. Greiner, arXiv:hep-ph/0104030.
3. A. J. Baltz and L. McLerran, Phys. Rev. **C58**, 1679 (1998) [arXiv:nucl-th/9804042].
4. A. J. Baltz, F. Gelis, L. D. McLerran, and A. Peshier, Nucl. Phys. A **695** (2001) 395 [arXiv:nucl-th/0101024].
5. Y. V. Kovchegov and D. H. Rischke, Phys. Rev. C **56**, 1084 (1997) [arXiv:hep-ph/9704201].
6. D. D. Dietrich, in preparation.

NUMERICAL STUDY OF THE EQUATION OF STATE FOR TWO FLAVOR QCD AT NON-ZERO BARYON DENSITY *

S. EJIRI,[a] C.R. ALLTON,[b] M. DÖRING,[a] S.J. HANDS,[b] O. KACZMAREK,[a] F. KARSCH,[a] E. LAERMANN,[a] and K. REDLICH[ac]

[a] *Fakultät für Physik, Universität Bielefeld, D-33615 Bielefeld, Germany*
[b] *Department of Physics, University of Wales Swansea, Singleton Park, Swansea, SA2 8PP, U.K.*
[c] *Institute of Theoretical Physics, University of Wroclaw,PL-50204 Wroclaw, Poland*

We discuss the equation of state (EoS) for two flavor QCD at non-zero temperature and density. Derivatives of $\ln Z$ with respect to quark chemical potential μ_q are calculated up to sixth order. From this Taylor series, the pressure, quark number density and associated susceptibilities are estimated as functions of temperature and μ_q. It is found that the fluctuations in the quark number density increase in the vicinity of the phase transition temperature and the susceptibilities start to develop a pronounced peak as μ_q is increased. This suggests the presence of a critical endpoint in the (μ_q, T) plane. Moreover, we comment on the hadron resonance gas model, which explains well our simulation results below T_c.

1. Introduction

It is important to study QCD at high temperature and density by numerical simulations of lattice QCD. In particular, studies of the equation of state (EoS) can provide basic input for the analysis of the experimental signatures for quark-gluon plasma formation. We study the EoS in the low but non-zero baryon number density regime, interesting for heavy-ion collisions. The simulation of QCD at non-zero μ_q is known to be difficult. However, studies based on a Taylor expansion with respect to μ_q turned out to be an efficient technique to investigate the low density regime [1,2].

Thermal fluctuations near the critical temperature T_c are important information for the understanding of the QCD phase diagram. Baryon number fluctuations are expected to diverge at the endpoint of the first order

*Presented by S. Ejiri. This work is supported by BMBF grant No.06BI102, DFG grant KA 1198/6-4, PPARC grant PPA/a/s/1999/00026 and KBN grant 2P03 (06925).

chiral phase transition line. The electric charge fluctuation in heavy-ion collisions is one of the most promising experimental observables to identify the critical endpoint. The fluctuations of quark number, isospin number and electric charge are estimated by the corresponding susceptibilities, χ_q, χ_I and χ_C, which are given by $\chi_q = \partial^2 p/\partial \mu_q^2$ and $\chi_I = \partial^2 p/\partial \mu_I^2$, where $\mu_q = (\mu_u + \mu_d)/2$, $\mu_I = (\mu_u - \mu_d)/2$, and $\mu_{u[d]}$ is the chemical potential for the $u[d]$ quark. For $\mu_u = \mu_d$, the charge susceptibility is $\chi_C = \chi_q/36 + \chi_I/4$.

In this study, we investigate the pressure p, quark number susceptibility χ_q and isovector susceptibility χ_I by calculating the Taylor expansion coefficients of p, c_n and c_n^I, up to sixth order in μ, which are defined by $p/T^4 = (\ln Z)/(VT^3) \equiv \sum_{n=0}^{\infty} c_n(\mu_q/T)^n$ and $\chi_q[\chi_I]/T^2 \equiv \sum_{n=2}^{\infty} n(n-1)c_n[c_n^I](\mu_q/T)^{n-2}$ for $\mu_u = \mu_d$. where $c_n = (\partial^n \ln Z/\partial \mu_q^n)/n!$ and $c_n^I = (\partial^n \ln Z/\partial \mu_q^{n-2} \partial \mu_I^2)/n!$. Moreover, we compare these results with the prediction from the hadron resonance gas model [3,4].

2. Quark gluon gas or hadron resonance gas

We expect the equation of state approaches that of a free quark-gluon gas (Stefan-Boltzmann (SB) gas) in the high temperature limit. The coefficients in the SB limit for $N_f = 2$, $\mu_I = 0$ are well known as $c_2 = c_2^I = 1, c_4 = c_4^I = 1/(2\pi^2)$ and $c_n = c_n^I = 0$ for $n \geq 6$.

On the other hand, in the low temperature phase QCD is well modelled by a hadron resonance gas. If the interaction between these hadrons can be neglected, the pressure is obtained by summing over the contributions from all resonance states of hadrons. The contribution to p/T^4 from a particle which has mass m_i, baryon number B_i and third component of isospin number I_{3i} is given by

$$\frac{p_{m_i}}{T^4} = \frac{1}{2\pi^2}\left(\frac{m_i}{T}\right)^2 \sum_{l=1}^{\infty} \frac{\eta^{l+1}}{l^2} K_2\left(\frac{l m_i}{T}\right) \exp\left[l \frac{3B_i\mu_q + 2I_{3i}\mu_I}{T}\right], \quad (1)$$

where K_2 is the modified Bessel function, and η is 1 for mesons and -1 for baryons. Moreover, since $m_i/T \gg 1$ for all baryons and $K_2(x) \approx \sqrt{\pi/2x}\exp(-x)$, the first term of l is dominant in the baryon sector. Therefore, the pressure can be written by $p/T^4 = G(T) + F(T)\cosh(3\mu_q/T)$ for $\mu_I = 0$, where $G(T)$ and $F(T)$ are the mesonic and baryonic components of p/T^4 at $\mu_q = 0$, respectively [4]. Similarly, we obtain $\chi_q/T^2 = 9F(T)\cosh(3\mu_q/T)$ and $\chi_I/T^2 = G^I(T) + F^I(T)\cosh(3\mu_q/T)$. Here, the mesonic component for χ_q is zero because mesons ($B_i = 0$) are independent of μ_q. Therefore, $c_4/c_2 = 3/4$, $c_6/c_4 = c_6^I/c_4^I = 3/10$ in the region where the non-interacting hadron resonance gas provides a good approximation.

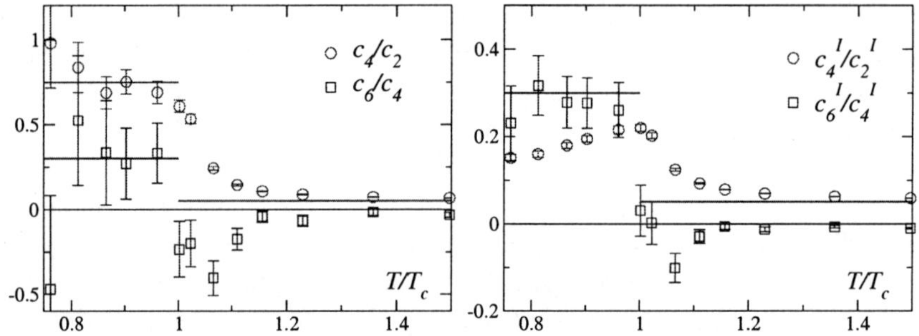

Figure 1. The ratio of Taylor expansion coefficients for χ_q (left) and χ_I (right).

We investigate these coefficients. Simulations are performed on a $16^3 \times 4$ lattice for 2 flavor QCD. The $p4$ improved action [5] is employed at $ma = 0.1$, which gives $m_{PS}/m_V = 0.7$. The number of configurations is 1000-5000 for each T. We use the random noise method [1,2]. The results for c_{n+2}/c_n and c_{n+2}^I/c_n^I are shown in Fig. 1. We find these results are consistent with the prediction from the hadron resonance gas model for $T/T_c \leq 0.96$ and approach the SB values, i.e. $c_4/c_2 = c_4^I/c_2^I = 1/2\pi^2$, $c_6/c_4 = c_6^I/c_4^I = 0$, in the high temperature limit. These results suggest that the models of free quark-gluon gas and hadron gas seem to explain the behavior of thermodynamical quantities well except in the narrow regime near T_c.

Moreover, if the data is consistent with the hadron resonance gas model, we can separate χ_I into the contributions from isotriplet mesons and baryons, since $G^I + F^I = 2c_2^I$ and $9F^I/2 = 12c_4^I$. Isosinglet mesons do not contribute to χ_I because $I_{3i} = 0$. Also, $9F = 2c_2$ from the equation of χ_q. Furthermore, for the case of $m_\pi \gg T$, the first term ($l = 1$) in Eq.(1) is dominant. This approximation holds for our present simulations where $m_\pi/T \simeq$ 4-5 but will no longer be valid with physical quark mass values. For our present simulation parameters, the ratio of $\partial^2(p_{mi}/T^4)/\partial(\mu_I/T)^2$ and p_{mi}/T^4 equals $(2I_{3i})^2$ for $I_{3i} = \{-1, 0, 1\}$. We thus find for the contribution of the triplet meson part: $\tilde{G} = 3G^I/8$. We then can estimate the contribution of the isosinglet mesons to p/T^4 by comparing $F + \tilde{G}$ with the total pressure. In Fig. 2 (left), we plot the baryon component $F = 2c_2/9$ (square), the triplet meson component $\tilde{G} = 3c_2^I/4 - c_4^I$ (circle) and the total (diamond). The solid line is the pressure obtained with the integral method in Ref. 6. The dashed lines in Fig. 2 (left) represent the predictions from the hadron resonance gas model for the mass parameter of our simulations. The resonance states for this quark mass are adjusted by the method

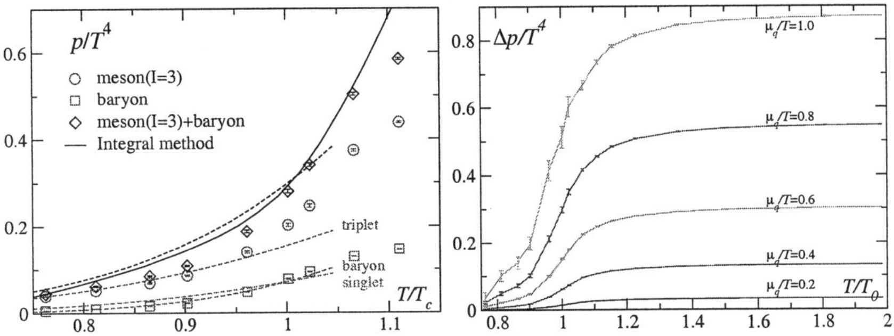

Figure 2. Pressure at $\mu_q = 0$ as a function of T (left), and difference of pressure at non-zero μ_q/T from $\mu_q = 0$ for each fixed μ_q/T (right). T_0 is T_c at $\mu_q = 0$.

described in Ref. 3. The simulation results for p/T^4 obtained with the integral method, together with the triplet meson and baryon contributions, are surprisingly well reproduced by this model calculation. The result of the singlet meson contribution may explain the difference in the pressure obtained from the integral method and that from the Taylor expansion which neglects the singlet part.

3. Pressure and susceptibilities at $\mu_q \neq 0$

Next, we calculate pressure and quark number susceptibility in a range of $0 \leq \mu_q/T \leq 1$, using the data of c_n and c_n^I; $\Delta(p/T^4) \equiv p(T,\mu_q)/T^4 - p(T,0)/T^4 = c_2(\mu_q/T)^2 + c_4(\mu_q/T)^4 + c_6(\mu_q/T)^6$, $\chi_q/T^2 = 2c_2 + 12c_4(\mu_q/T)^2 + 30c_6(\mu_q/T)^4$, and the corresponding equation for χ_I.

We draw $\Delta(p/T^4)$ for each fixed μ_q/T in Fig. 2 (right) and find that the difference from $p|_{\mu_q=0}$ is very small in the interesting regime for heavy-ion collisions, $\mu_q/T \approx 0.1$ (RHIC) and $\mu_q/T \approx 0.5$ (SPS), in comparison with the value at $\mu_q = 0$, e.g. the SB value for 2 flavor QCD at $\mu_q = 0$: $p^{SB}/T^4 \simeq 4.06$. The effect of a non-zero quark density on the pressure at $\mu_q/T = 0.1$ is only 1%. Comparing with previous result up to $O(\mu_q^4)$ [2], the effect from $O(\mu_q^6)$ seems to be small in the range we investigated. Also, the result is qualitatively consistent with that of Ref. 7 obtained by the reweighting method.

The data connected by solid lines in Fig. 3 are the results for χ_q (left) and χ_I (right) obtained by this method up to $O(\mu_q^4)$. We also calculated the susceptibilities by another two methods. Dot-dashed lines are from the hadron resonance gas model, using F, F^I and G^I we obtained. This turned out to be a good approximation for $T/T_c \lesssim 0.96$. Dashed lines are calculated

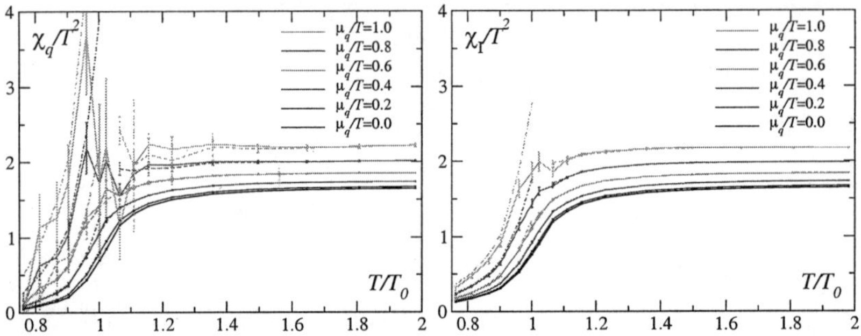

Figure 3. Quark number (left) and isovector (right) susceptibilities as a function of T for each fixed μ_q/T. T_0 is T_c at $\mu_q = 0$.

by the reweighting method using an approximation discussed in Ref. 1. Here the truncation error is $O(\mu_q^6)$ but the effect from higher order terms of μ_q is partially included. The sign problem in the reweighting method is serious when the complex phase fluctuations of the quark determinant are large at large μ_q/T. Hence we omitted data at which the standard deviation of the complex phase is larger than $\pi/2$. The difference among the three results is caused by the approximation in the higher order terms of μ_q/T.

Since the statistical error of c_6 is still large near T_c, the peak of χ_q is not clear. However, as seen in Fig. 1, c_6 changes its sign at T_c. This means the peak position of χ_q moves left, which is corresponding to the change of T_c as a function of μ_q. $T_c(\mu_q/T = 1)/T_c(\mu_q/T = 0)$ in Ref. 1 is about 0.93. Moreover, the strong enhancement of χ_q with increasing μ_q near T_c, obtained in Ref. 2, is observed also in the analysis up to $O(\mu_q^4)$. This suggests the presence of a critical endpoint in the (T, μ_q) plane. On the other hand, χ_I in Fig. 3 does not show any singular behavior. This is consistent with the sigma model prediction that only isosinglet degrees of freedom become massless at the critical endpoint [8].

References

1. C.R. Allton *et al.*, *Phys. Rev.* **D66**, 074507 (2002).
2. C.R. Allton *et al.*, *Phys. Rev.* **D68**, 014507 (2003).
3. F. Karsch, K. Redlich and A. Tawfik, *Eur. Phys. J.* **C29**, 549 (2003).
4. F. Karsch, K. Redlich and A. Tawfik, *Phys. Lett.* **B571**, 67 (2003).
5. U.M. Heller, F. Karsch, and B. Sturm, *Phys. Rev.* **D60**, 114502 (1999).
6. F. Karsch, E. Laermann and A. Peikert, *Phys. Lett.* **B478**, 447 (2000).
7. F. Csikor *et al.* *JHEP* **0405**, 046 (2004).
8. Y. Hatta and M.A. Stephanov, *Phys. Rev. Lett.* **91**, 102003 (2003).

PHASE CONVERSION AFTER A CHIRAL TRANSITION: EFFECTS FROM INHOMOGENEITIES AND FINITE SIZE

EDUARDO S. FRAGA

Instituto de Física, Universidade Federal do Rio de Janeiro
C.P. 68528, Rio de Janeiro, RJ 21941-972, Brazil
E-mail: fraga@if.ufrj.br

We discuss the kinetics of phase conversion, through the nucleation of bubbles and spinodal decomposition, after a chiral transition within an effective field theory approach to low-energy QCD. We study possible effects resulting from the finite size of the expanding system for both the initial and the late-stage growth of domains, as well as those effects due to inhomogeneities in the chiral field which act as a background for the fermionic motion.

To model the mechanism of chiral symmetry breaking present in QCD, and to study the dynamics of phase conversion after a temperature-driven chiral transition, one can resort to low-energy effective models. In particular, to study the mechanisms of bubble nucleation and spinodal decomposition in a hot expanding plasma, it is common to adopt the linear σ-model coupled to quarks, where the latter comprise the hydrodynamic degrees of freedom of the system[1,2,3,4,5,6]. The gas of quarks provides a thermal bath in which the long-wavelength modes of the chiral field evolve, and the latter plays the role of an order parameter in a Landau-Ginzburg approach to the description of the chiral phase transition. The gas of quarks and anti-quarks is usually treated as a heat bath for the chiral field, with temperature T. The standard procedure is then integrating over the fermionic degrees of freedom, using a classical approximation for the chiral field, to obtain a formal expression for the thermodynamic potential of an infinite system.

Let us consider a chiral field $\phi = (\sigma, \vec{\pi})$, where σ is a scalar field and π^i are pseudoscalar fields playing the role of the pions, coupled to two flavors of quarks according to the Lagrangian

$$\mathcal{L} = \overline{q}[i\gamma^\mu \partial_\mu + \mu_q \gamma^0 - M(\phi)]q + \frac{1}{2}\partial_\mu \phi \partial^\mu \phi - V(\phi) \; . \tag{1}$$

Here q is the constituent-quark field $q = (u, d)$ and $\mu_q = \mu/3$ is the quark chemical potential. The interaction between the quarks and the chiral field is given by $M(\phi) = g\left(\sigma + i\gamma_5\vec{\tau}\cdot\vec{\pi}\right)$, and $V(\phi) = \frac{\lambda^2}{4}\left(\sigma^2 + \vec{\pi}^2 - v^2\right)^2 - h_q\sigma$ is the self-interaction potential for ϕ.

The parameters above are chosen such that chiral $SU_L(2) \otimes SU_R(2)$ symmetry is spontaneously broken in the vacuum. The vacuum expectation values of the condensates are $\langle\sigma\rangle = f_\pi$ and $\langle\vec{\pi}\rangle = 0$, where $f_\pi = 93$ MeV is the pion decay constant. The explicit symmetry breaking term is due to the finite current-quark masses and is determined by the PCAC relation, giving $h_q = f_\pi m_\pi^2$, where $m_\pi = 138$ MeV is the pion mass. This yields $v^2 = f_\pi^2 - m_\pi^2/\lambda^2$. The value of $\lambda^2 = 20$ leads to a σ-mass, $m_\sigma^2 = 2\lambda^2 f_\pi^2 + m_\pi^2$, equal to 600 MeV. For $g > 0$, the finite-temperature one-loop effective potential also includes a contribution from the quark fermionic determinant. In what follows, we treat the gas of quarks as a heat bath for the chiral field, with temperature T and baryon-chemical potential μ. Then, one can integrate over the fermionic degrees of freedom, obtaining an effective theory for the chiral field ϕ. Using a classical approximation for the chiral field, one obtains the thermodynamic potential

$$\Omega(T, \mu, \phi) = V(\phi) - \frac{T}{\mathcal{V}} \ln\det\{[G_E^{-1} + W(\phi)]/T\} , \tag{2}$$

where G_E is the fermionic Euclidean propagator. ¿From the thermodynamic potential one can obtain all the thermodynamic quantities of interest.

To compute correlation functions and thermodynamic quantities, one has to evaluate the fermionic determinant within some approximation scheme. In $1D$ systems one can usually resort to exact analytical methods[7]. In practice, however, the determinant is usually calculated to one-loop order assuming a homogeneous and static background chiral field. Nevertheless, for a system that is in the process of phase conversion after a chiral transition, one expects inhomogeneities in the chiral field to play a role in driving the system to the true ground state.

We propose an approximation procedure to evaluate the finite-temperature fermionic determinant in the presence of a chiral background field, which systematically incorporates effects from inhomogeneities in the chiral field through a derivative expansion. The method is valid for the case in which the chiral field varies smoothly, and allows one to extract information from its long-wavelength behavior, incorporating corrections order by order in the derivatives of the field.

The Euler-Lagrange equation for static chiral field configurations contains a term which represents the fermionic density $\rho(\vec{x}_0) = (\nu_q/\mathcal{V})\left\langle \vec{x}_0 \left| (G_E^{-1} + M(\hat{x}))^{-1} \right| \vec{x}_0 \right\rangle$, where $|\vec{x}_0\rangle$ is a position eigenstate with eigenvalue $\vec{x}_0$, and $\nu_q = 12$ is the color-spin-isospin degeneracy factor. In momentum representation:

$$\rho(\vec{x}_0) = \nu_q T \sum_n \int \frac{d^3k}{(2\pi)^3} e^{-i\vec{k}\cdot\vec{x}} \frac{1}{\gamma^0(i\omega_n + \mu) - \vec{\gamma}\cdot\vec{k} + M(\hat{x})} e^{i\vec{k}\cdot\vec{x}} . \tag{3}$$

We can transfer the $\vec{x}_0$ dependence to $M(\hat{x})$ through a unitary transformation, expand $M(\hat{x}+\vec{x}_0)$ around $\vec{x}_0$, and use $\hat{x}^i = -i\nabla_{k_i}$ to write

$$\rho(\vec{x}_0) = \nu_q T \sum_n \int \frac{d^3k}{(2\pi)^3} \frac{1}{\gamma^0(i\omega_n + \mu) - \vec{\gamma}\cdot\vec{k} + M(\vec{x}_0)} \times$$
$$\left[1 + \Delta M(-i\nabla_{k_i}, \vec{x}_0) \frac{1}{\gamma^0(i\omega_n + \mu) - \vec{\gamma}\cdot\vec{k} + M(\vec{x}_0)} \right]^{-1} , \tag{4}$$

where $\vec{x}_0$ is a c-number, not an operator, and $\Delta M(\hat{x}, \vec{x}_0) = \nabla_i M(\vec{x}_0)\hat{x}^i + \frac{1}{2}\nabla_i\nabla_j M(\vec{x}_0)\hat{x}^i\hat{x}^j + \cdots$.

If we focus on the long-wavelength properties of the chiral field and assume that the static background, $M(\vec{x})$, varies smoothly and fermions transfer a small ammount of momentum to the chiral field, we can expand the expression above in a power series:

$$\rho(\vec{x}) = \nu_q T \sum_n \int \frac{d^3k}{(2\pi)^3} \frac{1}{\gamma^0(i\omega_n + \mu) - \vec{\gamma}\cdot\vec{k} + M(\vec{x})} \times$$
$$\sum_\ell (-1)^\ell \left[\Delta M(-i\nabla_{k_i}, \vec{x}) \frac{1}{\gamma^0(i\omega_n + \mu) - \vec{\gamma}\cdot\vec{k} + M(\vec{x})} \right]^\ell , \tag{5}$$

$$\Delta M(-i\nabla_{k_i}, \vec{x}) = \nabla_i M \left(\frac{1}{i}\right) \nabla_{k_i} + \frac{1}{2}\nabla_i\nabla_j M \left(\frac{1}{i}\right)^2 \nabla_{k_i}\nabla_{k_j} + \cdots , \tag{6}$$

which provides a systematic procedure to incorporate corrections brought about by inhomogeneities in the chiral field to the quark density, so that we can calculate $\rho(\vec{x}) = \rho_0(\vec{x}) + \rho_1(\vec{x}) + \rho_2(\vec{x}) + \cdots$ order by order in powers of the derivative of the background, $M(\vec{x})$. The leading-order term in this gradient expansion for $\rho(\vec{x})$ can be easily calculated and yields the well-known mean field result for ρ. The net effect of this leading term is to correct the potential for the chiral field to $V_{eff} = V(\phi) + V_q(\phi)$, where

$$V_q \equiv -\nu_q T \int \frac{d^3k}{(2\pi)^3} \ln\left(e^{[E_k(\phi) - \mu_q]/T} + 1 \right) + (\mu_q \to -\mu_q) , \tag{7}$$

where $E_k(\phi) = \sqrt{\vec{k}^2 + g^2\phi^2}$. This sort of effective potential is commonly used as the thermodynamic potential in a phenomenological description of the chiral transition for an expanding quark-gluon plasma created in a high-energy heavy-ion collision[2,3,4,5,6]. However, the presence of a non-trivial background field configuration, *e.g.* a bubble, can in principle dramatically modify the Dirac spectrum, hence the determinant[7,8]. Results for the correction of the Laplacian term will be presented elsewhere[9].

In the process of phase conversion through bubble nucleation in a QGP of finite size, the set of all supercritical bubbles integrated over time will eventually drive the entire system to its true vacuum. The scales that determine the importance of finite-size effects are the typical linear size of the system, the radius of the critical bubble and the correlation length. For definiteness, let us assume our system is described by a coarse-grained Landau-Ginzburg potential, $U(\phi, T)$, whose coefficients depend on the temperature. For the case to be considered, the scalar order parameter, ϕ, is *not* a conserved quantity, and its evolution is given by the time-dependent Landau-Ginzburg equation

$$\frac{\partial \phi}{\partial t} = \gamma \left[\nabla^2 \phi - U'(\phi, T) \right] \quad , \tag{8}$$

where γ is the response coefficient which defines a time scale for the system. The equation above is a standard reaction-diffusion equation, and describes the approach to equilibrium.

If $U(\phi, T)$ is such that it allows for the existence of bubble solutions (taken to be spherical for simplicity), then supercritical (subcritical) bubbles expand (shrink), in the thin-wall limit, with the following velocity:

$$\frac{dR}{dt} = \gamma(d-1) \left[\frac{1}{R_c} - \frac{1}{R(t)} \right] \quad , \tag{9}$$

where $R_c = (d-1)\sigma/\Delta F$ and ΔF is the difference in free energy between the two phases. The equation above relates the velocity of a domain wall to the local curvature. The response coefficient, γ, can be related to some characteristic collision time. One can measure the importance of finite-size effects for the case of heavy-ion collisions by comparing, for instance, the asymptotic growth velocity ($R >> R_c$) for nucleated hadronic bubbles to the expansion velocity of the plasma. In the Bjorken picture, the typical length scale of the expanding system is $L(T) \approx (v_z t_c)(T_c/T)^3 = L_0(T_c/T)^3$, where v_z is the collective fluid velocity and $L_0 \equiv L(T_c)$ is the initial linear scale of the system for the nucleation process which starts at $T \leq T_c$.

The relation between time and temperature provided by the cooling law that emerges from the Bjorken picture suggests the comparison between the following "velocities":

$$v_b \equiv \frac{dR}{dT} = -\left(\frac{3b\ell L_0}{2v_z\sigma T_c^2}\right)\left(\frac{T_c}{T}\right)^5\left(1-\frac{T}{T_c}\right) \quad , \tag{10}$$

the asymptotic bubble growth "velocity", and the plasma expansion "velocity" $v_L \equiv (dL/dT) = -(3L_0/T_c)(T_c/T)^4$. The quantity b is a number of order one to first approximation, and comes about in the estimate of the phenomenological response coefficient $\gamma(T) \approx b/2T$. Using the numerical values adopted previously and $\sigma/T_c^3 \sim 0.1$, we obtain[10]

$$\frac{v_b}{v_L} \approx \frac{20}{v_z}\left(\frac{T_c}{T}-1\right) \quad . \tag{11}$$

One thus observes that the bubble growth velocity becomes larger than the expansion velocity for a supercooling of order $\theta \approx v_z/20 \leq 5\%$. A simple estimate points to a critical radius larger than 1 fm at such values of supercooling[4]. Therefore, finite-size effects appear to be an important ingredient in the phase conversion process right from the start in the case of high-energy heavy-ion collisions[10].

Acknowledgments

Part of this paper is based on work done in collaboration with R. Venugopalan. E.S.F. is partially supported by CAPES, CNPq, FAPERJ and FUJB/UFRJ.

References

1. I. N. Mishustin and O. Scavenius, Phys. Rev. Lett. **83**, 3134 (1999).
2. O. Scavenius and A. Dumitru, Phys. Rev. Lett. **83**, 4697 (1999).
3. O. Scavenius, A. Mocsy, I. N. Mishustin and D. H. Rischke, Phys. Rev. C **64**, 045202 (2001).
4. O. Scavenius, A. Dumitru, E. S. Fraga, J. T. Lenaghan and A. D. Jackson, Phys. Rev. D **63**, 116003 (2001).
5. K. Paech, H. Stocker and A. Dumitru, Phys. Rev. C **68**, 044907 (2003).
6. C. E. Aguiar, E. S. Fraga and T. Kodama, nucl-th/0306041.
7. E. S. Fraga and C. A. A. de Carvalho, Phys. Rev. B **52**, 7448 (1995).
8. D. G. Barci, E. S. Fraga and C. A. A. de Carvalho, Phys. Rev. D **55**, 4947 (1997).
9. E. S. Fraga, in preparation.
10. E. S. Fraga and R. Venugopalan, Physica A (to appear), hep-ph/0304094.

COHERENT BARYOGENESIS AND NONTHERMAL LEPTOGENESIS: A COMPARISON

BJÖRN GARBRECHT, TOMISLAV PROKOPEC
AND MICHAEL G. SCHMIDT

Institut für Theoretische Physik, Heidelberg University, Philosophenweg 16, D-69120 Heidelberg, Germany
E-mail:B.Garbrecht-, T.Prokopec-, M.G.Schmidt@thphys.uni-heidelberg.de

We present a new mechanism for baryogenesis: at preheating after inflation fermions acquire a varying mass by their coupling to a time dependent field. Their CP-violating mass matrix can generate a charge asymmetry to be transformed into a lepton asymmetry through decay into standard model particles and heavy Majorana neutrinos. In a concrete model of hybrid inflation we compare "coherent baryogenesis" with nonthermal leptogenesis by perturbative decay of the inflation condensates.

1. Introduction

The baryon asymmetry of the universe $n_B/s \simeq 7 - 8 \times 10^{-11}$ (BAU), which is now deduced from early nucleosynthesis data and from WMAP data on microwave background fluctuations, is in surprisingly good agreement. Its theoretical explanation requires detailed knowledge on B, C/CP violation and nonequilibrium and this is an ideal testing ground for elementary particle physics models. Electroweak baryogenesis [1] in a strong first order phase transition requires a rather low Higgs mass. This and other ingredients can be tested in present and near future experiments - a great advantage. But this also led to the conclusion that the Standard Model (SM) is ruled out for such a mechanism and that the MSSM is close to this borderline. Out of equilibrium decay of heavy Majorana neutrinos - CP-violating at one loop level through a matrix of Yukawa couplings - has become very attractive again because such Majorana neutrinos also play an important role in explaining the newly found neutrino mass pattern. If one starts with thermalised Majorana neutrinos one can arrive at a simple and very restrictive picture [2] though in the SUSY case there are some problems with gravitino overproduction. We here present a new 'coherent baryogenesis' mechanism, which also contains lepton number violation by Majorana neutrinos and compare it with nonthermal leptogenesis by perturbative decay of the inflaton field and with nonthermal leptogenesis during preheating.

2. Coherent baryogenesis, formalism, and a concrete model

At the end of inflation, there is an oscillating condensate which couples to fermions and induces a varying mass matrix. This, in turn, yields fermion production, known as preheating. Moreover, the mass matrix induces CP-violating flavour oscillations, which is a tree level effect. Charge is produced and is frozen in when the scalar condensate settles to its minimum. Eventually, this charge gets converted to *(B-L)*. This mechanism we call "coherent baryogenesis" [3]. We use the following system of equations for charge and currents of a fermionic system derived in the Schwinger-Keldish formalism for two point functions after Wigner transform

$$\begin{aligned}
&\dot{f}_{0h} + i\,[M_H, f_{1h}] + i\,[M_A, f_{2h}] = 0\\
&\dot{f}_{1h} + 2h|\mathbf{k}|f_{2h} + i\,[M_H, f_{0h}] - \{M_A, f_{3h}\} = 0\\
&\dot{f}_{2h} - 2h|\mathbf{k}|f_{1h} + \{M_H, f_{3h}\} + i\,[M_A, f_{0h}] = 0\\
&\dot{f}_{3h} - \{M_H, f_{2h}\} + \{M_A, f_{1h}\} = 0
\end{aligned} \tag{1}$$

with f_{0h}: charge density, f_{3h}: axial charge density, f_{1h}: scalar density, f_{2h}: pseudoscalar density, h: helicity, k momentum, and hermitean and antihermitean part of the mass matrix, $M_H = \frac{1}{2}(M + M^+)$ and $M_A = \frac{1}{2i}(M - M^+)$. For pure quantum states this is equivalent to using the Dirac equation for mixing fermions, but it is also a generalization for mixed states.

In order to demonstrate our new proposal we implement it in a realistic model for hybrid inflation [4], the supersymmetric Pati-Salam model with gauge group $G_{PS} = SU(4)_c \times SU(2)_L \times SU(2)_R$ which after modifications does not suffer from the monopole problem as discussed in [5]. The relevant terms of the superpotential are

$$W = \kappa S(\bar{H}^c H^c - \mu^2) - \beta S\left(\frac{\bar{H}^c H^c}{M_s}\right)^2 + \zeta G H^c H^c + \xi G \bar{H}^c \bar{H}^c \tag{2}$$

with superfields $H^c = (\bar{4}, 1, 2)$, $\bar{H}^c = (4, 1, 2)$, $S = (1, 1, 1)$, $G = (6, 1, 1)$ and a notation $H^c = \begin{pmatrix} u^c_{H_1} & u^c_{H_2} & u^c_{H_3} & \nu^c_H \\ d^c_{H_1} & d^c_{H_2} & d^c_{H_3} & e^c_H \end{pmatrix}$. This leads to a scalar potential

$$V = 2\left|S\nu^{c*}_H\left(\kappa - 2\beta\frac{|\nu^c_H|^2}{M_S^2}\right)\right|^2 + \left|\kappa\left(|\nu^c_H|^2 - \mu^2\right) - \beta\frac{|\nu^c_H|^4}{M_S^2}\right|^2. \tag{3}$$

During inflation SUSY is broken, the sneutrino like scalar Higgses have an expectation value $\langle|\nu_H^c|\rangle = M_S\sqrt{\kappa/2\beta}$. This field and the inflaton field S begin to fall rapidly during the waterfall regime at the end of inflation (see figures below) before the SUSY vacuum is approached at $S = 0$.

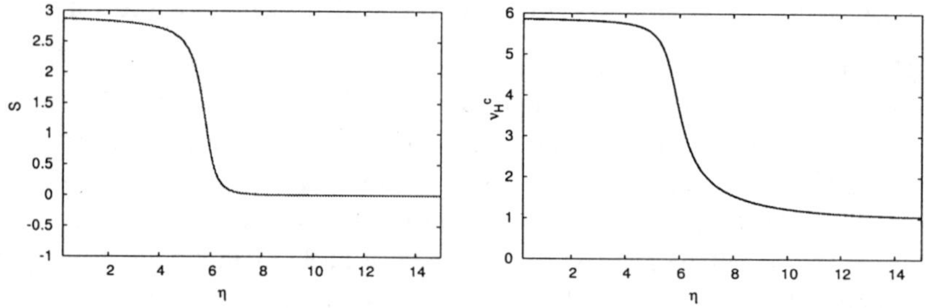

This induces a time dependent mass matrix for the Dirac fermions $\chi_{1j} = \left(-\Psi_{d^cH_j},\ \bar{\Psi}_{\bar{D}_j}\right)$, $\chi_{2_j} = \left(\Psi_{D_j},\ \bar{\Psi}_{d^cH_j}\right)$

$$(\bar{\chi}_{1j}, \bar{\chi}_{2j})\left[\begin{pmatrix} \Re\left[\langle\nu_H^c\rangle\xi\right] & \frac{1}{2}m_d \\ \frac{1}{2}m_d & \Re\left[\langle\nu_H^c\rangle\zeta\right] \end{pmatrix} + i\gamma_5\begin{pmatrix} -\Im\left[\langle\nu_H^c\rangle\xi\right] & -\frac{i}{2}m_d \\ \frac{i}{2}m_d & -\Im\left[\langle\nu_H^c\rangle\zeta\right] \end{pmatrix}\right]\begin{pmatrix} \chi_{1j} \\ \chi_{2j} \end{pmatrix},$$

where $m_d = \langle S\rangle\left(\kappa/2 - \beta\langle|\nu_H^c|^2\rangle/M_S^2\right)$ and we have chosen the parameters $\kappa = 0.007$, $\beta = 1$, $\xi = 0.12$, $\zeta = 0.12i$, $M_s = 100\mu$, $\mu = 3.9 \times 10^{16}$ GeV, $\gamma = \gamma_1 = 0.0001$, $\Gamma_S = \Gamma_\nu = 0.1\mu$ (a phenomenological damping term, which models tachyonic preheating). Note that the source of CP violation is here the phase between ξ and ζ, which enters at tree level.

3. Contribution from coherent baryogenesis

Resonant fermion production occurs whenever the fermion mode frequency changes nonadiabatically, which happens during the waterfall regime. The charges stored finally in the χ fermions are displayed in the figure below. Coupling to quarks and leptons is through the superpotential terms

$$\gamma F^c\bar{H}^cF^c\bar{H}^c/M_s \quad \text{and} \quad \delta F^cH^cF^cH^c/M_s \tag{4}$$

where $F^c = (\bar{4}, 1, 2)$. We then have the decay reactions

$$\chi_{1j} \to d^{c*} + \nu^{c*}, \qquad \chi_{2j} \to d^c + u^c \tag{5}$$

Because of the lepton number violating terms in (4) the charges hence get transformed to $q = \frac{1}{3}q_1 - \frac{2}{3}q_2$. This results in a baryon to entropy ratio

$$\frac{n_B}{s} = \frac{3}{4}\frac{n_B^{(0)}T_R}{V_0} \approx 1 \times 10^{-10}. \tag{6}$$

where $n_B^{(0)} \simeq 1.5 \times 10^{45}\,\mathrm{GeV}^3$ is the baryon density produced at preheating (the charge density q obtained from figure below multiplied by 3 colors and 1/3, the sphaleron conversion efficiency), $V_0 \simeq 3 \times 10^{64}\,\mathrm{GeV}^4$ is the energy density at the end of inflation, $T_R = [90/(\pi^2 g_*)]^{1/4}\sqrt{\Gamma M_P} \simeq 2.7 \times 10^9$ GeV is the reheat temperature, $g_* = 221.5$ is the number of relativistic degrees of freedom of the MSSM, $M_P \simeq 2.4 \times 10^{18}$ GeV the reduced Planck mass, $\Gamma \equiv H_R \simeq 15$ GeV is the perturbative inflaton decay rate, and $s = 2\pi^2 g^* T_R^3/45$ is the entropy density. This estimate of T_R is based on the assumption that tachyonic preheating thermalises the inflaton sector, and all other (light) species thermalise at $T \simeq T_R$. The decay rate of the inflaton Γ is dominated by the perturbative decay of the $\langle \nu_H^c \rangle$-condensate into the lightest Majorana neutrinos with mass $M_1 = \gamma_1 \langle \nu_H^c \rangle_0^2 / M_s \approx 4 \times 10^{10}$ GeV, such that $\Gamma \simeq (1/8\pi) m_{\nu_H^c} \left(\gamma_1 \langle \nu_H^c \rangle_0 / M_s\right)^2 \simeq 15$ GeV, where $\langle \nu_H^c \rangle_0 \simeq \mu$ and $m_{\nu_H^c} \simeq 4 \times 10^{14}$ GeV with our choice of parameters.

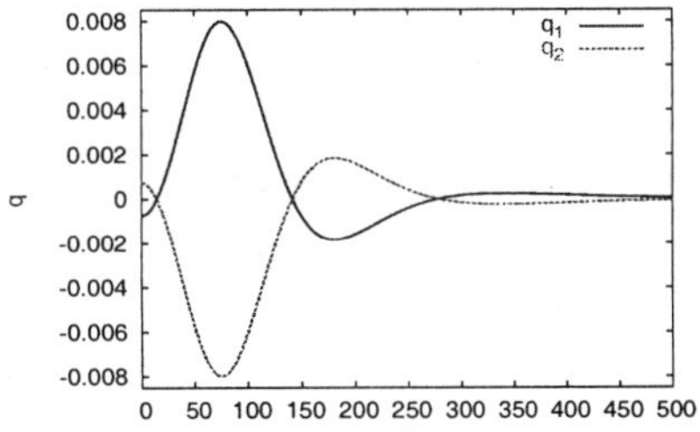

The baryon to entropy ratio (6) is somewhat larger than the observed value. It can be easily reduced by chosing a smaller CP violation (by a different choice of ξ and ζ), or by reducing the reheat temperature.

4. Reheating and nonthermal leptogenesis

In section 3 we have already mentioned the decay of the ν_H^c-condensate into the lightest Majorana neutrinos after preheating. The reheating temperature we obtained is $T_R \simeq 2.7 \times 10^9$ GeV, which meets the gravitino bound. It is much below the Majorana neutrino mass $M_1 \simeq 3.9 \times 10^{10}$ GeV. The two other masses $M_{2,3}$ we assume to be much heavier. Thus the produced Majorana neutrino is certainly nonthermal. For maximal mixing and CP violation via one-loop interference like in thermal leptogenesis, one obtains [6]

$$\frac{n_L}{s} \leq 3 \times 10^{-10} \frac{T_R}{m_{\nu_H^c}} \left(\frac{M_1}{10^6\mathrm{GeV}}\right)\left(\frac{m_{\nu^3}}{0.04\mathrm{eV}}\right) \simeq 8 \times 10^{-11}, \qquad (7)$$

where we assumed the last factor to be one, as suggested by atmospheric neutrino oscillations. This gives $n_B/s \leq 3 \times 10^{-11}$, which is significantly smaller than the coherent baryogenesis result (6). Increasing the cutoff scale M_s lowers the value (6), but reduces (7) at least equally fast.

Majorana neutrinos can also be produced nonperturbatively [7], just like the χ-particles in coherent baryogenesis. Nonperturbative production of the lightest Majorana neutrino N_1 is dominated by production of N_2 and N_3 with mass of order the inflaton mass, $M_{2,3} \simeq m_{\nu_H^c}$. Their decay asymmetry can be much larger than the asymmetry for N_1 [8]. We conclude that, like coherent baryogenesis, this production channel should also be regarded as a possible source for generating the BAU.

5. Conclusions

We considered two sources for baryogenesis during (p)reheating: (1) Direct charge production through coherent baryogenesis; (2) Perturbative decay of the inflaton into Majorana neutrinos.

For the given parameters, we found coherent baryogenesis to be the only viable baryogenesis mechanism. For this choice of parameters the gravitino bound is met. For a different choice of parameters (smaller M_s), coherent baryogenesis is further enhanced, making it thus robust baryogenesis mechanism. Since the CP-violating sources of coherent baryogenesis arise already at tree level and not at one-loop, as for leptogenesis, it is a competitive source of baryons over an ample section of phase space.

References

1. T. Prokopec, K. Kainulainen, M. G. Schmidt, S. Weinstock, SEWM 2002 Proceedings, hep-ph/0302192 and refs. therein;
   M. Carena, M. Quiros, C.E.M. Wagner Nucl. Phys. B **524** (1998) 3, hep-ph/9710401
2. W. Buchmüller, P. Di Bari, M. Plümacher, hep-ph/0401240 and refs. therein
3. B. Garbrecht, T. Prokopec, M. G. Schmidt, Phys. Rev. Let. **92**, 061302 (2004), hep-ph/0304088; hep-th/0211219, Europhys. J.
4. R. Jeannerot, S. Khalil, G. Lazarides, Q. Shafi, JHEP 0010, (2000) 012
5. V. N. Senoguz, Q. Shafi, Phys. Letters B**582**, (2004) 6, hep-ph/0309134
6. G. Lazarides, Q. Shafi, Phys. Lett **B258** (1991) 305
7. G. F. Guidice, M. Peloso, A. Riotto, I. Tkachev, JHEP 9908 (1999) 014, hep-ph/9905242
8. L. Covi, E. Roulet, F. Vissani, Phys. Lett. **B384** (1996) 169
   A. Pilaftsis, Nucl. Phys. **B504** (1997) 61

TWO ASPECTS OF COLOR SUPERCONDUCTIVITY: GAUGE INDEPENDENCE AND NEUTRALITY

A. GERHOLD*

Institut für Theoretische Physik, Technische Universität Wien,
Wiedner Hauptstr. 8-10, A-1040 Vienna, Austria
E-mail: gerhold@hep.itp.tuwien.ac.at

A formal proof is given that the fermionic quasiparticle dispersion laws in a color superconductor are gauge independent. It is shown that the gluon (photon) field acquires a non-vanishing expectation value in a color superconductor, which is related to color (electric) neutrality.

1. Introduction

It is well known that cold dense quark matter is a color superconductor.[1] Color superconducting phases are characterized by a non-vanishing gap function, which can be calculated self-consistently from the Schwinger-Dyson equation (gap equation). At asymptotic densities, calculations from first principles (QCD) should be possible. However, calculations of the gap function seem to give different results for covariant and Coulomb gauge.[2] This was the motivation to construct a formal proof of gauge independence, which is presented in Sec. 2. It turns out that this proof involves non-vanishing expectation values of the gauge fields. These expectation values act as effective chemical potentials for color and electric charge, which will be briefly discussed in Secs. 3-5.

*Work supported by the Austrian Science Foundation FWF, project no. 16387-N08.

2. Gauge independence of fermionic quasiparticle dispersion laws

The inverse quark propagator in the Nambu-Gor'kov basis is given by

$$\mathcal{S}^{-1} = \begin{pmatrix} \not{Q} + \mu\gamma_0 + \Sigma & \Phi^- \\ \Phi^+ & \not{Q} - \mu\gamma_0 + \bar{\Sigma} \end{pmatrix}$$
$$= \int d^4x \, e^{-iQ\cdot x} \frac{\delta^2\Gamma}{\delta\bar{\Psi}(x)\delta\Psi(0)}\Big|_{\psi=\bar{\psi}=A_i^a=0, A_0^a=\tilde{A}_0^a}, \tag{1}$$

where Φ^+ is the gap function, $\Phi^-(Q) = \gamma_0[\Phi^+(Q)]^\dagger\gamma_0$, Σ is the quark self energy, $\bar{\Sigma}(Q) = C[\Sigma(-Q)]^T C^{-1}$ with the charge conjugation matrix C, Γ is the effective action, and $\tilde{A}$ is the expectation value of the gluon field. The change in the effective action under the variation of some gauge parameter(s) is given by the following gauge dependence identity,[3,4]

$$\delta\Gamma = \Gamma_{,i}\delta X^i + \int dx \frac{\delta\Gamma}{\delta A^{a\mu}(x)} \delta X^{a\mu}_{(A)}(x). \tag{2}$$

We use the DeWitt notation for the fermions, $i = (\psi(x), \psi_c(x))^T$, $\bar{i} = (\bar{\psi}(x), \bar{\psi}_c(x))$. The δX's in Eq. (2) are quite complicated functionals, whose diagrammatic expansion is shown in Ref. 4. After taking the second derivative of Eq. (2), switching to momentum space,[a] and taking the determinant with respect to Nambu-Gor'kov, Dirac, color and flavor indices, we obtain[6]

$$\delta \det(\Gamma_{i\bar{j}}) + \delta\tilde{A}^{a0} \frac{\partial}{\partial\tilde{A}^{a0}} \det(\Gamma_{i\bar{j}}) = -\det(\Gamma_{i\bar{j}})[\delta X^k_{,k} + \delta X^{\bar{k}}_{,\bar{k}}]. \tag{3}$$

This identity means that the total variation of the determinant of the inverse propagator is proportional to the determinant of the inverse propagator itself. Therefore the fermionic quasiparticle dispersion laws are gauge independent, provided that the singularities of δX do not coincide with those of the quark propagator. In principle δX could develop mass shell singularities, as it is the case for the fermion damping rate in high temperature QCD.[7] Then a gauge dependent result could only be obtained by introducing an IR cutoff, to be lifted only at the very end of the calculation.[8] However, in the case of color superconductivity a simple IR cutoff turns out not to be sufficient to remove the gauge dependence of the gap in covariant gauges. Indeed, it has been argued that this gauge dependence will only disappear after the inclusion of vertex corrections in the gap equation.[9]

---

[a]Here we assume translation invariance, therefore our proof is not directly applicable for inhomogeneous color superconducting phases, which might appear at intermediate densities, see e.g. Ref. 5.

3. Gluon tadpole and color neutrality

In contrast to NJL models, where color neutrality has to be imposed as an external condition,[10] the color superconducting phases in full QCD are automatically color neutral. This can be demonstrated in the following way.[6,11] The fields A_0^a appear in the QCD action as Lagrange multipliers for the Gauss law constraint (at least in a gauge which does not involve A_0^a.) Therefore in the path integral the integration over the zero-momentum modes $A^a_{0,\vec{p}=0}$ produces delta functions, $\delta(N_a)$, where N_a are the color charges. This means that only color neutral field configurations contribute to the partition function.

In a color superconductor the gluon field acquires a non-vanishing expectation value.[6,12,13] The leading order contribution to this expectation value can be computed from the one-loop gluon tadpole diagram. For the 2SC phase we find[6]

$$\mathcal{T}^a \simeq -\delta^{a8} \frac{2g}{\sqrt{3}\pi^2}\, \mu\, \phi^2 \ln\left(\frac{\phi}{2\mu}\right). \tag{4}$$

The expectation value of A_0^a can be obtained by attaching the external gluon propagator, which gives[6,13]

$$\tilde{A}_0^8 \sim \phi^2/(g^2\mu). \tag{5}$$

This expectation value acts as an effective chemical potential for the color charge (via $\mu_8 = gA_0^8$), thus ensuring color neutrality.[13]

In Eq. (4) we have computed the quantity

$$\mathcal{T}^a \sim \left.\frac{\delta\Gamma}{\delta A_0^a}\right|_{A=0}. \tag{6}$$

We remark that at the order of our computation this calculation is completely equivalent to the self-consistent approach of Ref. 13, where $\tilde{A}$ has been computed from the Yang-Mills equation

$$\left.\frac{\delta\Gamma}{\delta A_0^a}\right|_{A=\tilde{A}} = 0. \tag{7}$$

The connection between the two approaches can be seen from the Taylor expansion

$$\left.\frac{\delta\Gamma}{\delta A_0^a}\right|_{A=0} = \left.\frac{\delta\Gamma}{\delta A_0^a}\right|_{A=\tilde{A}} - \left.\frac{\delta^2\Gamma}{\delta A_0^a \delta A_0^b}\right|_{A=\tilde{A}} \tilde{A}_0^b + \ldots \tag{8}$$

On the right hand side the first term vanishes, while the second term is essentially the inverse gluon propagator at zero momentum times the expectation value of A.

4. Gluon tadpole for CFL with $m_s \neq 0$

In the CFL phase one finds $\tilde{A}_0^a = 0$ at one-loop order for $m_s = 0$. For finite m_s the quark propagator in the Nambu-Gor'kov basis is given by

$$\mathcal{S} = \begin{pmatrix} [G_0^-]^{-1} & \Phi^- \\ \Phi^+ & [G_0^+]^{-1} \end{pmatrix}^{-1} \tag{9}$$

with

$$[(G_0^\pm)^{-1}]_{rs}^{ij} = [(\not{Q} \pm \mu\gamma_0)(\delta_{rs} - \delta_{r3}\delta_{s3}) + (\not{Q} \pm \mu\gamma_0 - m_s)\delta_{r3}\delta_{s3}]\delta^{ij}, \tag{10}$$

where the lower indices are flavor indices and the upper indices are color indices. At finite m_s the Dirac structure of the gap is more complicated than in the massless case,[14] but for $m_s \ll \mu$ we may assume that the additional terms are suppressed at least with m_s/μ,[15]

$$[\Phi^+]_{rs}^{ij} \simeq \phi_+ \Lambda_{\mathbf{q}}^+ \gamma_5 (\delta_r^i \delta_s^j - \delta_r^j \delta_s^i) + \mathcal{O}\left(\frac{m_s}{\mu}\right). \tag{11}$$

Assuming $m_s \gg \phi_+$ we find for the tadpole diagram at leading order in m_s

$$\begin{aligned} \mathcal{T}^a &\simeq \delta^{a8} \frac{g m_s^2}{2\sqrt{3}\pi^3} \int_{-\infty}^{\infty} dq_4 \int_0^\infty dq\, q\, \phi_+^2 \\ &\quad \times \frac{q_4^2 - 3(q-\mu)^2 - 8\phi_+^2}{\left(q_4^2 + (q-\mu)^2 + \phi_+^2\right)^2 \left(q_4^2 + (q-\mu)^2 + 4\phi_+^2\right)} \\ &\simeq \delta^{a8} \frac{g\mu m_s^2}{18\sqrt{3}\pi^2}(-21 + 8\ln 2), \end{aligned} \tag{12}$$

where we have neglected the momentum dependence of the gap function. Attaching the gluon propagator with the Debye mass[16] proportional to $g\mu$ we find as in Ref. 12

$$\tilde{A}_0^a \sim \delta^{a8} \frac{m_s^2}{g\mu}. \tag{13}$$

Therefore the effective chemical potential is $\mu_8 \sim m_s^2/\mu$, which is consistent with results from NJL model calculations.[10]

5. Photon tadpole and electric neutrality

If one takes into account also electrons and photons, electric neutrality will we achieved by a mechanism which is completely analogous to the one described in Sec. 3. For the expectation value of the photon field one finds in the 2SC phase (for $m_s = 0$)

$$A^0 \sim \frac{\phi^2}{e\mu} \ln\left(\frac{\phi}{2\mu}\right). \tag{14}$$

Acknowledgements

I would like to thank C. Manuel and A. Rebhan for very helpful discussions.

References

1. D. Bailin and A. Love, *Phys. Rept.* **107**, 325 (1984); M. G. Alford, K. Rajagopal and F. Wilczek, *Phys. Lett.* **B422**, 247 (1998); R. Rapp, T. Schäfer, E. V. Shuryak and M. Velkovsky, *Phys. Rev. Lett.* **81**, 53 (1998); K. Rajagopal and F. Wilczek, The condensed matter physics of QCD, hep-ph/0011333; M. G. Alford, *Ann. Rev. Nucl. Part. Sci.* **51**, 131 (2001).
2. D. T. Son, *Phys. Rev.* **D59**, 094019 (1999); R. D. Pisarski and D. H. Rischke, *Phys. Rev.* **D61**, 074017 (2000); T. Schäfer and F. Wilczek, *Phys. Rev.* **D60**, 114033 (1999); D. K. Hong, V. A. Miransky, I. A. Shovkovy and L. C. R. Wijewardhana, *Phys. Rev.* **D61**, 056001 (2000) [Erratum: *ibid.* **D62**, 059903 (2000)]; S. D. H. Hsu and M. Schwetz, *Nucl. Phys.* **B572**, 211 (2000); W. E. Brown, J. T. Liu and H.-c. Ren, *Phys. Rev.* **D61**, 114012 (2000); K. Rajagopal and E. Shuster, *Phys. Rev.* **D62**, 085007 (2000); R. D. Pisarski and D. H. Rischke, *Nucl. Phys.* **A702**, 177 (2002); D. K. Hong, T. Lee, D. P. Min, D. Seo and C. Song, *Phys. Lett.* **B565**, 153 (2003).
3. B. W. Lee and J. Zinn-Justin, *Phys. Rev.* **D7**, 1049 (1973); R. Fukuda and T. Kugo, *Phys. Rev.* **D13**, 3469 (1976). N. K. Nielsen, *Nucl. Phys.* **B101**, 173 (1975); I. J. R. Aitchison and C. M. Fraser, *Ann. Phys.* **156**, 1 (1984).
4. R. Kobes, G. Kunstatter and A. Rebhan, *Phys. Rev. Lett.* **64**, 2992 (1990); R. Kobes, G. Kunstatter and A. Rebhan, *Nucl. Phys.* **B355**, 1 (1991); A. Rebhan, *Lect. Notes Phys.* **583**, 161 (2002).
5. M. G. Alford, J. A. Bowers and K. Rajagopal, *Phys. Rev.* **D63**, 074016 (2001); R. Casalbuoni and G. Nardulli, *Rev. Mod. Phys.* **76**, 263 (2004).
6. A. Gerhold and A. Rebhan, *Phys. Rev.* **D68**, 011502 (2003).
7. R. Baier, G. Kunstatter and D. Schiff, *Phys. Rev.* **D45**, 4381 (1992).
8. A. Rebhan, *Phys. Rev.* **D46**, 4779 (1992).
9. D. f. Hou, Q. Wang and D. H. Rischke, *Phys. Rev.* **D69**, 071501 (2004).
10. P. Amore, M. C. Birse, J. A. McGovern and N. R. Walet, *Phys. Rev.* **D65**, 074005 (2002); M. Alford and K. Rajagopal, JHEP **0206**, 031 (2002); A. W. Steiner, S. Reddy and M. Prakash, *Phys. Rev.* **D66**, 094007 (2002); I. Shovkovy and M. Huang, *Phys. Lett.* **B564**, 205 (2003); M. Huang and I. Shovkovy, Neutral dense quark matter, hep-ph/0311155.
11. S. Y. Khlebnikov and M. E. Shaposhnikov, *Phys. Lett.* **B387**, 817 (1996).
12. A. Kryjevski, *Phys. Rev.* **D68**, 074008 (2003).
13. D. D. Dietrich and D. H. Rischke, *Prog. Part. Nucl. Phys.* **53**, 305 (2004).
14. R. D. Pisarski and D. H. Rischke, *Phys. Rev.* **D60**, 094013 (1999).
15. T. D. Fugleberg, *Phys. Rev.* **D67**, 034013 (2003).
16. D. H. Rischke, *Phys. Rev.* **D62**, 054017 (2000).

QCD PHASE DIAGRAM
IN NONLOCAL CHIRAL QUARK MODELS

D. GÓMEZ DUMM

IFLP, CONICET – Depto. de Física, Univ. Nac. de La Plata,
C.C. 67, 1900 La Plata, Argentina
E-mail: dumm@fisica.unlp.edu.ar

The QCD phase diagram is analyzed within chiral quark models with nonlocal separable interactions. For the case of two light flavors, we describe in detail the characteristics of the chiral phase transition in the $T - \mu$ plane, and analyze the color superconducting phase transition taking place at relatively large chemical potentials. We also consider the three-flavor version of the models, analyzing the predictions for masses and decay constants of light pseudoscalar mesons.

1. Introduction

The behavior of strongly interacting matter at finite temperature and/or density is a subject of fundamental interest, with applications in cosmology, astrophysics, and in the physics of relativistic heavy ion collisions. In view of the difficulty found when dealing with strong interactions in the nonperturbative regime, it has not been possible yet to obtain detailed information about the full $T - \mu$ phase diagram directly from QCD, and most of the theoretical work on the subject relies in low energy effective models. One of the most popular ones is that proposed long ago by Nambu and Jona-Lasinio (NJL)[1], in which quarks interact through local four fermion couplings. In the last few years, covariant nonlocal extensions of this model have been proposed[2], mainly motivated by some approaches to low energy quark dynamics (instanton liquid model, Schwinger-Dyson resummation techniques) in which nonlocality arises in a natural way. It has also been shown that these nonlocal theories present several advantages in comparison with the standard NJL scheme. Here we analyze, in this context, the structure of the QCD phase diagram. First, we consider the case of two-flavor models, studying the features of the chiral phase transition for finite T and μ and the presence of a possible superconducting phase in the region of large μ. Then we also address the three-flavor case, studying the predicted pattern of pseudoscalar meson masses and decay constants. The analysis summarized here is developed in detail in Refs.[3,4,5,6].

2. Formulation

We start by quoting the Euclidean quark effective action

$$S_E = \int d^4x \left\{ \bar{\psi}(x) \left(-i\partial\!\!\!/ + m_c\right) \psi(x) - \frac{G}{2} [j^S(x) j^S(x) + j_k^P(x) j_k^P(x)] \right\}, \quad (1)$$

where $\psi \equiv (u\ d)^T$, while the fermion currents $j^S(x)$ and $j_k^P(x)$ are given by

$$\left\{ \begin{array}{c} j^S(x) \\ j_k^P(x) \end{array} \right\} = \int d^4y\ d^4z\ r(y-x)\ r(x-z)\ \bar{\psi}(y) \left\{ \begin{array}{c} \mathbf{1} \\ i\gamma_5\tau_k \end{array} \right\} \psi(z) . \quad (2)$$

Here $r(x-y)$ is a soft covariant nonlocal regulator, which in the case of the instanton liquid model has a definite form in terms of modified Bessel functions. In the following we will consider the relatively simple case of a Gaussian regulator $r(p^2) = \exp(-p^2/2\Lambda^2)$, where Λ is a free parameter of the model, playing the rôle of an ultraviolet cutoff scale. In any case, our results are found to be qualitatively similar for other regulator shapes. Notice that the constituent quark mass m_c (we assume isospin symmetry) and the coupling constant G are also free parameters of the model.

We perform now a bosonization of the partition function of the theory, $Z = \int D\bar{\psi} D\psi \exp(-S_E)$. To do this we introduce scalar and pseudoscalar meson fields $\sigma(x)$, $\pi_k(x)$, integrating out the quark degrees of freedom. Then we carry out the mean field approximation (MFA), in which the meson fields are expanded around their translational invariant vacuum expectation values $\bar{\sigma}$ and $\bar{\pi}_k$ (the latter vanish due to charge and isospin symmetries). The extension to finite T and μ is obtained by performing the replacement

$$\int \frac{d^4p}{(2\pi)^4}\ F(p_4, \vec{p})\ \rightarrow \sum\!\!\!\!\!\!\int_p F(p_4, \vec{p}) \equiv T \sum_{n=-\infty}^{\infty} \int \frac{d^3p}{(2\pi)^3}\ F(\omega_n - i\mu, \vec{p}) \quad (3)$$

in the four-momentum integrals arising from the fermion determinant (here ω_n are Matsubara frequencies corresponding to fermion modes). The grand canonical thermodinamical potential per unit volume is then given by

$$\omega_{\rm MFA}(T, \mu, m_c) = -\frac{T}{V} \log Z_{\rm MFA} = \frac{\bar{\sigma}^2}{2G} - 4N_c \sum\!\!\!\!\!\!\int_p \log\left[p^2 + \Sigma^2(p^2)\right], \quad (4)$$

where $\Sigma(p^2) = m_c + \bar{\sigma}\ r^2(p^2)$ is the constituent (momentum-dependent) quark mass. Now $\bar{\sigma}(T, \mu, m_c)$ can be obtained from $\partial\omega_{\rm MFA}/\partial\bar{\sigma} = 0$ (gap equation), and one can easily derive the expressions for other relevant physical quantities, such as the quark condensate $\langle \bar{q}q \rangle$, quark density ρ, chiral susceptibility $\chi_{V,T,\mu} = -(\partial\langle \bar{q}q \rangle / \partial m_c)_{T,\mu}$, pressure, energy density, etc.

3. Chiral phase transition: analytical and numerical analysis

In order to study the physical quantities, the Matsubara sums can be worked out using complex plane analysis. While in the standard (local) NJL model the Euclidean quark propagator has just two simple poles at $p_4 = \pm i\sqrt{\vec{p}^2 + M^2}$, being M the quark constituent mass, in the case of nonlocal models the regulator leads in general to a complicated structure of poles and cuts in the complex plane. Depending on the parameters of the model and the shape of the regulator, there may be some poles on the real axis or not; it has been suggested that this last situation is a sort of realization of confinement, in the sense that the effective quark propagator has no poles at real energies[7], and one can talk about "confining" and "nonconfining" parameter sets. From our analysis it is seen[3,4] that using adequate integration paths, and introducing some auxiliary complex occupation number functions, the Matsubara sums can be converted into integrals. Moreover, we have found that in all considered cases the behavior of the physical quantities is strongly dominated by the contribution of the first pole of the quark propagator.

We have carried out numerical calculations to evaluate the dependence of the mentioned physical quantities with T and μ. To proceed, we have adjusted the free parameters Λ, m_c and G so as to reproduce the empirical values of the pion mass and decay constant, whereas as the third input we have chosen some fixed values of $\Sigma(0)$. Both confining and nonconfining parameter sets have been considered, studying in each case the features of the chiral phase transition. In general, the results have been found to be qualitatively similar (a detailed comparison can be found in Refs.[3,4]); for definiteness we present here the curves corresponding to one of these parameter sets, namely $\Lambda = 760$ MeV, $m_c = 7.7$ MeV and $G = 30$ GeV^{-2}.

Our main results are shown in Fig. 1, where we quote the values of the quark condensate, the density (relative to nuclear matter density ρ_0) and the chiral susceptibility as functions of T and μ. By looking at the curves for the quark condensate, it is seen that for $T = 0$ there is a first order phase transition at a given critical chemical potential μ_c. The latter goes down as T is increased, until one reaches an end point where the first order phase transition turns into a smooth crossover. For this parameter set the

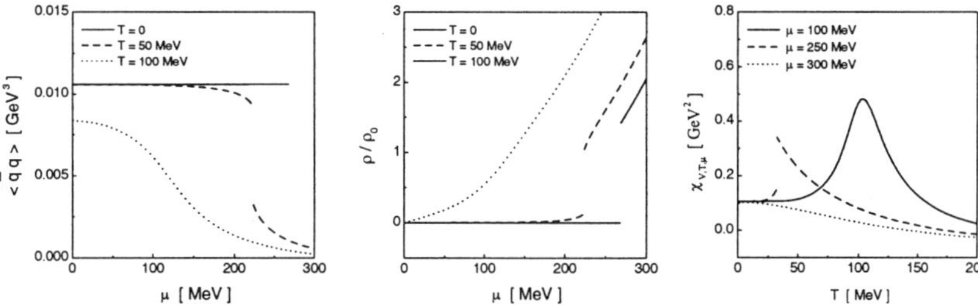

Figure 1. Behavior of the chiral condensate (left), the density (center) and the chiral susceptibility (right) with the temperature and the chemical potential.

end point is located at $(T_E, \mu_E) = (55 \text{ MeV}, 210 \text{ MeV})$. In the crossover region, the transition is also well defined by the presence of marked peaks in the chiral susceptibility, the sharpness of these peaks giving a measure of the crossover steepness. Thus we arrive at the phase diagrams shown in Fig. 2, in which we plot the phase transition temperature as a function of the chemical potential (left) and the density (right). The latter includes a "mixed phase" zone, which can be interpreted as a region where droplets containing light quarks of mass m_c coexist with a gas of constituent, massive quarks. It should be noticed that the model predicts a critical temperature at $\mu = 0$ of about 100 MeV, somewhat below the results obtained from lattice calculations which suggest this value to be $\sim 140 - 190$ MeV.

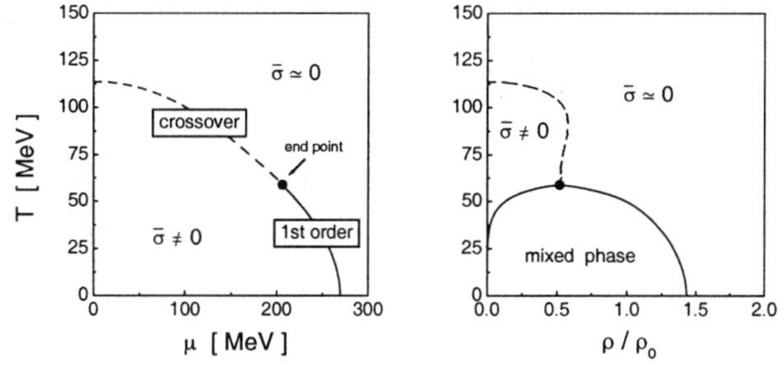

Figure 2. Phase transition temperature as a function of the chemical potential (left) and the density (right). Solid (dashed) lines stand for first order transition (crossover).

4. Extensions to diquark couplings and three flavors

The ability of this kind of models to allow for a two-flavor superconducting (2SC) phase at large chemical potentials is also under study[5]. This analysis considers an effective quark-quark interaction in the Euclidean action (1), carrying a new coupling constant h. Preliminary results show that for low T a first order phase transition into a 2SC phase is expected at some critical value of μ. A similar behavior is found when T is increased, until one reaches a "triple point" in which all three phases (hadronic, chiral and 2SC) coexist. The phase diagram is then completed with the inclusion of a second order phase transition line between the chiral and the 2SC phases. Unfortunately, the precise features of the phase diagram depend on the value of the diquark coupling h, which cannot be fixed from low energy phenomenology. A detailed discussion in this sense is given in Ref.[5].

Finally, we have considered[6] the inclusion of strange degrees of freedom, extending the flavor symmetry of the model to $SU(3)$ and incorporating the the $U(1)_A$ symmetry breaking through a nonlocal 't Hooft-like coupling. As a first step we have concentrated on the evaluation of meson masses and decay constants, finding our results in quite good agreement with empirical values, in particular in the case of the ratio f_K/f_π and the anomalous decay $\pi^0 \to \gamma\gamma$ (which are problematic in the standard NJL model). We have also obtained a reasonable description of the $\eta - \eta'$ phenomenology. Remarkably, this requires the presence of two significantly different state mixing angles.

Clearly, the two-flavor diquark channel analysis cannot be trusted beyond the strange quark mass threshold, and the $SU(3)$ version of the model should be addressed, leading presumably to the presence of a color flavor locking phase. We expect to report on these issues in forthcoming works.

Acknowledgments

I am indebted to N.N. Scoccola and A. Grunfeld for useful discussions, and to the organizers of SEWM04 for financial aid. This work has been partially supported by ANPCyT (Argentina) under grant PICT02-03-10718.

References

1. Y. Nambu and G. Jona-Lasinio, *Phys. Rev.* **122**, 345 (1961); *Phys. Rev.* **124**, 246 (1961). See also S. Klevansky, *Rev. Mod. Phys.* **64**, 649 (1992).
2. G. Ripka, *Quarks bound by chiral fields* (Oxford Univ. Press, Oxford, 1997).
3. I. General, D. Gómez Dumm and N.N. Scoccola, *Phys. Lett.* **B506**, 267 (2001).
4. D. Gómez Dumm and N.N. Scoccola, *Phys. Rev.* **D65**, 074021 (2002).
5. R. Duhau, A.G. Grunfeld and N.N. Scoccola, *work in progress.*
6. A. Scarpettini, D. Gómez Dumm and N.N. Scoccola, *Phys. Rev.* **D69**, 114018 (2004).
7. R.D. Bowler and M.C. Birse, Nucl. Phys. **A582**, 655 (1995).

QCD EQUATION OF STATE AND DARK MATTER

MARK HINDMARSH*AND OWE PHILIPSEN †
Dept. of Physics & Astronomy, University of Sussex, Brighton BN1 9QH, UK

The relic density of weakly interacting massive particles (WIMPs) depends on the entropy and the expansion rate of the Universe when they freeze out, at a temperature in the range 100 MeV – 10GeV. We use recent calculations of the QCD pressure to produce a new equation of state suitable for use in relic density calculations. We show that relic densities calculated by the dark matter package DarkSUSY receive corrections of several %, within the observational accuracy of current (WMAP) and future (Planck) CMB missions.

1. Introduction

WIMPS of mass m generically freeze out at a temperature of around $T_f \simeq m/25$ GeV, and so for masses in the range 10 – 100 GeV, T_f is not far from the QCD phase transition. The equation governing the relic density depends on both energy and entropy densities, and so the WIMP relic density is sensitive to the equation of state of the Universe in the range of a few hundred MeV to a few GeV.

We re-examine the equation of state for the Universe in this critical regime using recent work in both the deconfined[1] and confined [2] phases , and investigate the consequences for the WIMP relic density in one of the dark matter packages, DarkSUSY[3]. Replacing the standard-issue DarkSUSY equation of state, which is based on Refs. [4,5,6], with ours gives corrections to the relic densities calculated in the supplied test code of several percent. WMAP has already determined the matter density to better than 10% in single-field inflation models [7] and Planck promises to do much better, with one estimate at less than 1%[8], although other authors find that Sloan Digital Sky Survey (SDSS) data is required for this level of accuracy[9].

*email: m.b.hindmarsh@sussex.ac.uk
†email: o.philipsen@sussex.ac.uk, PPARC Advanced Fellow

2. Equation of state of QCD

Work on the pressure of non-Abelian gauge theories has a long history[10,11,12,13,14]. Perturbative expansions in the coupling constant g of quantities such as the pressure p converge badly, and particularly for a strongly-coupled theory like QCD. Strictly perturbative expansions, even when expanded to $O(g^6 \ln g)$[1], seem to work only at very high temperatures, above about $10^5 T_c$ (where T_c is the critical temperature).

However, by constructing a sequence of effective theories for the scales $2\pi T$, gT and $g^2 T$, one can match the calculated pressure reasonably well to pure-glue lattice data near the critical temperature by fitting for a non-perturbative and as yet unknown $O(g^6)$ term[1]. Below the phase transition the hadronic resonance gas model, which models the plasma as an ideal gas of mesons, baryons and their excited states, matches reasonably well to lattice data[2].

Near the transition, lattice calculations exist for the pressure and energy density for $N_f = 0$, 2, 2+1, and 3, although only the pure glue case has been extrapolated to the continuum. The critical temperatures (defined as the peak of the susceptibility) are currently given as $T_c(0) = 271 \pm 2$ MeV, $T_c(2) = 173 \pm 8$ MeV and $T_c(3) = 154 \pm 8$ MeV[15].

3. Relic density calculations

Consider a particle of mass m and number density n, undergoing annihilations $XX \to \ldots$ with total cross-section σ at temperature T.

It is convenient to convert the time variable to $x = m/T$, and to measure the relic abundance in terms of $Y = n/s$, where s is the entropy density. If the total entropy $S = sa^3$ is conserved, then we can write[16]

$$\frac{dY}{dx} = \frac{1}{3H}\frac{ds}{dx} \langle \sigma v_{\mathrm{M\o l}} \rangle \left(Y^2 - Y_{\mathrm{eq}}^2\right), \tag{3.1}$$

where Y_{eq} is the equilibrium value of Y, $H = \dot{a}/a$ is the Hubble parameter, a is the scalar factor of the expanding background, and $v_{\mathrm{M\o l}}$ is the Møller velocity which is a relativistic generalisation of the relative velocity of the annihilating particles[16]. The adiabaticity assumption is violated if the QCD transition is first order, but it is most likely to be a cross-over transition at the low chemical potentials which are relevant for the early Universe[12,13].

Using the Friedmann equation $H^2 = 8\pi G\rho/3$, and defining effective numbers of degrees of freedom for the energy and entropy densities through

$$\rho = \frac{\pi^2}{30} g_{\mathrm{eff}}(T) T^4, \quad s = \frac{2\pi^2}{45} h_{\mathrm{eff}}(T) T^3, \tag{3.2}$$

one finds that the relic density depends on the parameter

$$g_*^{1/2}(T) = \frac{h_{\rm eff}}{g_{\rm eff}^{1/2}}\left(1 + \frac{T}{3}\frac{d\ln h_{\rm eff}}{dT}\right). \tag{3.3}$$

4. Results: degrees of freedom and effect on relic densities

If the Standard Model at high temperature behaved as an ideal gas, the computation of $g_{\rm eff}$ and $h_{\rm eff}$ proceeds straightforwardly. The contributions from a particle of mass m_i to the energy density ρ and the entropy density $s = (\rho + p)/T$ are given by

$$g_{i,\rm eff} \equiv \frac{\rho_i}{\rho_0} = \frac{15}{\pi^4}\int_{x_i}^{\infty} \frac{(u^2 - x_i^2)^{\frac{1}{2}}}{e^u \pm 1} u^2 du, \tag{4.4}$$

$$h_{i,\rm eff} \equiv \frac{s_i}{s_0} = \frac{45}{12\pi^4}\int_{x_i}^{\infty} \frac{(u^2 - x_i^2)^{\frac{1}{2}}}{e^u \pm 1}(4u^2 - x_i^2) du, \tag{4.5}$$

where ρ_0 and s_0 are the ideal energy and entropy densities for a free boson. These contributions are summed. In practice there would still be some uncertainty, as the masses of the quarks are not accurately known (with the exception of the top quark).

Interactions in non-Abelian plasmas correct the ideal, with the dominant corrections come from the coloured degrees of freedom; electroweak corrections are small by comparison and will be ignored.

Near the transition, it seems that the correction to the ideal gas pressure p_0 for a coloured degree of freedom has a universal form[17]

$$p/p_0 = f_{\rm QCD}(T/T_c(N_f)). \tag{4.6}$$

It would be best to use lattice calculations here, but they as yet have been extrapolated to the continuum limit only in the $N_f = 0$ case. In using lattice calculations for $N_f = 3$, we also have the problem that the $O(g^6)$ fitting parameter for the analytic result has been calculated only for pure glue QCD. Hence we will take our correction factor from the resummed perturbative formula of Kajantie et al. [1] and scale the temperature dependence by $T_c(N_f = 3)/T_c(N_f = 0)$.

For our hadronic resonance model in the confining phase, we include all resonances listed in the Particle Data Group's table `mass_width_02.mc`, a total of 822 (counting spin states and antiparticles separately). We make a sharp switch between the high and low temperature calculations at the lattice value of the critical temperature of 154 MeV.

In Figs. 1 we plot our calculated effective numbers of degrees of freedom $h_{\rm eff}(T)$ and $g_*^{1/2}(T)$ defined in Eqs. 3.2 and 3.3, compared with those used in DarkSUSY[3], one of the leading packages for calculating the properties of supersymmetric dark matter. We see that there are differences, particularly near the QCD transition, which is perhaps not surprising as the QCD equation of state used in DarkSUSY[6] is quite old [4,5]. The spike in $g_*^{1/2}(T)$ is due to the sharp transition between the high and low temperature equations of state and is probably exaggerated.

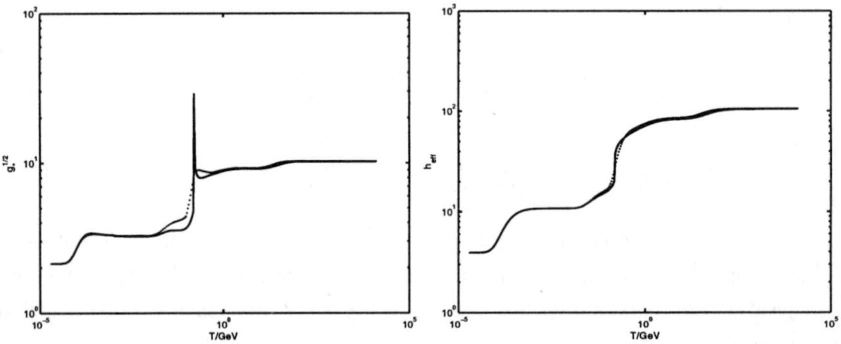

Figure 4.1. Degrees of freedom factors $g_*(T)$ (or rather its square root) and $h_{\rm eff}(T)$, defined in Eqs. 3.2 and 3.3, used in the current DarkSUSY (dotted) and modified by us (solid).

In Table 4.1 we exhibit the effect of the new equation of state on the density of relic neutralinos χ, for the models used to test DarkSUSY in the standard distribution. We see changes of about 3%.

5. Conclusions

By updating the equation of state of the Standard Model in the light of recent developments in high temperature QCD, we have found differences in WIMP relic density calculations of a few per cent for selected models in the DarkSUSY. Although the effect is small, it is not insignificant. In single field inflation models, the 68% confidence limit on the matter density is already $\Omega h^2 = 0.127 \pm 0.017$ from the WMAP First Year data[7], while Planck has been estimated to be able to reach a level of $\Delta\Omega h^2 = 0.0011$ in the most optimistic of circumstances[8]. Hence, in combination with measurements of the neutralino mass and annihilation cross-section at LHC, the

Table 4.1. The effect of a change in the equation of state on the dark matter density in benchmark supersymmetric models supplied with the DarkSUSY (DS) package.

Model	m_χ/GeV	$\Omega_m h^2$ (here)	$\Omega_m h^2$ (DS)
A′	242.83	0.0953	0.0929
B′	94.88	0.1247	0.1213
C′	158.09	0.1179	0.1149
D′	212.42	0.0886	0.0864
G′	147.98	0.1329	0.1294
H′	388.38	0.1673	0.1629
I′	138.08	0.1357	0.1319
J′	309.17	0.0990	0.0966
K′	554.19	0.0881	0.0863
L′	180.99	0.1005	0.0988

QCD equation of state may become experimentally accessible, motivating attempts to improve further on this work.

Acknowledgments

We are grateful to Mikko Laine and Paolo Gondolo for helpful discussions.

References

1. K. Kajantie, M. Laine, K. Rummukainen, and Y. Schroder, Phys. Rev. **D67**, 105008 (2003).
2. F. Karsch, K. Redlich, and A. Tawfik, Eur. Phys. J. **C29**, 549 (2003).
3. P. Gondolo *et al.*, (2004).
4. K. A. Olive, Nucl. Phys. **B190**, 483 (1981).
5. K. A. Olive, D. N. Schramm, and G. Steigman, Nucl. Phys. **B180**, 497 (1981).
6. M. Srednicki, R. Watkins, and K. A. Olive, Nucl. Phys. **B310**, 693 (1988).
7. H. V. Peiris *et al.*, Astrophys. J. Suppl. **148**, 213 (2003).
8. A. Balbi *et al.*, Astrophys. J. **588**, L5 (2003).
9. J. Lesgourgues, S. Pastor, and L. Perotto, (2004).
10. J.-P. Blaizot, E. Iancu, and A. Rebhan, (2003).
11. U. Kraemmer and A. Rebhan, Rept. Prog. Phys. **67**, 351 (2004).
12. F. Karsch, Lect. Notes Phys. **583**, 209 (2002).
13. E. Laermann and O. Philipsen, (2003).
14. J. O. Andersen and M. Strickland, (2004).
15. F. Karsch, E. Laermann, and A. Peikert, Nucl. Phys. **B605**, 579 (2001).
16. P. Gondolo and G. Gelmini, Nucl. Phys. **B360**, 145 (1991).
17. F. Karsch, E. Laermann, and A. Peikert, Phys. Lett. **B478**, 447 (2000).

ANALYTICAL APPROACH TO SU(2) YANG-MILLS THERMODYNAMICS

R. HOFMANN

Institut für Theoretische Physik, Universität Heidelberg, Philosophenweg 16, 69120 Heidelberg, Germany, E-mail: r.hofmann@thphys.uni-heidelberg.de

We propose an analytical approach to SU(2) Yang-Mills thermodynamics. The existence of a macroscopic and rigid adjoint Higgs field, generated by dilute trivial-holonomy calorons at large temperature T (electric phase), implies a twofold degeneracy of the ground state which signals a broken electric Z_2 symmetry. A finite energy density $\propto T$ of the ground state arises due to caloron interaction. An evolution equation for the effective gauge coupling, derived from thermodynamical self-consistency, predicts a second-order like transition (seen in lattice simulations) at T_c to a phase where monopoles are condensed and off-Cartan excitations decoupled. In this magnetic phase the ground state is unique and dominates the pressure (negative total pressure). While the magnetic phase has a massive, propagating 'photon' it confines fundamental matter (pre-confinement). The temperature dependence of the magnetic gauge coupling predicts the transition to the confining phase at $T_C \sim \frac{T_c}{1.9}$ where center-vortex loops condense and the 'photon' decouples. We believe that this transition is 'swallowed' by finite-size artefacts in lattice simulations. No thermodynamical connection exists between the confining and the magnetic phase.

The objective is to propose a macroscopic, effective theory for SU(2) Yang-Mills thermodynamics which can be generalized to SU(N) [1]. The approach is similar in spirit to the idea that superconductivity is macroscopically described by a U(1) Higgs theory [2,3]. We predict the phase structure of SU(2) Yang-Mills theory, the (quasiparticle) spectrum of its excitations, and the T dependence of thermodynamical quantities. As a result, the theory is shown to come in *three* rather than two phases. We predict a vanishing entropy density and an equation of state $\rho = -P$ at T_C and *negative* pressure P throughout the magnetic phase ($T_C \leq T \leq T_c$). An over-exponentially growing density of states (intersecting and single center-vortex loops) implies that the limiting temperature $T_C \sim \Lambda_{YM}$ of the confining phase can only then be exceeded if the spatial homogeneity of the system is sacrificed [4]. The situation of two disconnected thermodynamical regimes in a pure SU(2) gauge theory has an analogue in the $\mathbf{N}=1$ SUSY YM theory where a separating pole in the exactly known beta function exists [5].

The basic assumption is that a dilute-gas ensemble of calorons with trivial-holonomy (THC) [6] macroscopically forms a composite and adjoint Higgs field ϕ at high T (electric phase). By dilute gas we mean that in a minimal, local definition (lowest possible mass dimension) of ϕ,

$$\phi^a(x)|\phi(x)|^5 \equiv \langle \mathrm{tr}_{\mathrm{N}} F_{\mu\nu}(x)\, t^a\, F_{\nu\lambda}(x) F_{\lambda\mu}(x)\rangle_{A_\beta^{\mathrm{THC}}}\,, \qquad (1)$$

the average is performed over a single THC and its zero-mode deformations only. Fluctuations, that would lift the action of a given THC configuration A_β^{THC} above the BPS bound, are discarded in Eq. (1) because they mediate interactions between calorons to be considered later at the macroscopic level. The above assumption would be superfluous if it could be shown that at a given temperature T nontrivial solutions to the gap-equation (1) with $|\phi|=|\phi|(T)$ exist for a certain range of values of the fundamental gauge coupling $\bar{g}$ [a]. This is the objective of future research. While the (nonfluctuating, see below) field ϕ generates a T-dependent mass for topologically trivial modes $W^\pm$ on tree-level (thermal quasiparticles, solution to the infrared problem of thermal perturbation theory [7]) the 'photon' Z_0 remains tree-level massless in the electric phase. The 'condensate' $|\phi|$ represents a compositeness scale governing the maximal off-shellness of gauge-modes and the maximal center-of-mass energy flowing into a vertex. This yields a convering loop expansion of thermodynamical quantities despite a large value of the effective coupling constant e.

A fixed color orientation of ϕ forms a finite domain induced by a 'seed' caloron. At the points where at least four domains meet isolated zeros of ϕ exist with the associated magnetic monopoles being mappings from S_2 onto the coset spaces $\{\mathrm{SU(2)/U(1)}\}$ [8]. Microscopically, isolated monopoles are generated by nontrivial-holonomy calorons $(A_0(|\vec{x}| \to \infty) \neq 0$ and regular gauge copies) which form and dissociate during domain collisions [9,10,11,12,13], see [14] for an estimate of the surface tension of the spatial Wilson loop in terms of screened magnetic monopoles. A fluctuation A_β in the fundamental theory is decomposed into a BPS, topologically nontrivial part A_β^{THC}, building the ground-state, and topologically trivial a_β associated

[a] The integration over the instanton scale $\bar{\rho}$ in Eq. (1) is cut off in the UV at the compositeness scale $|\phi|^{-1}$.

with excitations: $A_\beta = A_\beta^{\rm THC} + a_\beta$. The macroscopic action is

$$S_E = \int_0^{1/T} d\tau \int d^3x \left(\frac{1}{2} \operatorname{tr}_{\rm N} G_{\mu\nu} G_{\mu\nu} + \operatorname{tr}_{\rm N} D_\mu \phi D_\mu \phi + V_E(\phi) \right) , \tag{2}$$

where $G^a_{\mu\nu} \equiv \partial_\mu a^a_\nu - \partial_\nu a^a_\mu - e f^{abc} a^b_\mu a^c_\nu$ denotes the field strength of a top. trivial mode a_β, e is the effective coupling constant, $D_\beta\phi \equiv \partial_\beta + ie[\phi, a_\beta]$, and $\operatorname{tr}_{\rm N} t^a t^b \equiv 1/2\, \delta^{ab}$. The potential $V_E(\phi) \equiv \operatorname{tr}_{\rm N} v_E^\dagger v_E$ is uniquely determined by the requirement of spatially homogeneous, periodic-in-euclidean-time τ solutions (τ independent modulus, $0 \le \tau \le 1/T$) to the BPS equation $\partial_\tau \phi = v_E$. Macroscopic BPS saturation derives from the zero-energy property of THCs (microscopically BPS) of which ϕ is composed (in absence of other gauge-field fluctuations), the other properties follow from thermodynamical equilibrium. The potential reads

$$V_E = \operatorname{tr} v_E^\dagger v_E \equiv \Lambda_E^6 \operatorname{tr} (\phi^2)^{-1} , \tag{3}$$

where Λ_E denotes a dynamically generated mass scale determined by a boundary condition to the thermodynamical evolution. Up to global gauge transformations the 'square-root' v_E is given as $v_E \equiv i\Lambda_E^3 \lambda_1 \phi/|\phi|^2$ where $\lambda_i\, (i = 1, 2, 3)$ denote the Pauli matrices and $|\phi| \equiv 1/2 \operatorname{tr} \phi^2$. Solutions to the BPS equation $\partial_\tau \phi = v_E$ are labelled by nonzero winding numbers $l \in \mathbf{Z}$:

$$\phi_l(\tau) = \sqrt{\frac{\Lambda_E^3}{2\pi T|l|}}\, \lambda_3 \exp(-2\pi i T l \lambda_1 \tau) . \tag{4}$$

On the solutions (4) we have $\partial^2_{|\phi_l|} V_E/T^2 = 12\pi^2\, l^2$ and $\partial^2_{|\phi_l|} V_E/|\phi_l|^2 = 3l^3\, \lambda_E^3$. We thus conclude that ϕ_l does not fluctuate thermodynamically and quantum mechanically ($\lambda_E \equiv 2\pi T/\Lambda_E$ is much larger than unity, see [1]). We restrict to lowest winding $|l| = 1$. Interactions between THC are accounted for macroscopically by taking ϕ_1 as a background in the equation of motion $D_\mu G_{\mu\beta} = 2ie[\phi, D_\beta \phi_1]$. On the macroscopic level there must not be a net field strength in the thermal ground state, and thus the only admissible solution $a_\beta^{bg,1}$ is pure gauge, $G_{\mu\beta}[a_\nu^{bg,1}] = 0 = D_\beta \phi_1$. We have $a_\beta^{bg,1} = \frac{\pi}{e} T \delta_{\beta 4} \lambda_1$. On the configuration $a_\beta^{bg,1}$ the Polyakov loop is $\mathbf{P} = -\mathbf{1}$. To asign (T dependent and τ independent) masses $m^2_{W^\pm} = -2e^2 \operatorname{tr} [\phi, t^{1,2}][\phi, t^{1,2}]$ to off-Cartan fluctuations a *singular* gauge transformation to unitary gauge needs to be performed [1]. The Polyakov loop transforms as $\mathbf{P} = -\mathbf{1} \rightarrow \mathbf{P} = +\mathbf{1}$. The singular gauge transformation does not change the *periodicity* of the fluctuations a_β. Thus it is *irrelevant* whether one integrates out a_β in winding or unitary gauge in a

loop expansion of thermodynamical quantities. The two distinct ground states $\mathbf{P} = \pm\mathbf{1} \neq 0$ together with the associated gauge-field fluctuations are physically equivalent. In unitary gauge there is a physical interpretation of a_β and integrating them out is simple. The existence of two ground states signals a broken electric Z_2 symmetry and thus deconfinement. By virtue of $D_\beta\phi = 0$ the vanishing energy density (pressure) of the hypothetical ground state, that is composed of noninteracting THCs only, is shifted to the finite energy density (pressure) (-)$V_E = 4\pi\Lambda_E^3 T$ of the physical ground state by THC interaction. As a consequence, the covariant BPS equation $D_\tau\phi = v_E$ is not satisfied by the above configurations mirroring that ground-state physics on its own is thermodynamically incomplete (gauge-field *excitations* are emitted and absorbed in addition).
An expression for the total pressure P in one-loop approximation is derived. Because of the compositeness scale $|\phi_1|$ quantum fluctuations can be neglected [1], and two-loop corrections are tiny (<2% of the one-loop value [15]). Thermodynamics is consistent (as in the underlying theory) if we impose $\partial_a P = 0$ where the mass parameter a is defined as $a = 2\pi e\lambda_E^{-3/2}$. This enforces an evolution of the effective gauge coupling with T which, in implicit form, is governed by

$$\partial_a \lambda_E = -\frac{24\,\lambda_E^4\,a}{(2\pi)^6} D(2a)\,. \tag{5}$$

The function $xD(x)$ denotes a scaled-out derivative w.r.t mass Ta of the thermal component of the one-loop pressure for a bosonic field. The right-hand side of Eq. (refeeq) is zero at $a = 0, \infty$ implying the existence of a highest and lowest attainable temperature. Shortly below $\lambda_{E,P}$ e shows a large bump of height $\sim \lambda_{E,P}$ making the assumed importance of THCs in the partition function self-consistent due to their small action. A plateau value $e_{\text{Plateau}} = 17.15$ (attractor) exists which is independent of the boundary condition imposed at large temperature T_P (IR $\leftrightarrow$ UV decoupling, renormalizability of the underlying theory). The constancy of e after the formation of ϕ_1 is consistent with the existence of isolated and locally conserved magnetic charge. We note here that the Stefan-Boltzmann limit is reached quickly (with three polarizations for the $W^\pm$ bosons!) due to a *power*-like approach.

For lack of space we only quote the results for the magnetic phase (see [1] for details) which sets in at T_c where $a \to \infty$: condensation of magnetic monopoles; decoupling of $W^\pm$ bosons; 'photon' Z_0 is massless at T_c and massive below; decoupling of Z_0, equation of state $\rho = -P$, and vanishing entropy density at $T_C \sim T_c/1.9$ (predicted from an evolution equation similar to Eq. (5) for the magnetic coupling g); negative total pressure for $T_C \leq T \leq T_c$. The Polyakov loop is +1 already in winding gauge. Thus only a single ground state exists and the electric Z_2 symmetry is restored (confinement of fundamental test charges but existence of a massive, propagating 'photon' Z_0). In the confining phase, setting in at $T_C \sim \frac{T_c}{1.9}$ by violent 'reheating' (entropy generation), no propagating gauge mode exists, excitations are made of crossed and uncrossed closed magnetic flux lines (center-vortex loops), and the ground-state pressure is precisely zero once the vortex condensate has settled to the minimum of its potential. Since excitations resemble vacuum diagrams in a ϕ^4 theory [16] and since the mass spectrum of these states is equidistant it is easy to estimate that the density of states in the confining phase is over-exponentially growing. This implies the existence of the Hagedorn temperature T_C and the thermodynamical disconnectedness of the confining phase from the other two phases.

Acknowledgments

The author would like to thank organizers and participants for a stimulating conference.

References

1. R. Hofmann, hep-ph/0404265.
2. V. L. Ginzburg and L. D. Landau, JETP **20**, 1064 (1950).
3. A. A. Abrikosov, Sov. Phys. JETP **5**, 1174 (1957).
4. R. Hagedorn, Nuovo Cim. Suppl. **3**, 147 (1965).
5. V. A. Novikov, M. A. Shifman, A. I. Vainshtein, and V. I. Zakharov, Phys. Lett. B **166**, 329 (1986).
6. B. J. Harrington and H. K. Shepard, Phys. Rev. **D17**, 105007 (1978).
7. A. D. Linde, Phys. Lett. B **108**, 389 (1982).
8. T. W. B. Kibble, J. Phys. A **9**, 1387 (1976).
9. W. Nahm, Phys. Lett. B **90**, 413 (1980).
   W. Nahm, Lect. Notes in Physics. 201, eds. G. Denaro, e.a. (1084) p. 189.
10. T. C. Kraan and P. van Baal, Nucl. Phys. B **533**, 627 (1998).
    T. C. Kraan and P. van Baal, Phys. Lett. B **435**, 389 (1998).
11. D. J. Gross, R. D. Pisarski, and L. G. Yaffe, Rev. Mod. Phys. **53**, 43 (1981).
12. D. Diakonov, N. Gromov, V. Petrov, and S. Slizovskiy, hep-th/0404042.
13. Ch. Hoelbing, C. Rebbi, and V. A. Rubakov, Phys. Rev. **D63**, 034506 (2001).
14. P. Giovannangeli and C. P. Korthals Altes, Nucl. Phys. B **608**, 203 (2001).
15. U. Herbst, R. Hofmann, J. Rohrer, work in progress.
16. C. M. Bender and T. T. Wu, Phys. Rev. **184**, 1231 (1969).

FREE ENERGIES OF STATIC THREE QUARK SYSTEMS

K. HÜBNER, O. KACZMAREK, F. KARSCH, O. VOGT

Fakultät für Physik

Universität Bielefeld, D-33615 Bielefeld, Germany

We study the behaviour of free energies of baryonic systems composed of three heavy quarks on the lattice in SU(3) pure gauge theory at finite temperature. For all temperatures above T_c we find that the connected part of the singlet (decuplet) free energy of the three quark system is given by the sum of the connected parts of the free energies of qq-triplets (-sextets). Using renormalized free energies we can compare free energies in different colour channels as well as those of qq- and qqq-systems on an unique energy scale.

1. Introduction

While existing studies on static baryonic systems focus on zero temperature simulations[1,2] or used maximal abelian gauge at finite temperature[3], we have calculated the free energies in different colour channels of heavy three quark systems at finite temperature using Coulomb gauge.

Here we restrict ourselves to the analysis of equilateral triangles above the critical temperature on a $32^3 \times 8$ lattice in SU(3) pure gauge theory. The qq-triplet and -sextet free energies have been calculated recently in Ref. 4 and also by us in this work.

2. Colour Channels of the Three Quark System

The state of a three quark system as the product of the irreducible representation of three quarks in colour space can be decomposed into symmetry states

$$3 \otimes 3 \otimes 3 = 1 \oplus 8 \oplus 8' \oplus 10, \tag{1}$$

where 3 is the irreducible representation of a quark in colour space and 1 denotes the singlet, 8 the first octet, $8'$ the second octet and 10 the decuplet state. The singlet is totally anti-symmetric, the first octet anti-symmetric in the first and second, the second octet in the second and third component and the decuplet is totally symmetric.

The derivation of the representation of free energies in these colour channels in terms of expectation values of Polyakov loop correlation functions is similiar to that for two quark systems[5], but more elaborate. Denoting the Polyakov-loop at $\mathbf{x}_i$ by L_i and $\beta = 1/T$, we find

$$\begin{aligned}\exp\left(-\beta F^{1}_{qqq}\right) = \frac{1}{6}\langle 27\,\mathrm{Tr}\,L_1\mathrm{Tr}\,L_2\mathrm{Tr}\,L_3 - 9\,\mathrm{Tr}\,L_1\mathrm{Tr}\,(L_2L_3)\\ - 9\,\mathrm{Tr}\,L_2\mathrm{Tr}\,(L_1L_3) - 9\,\mathrm{Tr}\,L_3\mathrm{Tr}\,(L_1L_2)\\ + 3\,\mathrm{Tr}\,(L_1L_2L_3) + 3\,\mathrm{Tr}\,(L_1L_3L_2)\rangle\end{aligned} \quad (2)$$

$$\begin{aligned}\exp\left(-\beta F^{8}_{qqq}\right) = \frac{1}{24}\langle 27\,\mathrm{Tr}\,L_1\mathrm{Tr}\,L_2\mathrm{Tr}\,L_3 + 9\,\mathrm{Tr}\,L_1\mathrm{Tr}\,(L_2L_3)\\ - 9\,\mathrm{Tr}\,L_3\mathrm{Tr}\,(L_1L_2) - 3\,\mathrm{Tr}\,(L_1L_3L_2)\rangle\end{aligned} \quad (3)$$

$$\begin{aligned}\exp\left(-\beta F^{8'}_{qqq}\right) = \frac{1}{24}\langle 27\,\mathrm{Tr}\,L_1\mathrm{Tr}\,L_2\mathrm{Tr}\,L_3 + 9\,\mathrm{Tr}\,L_3\mathrm{Tr}\,(L_1L_2)\\ - 9\,\mathrm{Tr}\,L_1\mathrm{Tr}\,(L_2L_3) - 3\,\mathrm{Tr}\,(L_1L_2L_3)\rangle\end{aligned} \quad (4)$$

$$\begin{aligned}\exp\left(-\beta F^{10}_{qqq}\right) = \frac{1}{60}\langle 27\,\mathrm{Tr}\,L_1\mathrm{Tr}\,L_2\mathrm{Tr}\,L_3 + 9\,\mathrm{Tr}\,L_1\mathrm{Tr}\,(L_2L_3)\\ +9\,\mathrm{Tr}\,L_2\mathrm{Tr}\,(L_1L_3) + 9\,\mathrm{Tr}\,L_3\mathrm{Tr}\,(L_1L_2)\\ + 3\,\mathrm{Tr}\,(L_1L_2L_3) + 3\,\mathrm{Tr}\,(L_1L_3L_2)\rangle.\end{aligned} \quad (5)$$

With this we obtain for the average free energy of the three quark system F^{av}_{qqq} the relation

$$\exp\left(-\beta F^{\mathrm{av}}_{qqq}\right) = \langle \mathrm{Tr}\,L_1\mathrm{Tr}\,L_2\mathrm{Tr}\,L_3\rangle \quad (6)$$

$$\begin{aligned}= \frac{1}{27}\exp\left(-\beta F^{1}_{qqq}\right) + \frac{8}{27}\exp\left(-\beta F^{8}_{qqq}\right)\\ + \frac{8}{27}\exp\left(-\beta F^{8'}_{qqq}\right) + \frac{10}{27}\exp\left(-\beta F^{10}_{qqq}\right).\end{aligned} \quad (7)$$

For the free energies of the qq-system we used the operators given in Ref. 5. These operators as well as those defined in (2)-(5) are gauge dependent and thus have to be evaluated in a fixed gauge. We used Coulomb gauge for our calculations.

3. Perturbation Theory and Renormalisation

Table 1 summarizes the Casimirs c_s found for the free energies in the different colour channels of the quark systems $q\bar{q}$, qq and qqq. Using this one obtains the leading order perturbative behaviour of the free energy in the symmetry state s as well as that of the average free energy for small distances or high temperatures

$$\beta F^s(R) = c_s \frac{\alpha\beta}{R} \quad \text{and} \quad \beta F^{\mathrm{av}}(R) = 1 + c_{\mathrm{av}} \frac{\alpha^2\beta^2}{R^2}, \tag{8}$$

where $\alpha = g^2/4\pi$. Here R is the usual Euclidian distance. For qqq-systems we restrict ourselves to equilateral triangles. In this case R denotes the edge length, which is an appropriate distance measure. In the case of the two octet free energies it is convenient to calculate the average of both. For small distances, the average free energies behave like the left most colour channels up to a T-dependent constant. In the following we will show results for

Table 1. Casimirs c_s and c_{av} for the leading order behaviour of F^s and F^{av}.

system	average	singlet	triplet	sextet	octet	decuplet
$q\bar{q}$	$-4/9$	$-4/3$			$+1/6$	
qq	$-4/9$		$-2/3$	$+1/3$		
qqq	$-4/3$	-2			$-1/2^{\star}$	$+1$

*This simple form only holds for equilateral triangles.

renormalized free energies. These are obtained from renormalized Polyakov loops. The relevant renormalization constants have been determined in Ref. 6 from an analysis of $q\bar{q}$ free energies and can directly be used also for the qqq-systems.

4. Colour Channels above T_c in qq- and qqq-systems

We compare the behaviour of the free energy in different colour channels for temperatures $T/T_c = 3, 6, 9$ on a $32^3 \times 8$ lattice and for equilateral triangular configurations in Figure 1a-c.

One can see clearly that the singlet free energies are strongly, the octet weaker attractive and the decuplet free energies are repulsive. Together with (7) this results in weakly attractive average free energies. All free energies at a given temperature approach a common T-dependent constant for large $R\sqrt{\sigma}$.

In Figure 1d we show the connected qqq-singlet free energies for equilateral triangles

$$\Delta F^1_{qqq}(R,T) = F^1_{qqq}(R,T) - F^1_{qqq}(\infty,T) \tag{9}$$

and compare it to the corresponding connected qq-triplet free energies, scaled by a factor of 3. Over the entire RT-interval we find that $\Delta F^1_{qqq}(R,T)$ and $3\Delta F^3_{qq}(R,T)$ coincide within errors for all temperatures, which suggests

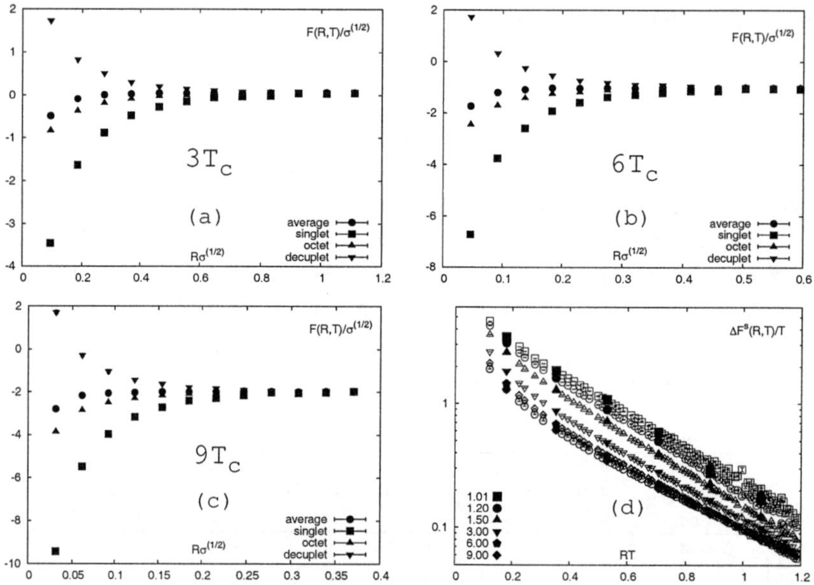

Figure 1. (a)-(c) Free energies of the qqq-system in different colour channels for temperatures $T/T_c = 3, 6, 9$. (d) Free Energies of the qq-(open symbols) and qqq-(full symbols) systems above T_c on a $32^3 \times 8$ lattice. Shown are $\Delta F^1_{qqq}(R,T)$ and $3\Delta F^3_{qq}(R,T)$ logarithmically.

that above T_c the interactions of the quarks in the qqq-singlet state can be decomposed into the pairwise interaction of three qq-pairs in a triplet state. The screening masses of both free energy channels are equal within errors. On the other hand we find that $F^1_{qqq}(\infty, T) = \frac{3}{2}F^3_{qq}(\infty, T)$, which shows that at large distances the static quark sources are screened independently by a gluon cloud. We find an analogous relation for the qqq-decuplet and qq-sextet free energies above T_c, the qqq-octet free energies show, however, some small deviations.

In Figure 2a we show $F^1_{qqq}(R,T)$ for several temperatures above T_c obtained on a $32^3 \times 8$ lattice renormalized by the procedure mentioned in section 3. For small distances we observe that for all temperatures the qqq-singlet free energies coincide, thus becoming T-independent.

5. Conclusion and Outlook

We calculated the free energies in different colour channels of heavy three quark systems at finite temperature. While the connected part of the qqq-

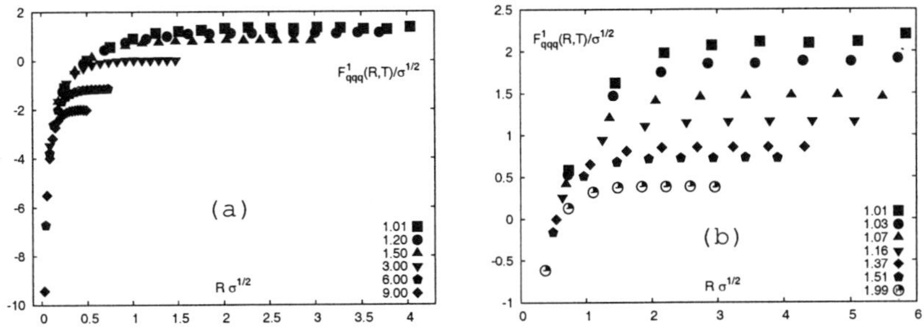

Figure 2. $F^1_{qqq}(R,T)$ (a) in pure gauge on a $32^3 \times 8$ lattice, (b) in 2-flavour QCD on a $16^3 \times 4$ lattice at $m/T = 0.4$.

singlet (-decuplet) free energies are found to be decomposable into three qq-triplet (-sextet) free energies for all distances calculated above T_c, the asymptotic large distance value of the free energies can be understood in terms of three independently screened quark sources. The qqq-octet free energies show deviations.

The approach used here for SU(3) gauge theory can also be applied to full QCD. In Figure 2b we show first results for $F^1_{qqq}(R)$ obtained on a $16^3 \times 4$ lattice in 2-flavour QCD. As in Figure 2a, the qqq-singlet becomes T-independent for small distances. Like the results in pure gauge simulations, we find that qqq-singlet (decuplet) free energies can be decomposed into qq-triplet (-sextet) free energies.

In the future we plan to perform a more detailed comparison between qq- and $q\bar{q}$-free energies in pure gauge and full QCD. Moreover, we are presently increasing the statistics for three quark free energies below T_c to be able to decide which flux tube geometry is realised close to T_c.

Acknowledgements

This work has been supported in parts by the German Research Foundation (DFG) under contracts GRK881 and KA1198/6-4.

References

1. T. T. Takahashi et al., *Phys. Rev.* **D65** (2002) 114509
2. C. Alexandrou, Ph. de Forcrand, *Nucl. Phys. B (Proc. Suppl.)* **129** (2004) 700
3. H. Ichie et al., *Nucl. Phys. B (Proc. Suppl.)* **119** (2003) 751-753
4. A. Nakamura, T. Saito, hep-lat/0406038
5. S. Nadkarni, *Phys. Rev.* **D34** 3904 (1986)
6. O. Kaczmarek et al., *Phys. Lett.* **B543** 41 (2002)

COLOR FERROMAGNETIC STATE OF DENSE QUARK MATTER

A. IWAZAKI

Department of Physics, Nishogakusha University, Shonan Ohi Chiba 277-8585, Japan.
E-mail: iwazaki@yukawa.kyoto-u.ac.jp

We discuss that dense quark matter possesses a new phase of color ferromagnetism which appears at a critical baryonic chemical potential around 200 MeV. Subsequently, color superconducting phase arises at a critical chemical potential larger than 1 GeV. The color ferromagnetic state is a stable state realized by the formation of a quantum Hall state of unstable gluons in the Savvidy vacuum

Quark matter is known or expected to have several phases, hadronic phase, quark gluon plasma phase and color superconducting phase[1]. Among the phases the hadronic phase and quark gluon plasma phase are observed. Although the color superconducting phase is very intriguing, present experiments could not produce the phase because chemical potentials of the quarks larger than 1 GeV is needed for the production of the phase.

We have recently discussed[2] a possibility of the stable Savvidy states (color ferromagnetic states) in dense quark matter. The state is caused by the condensation of the color magnetic field, not by the alignment of the quark's magnetic moments. We have discussed that the state is realized between the hadronic state and the color superconducting state when the chemical potential is varied. The state possesses a spontaneously generated color magnetic field in maximal Abelian sub-algebra and also involves a quantum Hall state of off-diagonal gluons. The gluons have been known to have unstable modes[3] in the color magnetic field B. The existence of these unstable modes implies that the naive ferromagnetic state (Savvidy vacuum [4]) is unstable. We have recently shown[2] that the formation of the gluon's quantum Hall state caused by the condensation of the unstable modes stabilizes the ferromagnetic state.

In this paper we give the brief discussion how the Savvidy vacuum is stabilized by the formation of the quantum Hall states[5] of the unstable gluons in dense quark matters. We treat only the case of zero temperature.

It has been known[4] in the gauge theory that one-loop effective potential for color magnetic field has non-trivial minimum; $V(gB) = \frac{11}{48\pi^2}g^2B^2\left(\log(gB/\Lambda^2) - \frac{1}{2}\right) - \frac{i}{8\pi}g^2B^2$, This apparently seems to imply spontaneous generation of the color magnetic field, namely, the realization of a ferromagnetic state. But it is not so simple since the imaginary part in $V(gB)$ is present when $gB \neq 0$. It means that the state with the magnetic field as well as the perturbative vacuum state with $gB = 0$ is not the state with the minimum energy. Namely, the unstable Savvidy vacuum should make transition to a new stable state by the condensation of unstable modes excited in the Savvidy vacuum. This unstable modes in the Savvidy vacuum can be seen easily in the following.

We decompose the gluon's Lagrangian with the use of the variables, "electromagnetic field" $A_\mu = A_\mu^3$, and "charged vector field" $\Phi_\mu = (A_\mu^1 + iA_\mu^2)/\sqrt{2}$ where indices $1 \sim 3$ denote color components,

$$L = -\frac{1}{4}\vec{F}_{\mu\nu}^2 = -\frac{1}{4}(\partial_\mu A_\nu - \partial_\nu A_\mu)^2 - \frac{1}{2}|D_\mu\Phi_\nu - D_\nu\Phi_\mu|^2$$
$$+ig(\partial^\mu A^\nu - \partial^\nu A^\mu)\Phi_\mu^\dagger\Phi_\nu + \frac{g^2}{4}(\Phi_\mu\Phi_\nu^\dagger - \Phi_\nu\Phi_\mu^\dagger)^2 \quad (1)$$

with $D_\mu = \partial_\mu + igA_\mu$. Using the Lagrangian we can derive that the energy E of the charged vector field $\Phi_\mu \propto e^{iEt}$ in the magnetic field, $A_\mu = A_\mu^B$, is given by $E^2 = k_3^2 + 2gB(n+1/2) \pm 2gB$ with a gauge choice, $A_j^B = (0, x_1B, 0)$ and $(\partial_\mu + igA_\mu^B)\Phi^\mu = 0$, where we have taken the spatial direction of $\vec{B}$ being along x_3 axis. $\pm 2gB$ (the integer $n \geq 0$) denotes the contribution from spin components of Φ_μ (Landau levels) and k_3 denotes momentum in the direction parallel to the magnetic field.

Then, we can find that the modes occupying the lowest Landau level ($n = 0$) with spin parallel to $\vec{B}$ are unstable; $E^2(n = 0) < 0$. The modes are present only under the color magnetic field and just the unstable modes in the Savvidy vacuum.

How do they form a new stable state by their condensation. The essence is the same as the case of the Higg potential. In the model of a scalar field with the potential a state with the vanishing expectation value of the field is unstable and the mass of the field in the state is imaginary. Unstable modes in the model are eigenstates of the field with small momenta; their energies become imaginary when their momenta are small enough. Among unstable modes with various momenta, the mode with the vanishing momentum is the most unstable one. It means that the energy of the mode has the largest imaginary part. Hence, the amplitude of the mode becomes large very rapidly. Eventually, this mode condenses and forms an uniform stable state with the nonvanishing expectation value of the field. This is the true groundstate of the Higgs model. We should mention that the mode to form the stable state is unique in the model of the scalar field.

A similar mechanism to form a stable state works in the Savvidy vacuum. Among the unstable modes the most unstable modes with $k_3 = 0$ are relevant to the formation of the true ground state also in the gauge theory. (In the Savvidy vacuum the most unstable modes are not unique and are degenerate in their momenta k_2. The degeneracy per unit area is given by $gB/2\pi$. This non uniquness of the modes makes the problem of the stabilization of the Savvidy vacuum become non trivial.) Since they have no x_3 dependence, they are two-dimensional objects occupying the lowest Landau level. The problem is how they form a new stable state. The situation is quite similar to the case in the two-dimensional electrons occupying Lowest Landau level under strong magnetic field. It is well known that they form stable fractional quantum Hall states (Laughlin states). The only difference is that in the gauge theory gluons are bosons, while electrons are fermions.

In order to find the stable state in the gauge theory, we extract only the most unstable modes from the Lagrangian, eq. (1), ignoring the other modes coupled with them and obtain a two-dimensional Lagrangian,

$$L_{\text{unstable}} = |(i\partial_\nu - gA_\nu^B)\phi_{\text{u}}|^2 + 2gB|\phi_{\text{u}}|^2 - \frac{\lambda}{2}|\phi_{\text{u}}|^4, \qquad (2)$$

with $\lambda = g^2/\ell$, where the field $\phi_{\text{u}} = (\Phi_1 - i\Phi_2)\sqrt{\ell/2}$ denotes the unstable modes in the lowest Landau level. ℓ denotes the coherent length of the magnetic field, namely, its extension in the direction of the field. Here we consider the quark matter with its length scale ℓ. Then, obviously, a condition of $\ell \gg \ell_B = 1/\sqrt{gB}$ must be satisfied for the consistency. The index ν runs from 0 to 2.

The presence of the negative mass square term implies that the Savvidy vacuum, which has no condensation of the field, is unstable. Physically, the Lagrangian L_{unstable} describes such a system that the particles of ϕ_{u} move in the magnetic field and interact with each other through a repulsive potential of a delta function. A numerical simulation[6] indicates that the nonrelativistic particles with such an interacting potential can form a stable Laughlin state even if they are bosons. Thus, the gluons represented by ϕ_{u}

are also expected to form a quantum Hall state.

In order to discuss a new stable state formed by the field, $\phi_{\rm u}$, namely, a quantum Hall state, we introduce Chern-Simons gauge field to make composite gluons; gluons attached with the Chern-Simons flux. Then, a Lagrangian[7] describing equivalently the original system $L_{\rm unstable}$ is given by

$$L_a = |(i\partial_\nu - gA_\nu + a_\nu)\phi_a|^2 + 2gB|\phi_a|^2 - \frac{\lambda}{2}|\phi_a|^4 + \frac{\epsilon^{\mu\nu\lambda}}{4\alpha}a_\mu\partial_\nu a_\lambda, \qquad (3)$$

where the statistical factor α should be taken as $\alpha = 2\pi\times$integer to keep the equivalence of the system described by L_a to that of $L_{\rm unstable}$. The field ϕ_a is called as composite gluons attached with the Chern-Simons flux a_i.

Solving equations of motion derived from the Lagrangian we can see that the condensed state of ϕ_a can occurs ($\langle\phi_a\rangle \neq 0$) and that the state is a fractional quantum Hall state. The point is that such a condensation can arise only when Chern-Simons gauge field acting on ϕ_a cancels with the color magentic field A_ν. Then, since there is no total "magnetic field" acting on ϕ_a, a nontrivial uniform solution of ϕ_a can exist. The above cancellation occurs when the following condition is satisfied,

$$2a_0|\phi_a|^2 = \frac{gB}{2\alpha} \quad \text{and} \quad a_0^2 + 2gB = \lambda|\phi_a|^2. \qquad (4)$$

That is, the quantum Hall state represented by the condensed state, $\langle\phi_a\rangle = v \neq 0$, arises only when the magnetic field is canceled by the Chern-Simons field, $-\epsilon_{ij}\partial_i a_j = gB = 2\alpha\rho_{\rm c}$; $\rho_{\rm c}$ is given by the left hand side of eq. (4), i.e. $\rho_{\rm c} = 2a_0v^2$. This $\rho_{\rm c}$ represents color charge density possessed by the gluons ϕ_a. The composite gluons condense to form the quantum Hall state only when the filling factor $\nu = 2\pi\rho_{\rm c}/gB$ is equal to π/α. This is quite similar to the case of the electron's quantum Hall state. Therefore, we can show in a similar way to the case of electrons[8] that the condensed state of ϕ_a describes quantum Hall effects, namely the state is the quantum Hall state. More importantly, we can show that the state is stable; there no unstable modes appear in the state. Consequently, the Savvidy vacuum with the color magnetic field is stabilized by the formation of the gluon's quantum Hall state.

The quantum Hall state of gluons is realized in a sector with nonzero color charge, not in the vacuum sector; the condensed state of ϕ_a possesses a color charge. Such a state can arise in dense quark matter where the color charge of quarks is transmitted to the condensate. Therefore, the stable Savvidy vacuum with the color magnetic field and with the gluon condensation is a possible state of quark matter. The state is called a color ferromagnetic state.

As we have mentioned, the quark matter is known to possess the hadronic phase and the quark gluon plasma phase. These phases arise in a very low density of baryons or in high temperature. On the other hand, in a very high density of baryons the matter possibly shows a color superconducting phase. Then, a significant question is where or not there is a window of the baryon density in which the stable color ferromagnetic phase can arise.

We have shown in the paper[9] that the color superconducting phase is favored in much larger chemical potential of the baryon number than 1 GeV. On the other hand, the color ferromagentic phase is favored in the chemical potential about several hundred MeV. More precisely we can show that a phase transition from the hadronic phase to the color ferromagnetic phase occurs in a critical chemical potential of the baryon number such as 180MeV $\sim$ 240MeV even if taken into account ambiguities in several parameters of the theory, e.g. strength of the color magnetic field, the mass of the quarks in dense matter, etc. Furthermore, a phase transition from the color ferromagnetic phase to the color superconducting phase occurs in a critical baryonic chemical potential such as 1 $\sim$ 3 GeV. Therefore, the quark matter possibly shows the color ferromagnetic properties in baryonic chemical potentials of several hundred MeV. The region of the chemical potential is accessible at present experiments in SPS.

References

1. K. Rajagopal and F. Wilczek, hep-ph/0011333.
2. A. Iwazaki and O. Morimatsu, Phys. Lett. **B571**, 61 (2003); A. Iwazaki, O Morimatsu, T. Nishikawa and M Ohtani, Phys. Lett **B579**, 347 (2004).
3. N.K. Nielsen and P. Olesen, Nucl. Phys. **B144**, 376 (1978); Phys. Lett. **79B**, 304 (1978).
4. G.K. Savvidy, Phys. Lett. **71B**, 133 (1977).
   H. Pagels, Lecture at Coral Gables, Florida, 1978.
5. *The Quantum Hall Effect, 2nd Ed.*, edited by R.E. Prange and S.M. Girvan (Springer-Verlag, New York, 1990).
6. T. Nakajima and M. Ueda, Phys. Rev. Lett. **91** 140401 (2003).
7. G.W. Semenoff, Phys. Rev. Lett. **61**, 516 (1988); G.W. Semenoff and P. Sodano, Nucl. Phys. **B328**, 753 (1989).
8. Z.F. Ezawa, M. Hotta and A. Iwazaki, Phys. Rev. **B46**, 7765 (1992); Z.F. Ezawa and A. Iwazaki, J. Phys. Soc. Jpn. **61**, 4133 (1990).
9. A. Iwazaki, O Morimatsu, T. Nishikawa and M Ohtani; hep-ph/0404201.

AXIAL CURRENTS FROM CKM MATRIX CP VIOLATION AND ELECTROWEAK BARYOGENESIS

T. KONSTANDIN
Institut für Theoretische Physik,
Philosophenweg 16,
69120 Heidelberg, Germany
E-mail: T.Konstandin@ThPhys.Uni-Heidelberg.de

The first principle derivation of kinetic transport equations suggests, that a CP-violating mass term during the electroweak phase transition can induce axial vector currents. Since the important terms are of first order in gradients there is a possibility to construct new rephasing invariants that are proportional to the CP phase in the Cabibbo-Kobayashi-Maskawa matrix and to circumvent the upper bound of CP-violating contributions in the Standard Model, the Jarlskog invariant.

1. Introduction

All models that intend to describe the baryon asymmetry of the universe (BAU) by electroweak baryogenesis (EWB)[1] depend on extensions of the Standard Model (SM) since the SM fails on the following grounds:

A) First order phase transition: Sakharov[2] pointed out that baryogenesis necessarily requires non-equilibrium physics. The expansion of the universe is too slow at the electroweak scale and one needs bubble nucleation during a first order EWPT. The phase diagram of the Standard Model is studied in detail[3], and it is well known that there is no first order phase transition in the Standard Model for the experimentally allowed Higgs mass.

B) Sphaleron bound: To avoid washout after the phase transition, the *vev* of the broken Higgs field has to meet the criterion $\langle \Phi \rangle \gtrsim T_c$, i.e. a strong first order phase transition. This results in the Shaposhnikov bound on the Higgs mass[4].

C) Lack of CP violation: Since the only source of CP violation in the Standard Model is the Cabibbo-Kobayashi-Maskawa (CKM) matrix (apart from the neutrino mass matrix, which provides an even tinier source of CP violation) one has to face that it is too weak to account for the observed magnitude of BAU.

In the following we will address the last point, the lack of sufficient CP violation, in the framework of kinetic theory.

The strong first order phase transition is assumed to occur at about $T_c \simeq 100$ GeV and is parametrised by the velocity of the phase boundary (wall velocity) v_w and its thickness l_w.

In this article we will focus on the following main points:

- We will demonstrate how CP violating sources can arise in semi-classical Boltzmann type equations.
- We argue that the Jarlskog determinant as an upper bound of CP violation in the SM is not valid during the EWPT.
- We give a rough estimate of the CP violating sources during the EWPT and conclude that the source is by orders larger than considering the Jarlskog determinant but still insufficient to explain the magnitude of the BAU.

2. Axial Currents in Kinetic Theory

Starting point in kinetic theory are the exact Schwinger-Dyson equation for the two point functions in the closed time-path (CTP) formalism.

$$\begin{aligned} &e^{-i\diamond}\{S_0^{-1} - \Sigma_R, S^<\} - e^{-i\diamond}\{\Sigma^<, S_R\} \\ &\quad = \frac{1}{2}e^{-i\diamond}\{\Sigma^<, S^>\} - \frac{1}{2}e^{-i\diamond}\{\Sigma^>, S^<\}, \\ &e^{-i\diamond}\{S_0^{-1} - \Sigma_R, A\} - e^{-i\diamond}\{\Sigma_A, S_R\} = 0, \end{aligned} \tag{1}$$

where we have used the definitions and relations

$$\begin{aligned} &S^{\bar{t}} := S^{--},\ S^t := S^{++},\ S^< := S^{+-},\ S^> := S^{-+}, \\ &A := \frac{i}{2}(S^< - S^>), S_R := S^t - \frac{i}{2}(S^< + S^>), \\ &\diamond\{A, B\} := \partial_{X^\mu} A \partial_{k_\mu} B - \partial_{k_\mu} A \partial_{X^\mu} B. \end{aligned} \tag{2}$$

S denotes the Wightman function, S_0^{-1} the free inverse propagator and Σ the selfenergy. $S^{\pm\pm}$ denotes the entries of the 2×2 Keldysh matrices and all functions depend on the average coordinate X and the momentum k. To simplify the equations, one can perform a gradient expansion. The terms on the left hand side will be expanded up to first order, whereas the collision terms on the right hand side vanish in equilibrium and are just taken up to zeroth order. The expansion parameter is formally ∂_X/k, which close to equilibrium and for typical thermal excitations reduces to $(l_w * T_c)^{-1}$.

We will not solve the full transport equations, but only look for the appearance of CP-violating source terms.

To start with, we consider a toy model with only one flavour and a mass term that contains a space dependent complex phase[5]. The inverse propagator in a convenient coordinate system reads

$$k_0\gamma_0 + k_3\gamma_3 + m_R(X_3) + im_I(X_3)\gamma_5. \tag{3}$$

Using the spin projection operator $P_s = \frac{1}{2}(1 + s\gamma_0\gamma_3\gamma_5)$ the Schwinger-Dyson equations can be decoupled and finally yield ($me^{i\theta} = m_R + i\,m_I$)

$$\left(k_0^2 - k_3^2 - m^2 + s\frac{m^2\theta'}{k_0}\right) Tr(\gamma_0 S_s^<) = 0 \tag{4}$$

$$\left(k_3\partial_{X_3} - \frac{(m^2)'}{2}\partial_{k_3} + s\frac{(m^2\theta')'}{2k_0}\partial_{k_3}\right) Tr(\gamma_0 S_s^<) = Coll. \tag{5}$$

We see, that in our approximation, the quasi-particle picture is still valid, since the Wightman function fulfills an algebraic constraint. Furthermore, the first order corrections lead to some source term, that is proportional to the complex phase of the mass and therefore CP violating.

Performing the calculation with several flavours, one finds the generalization of this CP violating term, reading $Tr(m^{\dagger\prime}m - m^\dagger m')$.

3. Enhancement of CP Violation in the SM

Jarlskog proposed an upper bound for CP violating effects in the Standard Model. Following her argument of rephasing invariants, the first CP violating quantity constructed out of the Yukawa couplings is the Jarlskog determinant[7]

$$\begin{aligned} &Im\,det[\tilde{m}_u\,\tilde{m}_u^\dagger, \tilde{m}_d\,\tilde{m}_d^\dagger] \\ &\quad = Tr(Cm_u^4C^\dagger m_d^4 Cm_u^2C^\dagger \tilde{m}_d^2) \\ &\quad \approx -2J \cdot m_t^4 m_b^4 m_c^2 m_s^2, \end{aligned} \tag{6}$$

When applied to the case of electroweak baryogenesis, one finds the upper bound of the BAU[8]

$$[\tfrac{g_W^2}{2M_W^2}]^7 J \cdot m_t^6 m_b^4 m_c^2 m_s^2 \approx 10^{-22}, \tag{7}$$

Though, two assumptions, that are needed for this argument are not fulfilled during the electroweak phase transition[6].

A) Since the mass matrix is space dependent, one needs space dependent diagonalization matrices to go to the mass eigenbasis. This leads to new physical relevant quantities, that can as well be CP violating. As a generalization of the CP violating source term in the kinetic toy model above, we found $Tr(m^{\dagger\prime}m - m^{\dagger}m')$. However in the Standard model this term vanishes at tree level, for the mass matrix is proportional to its derivative.

B) The argument of Jarlskog is based on the fact, that the examined quantity is perturbative in the Yukawa coupling. The calculation of the selfenergy in a thermal plasma involves integrations over divergent logarithms, of the form

$$h_2(\omega,\kappa) = \frac{1}{\kappa}\int_0^\infty \frac{d|\mathbf{p}|}{2\pi}\Big(\frac{|\mathbf{p}|}{\epsilon_h}L_2(\epsilon_h,|\mathbf{p}|)f_B(\epsilon_h) - \frac{|\mathbf{p}|}{\epsilon_u}L_1(\epsilon_u,|\mathbf{p}|)f_F(\epsilon_u)\Big).$$

$$L_{1/2}(\epsilon,|\mathbf{p}|) = \log\left(\frac{\omega^2-\kappa^2\pm\Delta+2\epsilon\omega+2\kappa|\mathbf{p}|}{\omega^2-\kappa^2\pm\Delta+2\epsilon\omega-2\kappa|\mathbf{p}|}\right) + \log\left(\frac{\omega^2-\kappa^2\pm\Delta-2\epsilon\omega+2\kappa|\mathbf{p}|}{\omega^2-\kappa^2\pm\Delta-2\epsilon\omega-2\kappa|\mathbf{p}|}\right),$$

that lead to a significant space dependence of the selfenergy. Since the

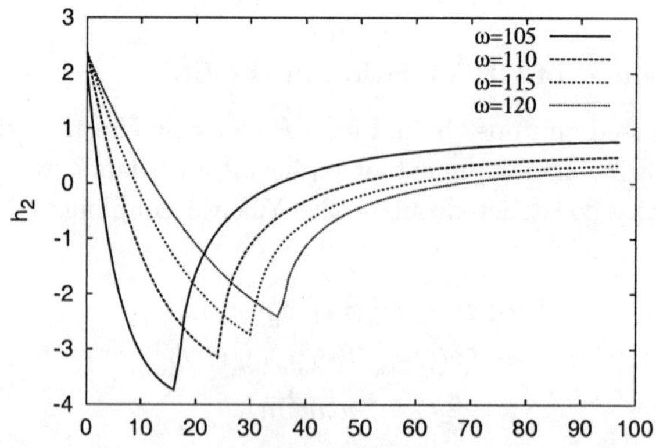

Figure 1. Dependence of h_2 on the Higgs vev Φ in % of its value $\Phi^0 = 246$ GeV at T=0. The external energies and momenta are fixed at $\omega = 105$ GeV to $\omega = 120$ GeV, k=100 GeV, the mass of the quark in the loop is $m_u = 100$ GeV.

space dependence is due to a resonance with the plasma particles, the selfenergy is highly sensitive to the quark masses and the W mass, that both change continously in the wall profile. However since CP violating effects

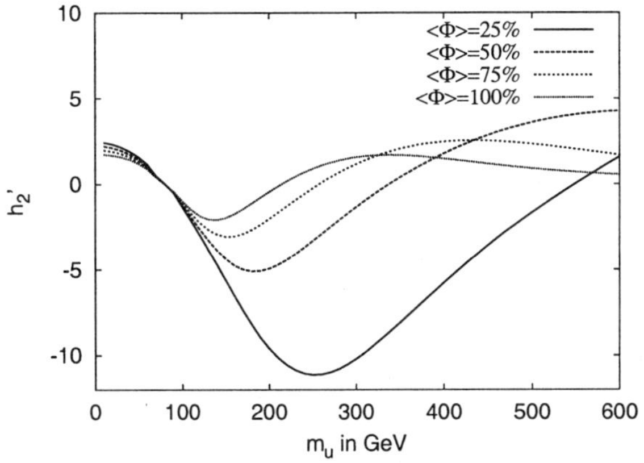

Figure 2. Dependence of h_2' on the mass of the quark in the loop with an on-shell external quark of mass $m_e = 4$ GeV. The Higgs vev is chosen in a range of 25% to 100% of its value in the broken phase at T=0.

only appear as an interference of the two loop and the one loop term, an estimation of the source term leads to the upper bound[6]

$$\frac{\delta\omega}{\omega} \sim J \cdot m_t^4 m_s^2 m_b^2 m_c^2 \frac{\alpha_w^3 h_2'}{m_W^8 l_w T^3} \approx 10^{-15}.$$

We conclude, that the axial current is enhanced seven orders in magnitude. Still the CP-violating source due to the CKM matrix might be too weak to account for the BAU.

Acknowledgements. I would like to thank M.G. Schmidt and T. Prokopec for the nice collaboration.

References

1. V. A. Kuzmin, V. A. Rubakov and M. E. Shaposhnikov, Phys. Lett. B **155** (1985) 36.
2. A. D. Sakharov, Pisma Zh. Eksp. Teor. Fiz. **5** (1967) 32
3. K. Kajantie, M. Laine, K. Rummukainen and M. E. Shaposhnikov, Nucl. Phys. B **493**, 413 (1997)
4. G. R. Farrar and M. E. Shaposhnikov, Phys. Rev. D **50** (1994) 774
5. K. Kainulainen, T. Prokopec, M. G. Schmidt and S. Weinstock, JHEP **0106** (2001) 031
6. T. Konstandin, T. Prokopec and M. G. Schmidt, Nucl. Phys. B **679** (2004) 246
7. C. Jarlskog and R. Stora, Phys. Lett. B **208** (1988) 268.
8. G. R. Farrar and M. E. Shaposhnikov, Phys. Rev. Lett. **70** (1993) 2833

DILUTE MONOPOLE GAS, AND K-TENSIONS IN GLUODYNAMICS

C. P. KORTHALS ALTES AND P. GIOVANNANGELI

Centre Physique Théorique au CNRS, Case 907, Luminy, F13288, Marseille, France

E-mail:altes@cpt.univ-mrs.fr

A dilute monopole gas explains, in quarkless gluodynamics, the small ratio δ between the square of magnetic screening mass m_M and spatial Wilson loop tension. This ratio is 0.0895 for $T = 0$ to 0.0594 at $T = \infty$ for any number of colours with order N^{-2} corrections and equals up to a numerical factor of O(1) the diluteness. The monopoles have a size $l_M = m_M^{-1}$. The GNO classification tells us they are in a representation of the magnetic SU(N) group. Choosing the adjoint for the dilute gas predicts the k-tensions to scale as $k(N-k)$, within a percent for high T, and a few percent for low T for the seven ratios determined by lattice simulation. The transition is determined by the transition of the dilute Bose gas at $T_c = 0.174 m_M$, and the transition is that of a non- or nearly-relativistic Bose gas.

1. Introduction

An ancient idea in QCD [1] is that monopoles are responsible for flux tube formation through a dual superconductor mechanism. Another mechanism, that of the "Copenhagen vacuum" of the early eighties [2], proposes macroscopic Z(N) Dirac strings or Z(N) vortices.

Here we will discuss a specific model of the first type [4], that works surprisingly well in its most direct applications. A straightforward way to see its workings is to start from the plasma phase.

We know that electric quasi-particles, the gluons, are approximately free at very high temperature. This follows from the Stephan-Boltzmann law.

We may guess that the same happens to the magnetic monopoles that were condensed in the cold phase.

What kind of monopoles can we expect? In the absence of any spontaneous breaking the GNO classification [5] applies. For a theory with only Z(N) neutral fields, like the adjoint, the magnetic group is $SU(N)$. So we can have monopoles in the fundamental, adjoint or any representation we like.

One of the consequences of the idea of a monopole condensate is the screening of the magnetic Coulomb force between two static magnetic sources. This is a straightforward generalization of the screening of the Coulomb force between two heavy quarks. There is now ample affirmation of screening [11] [16] [6] from lattice simulations.

2. Electric flux loops

Here we will keep the factor 2.

Take a Z(N) vortex of strength k, the 't Hooft loop[3]. We introduce in the Lie algebra of traceless hermitean $N \times N$ matrices a basis for the Cartan subalgebra, the $(N-1) \times (N-1)$ dimensional subspace of diagonal matrices. This basis of $N-1$ diagonal matrices Y_k is chosen such that in exponentiated form $\exp\big(-i2\pi Y_k\big)$ it gives the $N-1$ centergroup elements $\exp ik\frac{2\pi}{N}$. A simple choice is

$$Y_k = \frac{1}{N} diag(N-k, \ldots\ldots, N-k, -k, \ldots\ldots, -k). \tag{1}$$

The entry $N-k$ comes in k times, and $-k$ comes in $N-k$ times, so the trace is 0.

The flux operator becomes in this notation:

$$V_k(L) = \exp i\frac{4\pi}{g}\int d\vec{S}.Tr\vec{E}Y_k. \tag{2}$$

It does correspond to the vortex operator of 't Hooft with strength k.

The next question is: does the gas of deconfined gluons induce an area law in this operator?

The charge of a gluon with respect to the charge Y_k is found from the form Y_k takes in the adjoint representation. We have $2k(N-k)$ gluon species in the gluon multiplet with charge ± 1. All other gluons have vanishing charge so do not contribute to the flux. Since we take the gluons to be statistically independent, the charged ones all contribute a factor to $\langle V_k \rangle$, and this factor is the same by $SU(N)$ symmetry: $\exp -2l_D n$.

Conclusion: the expectation value of the loop is

$$\langle V_k \rangle = \exp\big(-\rho_k(T)A(L)\big) \tag{3}$$

with the tension

$$\rho_k = 4l_D nk(N-k). \tag{4}$$

So this is the k-scaling law for the electric flux loop.

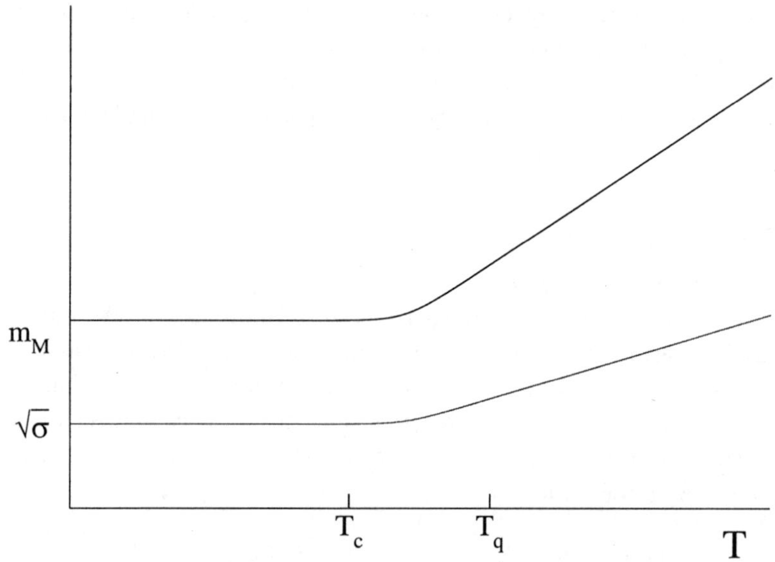

Figure 1. Magnetic mass m_M and tension σ as function of temperature, schematically. $m_M(0) = m_{0++}$ and from ref. (13): $T_c = 0.174\ m_{0++}$. The temperature $T_q \leq m_{0++}$ is where the de Broglie thermal wave length becomes equal to the magnetic screening length. For the calculation of the tension it is below T_q that quantum statistics applies, above classical statistics applies as in section 5. The dimensionless ratio of the spatial tension in the fundamental representation and the square of the magnetic mass is small, on the order of a few percent. In two extreme cases the ratio $\delta(T)$ is accurately known from simulations : at $T = \infty$ in the 3d theory, and at $T = 0$. In between there are simulations that are consistent with a slowly varying $\delta(T)$.

2.1. *Flux representation of the Wilson loop*

If we want to proceed with the Wilson loops, as we did with the 't Hooft loops, we need clearly a representation similar to eq.(2).

Of course a natural guess is to take this electric flux formula, replace $\vec{E}$ by $\vec{B}$, $\alpha = g^2/4\pi$ by $1/\alpha$ to get:

$$W_R = \exp{(ig \int d\vec{S}.Tr\vec{B}Y_k)}. \tag{5}$$

This relation holds in a dynamical sense as an average.

3. Predictions for the Wilson loops

For the dilute gas in the adjoint representation the computation is not different from the one with gluons. The adjoint monopoles have a magnetic charge equal to $\pm\frac{2\pi}{g}$, or 0. Only the former contribute and their individual contribution to the loop in eq. (5) is -1. The diluteness, and classical statistics (as the thermal wave length is T^{-1} and the typical interparticle distance is $(g^2T)^{-1}$ this is justified, see also fig. 1.) then give for every charged species the same contribution $\exp -2l_M n_M$, n_M being the common density of a given species. As there are $2k(N-k)$ charged species in the adjoint:

$$\sigma_k = 4k(N-k)l_M n_M. \tag{6}$$

One can do a similar calculation for monopoles in the fundamental representation or in any other.

4. Comparison to lattice simulations

The ratios found [10] for the totally antisymmetric irreps are close — within a percent for the central value — as far as the adjoint multiplet of magnetic quasi-particles is concerned :

$$\begin{aligned} \mathrm{SU}(4) : \sigma_2/\sigma_1 &= 1.3548 \pm 0.0064 \quad \text{adjoint} : 1.3333 \quad \text{fundamental} : 1.8182 \\ \mathrm{SU}(6) : \sigma_2/\sigma_1 &= 1.6160 \pm 0.0086 \quad \text{adjoint} : 1.6000 \quad \text{fundamental} : 1.9686 \\ \sigma_3/\sigma_1 &= 1.808 \ \ \pm 0.025 \quad\ \ \text{adjoint} : 1.8000 \quad \text{fundamental} : 2.3635 \end{aligned}$$

There is a less precise determination of the ratio $\sigma_2/\sigma_1 = 1.52 \pm 0.15$ in $SU(5)$ [13]. But the central value is within 1 to 2% of the predicted value 3/2 from the adjoint. The fundamental gives a ratio 1.8231.

The $SU(8)$ ratios are known on a rather course lattice [13] and using a different algorithm:

$$\begin{aligned} \sigma_2/\sigma_1 &= 1.692(29) \quad \text{adjoint} : 1.714 \quad \text{fundamental} : 2.106 \\ \sigma_3/\sigma_1 &= 2.160(64) \quad \text{adjoint} : 2.143 \quad \text{fundamental} : 2.958 \\ \sigma_4/\sigma_1 &= 2.26(12) \quad\ \ \text{adjoint} : 2.286 \quad \text{fundamental} : 3.256 \end{aligned}$$

The fundamental monopoles are clearly disfavoured by the data.

5. Conclusions

There is remarkable agreement with an accuracy of about a percent between numerical simulation at high T and the dilute adjoint monopole gas. This diluteness appears as a small parameter for *every* SU(N) theory. Some parameters, like the mass of the magnetic quasi-particle, are still within a large range, although its size is l_M. Is it heavy, on the order of the lowest glueball mass (i.e. m_M), then the transition is that of a non-relativistic Bose gas. Is it light (with respect to its size) then the transition is that of a near-relativistic BE transition.

Perhaps the best way to summarize our approach is to return to fig. 1. It is clear that the relation:

$$\sigma \sim l_M n_M \tag{7}$$

implies that $l_M^2\sigma$ is the diluteness δ, and must be small for consistency. And lattice data have borne that out!

Of course for most observables in the low T phase the Bose statistics is all-important. Thus the string tension will become non-zero below T_c, the character of the transition, i.e. a jump or continous behaviour in the occupation fraction of the $\vec{p}=0$ states will be crucial to know and and is calculable in this model.

References

1. G. 't Hooft, in High Energy Physics, ed. A. Zichichi (Editrice Compositori Bologna, 1976);S. Mandelstam, Phys. Rep. 23C (1976), 245
2. H. B. Nielsen, P. Olesen, Nucl. Phys.B61, 45,(1973); Nucl. Phys.B160, 380 (1979).
3. G.'t Hooft, Nucl.Phys.B138, 1 (1978).
4. P. Giovannangeli, C.P. Korthals Altes, Nucl.Phys.B608:203-234,2001; hep-ph/0102022.
5. P. Goddard, J. Nuyts, D. Olive, *Nucl.Phys* **B125**, 1 (1977). S.Coleman, Erice lectures 1981.
6. P. de Forcrand, C. P. Korthals Altes, O. Philipsen, in preparation
7. M. Teper, Phys.Rev.D59, 014512, 1999; hep-lat/9804008.
8. B. Lucini, M. Teper, JHEP 0106,050 (2001), M. Teper, hep-th/9812187 .
9. B. Lucini, M. Teper, U. Wenger, JHEP 0401:061,2004, hep-lat/0307017.
10. M. Teper, B. Lucini, Phys.Rev. D64 (2001) 105019.
11. C. Hoelbling, C. Rebbi, V.A. Rubakov, Phys. Rev.D63 :034506,2001; hep-lat/0003010.
12. D. Diakonov, V. Petrov, Phys. Lett.B242, 425 (1990); see also hep-lat/0008004.
13. H. Meyer, private communication and hep-lat/0312034.
14. T. Bhattacharya, A. Gocksch, C. P. Korthals Altes and R. D. Pisarski, Phys.Rev.Lett.66, 998 (1991); Nucl.Phys.B 383 (1992), 497.
15. C.P. Korthals Altes, A. Kovner and M. Stephanov, Phys.Lett.B469, 205 (1999), hep-ph/9909516.
16. Ph. de Forcrand, M. D'Elia, M. Pepe, Phys.Rev.Lett.86:1438,2001; hep-lat/0007034.

INFRARED QCD AND THE RENORMALISATION GROUP*

DANIEL F. LITIM

CERN, Theory Group, CH – 1211 Geneva 22, and
School of Physics, Univ. Southampton, Southampton SO17 1BJ, U.K.

JAN M. PAWLOWSKI, SERGEI NEDELKO, LORENZ V. SMEKAL

Inst. f. Theoretische Physik III, Universität Erlangen,
Staudtstr. 7, 91058 Erlangen, Germany

We study the infrared regime of QCD by means of a Wilsonian renormalisation group. We explain how, in general, the infrared structure of Green functions is deduced in this approach. Our reasoning is put to work in Landau gauge QCD, where the leading infrared terms of the propagators are computed. The results support the Kugo-Ojima scenario of confinement. Possible extensions are indicated.

1. Introduction

Many aspects of QCD are well understood in its ultraviolet limit where the gauge coupling is small as a consequence of asymptotic freedom. Therefore, perturbation theory is expected to provide a viable description of QCD phenomena at high energies or high temperature. On the other hand, in the infrared limit, quarks and gluons are confined to hadronic states and the gauge coupling is expected to grow large. In consequence, a reliable description of low energy phenomena -like confinement, the physics of bound states, chiral symmetry breaking or the confinement-deconfinement transition- call for a non-perturbative analysis.

Renormalisation group methods provide important analytical tools in the study of non-perturbative phenomena. In this contribution, we study Landau gauge QCD in the non-perturbative infrared regime [1] by means of an exact (or functional) renormalisation group [2,3]. The strength of the approach is its flexibility when it comes to approximations. Furthermore, efficient optimisation procedures are available [4], increasing the domain of

*Presented by DFL. This work is supported by EPSRC.

validity and the convergence of the flow. We discuss how, in general, the momentum structure of Green functions in the infrared is deduced from non-perturbative renormalisation group equations. As an application we compute the leading non-perturbative infrared coefficients (or anomalous dimensions) for the gluon and ghost propagator in Landau gauge. Our results are discussed in the light of the Kugo-Ojima confinement criterion and earlier findings based on other non-perturbative methods. For technical details we refer to the original publication [1].

2. Renormalisation group for QCD

The exact renormalisation group is based on the Wilsonian idea of integrating-out momentum degrees of freedom within a path integral representation of quantum field theory. Central to this approach is the effective action Γ_k, where quantum fluctuations with momenta $q^2 > k^2$ are already integrated out. The renormalisation group equation for Γ_k is given by

$$\partial_t \Gamma_k = \frac{1}{2} \,\mathrm{Tr}\, \frac{1}{\Gamma_k^{(2)} + R} \,\partial_t R\,. \tag{1}$$

Here, $t = \ln k$ is the logarithmic scale parameter, Tr denotes a trace over loop momenta q and a sum over indices and fields, and $R(q^2)$ is the infrared momentum cutoff at momentum scale k. R obeys a few restrictions which ensure that the flow is well-defined and finite both in the ultraviolet and the infrared. The flow interpolates between the initial (classical) action in the ultraviolet and the full quantum effective action in the infrared where the cutoff is removed. For gauge theories, the flow is ammended by a set of modified Slavnov-Taylor identities [5,6]. They ensure the requirements of gauge symmetry for Green functions in the physical limit $k \to 0$. Since this approach allows even for truncations which are non-local in momenta and the fields it is particularly useful for gauge theories. Furthermore, it is worth emphasising that the flow only involves fully dressed propagators and vertices. Therefore, the correct RG scaling properties are represented by the full flow and truncations with the correct symmetry properties.

So far the flow equation (1) has been applied to Landau gauge QCD for a determination of the heavy quark effective potential and effective quark interactions above the confinement scale [7]. For an implementation in general axial gauges and further applications in Yang-Mills theories see Refs. [8].

3. Infrared regime of QCD

In quantum field theory, important physical information for low energy phenomena is given through the momentum structure of Green functions in the deep infrared regime. In QCD, the characteristic scale differentiating between strong and weak coupling is $\Lambda_{\rm QCD} \approx 200$ MeV. The strongly coupled deep infrared regime is defined as

$$p^2 \ll \Lambda^2_{\rm QCD} \tag{2}$$

and p denotes a momentum argument of a QCD Green function. As a consequence of confinement, it is expected that gluon and ghost propagators display strong deviations from a simple particle pole for sufficiently small momenta. Within covariant linear gauges, the necessary conditions for confinement in terms of local fields were formulated by Kugo and Ojima [9]. In Landau gauge, they state that the gluonic correlations are suppressed in the infrared as a consequence of a mass gap, while the ghost correlations are infrared enhanced and dominant. This type of behaviour has already been detected in solutions of truncated Schwinger-Dyson equations [10] and stochastic quantisation [11], and lattice simulations [12].

Now we turn to the infrared analysis based on (1), where two-point functions and propagators depend additionally on the cutoff scale k. As long as $k^2 \gg p^2$, the propagators barely differ from the classical ones, given that no quantum fluctuations with momenta p or smaller have yet been integrated out. In turn, as soon as

$$k^2 \ll p^2 \tag{3}$$

all quantum fluctuations have been integrated out and physical quantities as well as general vertex functions upon appropriate rescaling, are no longer affected by the infrared cutoff. Consequently, the non-trivial infrared behaviour of physical Green functions resides in the momentum regime given by (2) and (3). It can be shown that Green functions in the regime (2) depend on the cutoff scale k only parametrically through dimensionless ratios, or through k-dependent renormalisation group factors that leave the full action invariant [1]. This property implies a fixed point behaviour and strongly facilitates the evaluation of (1) in given truncations.

In order to integrate the flow (1) in the infrared regime, we introduce an appropriate truncation of Γ_k and retain the full momentum dependence of the QCD propagators. The truncation is ammended by vertices which fulfil the truncated Slavnov-Taylor identities. The gluon and ghost two-point function are parametrised as

$$\Gamma^{(2)}_{k,A/C}(p^2) = z_{A/C} \cdot Z_{A/C}(x) \cdot p^2 \,. \tag{4}$$

Here, $x = p^2/k^2$, z denotes a possibly k-dependent RG factor, and $Z(x) = x^\kappa(1+\delta Z(x))$ parametrises the non-trivial modifications of the momentum structure through quantum fluctuations. The indices A/C refer to the gluon/ghost fields, respectively. In (4), we have suppressed the trivial colour structure and the transversal projector for the gluons. The longitudinal modes do not contribute to the flow in Landau gauge. Note that (4) is the most general parametrisation of the propagators and no assumptions have been made upon their structure. In the deep infrared region the behaviour of (4) is dominated by the term x^κ, where $\kappa_{A/C}$ is a non-perturbative anomalous dimension of the gluon and ghost propagator in Landau gauge. The function δZ constitutes the transition behaviour between the deep infrared regime and the cutoff regime. In addition, non-renormalisation of the ghost-gluon vertex implies [13]

$$\kappa_A = -2\kappa_C \,, \qquad \alpha_s = \frac{g^2}{4\pi}\frac{1}{z_A z_C^2} \,. \tag{5}$$

The coefficients $\kappa_{A/C}$ and the value for α_s in the deep infrared regime are deduced from integrating the flow for the propagators using (2) and (3). This leads to two integral equations for $\delta Z_{A/C}$ of the form

$$\delta Z_{A/C}(x) = F_{A/C}[\delta Z_{A/C}, \kappa_{A/C}, \alpha_s] \,. \tag{6}$$

Explicit expressions for the integrals F are given in [1]. The simultaneous solutions of (6) lead to explicit solutions for the infrared coefficients κ, α_s and δZ. To leading order, the back-coupling of δZ on the right hand side of (6) can be neglected. In this limit, we find

$$\kappa_C = 0.59535\cdots \qquad \alpha_s = 2.9717\cdots \tag{7}$$

independently of the cutoff function R. Furthermore, the numerical values agree with the most advanced results obtained within the Dyson-Schwinger approach and stochastic quantisation. Iterating (6) beyond leading order ($\delta Z \neq 0$), we find that κ_C varies slightly with the cutoff, ranging between 0.539 for the sharp cutoff and (7) for appropriately optimised ones. More recently, the infrared region has also been studied in [14].

4. Discussion

We have shown how, in general, the momentum dependence of Green functions is extracted in the infrared regime based on an exact renormalisation group. Applied to QCD in the Landau gauge, we detected a strong enhancement of the ghost propagator while the gluon propagator develops a mass gap. Our results provide a further independent evidence for the Kugo-Ojima scenario of confinement. In the simplest possible truncation with dressed propagators our results match the state-of-the-art within both the Dyson-Schwinger and the stochastic quantisation approach. This is quite remarkable, in particular in view of the conceptual and technical differences between these methods. Current work deals with straightforward extensions of the present analysis and covers vertex corrections, investigations of QCD Green functions at finite temperature, and dynamical quarks. In either case, the correct RG scaling as well as the inherent finiteness of the integrated flow are most important.

References

1. J. M. Pawlowski, D. F. Litim, S. Nedelko and L. von Smekal, Phys. Rev. Lett. **93** (2004) 152002.
2. F. Wegner, A. Houghton, Phys. Rev. **A 8** (1973) 401; K. G. Wilson and J. B. Kogut, Phys. Rept. **12** (1974) 75; J. Polchinski, Nucl. Phys. **B231** (1984) 269.
3. C. Wetterich, Phys. Lett. B **301** (1993) 90; M. Bonini, M. D'Attanasio and G. Marchesini, Nucl. Phys. B **409** (1993) 441; U. Ellwanger, Z. Phys. C **62** (1994) 503; T. R. Morris, Int. J. Mod. Phys. A **9** (1994) 2411.
4. D. F. Litim, Phys. Lett. B **486** (2000) 92; Phys. Rev. D **64** (2001) 105007; JHEP **0111** (2001) 059; Nucl. Phys. B **631** (2002) 128.
5. D. F. Litim and J. M. Pawlowski, hep-th/9901063.
6. F. Freire, D. F. Litim and J. M. Pawlowski, Phys. Lett. B **495** (2000) 256.
7. U. Ellwanger, M. Hirsch and A. Weber, Z. Phys. C **69** (1996) 687; Eur. Phys. J. C **1** (1998) 563; B. Bergerhoff and C. Wetterich, Phys. Rev. D **57** (1998) 1591.
8. D. F. Litim and J. M. Pawlowski, Phys. Lett. **B435** (1998) 181; JHEP **0209** (2002) 049; J. M. Pawlowski, Phys. Rev. D **58** (1998) 045011; Int. J. Mod. Phys. A **16** (2001) 2105; Acta Phys. Slov. **52** (2002) 475; M. Reuter and C. Wetterich, Phys. Rev. D **56** (1997) 7893; U. Ellwanger, Nucl. Phys. B **560** (1999) 587; F. Freire, Phys. Lett. B **526** (2002) 405; H. Gies, Phys. Rev. D **66** (2002) 025006; H. Gies and C. Wetterich, hep-th/0209183.
9. T. Kugo and I. Ojima, Prog. Theor. Phys. Suppl. **66** (1979) 1.
10. R. Alkofer and L. von Smekal, Phys. Rep. **353** (2001) 281.
11. D. Zwanziger, Phys. Rev. D **65** (2002) 094039.
12. A. Cucchieri, Phys. Lett. B **422** (1998) 233; F. D. R. Bonnet, P. O. Bowman, D. B. Leinweber and A. G. Williams, Phys. Rev. D **62** (2000) 051501; K. Langfeld, H. Reinhardt and J. Gattnar, Nucl. Phys. B **621** (2002) 131.
13. C. Lerche and L. von Smekal, Phys. Rev. D **65** (2002) 125006.
14. J. Kato, hep-th/0401068, C. S. Fischer and H. Gies, hep-ph/0408089.

RESIDUAL CONFINEMENT IN HIGH-TEMPERATURE YANG-MILLS THEORY*

A. MAAS, J. WAMBACH

Darmstadt University of Technology, Schloßgartenstraße 9, D-64289 Darmstadt, Germany

B. GRÜTER, R. ALKOFER

Tübingen University, Auf der Morgenstelle 14, D-72076 Tübingen, Germany

The infrared behavior of Landau gauge gluon and ghost propagators are investigated in Yang-Mills theory at non-vanishing temperatures. Self-consistent solutions are presented for temperatures below the presumed phase transition and in the infinite temperature limit. Gluon confinement is manifest in the infrared behavior of these propagators. As expected confinement prevails below the phase transition. In the infinite-temperature limit a qualitative change is observed: the chromoelectric sector exhibits a near-perturbative behavior while long-range chromomagnetic interactions, mediated by soft ghost modes, are still present. The latter behavior is in agreement with corresponding lattice results. It furthermore implies that part of the gluons are still confined.

1. Introduction

It is by now well established that QCD undergoes a phase transition or at least a rapid crossover at some critical temperature. The naive expectation is that the novel kind of matter formed above the phase transition is a plasma of deconfined gluons and quarks reaching ideal gas behavior at very high temperatures. Recent lattice calculations (see *e.g.* Cucchieri et al.[1]) have raised doubts about this simple picture. In addition, it is known that the dimensionally reduced theory, to which the equilibrium state evolves in the infinite temperature limit, is confining. The present work further investigates the properties of finite- and high-temperature QCD within the Dyson-Schwinger approach[2].

*This work is supported by the BMBF under grant number 06DA917, by the European Graduate School Basel-Tübingen (DFG contract GRK683), and by the Helmholtz association (Virtual Theory Institute VH-VI-041).

The following investigations are restricted to pure Yang-Mills theory as substantial evidence exists that the non-perturbative features of QCD are generated in the gauge sector. Hence, the theory studied here is an equilibrium Yang-Mills theory governed by the Euclidean Lagrangian

$$\begin{aligned} \mathcal{L} &= \tfrac{1}{4}F^a_{\mu\nu}F^a_{\mu\nu} + \bar{c}^a\partial_\mu D^{ab}_\mu c^b \\ F^a_{\mu\nu} &= \partial_\mu A^a_\nu - \partial_\nu A^a_\mu - gf^{abc}A^b_\mu A^c_\nu \\ D^{ab}_\mu &= \delta^{ab}\partial_\mu + gf^{abc}A^c_\mu \,, \end{aligned}$$

with the field strength tensors $F^a_{\mu\nu}$ and the covariant derivative D^{ab}_μ. A^a_μ denotes the gluon field and $\bar{c}^a$ and c^a are the ghost fields describing part of the quantum fluctuations of the gluon field. The gauge chosen is the Landau gauge which is best suited for the purpose at hand.

The gluon and ghost fields are described by their respective two-point Green's functions or propagators. Their infrared properties are linked to the presence of confinement. Especially, a particle is absent from the physical spectrum if its propagator $D(q^2)$ vanishes in the infrared[2]

$$\lim_{q\to 0} D(q^2) = 0. \tag{1}$$

2. Results

To present results we first define dressing functions from the propagators:

$$\begin{aligned} D_G(q) &= -\frac{G(q_0^2,q_3^2)}{q^2}\,, \\ D_{\mu\nu}(q) &= P^T_{\mu\nu}(q)\frac{Z(q_0^2,q_3^2)}{q^2} + P^L_{\mu\nu}(q)\frac{H(q_0^2,q_3^2)}{q^2}\,. \end{aligned}$$

Here G denotes the ghost dressing function, Z and H the ones for gluons being transverse or longitudinal with respect to the heat bath. To obtain these functions the truncated set of Dyson-Schwinger equations displayed in Fig. 1 have been solved[3,4,5]. The necessity to truncate the infinite set of coupled Dyson-Schwinger equations generates several problems with gauge invariance. These have been dealt with and the errors are, at least qualitatively, under control.

Calculations have been performed at $T = 0$[3], $T < T_c$[4] and $T \to \infty$[5]. Those at $T < T_c$ have used a toroidal discretized space while the others were done in the continuum. Note that, for $T \to \infty$, the theory becomes dimensionally reduced to a three-dimensional (3d) Yang-Mills theory plus an adjoint Higgs field, where Z corresponds to the 3d gluon (the chromomagnetic sector of the original 4d theory) and H to the Higgs (the chromoelectric sector of the original 4d theory).

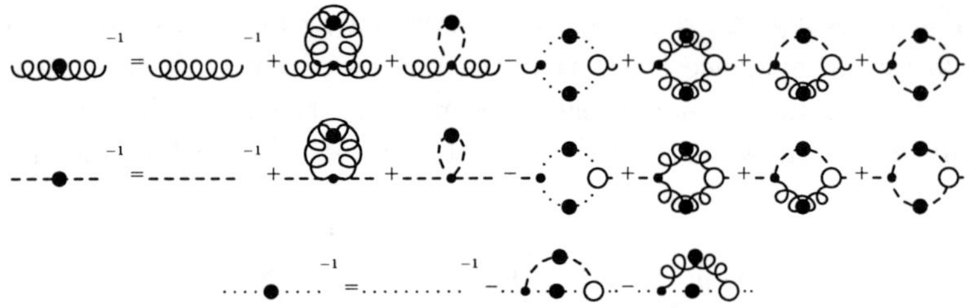

Figure 1. The truncated Dyson-Schwinger equations at finite temperature. The dotted lines denote ghosts, the dashed lines longitudinal gluons and the wiggly lines transverse gluons. Lines with a full dot represent self-consistent propagators and small dots indicate bare vertices. The open circled vertices are full and must be constructed in a given truncation scheme. A bare ghost-gluon vertex and slightly modified bare gluon vertices have been used. Note that for soft modes the ghost-longitudinal and the 3-point coupling of three longitudinal and of one longitudinal and two transverse gluons vanish.

Solutions for the ghost dressing function are shown in Fig 2. It is found that they are not affected by temperature qualitatively. The divergence at zero momentum indicates the mediation of long-range forces which are still present at $T \to \infty$. In addition, this divergence is connected to the confinement mechanism[2] and thus indicates the presence of some residual gluon confinement in the high-temperature phase.

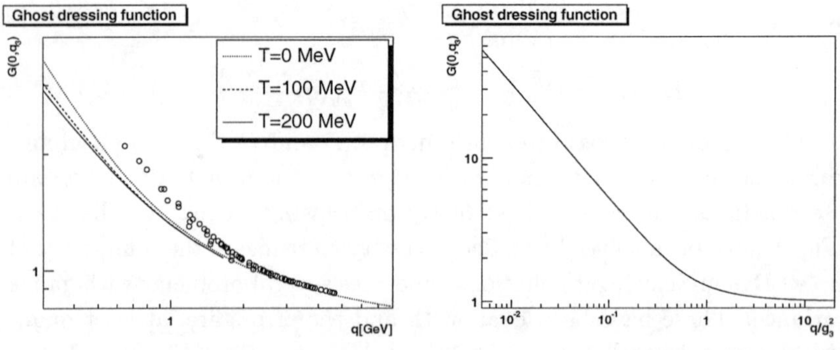

Figure 2. The ghost dressing function compared to lattice results[6]: The left panel is for $T < T_c$ while the right panel shows $T \to \infty$. g_3 is the dimensionful coupling constant in the dimensionally reduced theory.

The results for the transverse gluon are shown in Fig. 3. Below the phase transition, the propagator exhibits confinement by virtue of condition (1). It becomes steeper in the infrared with increasing temperature.

In the high-temperature phase there is no qualitative difference. Hence the transverse gluon propagator also shows explicitly the presence of residual confinement which can be interpreted as over-screening. In fact, the temperature dependence is very slight. The lattice data at $T = 0$ show strong finite-volume effects in the infrared and are expected to bend over for larger volumes as is the case in the dimensionally reduced theory, where significantly larger lattices can be employed.

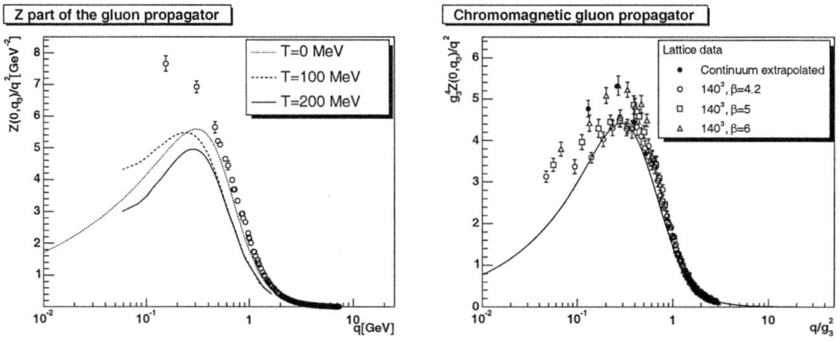

Figure 3. The transverse gluon propagator: The left panel displays $T < T_c$ results compared to $T = 0$ lattice data[7] while the right panel shows the $T \to \infty$ case compared to corresponding lattice data[1,8].

For the longitudinal gluon propagator the results are shown in Fig. 4. At $T = 0$, $H = Z$ and thus longitudianl gluons are confined as well. At non-zero temperature, H becomes shallower in the infrared, and is thus severely affected by finite-volume effects. It, however, still exhibits confinement according to volume-scaling studies[4]. In the high-temperature phase, the situation is much different. In contrast to the transverse sector, it shows screening and behaves similar to a massive particle. Nevertheless a comparison to lattice results and perturbation theory indicates that the Higgs propagator contains sizeable non-perturbative effects[5].

3. Conclusions

We have analyzed the infrared behavior of gluon and ghost propagators employing Dyson-Schwinger equations. At vanishing and small temperatures the results show manifestly gluon confinement. It is found that, even in the $T \to \infty$ limit, strong long-range correlations are present, leading to a non-perturbative behavior of the soft modes. Part of the gluons are

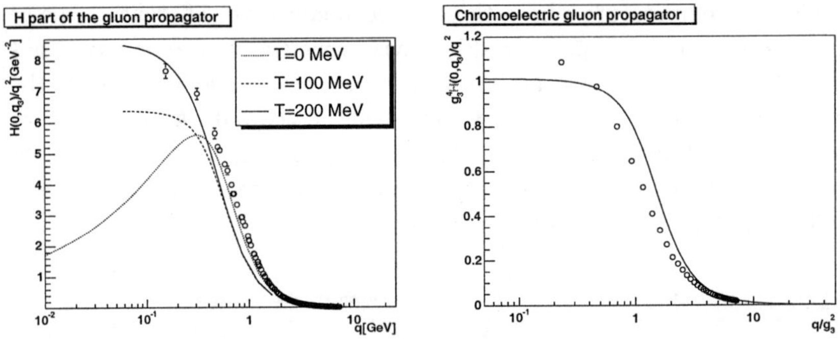

Figure 4. The longitudinal gluon propagator: The left panel is for $T < T_c$ compared to $T = 0$ lattice data[7] while in the right panel $T \to \infty$ results are compared to corresponding lattice data[1].

still confined: The Gribov-Zwanziger scenario (see *e.g.* Zwanziger[9]) applies at **all** temperatures. These results, together with the one of lattice calculations, demonstrate the presence of strong non-perturbative effects even at high temperatures. In the infrared sector of a Yang-Mills theory soft quantum fluctuations win over thermal fluctuations at all temperatures.

Acknowledgments

A.M. thanks the organizers of the Strong and Electroweak Matter 2004 for the very inspiring conference and the opportunity to present this work.

References

1. A. Cucchieri, F. Karsch and P. Petreczky, *Phys. Rev.* **D64**, 036001 (2001).
2. R. Alkofer and L. von Smekal, *Phys. Rept.* **353**, 281 (2001).
3. C. S. Fischer and R. Alkofer, *Phys. Lett.* **B536**, 177 (2002); *Phys. Rev.* **D67**, 094020 (2003).
4. B. Grüter, R. Alkofer, A. Maas and J. Wambach, arXiv:hep-ph/0408282
5. A. Maas, J. Wambach, B. Grüter and R. Alkofer, arXiv:hep-ph/0408074.
6. K. Langfeld, J. C. R. Bloch, J. Gattnar, H. Reinhardt, A. Cucchieri and T. Mendes, arXiv:hep-th/0209173.
7. P. O. Bowman, U. M. Heller, D. B. Leinweber, M. B. Parappilly and A. G. Williams, arXiv:hep-lat/0402032.
8. A. Cucchieri, T. Mendes and A. R. Taurines, *Phys. Rev.* **D67**, 091502 (2003).
9. D. Zwanziger, *Phys. Rev.* **D69**, 016002 (2004) and references therein.

SCALAR O(N) MODEL AT FINITE TEMPERATURE — 2PI EFFECTIVE POTENTIAL IN DIFFERENT APPROXIMATIONS

JÜRGEN BAACKE and STEFAN MICHALSKI

Universität Dortmund
Institut für Physik, Otto-Hahn-Str. 4
D-44221 Dortmund, Germany

We calculate the two-particle irreducible (2PI) effective potential of the $O(N)$ linear sigma model in 1+1 dimensions. The approximations we use are the next-to-leading order of a $1/N$ expansion (for arbitrary N) and a kind of "resummed loop approximation" for $N = 1$. We show that the effective potential of the $1/N$ expansion is convex for $N = 4$ and $N = 10$ whereas it is not for the "loop" expansion and the case $N = 1$ of the $1/N$ expansion.

1. Introduction

Inspired by earlier analyses out of equilibrium and in thermal equilibrium [1] we are interested in the phase structure of the $O(N)$ linear sigma model in different approximations. Since renormalization is quite a task in 3+1 dimensions we first carry out this analysis in 1+1 dimensions. For all details the reader is referred to a more comprehensive analysis published recently [2].

2. Basic Equations

2.1. *Classical action*

We take the following classical action for the $O(N)$ linear sigma model with spontaneous symmetry breaking in 1+1 dimensions

$$\mathcal{S}[\Phi] = \int \mathrm{d}^2x \, \frac{1}{2}(\partial_\mu \vec{\Phi})^2 - \frac{\lambda}{4N}\left(\vec{\Phi}^2 - Nv^2\right)^2 \quad \text{where} \quad \vec{\Phi} = (\Phi_1, \ldots, \Phi_N) \, . \tag{1}$$

In order to consider finite temperature we use the Matsubara formalism where the Minkowskian integral over momentum space is transformed to a sum and an integral over spatial momenta. We will use the short-hand notation $T\sum_{\omega_n}\int \frac{\mathrm{d}p}{2\pi} \equiv \sum\!\!\!\!\!\!\int_p$. For the effective action we consider a homogeneous background condensate $\vec{\phi} = \langle\vec{\Phi}\rangle = (\phi, \phi, \ldots, \phi)$ which can be $O(N)$-rotated such that it points only in the 1-direction $\vec{\phi} = \left(\sqrt{N}\phi, 0, \ldots, 0\right)$.

2.2. *2PI effective action*

The two-particle irreducible (2PI) effective action [4] reads

$$\Gamma[\phi, G] = \mathcal{S}\left[\sqrt{N}\phi\right] + \frac{1}{2}\mathrm{Tr}\ (i\mathcal{D}^{-1}G - 1) + \frac{i}{2}\ln\det G^{-1}D_0 + \Gamma_2[G, \phi]\ . \quad (2)$$

Here the two-point function G is a $N \times N$ matrix which is $O(N)$ symmetric, $G = \mathrm{diag}\,[G_\sigma, G_\pi, \ldots, G_\pi]$, and $\mathcal{D}$ denotes the analogous matrix of classical Green functions with $i\mathcal{D}^{-1}_{\sigma,\pi}(k) = k^2 - \lambda(f_{\sigma,\pi}\ \phi^2 - v^2)$ and $f_{\sigma,\pi} = 3, 1$. The explicit form of all higher-order corrections denoted by $\Gamma_2[G, \phi]$ is related to the type of approximation used.

2.2.1. *Next-to-leading order of* $1/N$ *expansion*

The $1/N$ expansion is a systematic expansion of the effective action in powers of $1/N$. The classical and one-loop part of the effective action is of leading order $\mathcal{O}(N)$ whereas all further contributions (except for the double-bubble graph) are of higher order. Here we will only take into account the contributions of next-to-leading order (NLO) of the 2PI-$1/N$ expansion[a] Therefore, higher loop contributions are separated into three parts

$$\Gamma_2[\phi, G] = \Gamma_2^{\mathrm{db}}[G] + \Gamma_2^{\mathrm{pearls}}[G] + \Gamma_2^{\mathrm{sunset}}[\phi, G]\ .$$

The first term (so called double-bubble) contributes both at leading and next-to-leading order of $1/N$ (cf. larger articles for details [2,3]) To sum up all contributions of next-to-leading order (NLO) we use the functional [3]

$$\Gamma_2^{\mathrm{pearls}}[G] = \frac{i}{2}\sum\!\!\!\!\!\!\int_p \left\{\ln\left[1 + i\frac{\lambda}{N}\mathrm{Tr}\,\mathcal{F}(p)\right] - i\frac{\lambda}{N}\mathrm{Tr}\,\mathcal{F}(p)\right\}\ , \quad (3)$$

where $\mathrm{Tr}\,\mathcal{F}$ is the trace of all fish graphs

$$\mathrm{Tr}\,\mathcal{F}(p) = \mathcal{F}_\sigma(p) + (N-1)\mathcal{F}_\pi(p) \quad \text{with} \quad \mathcal{F}_*(p) = \sum\!\!\!\!\!\!\int_p G_*(k)\,G_*(k+p)\ . \quad (4)$$

The subtracted graph in Eq. (3) is the NLO part of the double-bubble which is dealt with separately [2].

---

[a] Due to resummation contributions of all powers of $1/N$ contribute as well. Though the order is determined from the 2PI graphs of the action.

The generalization of sunset diagrams in the effective action is achieved by cutting a sigma line of $\Gamma_2^{\rm pearls}$ and pinning the two open legs to the background by multiplying by a factor of $N\phi^2$

$$\Gamma_2^{\rm sunset}[\phi, G] = i\lambda\, \phi^2 \sum\!\!\!\!\!\!\int_p G_\sigma(p) \frac{i\frac{\lambda}{N}{\rm Tr}\,\mathcal{F}(p)}{1 + i\frac{\lambda}{N}{\rm Tr}\,\mathcal{F}(p)}\ . \tag{5}$$

Expanding this in powers of $\mathcal{F}$ one finds graphs of the generalized sunset kind (see again our larger article [2] for figures and more details).

2.2.2. *"Loop" expansion for $N = 1$*

Since for $N = 1$ the $1/N$ expansion is somewhat meaningless, one could have the idea to improve a loop expansion by summing up all pearls and generalized sunset graphs as in Eqs. (3) and (5). We will call this "loop expansion" although this is not literally correct. We take into account the same graphs (topologically) as in the $1/N$ expansion but with a combinatorical factor that disregards their order of $1/N$. The respective expression for the resummation of pearls is (note the additional combinatorical factor)

$$\Gamma_2^{\rm pearls}[G] = \frac{i}{2} \sum\!\!\!\!\!\!\int_p \left\{ \ln\left[1 + 3i\lambda\mathcal{F}(p)\right] - i3\lambda\mathcal{F}(p) + 3\lambda^2 \left[\mathcal{F}(p)\right]^2 \right\}\ . \tag{6}$$

The subtracted graphs in Eq. (6) have to be dealt with separately due to different combinatorical factors. The last term in Eq. (6) is the basketball graph which has a different topology than the graphs with more than two vertices. The sum of generalized sunset graphs is obtained in an analogous way to the $1/N$ expression (5). It reads

$$\Gamma_2^{\rm sunset}[\phi, G] = i\lambda\, \phi^2 \sum\!\!\!\!\!\!\int_p G_\sigma(p) \frac{3i\lambda\mathcal{F}(p)}{1 + 3i\lambda\mathcal{F}(p)} + 6\lambda^2\mathcal{F}(p)G(p)\ . \tag{7}$$

2.3. *Dyson–Schwinger Equation*

In order to calculate the 1PI effective potential from the 2PI effective action we have to solve the Dyson–Schwinger equation $\delta\Gamma/\delta G = 0$ for the 2-point function. Using the convention $iG_*^{-1}(p) = i\mathcal{D}_*^{-1} - \Sigma_*(p)$, where $* = \sigma, \pi$, we can express the Dyson–Schwinger equation in terms of the self-energy $\Sigma(p)$

$$\Sigma_\sigma(p) = 3\frac{\lambda}{N}\, \mathcal{B}_\sigma + (N-1)\frac{\lambda}{N}\mathcal{B}_\pi - 2\,\frac{\delta}{\delta G_\sigma(p)}\left(\Gamma_2^{\rm pearls} + \Gamma_2^{\rm sunset}\right) \tag{8a}$$

$$\Sigma_\pi(p) = \frac{\lambda}{N}\, \mathcal{B}_\sigma + (N+1)\frac{\lambda}{N}\mathcal{B}_\pi - 2\,\frac{\delta}{\delta G_\pi(p)}\left(\Gamma_2^{\rm pearls} + \Gamma_2^{\rm sunset}\right)\ . \tag{8b}$$

The 1PI effective action is obtained by substituting a solution $G(\phi)$ of Eqs. (8) into $\Gamma[\phi, G]$. We will plot the 1PI effective potential that differs from that only by a total factor of volume times temperature.

3. Numerical Results

For a given temperature T and different values of ϕ we numerically solve the Dyson–Schwinger equation (8) by iteration.

3.1. *Next-to-leading order of 2PI* $1/N$ *expansion*

We take here a value of $\lambda = 0.5$ for the coupling constant and show results for $N = 1$, $N = 4$ and $N = 10$. Figure 1 indicates that there is only one

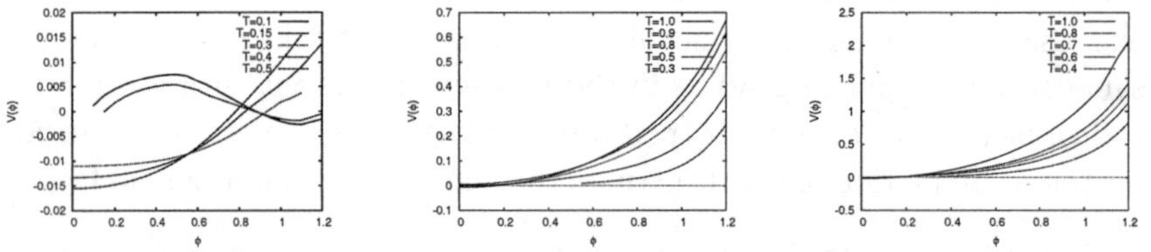

Figure 1. Shape of the effective potential in the 2PI $1/N$ approximation at NLO. From left to right: $N = 1$ and $\lambda = 0.1$ (not normalized), $N = 4$ and $\lambda = 0.5$, $N = 10$ and $\lambda = 0.5$.

vacuum at $\phi = 0$ for the considered temperatures and the cases $N = 4$ and $N = 10$. The potential for $N = 1$ shows signs of a false vacuum which is actually not expected in a $1/N$ expansion. Though for $N = 1$ this expansions is obviously not meaningful. For the sake of direct comparison we mention — without showing a plot — that for $N = 1$ and $\lambda = 0.5$ the $1/N$ potential is convex at temperatures $T = 0.7$ and $T = 0.8$, e.g., whereas it is not when using the "loop expansion" (see below and Fig. 2).

3.2. *"Loop expansion" for* $N = 1$

We display the effective potential in the resummed "loop expansion" (cf. Eqs. (6) and (7)) in Fig. 2 at different temperatures and for three values of the coupling constant. The potential exhibits a typical structure with a false vacuum — a clear sign of a first-order phase transition. For higher temperatures, e.g. $T = 1.2$ and $\lambda = 1$ or $T = 0.8$ and $\lambda = 0.5$, the false vacuum has disappeared but a "relic" consisting of two inflection points remains.

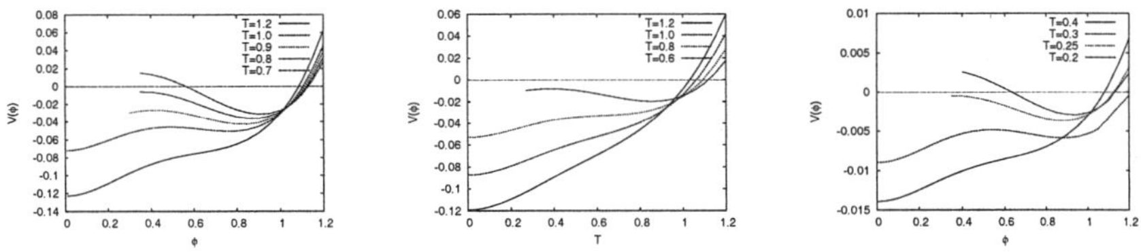

Figure 2. Effective potential of the "resummed loop expansion" at different temperatures for $\lambda = 1$, $\lambda = 0.5$ and $\lambda = 0.1$.

4. Conclusion and Outlook

We have solved the Dyson–Schwinger equation to compute the effective potential of the $O(N)$ linear sigma model in 1+1 dimensions both in a "resummed loop" expansion for $N = 1$ and at NLO of a $1/N$ expansion for arbitrary N. For $N = 4$ and $N = 10$ the effective potential is convex for all parameters we chose as expected from very old arguments [5].

For $N = 1$ we find (indications of) false vacua in both approximations. The $1/N$ expansion seems to be meaningless for $N = 1$ concerning the shape of the effective potential. For the "resummed loop expansion" one has to admit that this approximation only serves as an example of a non-systematic expansion and therefore the effective potential has a non-physical shape. As stated above, further results can be found in a more comprehensive publication[2].

Acknowledgements. S.M. thanks all the organizers of *SEWM 2004* for a wonderful meeting in Helsinki. S.M. was supported by *Deutsche Forschungsgemeinschaft* as a member of *Graduiertenkolleg 841*.

Bibliography

1. J. Baacke and S. Michalski, Phys. Rev. D **67**, 085006 (2003) [arXiv:hep-ph/0210060]; Phys. Rev. D **65**, 065019 (2002) [arXiv:hep-ph/0109137]; J. Baacke and A. Heinen, Phys. Rev. D **68**, 127702 (2003) [arXiv:hep-ph/0305220]. Phys. Rev. D **67**, 105020 (2003) [arXiv:hep-ph/0212312].
2. J. Baacke and S. Michalski, arXiv:hep-ph/0407152 (to be published in PRD).
3. G. Aarts, D. Ahrensmeier, R. Baier, J. Berges and J. Serreau, Phys. Rev. D **66**, 045008 (2002) [arXiv:hep-ph/0201308]; J. Berges, Nucl. Phys. A **699**, 847 (2002) [arXiv:hep-ph/0105311].
4. see e.g. J. M. Cornwall, R. Jackiw and E. Tomboulis, Phys. Rev. D **10**, 2428 (1974).
5. S. R. Coleman, *"There Are No Goldstone Bosons In Two-Dimensions"*, Commun. Math. Phys. **31**, 259 (1973).

CUTOFF EFFECTS IN MESON SPECTRAL FUNCTIONS

T. BLUM

Physics Department, University of Connecticut, Storrs, CT 06269-3046 and RIKEN-BNL, Brookhaven National Laboratory, Upton, NY 11973 USA

P. PETRECZKY

Physics Department and RIKEN-BNL, Brookhaven National Laboratory, Upton, NY 11973 USA

We study the lattice spacing dependence of meson spectral functions calculated in quenched QCD with domain wall fermions as well as clover Wilson fermions in quenched and partially-quenched QCD. We conclude that for lattice spacing $a \leq 3$ GeV all excited states appearing in the spectral functions are lattice artifacts.

1. Introduction

Temporal meson correlation functions in Euclidean time can be related to meson spectral functions by analytic continuation. Furthermore using the Maximum Entropy Method one can extract the spectral functions from meson correlators calculated on a finite lattice [1]. The method was applied at zero [1,2] as well as non-zero temperature [3,4,5,6]. In such an analysis one usually uses all time slices in the correlation function not just the long distance part. Therefore one is sensitive to short distance physics which may be affected by lattice artifacts. The lattice distortion of the correlators at short distances translates into lattice artifacts in the spectral functions for large energies. So far only Wilson gauge action and Wilson (clover) fermion action have been used to study meson spectral functions. The lattice artifacts seen in numerical simulations for this combination of actions turns out to be quite different from what is expected in the free theory [7]. The purpose of this paper is to study cutoff effects of meson spectral functions using different gauge and fermion actions.

2. Numerical results

We consider the temporal correlators of meson currents

$$G(\tau, T) = \sum_{\vec{x}} \langle J(\vec{x}, \tau) J(\vec{0}, 0) \rangle \ , \tag{1}$$

with $J = \bar{q}(\vec{x}, \tau)\Gamma q(\vec{x}, \tau)$ and $\Gamma = \gamma_5, \gamma_\mu$ for pseudo-scalar and vector channels. For temperature T the spectral function is related to the meson correlator by the integral relation

$$G(\tau, T) = \int_0^\infty d\omega \sigma(\omega, T) \frac{\cosh \omega(\tau - 1/(2T))}{\sinh \omega/(2T)} \ . \tag{2}$$

Although this relation is valid *a priori* only in the continuum, it has been

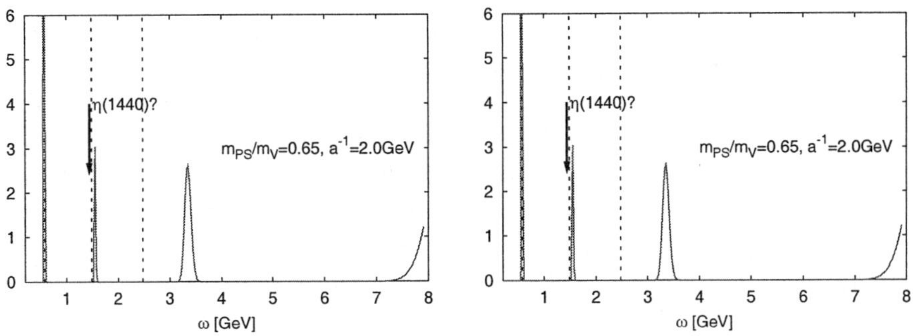

Figure 1. The pseudo-scalar (left) and vector (right) spectral functions with DWF at $a^{-1} = 2.0$ GeV. The vertical line and band show the mass of the ground and first excited states.

shown recently that in the limit of a non-interacting theory it holds on the lattice [7] as well.

We use quenched domain wall fermon correlators on DBW2 gauge configurations from the RBC collaboration. Furthermore, we calculate tree-level clover fermion correlators on quenched 1-loop tadpole-improved Symanzik lattices and on partially-quenched 2+1 flavor Asqtad fermion lattices from the MILC collaboration.

To extract the spectral function we use the Maximum Entropy Method. We also parametrize the spectral functions as a sum of delta functions $\sigma(\omega) = \sum_i F_i \delta(\omega^2 - m_i^2)$ and determine the parameters F_i and m_i using constrained curve fitting techniques[8] which allows for better systematic control of the results.

We start the discussion of our numerical results with the case of domain wall fermions. Here we performed calculations at three different lattice spacings, $a^{-1} = 1.3,\ 2.0,\ 3.0$ GeV and valance quark mass around the strange quark mass. For the coarsest lattice we see a two peak structure in the spectral functions with the second peak located around 2GeV. The positions of the peaks are reasonably reproduced using a two exponential fit. In Fig. 1 we show our results for finer lattices, $a^{-1} = 2.0$ GeV for the pseudo-scalar and vector spectral functions. We see a three peak structure, with the first excited state being close to the position of the physical excited states of the η and ϕ mesons. The third peak is likely to be a lattice artifact. We also performed a simple two exponential fit of the corresponding meson correlators. The results of the fit are shown as the vertical line and as a band. The ground state mass from the fit (statisitcal error $\sim 1\%$ not shown) is in reasonable agreement with the MEM analysis while for first excited state there is larger statistical uncertainty indicated by the band. To see whether or not the second and third peak correspond to physical resonances or lattice artifacts, we calculated the spectral functions at $a^{-1} = 3.0$ GeV. The results are shown in Fig. 2 which shows that the ground state peak did not shift (apart from a small shift due to slightly larger quark mass) while the second and the third peaks have moved roughly as a^{-1}. This indicates that these peaks are lattice artifacts. Similar results were found for the Wilson action [1,2]. In Ref. [2] they were interpreted as bound states of Wilson doublers. For DWF, however, such an interpretation is more difficuilt to apply. For this lattice spacing we have also performed a constrained fit with three exponentials. The ground state mass shown as the vertical line in Fig. 2 was obtained from a simple uncostrained fit which was then used as a prior for the constrained fit. The result for the second and third peak from the constrained fit is shown as bands in Fig. 2. Reasonable agreement between the MEM and constrained curve fits is found.

One may wonder whether the second and third peaks are artifacts of quenching. Absent dynamical quarks, all hadrons are absolutely stable (i.e. have zero width) no matter how large their mass is. If dynamical quarks are present, hadrons may decay, and the higher the mass the larger the width. Therefore higher lying resonances merge into a continuum. To check this we performed calculations in quenched and partially-quenched QCD using the clover action. More precisely we used gauge configurations with Symanzik action generated by the MILC collaboration as well the dynamical 2+1 Asqtad configurations. In both cases the lattice spacing was $a \approx 0.09 fm$. We use the tree-level value for the clover coefficient, $c_{sw} = 1$, and $\kappa = 0.14$

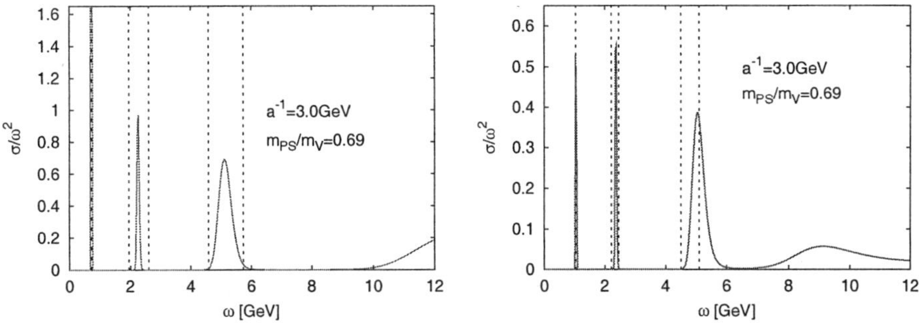

Figure 2. The pseudo-scalar (left) and vector (right) spectral functions with DWF at $a^{-1} = 3.0$ GeV. The vertical bands indicate the error on the peak location as determined from a constrained fit

which corresponds to $m_{PS}/m_V = 0.68$. Results for the spectral functions are shown in Fig. 3. As one can see, the spectral functions for quenched and partially quenched QCD are very similar, indicating that for quark masses around the strange quark mass effects due to quenching are quite small. In addition we have also calculated meson spectral functions wit HYP smeared clover action in $2+1$ partially-quenched QCD, i.e. we have applied HYP smearing on the dynamical configuations before calculating the meson correlators. We have performed calculations at several quark masses. From extrapolation of the pseudo-scalar mass squared to zero we determine the critical hopping parameter to be $\kappa_c = 0.12718(2)$ (statistical error only). The purpose of calculation with HYP smearing was to see whether the smoothing of the gauge fields removes the artificial peaks, and the corresponding results are shown in Fig. 3. As one can see from the figure, the lattice artifacts are reduced but not removed completely by HYP smearing. However, as a by-product we observe that number of exceptional configurations as well as the number of CG in the fermions matrix inversions are greatly reduced. In fact we did not observe any exceptional configurations in a sample of 59 configurations down to quark masses $\sim m_s/8$ (m_s is the strange quark mass).

3. Conclusions

In conclusions, we have studied cutoff effects in meson spectral functions at zero temperature using improved gauge actions and Domain Wall as well as clocer fermions. The spectral functions show cutoff effects which are similar to those obtained with standard gauge action and Wilson fermions. We have seen that for lattice spacings $a^{-1} < 3.0GeV$ it is not possible to

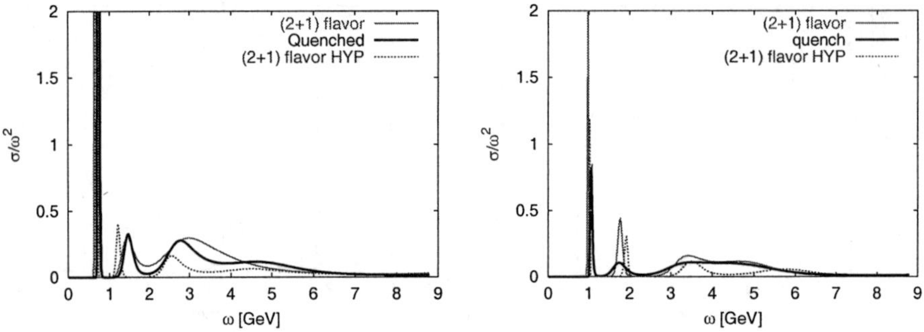

Figure 3. Pseudo-scalar (left) and vector (right) spectral functions for clover action in quenched and partially-quenched QCD.

reliable extract the properties of excited meson states.

Acknowledgment

This work has been authored under the contract DE-AC02-98CH10886 with the U.S. Department of energy. TB was partly supported by the LDRD program of Brookhaven National Laboratory, Project No. 04-041. We thank the RBC collaboration for the DWF correlators. The program for the MEM analysis was provided by Ines Wetzorke to whom we are very grateful. The calculations were performed on the IBM-SP supercomputer at NERSC using gauge configurations from the MILC collaboration and the MILC code for which we are grateful.

References

1. Y. Nakahara *et al.*, Phys. Rev. D **60** (1999) 091503; M. Asakawa *et al.*, Prog. Part. Nucl. Phys. **46** (2001) 459
2. Yamazaki, *et al.*., Phys. Rev. D65 (2002) 014501
3. F. Karsch, *et al.*, Phys. Lett. **B530** (2002) 147
4. T. Umeda *et al.*, hep-lat/0211003
5. M. Asakawa and T. Hatsuda, Phys.Rev.Lett. 92 (2004) 012001
6. S. Datta, *et al.*, Phys.Rev. D69 (2004) 094507
7. F. Karsch, *et al.*, Phys. Rev. D **68** (2003) 014504
8. G.P. Lepage, *et al.*., Nucl.Phys.Proc.Suppl. 106 (2002) 12-20

ANOMALOUS SPECIFIC HEAT IN ULTRADEGENERATE QED AND QCD

A. GERHOLD, A. IPP, A. REBHAN

Institut für Theoretische Physik, Technische Universität Wien, Wiedner Haupstr. 8-10, A-1040 Vienna, Austria

We discuss the origin of the anomalous $T \ln T^{-1}$ behavior of the low-temperature entropy and specific heat in ultradegenerate QED and QCD and report on a recent calculation which is complete to leading order in the coupling and which contains an infinite series of anomalous terms involving also fractional powers in T. This result involves dynamical hard-dense-loop resummation and interpolates between Debye screening effects at larger temperatures and non-Fermi-liquid behavior from only dynamically screened magnetic fields at low temperature.

1. Introduction

Cold dense quark matter has important deviations from Fermi-liquid behavior: below $T_c^{CSC} \sim 6$ to 60 MeV there is color superconductivity and long-range magnetic interactions are responsible for the fact that the energy gap is not proportional to $\exp(-c/g^2)$ as with short-range interactions, but instead to $\exp(-c'/g)$.[1,2] Above T_c^{CSC}, and for unpaired quarks also below, the only weakly (dynamically) screened magnetic interactions are also responsible for an anomalous behavior of entropy and specific heat, with a behavior $C_v \sim \alpha_s N_g N_f \mu^2 T \ln T^{-1}$ first discovered in the context of nonrelativistic QED by Holstein, Norton, and Pincus[3,4].

In QED this effect is probably unobservably small (though it may arise also from effective gauge field dynamics in systems of strongly correlated electrons), but in QCD it can be orders of magnitude larger because there are $N_g = 8$ gauge bosons instead of only one, and also α_s is much larger than α. However, more recently the existence of this effect had been questioned by Boyanovsky and de Vega[5], who instead found a $\alpha T^3 \ln T$ (when their renormalization-group summation of log's is undone).

In Ref.[6], we have recently confirmed the correctness of the original result and succeeded in calculating higher terms of the low-temperature expansion, for which only the coefficient of the leading log was known before.

The higher terms involve cubic roots of temperature, which can be traced to the fact that the frequency-dependent screening length of quasi-static magnetic modes is given by[7] $\kappa \simeq (\pi m_D^2 \omega/4)^{1/3}$, where m_D is the (electric) Debye mass.

2. Origin of the $T \ln T$ term

The non-Fermi-liquid behavior is usually discussed in terms of the spectral properties of the fermions. The leading fermionic contribution to the entropy density can be written as

$$\mathcal{S}_f \simeq -4NN_f \int \frac{d^4K}{(2\pi)^4} \frac{\partial n_f(\omega)}{\partial T} \left(\mathrm{Im} \ln S_+^{-1} + \mathrm{Im}\, \Sigma_+ \,\mathrm{Re}\, S_+ \right), \tag{1}$$

where it suffices to consider the particle (+) as opposed to antiparticle contribution. The $T \ln T$ behavior can then be obtained[3] from the singular behavior of the fermion self-energy at the Fermi surface[3,8,9],

$$\Sigma_+ \simeq \frac{g^2 C_f}{24\pi^2}(\omega - \mu) \ln \left(\frac{M^2}{(\omega-\mu)^2} \right) + i \frac{g^2 C_f}{12\pi} |\omega - \mu|. \tag{2}$$

However, it is not a priori justified to leave out the contributions from the gauge bosons. The entropy contributed by transverse modes can be written to two-loop accuracy as

$$\mathcal{S}_T \simeq -2N_g \int \frac{d^4K}{(2\pi)^4} \frac{\partial n_b(\omega)}{\partial T} \Big(\underbrace{\mathrm{Im} \ln D_T^{-1}}_{(A)} \underbrace{- \mathrm{Im}\, \Pi_T \,\mathrm{Re}\, D_T}_{(B)} \Big). \tag{3}$$

The fact that quasi-static transverse gauge bosons are only weakly screened by $\Pi_T \simeq -i\pi m_D^2 \omega/(4k)$ leads to nonanalytic behavior in T, which is exactly the same as that of the interaction part of $\mathcal{S}_f$:

$$\mathcal{S}_f^{\mathrm{int}} \simeq \mathcal{S}_{(A)} \simeq -\frac{N_g m_D^2 T}{36} \ln T^{-1} \tag{4}$$

It is only because $\mathcal{S}_{(B)} \simeq -\mathcal{S}_{(A)}$ that $\mathcal{S}_f$ already gives the complete result to leading order.

If one organizes the calculation differently, as done in Ref.[5] where the specific heat is extracted from the internal energy, one in fact finds that all $T \ln T$ terms as contributed by the fermions cancel out. However, as we have shown in Ref.[10], the complete result is then coming from the internal energy of the gauge boson sector, explicitly neglected in Ref.[5], which resolves the contradiction with the original results.

3. Complete leading-order results

It turns out that it is in fact advantageous to reorganize the calculation such that all anomalous contributions come from the gauge boson sector, by integrating out the fermionic degrees of freedom first. This allows one to systematically calculate beyond the leading-log approximation without having to calculate the fermionic spectral densities beyond leading order. In Ref.[10] we have most recently shown that the infinite series of anomalous contributions is contained in the following hard-dense-loop (HDL)[11] resummed expression valid for $T \ll \mu$:

$$\frac{1}{N_g}(\mathcal{S}-\mathcal{S}^0) = -\frac{g_{\rm eff}^2\mu^2 T}{24\pi^2} - \frac{1}{2\pi^3}\int_0^\infty dq_0\, \frac{\partial n_b(q_0)}{\partial T}\int_0^\infty dq\, q^2 \Bigg[2\,\mathrm{Im}\,\ln\left(\frac{q^2-q_0^2+\Pi_T^{\rm HDL}}{q^2-q_0^2}\right) + \mathrm{Im}\,\ln\left(\frac{q^2-q_0^2+\Pi_L^{\rm HDL}}{q^2-q_0^2}\right)\Bigg] + O(g_{\rm eff}^4\mu^2 T). \quad (5)$$

Here S^0 is the ideal-gas value of the entropy density, and $g_{\rm eff}^2 = g^2 N_f/2$ in QCD, but $e^2 N_f$ in QED.

At low temperature $T \ll g_{\rm eff}\mu$, one finds that the Landau damping cuts of the HDL propagators give rise to a series of the form

$$\frac{\mathcal{S}-\mathcal{S}_0}{N_g} = \frac{g_{\rm eff}^2\mu^2 T}{36\pi^2}\left(\ln\frac{4g_{\rm eff}\mu}{\pi^2 T} - 2 + \gamma_E - \frac{6}{\pi^2}\zeta'(2)\right) + c_1 T^{5/3} + c_2 T^{7/3} + c_3 T^3(\ln(g_{\rm eff}\mu/T) + c_4) + O(T^{11/3}), \quad (6)$$

where the coefficients up to and including c_4 can be found in Ref.[10].

At $g_{\rm eff}\mu \lesssim T \ll \mu$, one has also important quasiparticle contributions, most of which are invisible in the low-T expansion since they involve terms suppressed by factors of $e^{-m_D/\sqrt{3}T}$. For $g_{\rm eff}\mu \ll T \ll \mu$, one finally makes contact with more familiar results from thermal perturbation theory, as $\mathcal{S}-\mathcal{S}_0 \to N_g(-g_{\rm eff}^2\mu^2T^2/(12\pi^2) + g_{\rm eff}^3\mu^3/(12\pi^4) + \ldots)$, where it is the longitudinal plasmon term $\propto m_D^3$ which makes (HDL) resummation necessary. The expression (5) thus interpolates between two physically rather different collective phenomena: the plasmon effect from (electric) Debye screening, which only arises at $T \gg g_{\rm eff}\mu$, and non-Fermi-liquid effects from the only dynamically screened magnetic interactions at $T \ll g_{\rm eff}\mu$, see Fig. 1.

4. Results for large and finite N_f

The above result (5) is the leading-order result for both QCD and QED at $T \ll \mu$. Higher-order terms either involve extra powers of g^2 or T^2/μ^2 (but not at the same time higher powers of $\ln T$)[4,12].

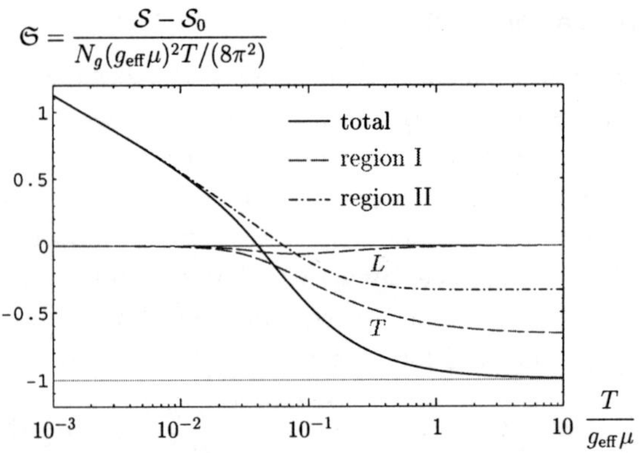

Figure 1. The function $\mathfrak{S}(T/(g_{\rm eff}\mu))$ which determines the leading-order interaction contribution to the low-temperature entropy. The normalization is such that $\mathfrak{S} = -1$ corresponds to the result of ordinary perturbation theory. The dash-dotted line shows the contribution from spacelike momenta (region II), comprising HDL Landau damping and hard contributions; the two dashed lines give the transverse (T) and longitudinal (L) quasiparticle pole contributions (region I).

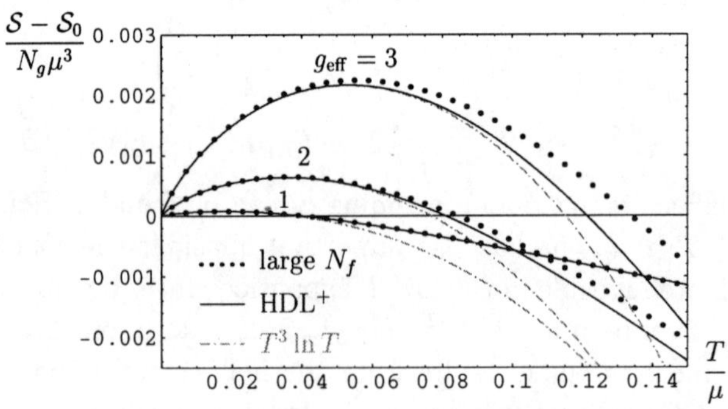

Figure 2. Complete entropy density in the large-N_f limit for the three values $g_{\rm eff}(\bar{\mu}_{\rm MS} = 2\mu) = 1, 2, 3$ (heavy dots), compared with the full HDL result (solid line). Also given is the low-temperature series up to and including the $T^3 \ln T$ contributions.

One case where we can actually investigate quantitatively the importance of higher-order terms is in the exactly solvable limit of large flavor number, which has been worked out for finite chemical potential in Ref. [13]. Figure 2 compares with the exact large-N_f result for $g_{\rm eff}(\bar{\mu}_{\rm MS} = 2\mu) = 1, 2, 3$, and we find that the HDL resummed result works well in the range where the entropy exceeds its ideal-gas value. The exact large-N_f result turns out to have even a slightly larger anomalous entropy.

In Fig. 3 we finally give results for the specific heat at finite N_f, which shows its anomalous behavior for a potentially interesting range in T/μ.

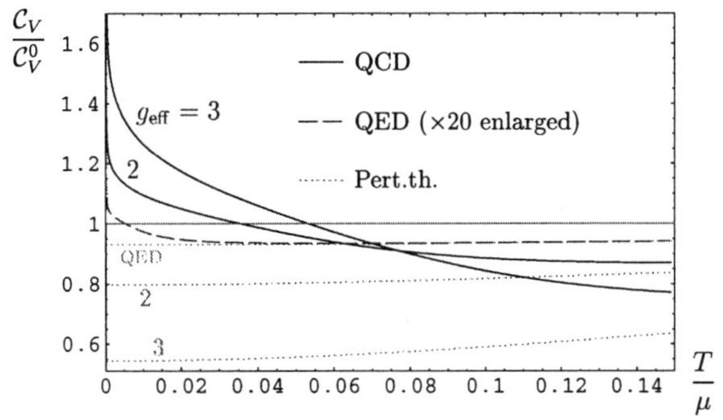

Figure 3. The HDL-resummed result for the specific heat $\mathcal{C}_v$, normalized to the ideal-gas value for $g_{\text{eff}} = 2$ and 3 corresponding to $\alpha_s \approx 0.32$ and 0.72 in two-flavor QCD, and $g_{\text{eff}} \approx 0.303$ for QED. The deviation of the QED result from the ideal-gas value is enlarged by a factor of 20 to make it more visible.

Acknowledgments

This work has been supported by the Austrian Science Foundation FWF, project no. 16387-N08 and by the ÖAD, project no. Amadée-16/2003.

References

1. D. T. Son, *Phys. Rev.* **D59**, 094019 (1999).
2. D. H. Rischke, *Prog. Part. Nucl. Phys.* **52**, 197 (2004).
3. T. Holstein, R. E. Norton and P. Pincus, *Phys. Rev.* **B8**, 2649 (1973).
4. S. Chakravarty, R. E. Norton and O. F. Syljuåsen, *Phys. Rev. Lett.* **74**, 1423 (1995).
5. D. Boyanovsky and H. J. de Vega, *Phys. Rev.* **D63**, 114028 (2001).
6. A. Ipp, A. Gerhold and A. Rebhan, *Phys. Rev.* **D69**, 011901 (2004).
7. H. A. Weldon, *Phys. Rev.* **D26**, 1394 (1982).
8. W. E. Brown, J. T. Liu and H.-c. Ren, *Phys. Rev.* **D61**, 114012 (2000).
9. C. Manuel, *Phys. Rev.* **D62**, 076009 (2000).
10. A. Gerhold, A. Ipp and A. Rebhan, hep-ph/0406087.
11. E. Braaten and R. D. Pisarski, *Nucl. Phys.* **B337**, 569 (1990); T. Altherr and U. Kraemmer, *Astropart. Phys.* **1**, 133 (1992); H. Vija and M. H. Thoma, *Phys. Lett.* **B342**, 212 (1995); C. Manuel, *Phys. Rev.* **D53**, 5866 (1996).
12. T. Schäfer and K. Schwenzer, hep-ph/0405053.
13. A. Ipp and A. Rebhan, *JHEP* **0306**, 032 (2003).

COLOR-SUPERCONDUCTING PHASES IN COLD AND DENSE QUARK MATTER

ANDREAS SCHMITT

Institut für Theoretische Physik, J.W. Goethe-Universität
D-60054 Frankfurt/Main, Germany
E-mail: aschmitt@th.physik.uni-frankfurt.de

A discussion of several phases of color-superconducting quark matter is presented. After briefly discussing color superconductors where quarks of different flavors form Cooper pairs, we focus on color superconductors with Cooper pairs consisting of quarks of the same flavor. These Cooper pairs, contrary to the former ones, carry total spin one. Possible spin-one phases are classified, and the magnitude as well as the angular structure of their energy gap is discussed. It is argued that spin-one phases, contrary to spin zero-phases, exhibit an electromagnetic Meissner effect.

It is well-known that certain metals and alloys exhibit a superconducting phase below a critical temperature T_c. In this phase, electrons in the vicinity of the Fermi surface form Cooper pairs which leads to a gap ϕ in the quasiparticle excitation spectrum [1]. An attractive interaction between the electrons is provided by the exchange of virtual phonons, and the electromagnetic gauge group $U(1)_{em}$ is spontaneously broken. A similar mechanism occurs in sufficiently cold and dense quark matter [2]. In this case, the attractive color-antitriplet channel is responsible for the formation of quark Cooper pairs. Due to the intrinsic properties of quarks (color, flavor, electric charge), many pairing patterns seem to be theoretically possible. In other words, besides the electromagnetic gauge group, also the color gauge group $SU(3)_c$, the flavor group $SU(N_f)_f$, and the baryon number conservation group $U(1)_B$ may be broken completely or to a certain residual subgroup. In recent years there have been interesting works studying the ground state of cold and dense quark matter, i.e., it has been investigated which color-superconducting phase is favored for certain ranges of the quark chemical potential μ. This question is also of phenomenological interest, since matter in the interior of neutron stars can reach densities up to an order of magnitude larger than the nuclear ground state density while the temperature can be of the order of keV. Therefore, the core of a neutron star can be expected to be a color superconductor.

For very large densities, where the quark masses of the u, d, and s quarks are degenerate and much smaller than the chemical potential, $m_u \simeq m_d \simeq m_s \ll \mu$, the ground state is the so-called color-flavor-locked (CFL) phase [3]. In this phase, quarks of all flavors and all colors form Cooper pairs, breaking spontaneously $SU(3)_c \times SU(3)_f \times U(1)_{em} \times U(1)_B$ to a subgroup $SU(3)_{c+f} \times U(1)_{c+em}$. Going to smaller densities renders the situation more complicated because a larger strange mass, $m_u \simeq m_d \ll m_s$, as well as the conditions of β-equilibrium and electric charge neutrality impose restrictive conditions on the system. The simplest solution seems to be a pairing of only u and d quarks [2], the so-called 2SC phase. In this phase, quarks of one color remain unpaired, and the symmetry breaking pattern here is $SU(3)_c \times SU(2)_f \times U(1)_{em} \times U(1)_B \rightarrow SU(2)_c \times SU(2)_f \times U(1)_{c+em} \times U(1)_{em+B}$. In both CFL and 2SC phases the Fermi momenta of the quark flavors participating in pairing are assumed to be equal. This is a necessary condition for the conventional BCS pairing mechanism, since in both cases quarks of different flavor form Cooper pairs. However, for moderate densities, this assumption is not valid. Therefore, the ground state is neither the (pure) CFL nor the (pure) 2SC state. Several other possibilities have been discussed. In principle, there are two classes of alternatives. In the first class, the difference in Fermi momenta is nonzero but small enough to still allow for pairing of quarks of different flavor. The second class accounts for cases in which the difference in Fermi momenta is so large, that, besides the normal-conducting state, pairing of quarks of the same flavor is the only possibility.

Let us first mention some options for the first class. First, there might be a phase in which the Cooper pairs carry nonzero total momentum. In this case, the system exhibits a crystalline structure due to a spatially varying energy gap [4]. This kind of superconductivity is called LOFF phase and was first discussed in solid-state physics [5]. Second, there have been studies about the "gapless" 2SC [6] and CFL [7] phases. In these phases, at least a part of the quasiparticle excitations are ungapped, although the gap parameter is nonzero. This feature obviously can have enormous physical consequences, for instance for the specific heat of the system or the cooling of a neutron star with a core in a gapless color-superconducting phase. Other possible phases are derived from the CFL phase and contain kaon and/or eta condensates [8]. Moreover, besides a deplacement (LOFF phase), also a deformation of the Fermi surface has been discussed [9].

Let us now discuss pairing of quarks of the same flavor [2,10]. The simplest situation is a system of only one quark flavor. More realistic scenarios are two- or three-flavor systems where each quark flavor separately forms Cooper pairs. Another possibility is a system where u and d quarks form a (gapless) 2SC superconductor while the s quarks pair separately. In either case, since the Cooper pairs carry spin one, $J = 1$, rotational symmetry is broken. In general, the group $SU(3)_c \times SU(2)_J \times U(1)$ is broken spontaneously by an order parameter which has a 3×3 structure, arising from the color and spin (anti)triplets. A systematic group theoretical classification, analogous to that of the similar situation in superfluid ^{3}He, yields a list of possible symmetry breaking patterns, each corresponding to a special order parameter [12], cf. Fig. 1. From Fig. 1 one can for instance read off that the color-spin-locked (CSL) phase exhibits an electromagnetic Meissner effect due to pure symmetry reasons. An explicit calculation of the (rotated) photon Meissner mass shows that in the case of a many-flavor system also the polar phase (and most likely all other spin-one phases) expel external magnetic fields [13].

For each of the phases, the corresponding energy gap ϕ can be computed. This has been done with the help of the QCD gap equation, which has been formalized in a universal way in order to easily insert the different order parameters [14]. In Fig. 2 we show the angular structure of the gaps for the polar, planar, A, and CSL phases. For each order parameter we distinguish the cases where quarks of the same ("longitudinal") and opposite ("transverse") chirality form Cooper pairs and where a combination of these two pairing patterns is realized ("mixed"). In some of the phases (e.g., the transverse polar phase) the gap has nodal points at the north and south poles of the Fermi sphere, while in the longitudinal polar phase, the gap has a nodal line at the equator of the Fermi surface. Two of the phases, the mixed CSL and transverse A phases, have two different gaps. In these cases, one computes a deviation from the BCS relation $T_c = 0.57\phi$ [14].

The magnitude of these gaps is in the range of tens to hundreds of keV, i.e., they are roughly three orders of magnitude smaller than those in the 2SC and CFL phases. This suppression is produced by a factor originating from the angular structure, which is e^{-6} for all longitudinal and $e^{-4.5}$ for all transverse gaps. In the case of the mixed gaps, this factor depends on the order parameter but is always larger than e^{-6} and smaller than $e^{-4.5}$ [12]. This indicates that the transverse phases are the energetically preferred ones.

phase	$SU(3)_c$ x $SU(2)_J$ x $U(1)$ broken to	order parameter
polar	SU(2) x U(1) x U(1)	$\begin{pmatrix} 0 & 0 & 0 \\ 0 & 0 & 0 \\ 0 & 0 & 1 \end{pmatrix}$
planar	U(1) x U(1)	$\begin{pmatrix} 1 & 0 & 0 \\ 0 & 1 & 0 \\ 0 & 0 & 0 \end{pmatrix}$
A	SU(2) x U(1) x U(1)	$\begin{pmatrix} 0 & 0 & 0 \\ 0 & 0 & 0 \\ 1 & i & 0 \end{pmatrix}$
CSL	SU(2)	$\begin{pmatrix} 1 & 0 & 0 \\ 0 & 1 & 0 \\ 0 & 0 & 1 \end{pmatrix}$
	SU(2) x U(1)	$\begin{pmatrix} 0 & 0 & 0 \\ 0 & 0 & 0 \\ \Delta_1 & \Delta_2 & \Delta_3 \end{pmatrix}$
	U(1) x U(1)	$\begin{pmatrix} 0 & 0 & \Delta_1 \\ 0 & 0 & \Delta_2 \\ 0 & 0 & 0 \end{pmatrix}$
	U(1) x U(1)	$\begin{pmatrix} \Delta_1 & i\Delta_1 & 0 \\ \Delta_2 & i\Delta_2 & 0 \\ 0 & 0 & 0 \end{pmatrix}$
	U(1)	$\begin{pmatrix} \Delta_1 & \Delta_2 & \Delta_3 \\ \Delta_4 & \Delta_5 & \Delta_6 \\ 0 & 0 & 0 \end{pmatrix}$
	U(1)	$\begin{pmatrix} \Delta_1 & i\Delta_1 & 0 \\ \Delta_2 & i\Delta_2 & 0 \\ \Delta_3 & i\Delta_3 & 0 \end{pmatrix}$
	U(1)	$\begin{pmatrix} 0 & 0 & \Delta_1 \\ 0 & 0 & \Delta_2 \\ 0 & 0 & \Delta_3 \end{pmatrix}$
	U(1)	$\begin{pmatrix} \Delta_1 & i\Delta_1 & 0 \\ \Delta_2 & i\Delta_2 & 0 \\ 0 & 0 & \Delta_3 \end{pmatrix}, \begin{pmatrix} 0 & 0 & \Delta_1 \\ 0 & 0 & \Delta_2 \\ \Delta_3 & i\Delta_3 & 0 \end{pmatrix}$

Figure 1. Symmetry breaking patterns and order parameters for a spin-one color superconductor. The original symmetry group (first line) is given by the color gauge group $SU(3)_c$ (blank background), the spin group $SU(2)_J$ (grey background), and a $U(1)$ (hatched background). The latter can be regarded as the electromagnetic gauge group or the baryon number conservation group. The backgrounds of the residual groups illustrate the symmetry breaking pattern. For instance, the blank $SU(2)$ occurring in the residual group of the polar phase is generated solely by generators of the original $SU(3)_c$ while the blank/hatched $U(1)$ in the same line is generated by a linear combination of the generators of the original $SU(2)_J$ and $U(1)$ groups.

Acknowledgments

I would like to thank Q. Wang and D.H. Rischke who substantially contributed to this work. Moreover, I thank the organizers of SEWM 2004 for

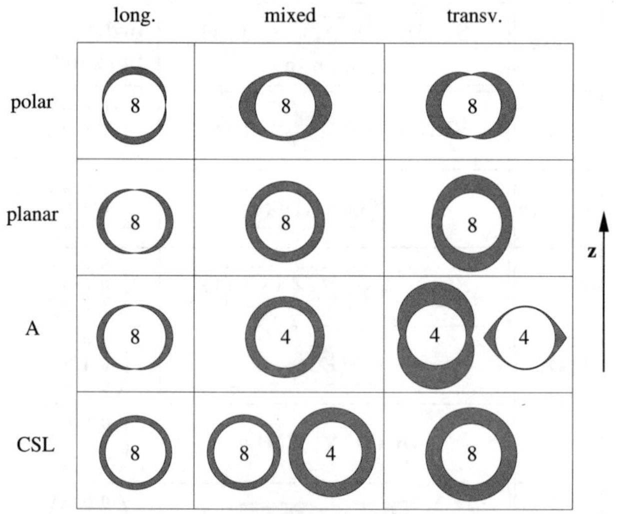

Figure 2. Angular structure of the gaps for several spin-one phases. In each case, the gap is symmetric with respect to rotations around the z-axis. The numbers indicate the degeneracy of the respective quasiparticle excitation branch. Only the gapped branches (out of 12 branches) are shown.

the chance to participate in a perfectly organized conference.

References

1. J. Bardeen, L.N. Cooper, and J.R. Schrieffer, Phys. Rev. **108**, 1175 (1957).
2. D. Bailin and A. Love, Phys. Rep. **107**, 325 (1984).
3. M. Alford, K. Rajagopal, and F. Wilczek, Nucl. Phys. **B537**, 443 (1999).
4. M.G. Alford, J.A. Bowers, and K. Rajagopal, Phys. Rev. D **63**, 074016 (2001).
5. A.I. Larkin and Y.N. Ovchinnikov, Zh. Eksp. Teor. Fiz. **47**, 1136 (1964); P. Fulde and R.A. Ferrell, Phys. Rev. **135**, A550 (1964).
6. I. Shovkovy and M. Huang, Phys. Lett. B **564**, 205 (2003); Nucl. Phys. **A729**, 835 (2003).
7. M. Alford, C. Kouvaris, and K. Rajagopal, Phys. Rev. Lett. **92**, 222001 (2004).
8. P.F. Bedaque and T. Schäfer, Nucl. Phys. **A697**, 802 (2002); A. Kryjevski and D.B. Kaplan, hep-ph/0404290.
9. H. Müther and A. Sedrakian, Phys. Rev. D **67**, 085024 (2003).
10. T. Schäfer, Phys. Rev. D **62**, 094007 (2000).
11. D. Vollhardt and P. Wölfle, *The Superfluid Phases of Helium 3* (Taylor & Francis, London, 1990).
12. A. Schmitt, *Ph.D. thesis*, nucl-th/0405076.
13. A. Schmitt, Q. Wang, and D.H. Rischke, Phys. Rev. Lett. **91**, 242301 (2003); Phys. Rev. D **69**, 094017 (2004).
14. A. Schmitt, Q. Wang, and D.H. Rischke, Phys. Rev. D **66**, 114010 (2002).

NON FERMI LIQUID EFFECTS IN DENSE MATTER AND COMPACT STAR COOLING

KAI SCHWENZER[1] AND THOMAS SCHÄFER[1,2]

[1]*Department of Physics, North Carolina State University, Raleigh, NC 27695*
[2]*Riken-BNL Research Center, Brookhaven National Laboratory, Upton, NY 11973*

We study non Fermi liquid effects due to the exchange of unscreened magnetic gluons in the normal phase of high density QCD. We show that the fermionic self energy is one-loop exact in the weak coupling limit. The corresponding dispersion relation is employed to study the neutrino emission of quark matter and the resulting cooling behavior.

1. Introduction

Quark matter at high density is known to exhibit non Fermi liquid behavior due to the appearance of a color superconducting phase with gapped fermionic excitations[1]. The consideration of physical requirements as the finite quark masses and charge neutrality, leads to different phases which include ungapped excitations resembling the quasiparticles of a Fermi liquid. In the presence of unscreened color magnetic interactions, however, the Fermi liquid description ultimately breaks down at very small scales. Contrary to a previous analysis we find that the logarithmic non Fermi liquid enhancement at one-loop level is not changed by higher order fluctuations and thereby perturbation theory remains valid.

Compact stars represent a unique opportunity to observe ultradense matter in nature and their cooling behavior provides one of the main sources of information on their decomposition[2]. In particular, the neutrino emission from ungapped quarks via the direct URCA process would constitute by far the most effective cooling mechanism. Due to the low temperatures of compact stars, however, non Fermi liquid effects, which determine the infrared behavior of the fermionic excitations, have to be considered in the cooling analysis. We will exemplarily include these effects for the idealized case of fermionic excitations in the normal phase of dense matter.

2. Non Fermi liquid dispersion relation

The influence of the medium on the propagation of fermionic excitations is described by the fermion self energy. We compute this quantity in the high density effective theory[3] for QCD

$$L = \sum_{v} \psi_v^\dagger (iv \cdot D)\psi_v - \frac{1}{4} G^a_{\mu\nu} G^a_{\mu\nu} + \ldots , \tag{1}$$

which includes only the relevant particle and hole excitations in the vicinity of the Fermi surface. The leading logarithmic term in the one-loop self energy depends only on the energy $\omega = p_0 - \mu$ and is given by[4]

$$\Sigma(\omega) = \frac{g^2}{9\pi^2} \omega \log\left(\frac{\Lambda}{\omega}\right), \tag{2}$$

where Λ is a cutoff scale. This result implies that for $\omega \sim \Lambda \exp(-9\pi^2/g^2)$ the perturbative correction is of order 1, and higher order terms of the form $g^{2n} \log^n(\Lambda/\omega)$ may have to be included. In order to study this problem we consider the Dyson-Schwinger equation for the fermion self energy

$$-i\Sigma(p) = -g^2 \int \frac{d^4k}{(2\pi)^4} v_i \frac{\lambda^a}{2} S(p+k) v_j \frac{\lambda^b}{2} D^{ab}_{ij}(k), \tag{3}$$

where D^{ab}_{ij} is the HDL resummed transverse gluon propagator which dominates in the infrared regime. We have neglected vertex corrections since they are not logarithmically enhanced in the considered kinematic regime[5]. We also note that the Dyson-Schwinger equation for the fermion propagator does not contain hard loops. This implies that we do not have to solve coupled Dyson-Schwinger equations for the quark and gluon propagators. Since the infrared divergence in the self energy depends only on the energy[6] we can neglect the subleading momentum dependence which leads to

$$-i\Sigma(p_0) = \frac{4}{3} g^2 \int \frac{d^4k}{(2\pi)^4} \frac{1-(\vec{v}\cdot\hat{k})^2}{p_0+k_0-\vec{v}\cdot\vec{k}+\Sigma(p_0+k_0)} \frac{1}{k_0^2-\vec{k}^2+i\frac{\pi}{4}m^2\frac{|k_0|}{|\vec{k}|}} \tag{4}$$

with the dynamical gluon mass $m^2 = (N_f g^2 \mu^2)/(4\pi^2)$. After analytic continuation to euclidean space and angular integrations we find

$$\Sigma(p_4) = \frac{g^2}{3\pi^3} \int dk_4 \int k dk \left(\frac{(p_4+k_4+\Sigma(p_4+k_4))^2+k^2}{k^2} \right. \tag{5}$$

$$\left. \cdot \arctan\left(\frac{k}{p_4+k_4+\Sigma(p_4+k_4)}\right) - \frac{p_4+k_4+\Sigma(p_4+k_4)}{k} \right) \frac{1}{k_4^2+k^2+\frac{\pi}{4}m^2\frac{|k_4|}{k}},$$

In order to derive an approximate analytic expression we note that only the integration over the cut of the arc tangent yields a logarithmic singularity[7].

Thereby, we find in first approximation exactly the 1-loop result eq. (2). The full numeric solution of the Dyson-Schwinger equation (5) shown in fig. 1 supports this infrared result and contradicts a previous analysis that found a scaling behavior with an anomalous dimension[8]. A consistent RG analysis which respects the correct scaling behavior in the vicinity of the Fermi surface clarifies this discrepancy and confirms the one-loop exactness[7].

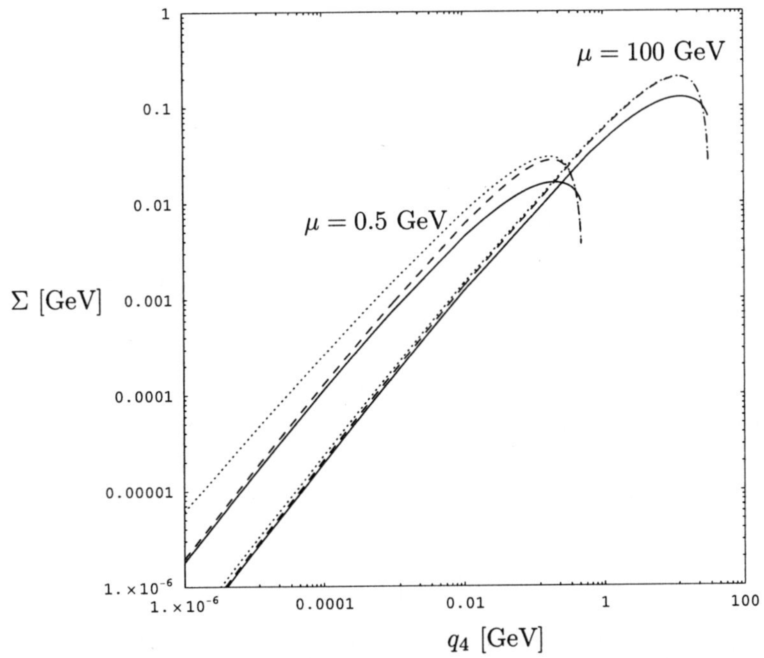

Figure 1. Numerical solution of the Dyson-Schwinger equation for the fermion self energy. We show the euclidean self energy $\Sigma(p_4)$ as a function of p_4 for two different chemical potentials, $\mu = 0.5$ GeV and $\mu = 100$ GeV. The solid lines show the numerical solution, the dashed curves show the analytic result for the leading logarithm, and the dotted curves show the previously proposed anomalous scaling behavior[8].

This implies that as long as g is small the one-loop self energy is a solution of the Dyson-Schwinger equation even in the regime $g^2 \log(\Lambda/p_4) \sim 1$ and perturbation theory remains well defined. However, the excitations in the vicinity of the Fermi surface are no longer described by a Breit-Wigner distribution function. Since the relation between the chemical potential and the Fermi momentum is not influenced by low energy gluonic interactions, using the Fermi liquid result[9] we finally obtain for the infrared dispersion relation of a fermion in the normal phase of dense matter

$$p = \left(1 - \frac{2\alpha_s}{3\pi}\right)\mu + \left(1 + \frac{4\alpha_s}{9\pi}\log\left(\frac{\Lambda}{\omega}\right)\right)\omega\,. \tag{6}$$

3. Neutrino emission and compact star cooling

We will now discuss the influence of the logarithmically enhanced dispersion relation eq. (6) on the cooling of quark matter in the normal phase due to neutrino emission. The dominant contribution to the emission of neutrinos arises from the quark URCA processes $d \to u + e^- + \bar{\nu}_e$ and $u + e^- \to d + \nu_e$. The quarks in these expressions are quasi particles dressed by medium fluctuations. The neutrino emissivity is given by the total energy loss due to neutrino emission averaged over the initial quark helicities and summed over the final state phase space. We find to leading order in T/μ

$$\epsilon \approx \frac{457}{630}\, G_F^2 \cos^2\theta_c\, \alpha_s\, \mu_q^2\, \mu_e\, T^6 \left(1 + \frac{8\alpha_s}{9\pi}\log\left(\frac{\Lambda}{T}\right) + \frac{16\alpha_s^2}{81\pi^2}\log^2\left(\frac{\Lambda}{T}\right)\right). \tag{7}$$

The first term is the standard result by Iwamoto[12], whereas the anomalous logarithmic terms are non Fermi liquid corrections. Note that in the infrared regime $T \ll \Lambda = O(\mu)$ relevant for the physics of compact stars these terms are logarithmically enhanced and have to be considered although they are formally of higher order in the strong coupling.
Besides the neutrino emissivity, the other relevant quantity that determines the cooling process is the specific heat. The inclusion of magnetic interactions to this quantity yields the infrared enhanced result[10,11]

$$c_v = \frac{N_c N_f}{3}\mu^2 T + \frac{(N_c^2 - 1)N_f g^2}{72\pi^2}\mu^2 T \log\left(\frac{\Lambda}{T}\right). \tag{8}$$

The cooling is determined by the time variation of the energy density

$$\frac{\partial u}{\partial t} = \frac{\partial u}{\partial T}\frac{\partial T}{\partial t} = c_v(T)\frac{\partial T}{\partial t} = -\epsilon(T)\,. \tag{9}$$

Whereas without non Fermi liquid corrections this equation results in a simple decay $T \propto 1/t^{\frac{1}{4}}$, the gluonic interactions introduce a scale dependence and prevent an analytic solution employing eqs. (7,8). Therefore we will give a numeric solution for the cooling behavior using the generic values for the input parameters. We choose $\mu_q = 500$ MeV, compute μ_e from the condition of charge neutrality and use for the cutoff scale Λ the expression derived for the specific heat[11]. The strong coupling is taken from a one-loop

RG evolution at the scale μ_q which yields a value of the order 1, so that our perturbative result should be viewed as an estimate for the physical case only. The cooling curve is shown in fig. 2. Although the non Fermi liquid corrections dominate both the neutrino emissivity and the specific heat in the relevant temperature range, the leading logarithmic correction just cancels in the cooling relation eq. (9) . Therefore, the cooling behavior of the non Fermi liquid result is only slightly changed compared to the Fermi liquid case and supports the previous result[12], that the quark URCA process a very effective cooling mechanism.

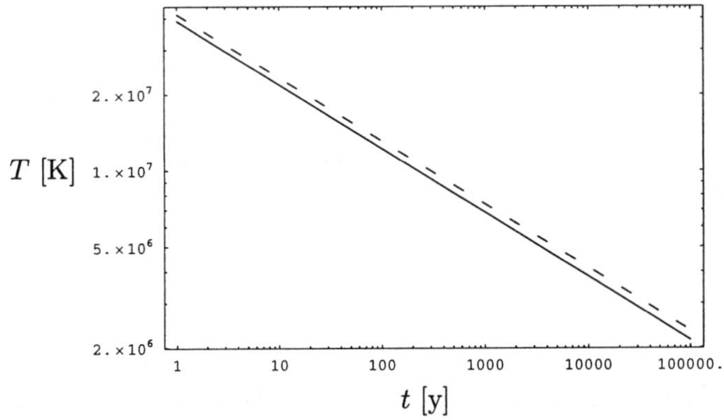

Figure 2. The cooling behavior of ungapped quark matter. The dashed line shows the Fermi liquid result, whereas the solid line shows the result including non Fermi liquid effects. Although the non Fermi liquid corrections to both the specific heat and the neutrino emissivity are significant, they nearly cancel out in the cooling rate.

References

1. D. Bailin and A. Love, Phys. Rept. **107**, 325 (1984);
2. J. M. Lattimer and M. Prakash, [astro-ph/0405262].
3. D. K. Hong, Phys. Lett. B **473**, 118 (2000).
4. C. Manuel, Phys. Rev. D **62**, 076009 (2000).
5. W. E. Brown, J. T. Liu and H. c. Ren, Phys. Rev. D **62**, 054013 (2000).
6. J. Polchinski, Nucl. Phys. B **422**, 617 (1994).
7. T. Schäfer and K. Schwenzer, [hep-ph/0405053].
8. D. Boyanovsky and H. J. de Vega, Phys. Rev. D **63**, 034016 (2001).
9. G. Baym and S. A. Chin, Nucl. Phys. A **262**, 527 (1976).
10. T. Holstein, A. E. Norton, P. Pincus, Phys. Rev. **B8**, 2649 (1973);
11. A. Gerhold, A. Ipp and A. Rebhan, [hep-ph/0406087].
12. N. Iwamoto, Phys. Rev. Lett. **44**, 1637 (1980).

PRETHERMALISATION AND THE BUILD-UP OF THE HIGGS EFFECT

D. SEXTY AND A. PATKÓS*

Department of Atomic Physics
Eötvös University
H-1117 Pázmány Péter sétány 1/A
E-mail: denes@achilles.elte.hu, patkos@ludens.elte.hu

Real time field excitations in the broken symmetry phase of the classical abelian Gauge+Higgs model are studied numerically in the unitary gauge, for systems starting from the unstable maximum of the Higgs potential.

1. Introduction

Tracking the transmutation of the angular component of a complex Higgs field into the longitudinal polarisation state of the gauge field during the termination of inflation in hybrid models might reveal interesting details of the real time Higgs effect and of the electroweak dynamics [1]. The excitation rate of the different polarisation states might be different as well as the thermal relaxation rates [2]. The production of gauged cosmic strings is another important aspect of this process[3]. Our numerical study concentrates on the non-equilibrium phase transition aspects.

2. Partial Pressures and Energy Densities of the Model

The main observables studied in the present investigation of the classical abelian Gauge+Higgs model,

$$L = -\frac{1}{4}F_{\mu\nu}F^{\mu\nu} + \frac{1}{2}D_\mu\Phi(D^\mu\Phi)^* - V(\Phi), \quad V(\Phi) = \frac{1}{2}m^2\Phi^2 + \frac{\lambda}{24}\Phi^4 \quad (1)$$

*Work supported by grant T-037689 of the Hungarian National Science Foundation.

are the partial energy densities and pressures:

$$\begin{aligned}
\epsilon &= \epsilon_\rho + \epsilon_T + \epsilon_L, \quad p = p_\rho + p_T + p_L, \\
\epsilon_\rho &= \frac{1}{2}\Pi_\rho^2 + \frac{1}{2}(\nabla\rho)^2 + V(\rho), \quad p_\rho = \frac{1}{2}\Pi_\rho^2 - \frac{1}{6}(\nabla\rho)^2 - V(\rho), \\
\epsilon_T &= \frac{1}{2}[\mathbf{\Pi}_T^2 + (\nabla \times \mathbf{A}_T)^2 + e^2\rho^2\mathbf{A}_T^2], \\
p_T &= \frac{1}{6}[\mathbf{\Pi}_T^2 + (\nabla \times \mathbf{A}_T)^2 - e^2\rho^2\mathbf{A}_T^2], \\
\epsilon_L &= \frac{1}{2}\left[\mathbf{\Pi}_L^2 + e^2\rho^2\left(\mathbf{A}_L^2 + \frac{1}{(e^2\rho^2)^2}(\nabla\Pi_L)^2\right)\right], \\
p_L &= \frac{1}{6}[\mathbf{\Pi}_L^2 - e^2\rho^2\mathbf{A}_L^2] + \frac{1}{2}\frac{1}{e^2\rho^2}(\nabla\mathbf{\Pi}_L)^2.
\end{aligned} \tag{2}$$

Index T refers to the transversal, L to the longitudinal part of the gauge field $\mathbf{A}$. The expressions are valid in the unitary gauge: $\rho = |\Phi|$. A_0 was eliminated with the Gauss constraint. It was checked numerically that in equilibrium $\epsilon_\rho : \epsilon_L : \epsilon_T = 1 : 1 : 2$ within statistical fluctuations [4].

The equations of motion derived from (1) were solved in the $A_0 = 0$ gauge and the solutions were transformed to the unitary gauge for measurements. Initially $\rho(\mathbf{x}, t = 0) = 0$ was put uniformly. Inhomogenuous modes beyond the region of spinodal instability (up to $|\mathbf{k}| \leq 5|m|$) were filled with low amplitude white noise. The spatial lattice spacing was chosen to $a_s = 0.35|m|^{-1}$ and the time step $\delta t = 0.04|m|^{-1}$.

3. Spectral Analysis of the Process of Excitation

Spectral representation of the energy densities $\epsilon(\mathbf{k})$ (pressures $p(\mathbf{k})$) were defined using the Fourier transform of the square root of various pieces of the local energy densities (pressures) and taking their absolute squares.

In **Fig.1** the degrees of excitation $\epsilon_T(\mathbf{k})/\epsilon_\rho(\mathbf{k})$ and $\epsilon_L(\mathbf{k})/\epsilon_\rho(\mathbf{k})$ are displayed as functions of $|\mathbf{k}|$ at different times. The low frequency part of $\epsilon_L(\mathbf{k})$ develops an over-excited peak, while $\epsilon_T(\mathbf{k})$ is excited weakly at early times. Their slow evolution towards equipartition defines an extremely long thermalisation time-scale ($\tau > 10^5|m|^{-1}$).

4. Signals for Prethermalisation

The dispersion relations $\omega^2(\mathbf{k})$ of the modes $(\rho, \mathbf{A}_T)$ are obtained as $|\Pi_\rho(\mathbf{k})|^2/|\rho(\mathbf{k})|^2$ and $|\Pi_T(\mathbf{k})|^2/|\mathbf{A}_T(\mathbf{k})|^2$. For the longitudinal mode inspection of ϵ_L in (2) suggests $\omega_L^2(\mathbf{k}) \equiv |[e^2\rho^2\mathbf{A}_L](\mathbf{k})|^2/|\Pi_L(\mathbf{k})|^2$. The masses of

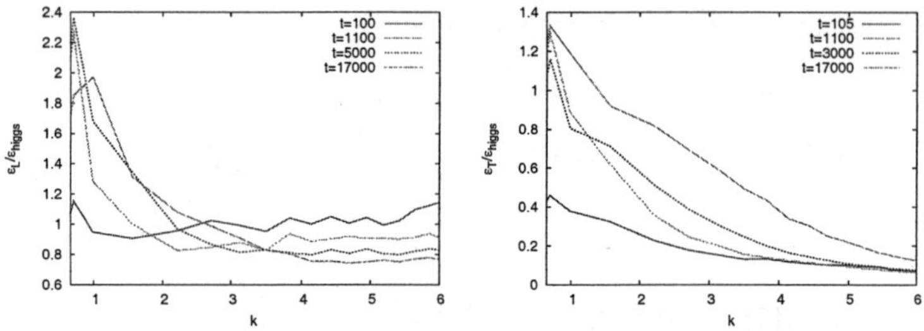

Figure 1. Evolution of the longitudinal (left) and transversal (right) energy excitations relative to the $\mathbf{k} \neq 0$ Higgs modes.

the longitudinal and transversal modes become degenerate early when calculated from modes $1 \leq |\mathbf{k}| \leq 5$; evolutionary effects are seen only at high k (see **Fig. 2**).

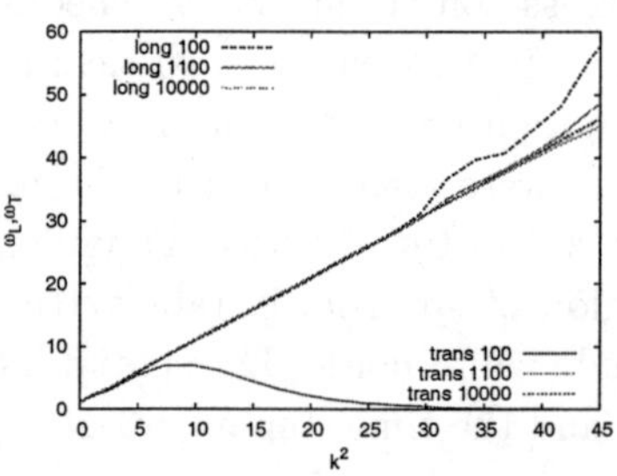

Figure 2. Evolution of the longitudinal and transversal dispersion relations

Spectral equations of state for the different fields can be defined assuming the mode-by-mode equality of the kinetic and potential spectral energy densities. Using this in (2) one arrives at the following equation for $\mathbf{A}_T$:

$$w_T(\mathbf{k}) \equiv p_T(\mathbf{k})/\epsilon_T(\mathbf{k}) = \mathbf{k}^2|\mathbf{A}_T(\mathbf{k})|^2/3|\Pi_T(\mathbf{k})|^2 = \mathbf{k}^2/3\omega_T^2(\mathbf{k}). \quad (3)$$

Similar relation holds for ρ. Using $\omega_L^2(\mathbf{k})$ for the longitudinal mode the same formula appears on the right end of the chain if the definition $w_L(\mathbf{k}) \equiv [e^2\rho^2 p_L](\mathbf{k})/[e^2\rho^2\epsilon_L](\mathbf{k})$ is applied. The expected functional form (with newly fitted squared mass values) was compared with the measured $w_{T,L}(|\mathbf{k}|)$. The modes filled initially almost instantly obey the expected form (3). Higher $|\mathbf{k}|$ modes are gradually filled and $w(\mathbf{k})$ "climbs up" to the stable free particle behavior. These prompt features illustrate the phenomenon of "prethermalisation" [5] in a gauge system.

5. Gauge-Higgs Cross-Correlations

The intiuitive quasi-particle picture in the unitary gauge conjectures that at low enough temperature and at moderate couplings the statistically independent field variables are just $\rho, \mathbf{A}_L, \mathbf{A}_T$. In the analysis of the equations of state above we avoided to rely on the statistical independence of these three variables which we are going to test next.

The transverse polarisation. The correlation coefficient between the quadratic spatial averages of ρ and $\mathbf{A}_T$ is defined as

$$\Delta[\mathbf{A}_T, \rho] \equiv \left(\overline{\rho^2(\mathbf{x},t)\mathbf{A}_T^2(\mathbf{x},t)} - \overline{\rho^2(\mathbf{x},t)}\ \overline{\mathbf{A}_T^2(\mathbf{x},t)}\right) / \overline{\rho^2(\mathbf{x},t)\mathbf{A}_T^2(\mathbf{x},t)}. \quad (4)$$

In **Fig.3a** the time evolution of this quantity is displayed in two characteristic runs. In the first a large negative value is reached almost instantly after the Higgs-field rolls down. After a longer time interval $\Delta[\mathbf{A}_T, \rho]$ suddenly jumps to a value compatible with zero. In the other run one can observe negative "needles" occuring on the background of near-zero fluctuations.

A large negative value of the correlation coefficient (4) represents a very sensitive indicator for the presence of relativistic Abrikosov-strings. The location of the points where $\rho/|m| < 0.3$ displays a vortex network very nicely (see **Fig. 3b**), but $\Delta \approx 0$ excludes the presence of vortices.

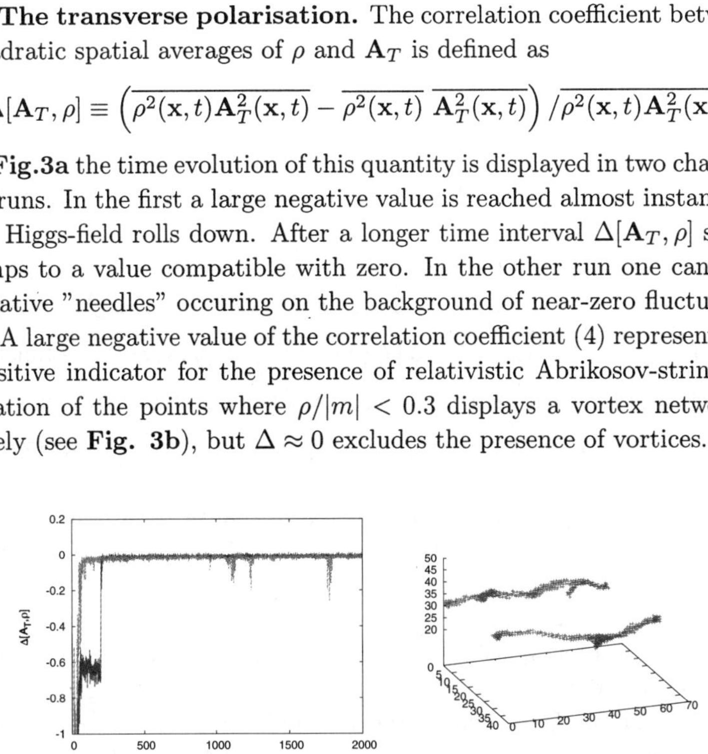

Figure 3. Non-smooth $\mathbf{A}_T - \rho$ correlation histories, and the corresponding vortex pair.

The longitudinal polarisation. The quantity $\Delta[\mathbf{A}_L, \rho]$ always displays instantly after the roll-down values significantly different from zero. and shows only a very mild time variation. It was checked that a non-zero value is present also in equilibrium [4]. This correlation coeficient linearly increases with the temperature. These observations point to the fact that the true quasiparticle field is a composite of ρ and $\mathbf{A}_L$. This is not a very great surprise for relatively strongly coupled systems, still it should be confronted with the fact that (without vortices) $\mathbf{A}_T(\mathbf{k},t)$ and $\rho(\mathbf{k},t)$ perform statistically independent and Gaussian small oscillations.

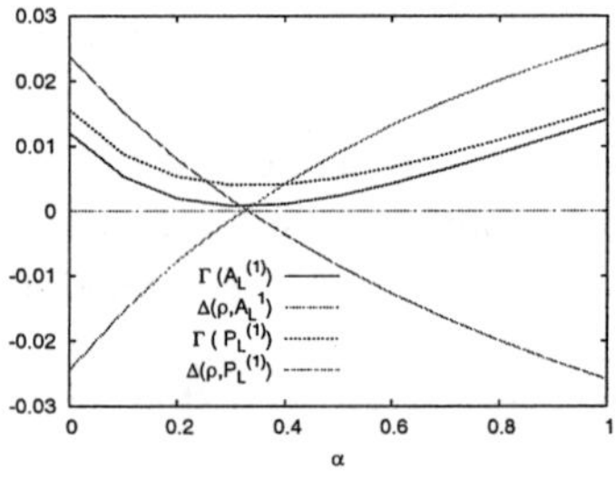

Figure 4. Correlations testing independence and Gaussianity of (5) in function of α

We have tested several trial composite fields at equilibrium [4]. In **Fig. 4** we show the variation of $\Delta[P_L,\rho],\Delta[Q_L,\rho]$ and of $\Gamma[P_L],\Gamma[Q_L]$ defined by the ratio $\Gamma[Q_L]\equiv\left(\overline{(Q_L^2(\mathbf{x},t))^2}-3\left(\overline{Q_L^2(\mathbf{x},t)}\right)^2\right)/\overline{(Q_L^2(\mathbf{x},t))}^2$ as a function of α which characterizes the compositeness of the conjugate variables

$$Q_L(\mathbf{x},t)=(1+\alpha\rho^2(\mathbf{x},t))\mathbf{A}_L(\mathbf{x},t),\quad P_L(\mathbf{x},t)=\Pi_L(\mathbf{x},t)/(1+\alpha\rho^2(\mathbf{x},t)). \tag{5}$$

The coefficients Γ signal deviations from Gaussianity. The nice surprise is that in equilibrium there exists a single optimal choice α_{opt} where $\Delta[P_L,\rho],\Delta[Q_L,\rho]$ vanish and both Γ's are minimal. It turns out, however, that during the non-equilibrium phase transition no such α_{opt} appears to exist, the "longitudinal quasiparticle" coordinate presumably emerges only on the thermalisation scale.

In conclusion, we found that in the abelian Higgs model an early quasi-particle characterisation works well for the dispersion relations and the equations of state just after the symmetry breaking is completed. The non-linear quasi-particle field containing the longitudinal vector component builds up much more slowly in the process of complete thermalisation.

References

1. J. Smit, Cold Electroweak Baryogenesis, to appear in Proc. of SEWM'04
2. J.-I. Skullerud, J. Smit and A. Tranberg, JHEP **0308** (2003) 045
3. M. Hindmarsh and A. Rajantie, Phys. Rev. **D64** (2001) 065016
4. D. Sexty and A. Patkós, hep-ph/0404235
5. J. Berges, Sz. Borsányi, C. Wetterich,hep-ph/0403234

VECTOR MESON AT NON-ZERO BARYON DENSITY AND ZERO SOUND

S.J. HANDS

Department of Physics, University of Wales Swansea,
Singleton Park, Swansea SA2 8PP, U.K.

C.G. STROUTHOS

Division of Science and Engineering,
Frederick Institute of Technology, Nicosia 1036, Cyprus

We present simulation results of the $(2+1)d$ four-fermion model with a baryon chemical potential μ. We examine temporal correlation functions of the vector meson, and find evidence of phonon-like behavior characterised by a linear dispersion relation in the long wavelength limit. We also discuss the consistency of our numerical results with analytical solutions to the Boltzmann equation corresponding to zero sound. We argue that our results provide the first evidence for a collective excitation in a lattice simulation.

We study the $(2+1)d$ Gross-Neveu (GN) model with a Z_2 chiral symmetry describing N_f flavors of self-interacting fermions, with continuum lagrangian

$$\mathcal{L} = \bar{\psi}(\partial\!\!\!/ + \mu\gamma_0)\psi - \frac{g^2}{2N_f}(\bar{\psi}\psi)^2. \tag{1}$$

As μ is increased, the model exhibits a sharp first-order transition at a critical μ_c from a phase where a fermion mass m_f is dynamically generated but baryon density $n_B = \langle\bar{\psi}\gamma_0\psi\rangle$ vanishes to one where $m_f = 0$ and $n_B \propto \mu^2$ [1]. The main new feature emerging for $\mu > \mu_c$ is the existence of a new scale, the Fermi momentum k_F. Since the lowest energy excitations of the ground state have $|\vec{k}| \approx k_F$, measurements of Euclidean timeslice correlators with $\vec{k} \neq 0$ are mandatory.

If certain conditions are fulfilled, such as an effective interaction between quasiparticles which is both short-ranged and repulsive, then there is a massless bosonic excitation in the spectrum as $T \to 0$. Carrying zero baryon number, it is a *phonon*, ie. a quantum of a collective excitation called *zero sound* [2]. Sound propagation occurs in any elastic medium; zero sound happens when the elasticity originates not from collisions between individual particles, but from the force on a single particle due to its coherent interaction with all others present in the medium. It can be pictured as a propagating distortion in the local shape of the Fermi surface, and its speed of propagation exceeds that of conventional "first" sound. In [3] we solved the self-consistent equation for zero sound in the framework of Fermi liquid theory in the case of the $3d$ GN model to leading non-trivial order in $1/N_f$. We found solutions for sound propagation with speed $\beta_0 > \beta_F$, the Fermi velocity. Here we discuss our numerical results and compare them with the analytical solutions presented in detail in [3].

The lattice regularized action of our model uses staggered fermions. Details of the formulation and simulation algorithm are given in [1,4]. The simulation parameters are $N_f = 4$ and $a/g^2 = 0.75$, corresponding to a physical fermion mass at $\mu = 0$ of $m_f \simeq 0.17a^{-1}$. All results are taken in the chirally-restored phase, ie. with $\mu > \mu_c \simeq 0.16a^{-1}$.

By analysing the decay of the appropriate correlation function with Euclidean time, we calculated the dispersion relations $E(k)$ for the spin-$\frac{1}{2}$ quasiparticle, which carries a baryon charge, and $\omega(k)$ for various meson states of the form $\bar{\psi}\Gamma\psi$ which are best thought of as excitations formed from a particle-hole pair. We simulated $L_s^2 \times 48$ systems with $L_s = 32$ and 48, with μ ranging from 0.2 to 0.6, and measured $E(k), \omega(k)$ for $\vec{k} = (k, 0)$ with $k = 0, 2\pi/L_s, \ldots, \pi/2$. The increased momentum resolution offered by $L_s = 48$ has proved to be important. Note that the staggered fermion action is only invariant under translations of two lattice spacings, which restricts the space of accessible momenta. In [3,5] we measured the quasiparticle propagator and showed that the transition between particle and hole behavior is rather sudden, characteristic of a well-defined Fermi surface. We also presented [5] the theoretical predictions of Fermi liquid properties to leading non-trivial order in $1/N_f$ for $\mu \gg \mu_c$. The results for the Fermi velocity β_F and the Fermi momentum k_F are given by:

$$\beta_F = 1 - \frac{1}{16N_f} \simeq 0.984 \quad ; \quad \frac{k_F}{\mu\beta_F} = 1 - \frac{1}{8N_f} \simeq 0.969. \tag{2}$$

As discussed in [3,5] the agreement between the analytical predictions and the numerical results on 48^3 lattices for $\mu = 0.2, ..., 0.6$ and with discretization effects taken into account is at best qualitative. It should be noted that analytic predictions of β_F and k_F to $O(1/N_f)$ for finite μ strictly requires a knowledge of the gap equation to two loops, which is not yet available.

For the remaining analysis we will therefore assume the free field values $k_F = \mu$, $\beta_F = 1$.

Next we consider the meson sector, ie. correlators of the form

$$C_\Gamma(\vec{k},t) \equiv \sum_{\vec{x}} \langle \bar{\psi}\Gamma\psi(\vec{0},0)\ \bar{\psi}\Gamma\psi(\vec{x},t)\rangle e^{i\vec{k}.\vec{x}}, \tag{3}$$

which carry zero baryon number. In [5] we showed that the generic large-distance behaviour in any given channel is dominated by zero-energy particle-hole pairs, and as a result the decay is algebraic, ie. $C_\Gamma(t) \propto t^{-\lambda(\vec{k})}$, with the exponent λ determined by the geometry of the overlap between two Fermi disks with relative displacement $\vec{k}$ between their centres. A particularly favourable configuration occurs for $|\vec{k}| \approx 2\mu$, in which case the Fermi surfaces just kiss, and $\lambda = \frac{3}{2}$. In this paper we focus on point-split meson operators corresponding to spatial components of the conserved vector current, ie. of the form

$$j_i(x) = \frac{\eta_i(x)}{2}\left[\bar{\chi}(x)\chi(x+\hat{\imath}) + \bar{\chi}(x+\hat{\imath})\chi(x)\right], \tag{4}$$

where $\chi, \bar{\chi}$ are staggered fermion fields, $\eta_1(x) = (-1)^t$, and $\eta_2(x) = (-1)^{t+x_1}$. We set $\vec{k} = (k,0)$, and define $C_\parallel$ in terms of the correlator $\langle j_1(0)j_1(x)\rangle$ and $C_\perp \sim \langle j_2(0)j_2(x)\rangle$. In Fig. 1 we show $|C_{\parallel,\perp}|(k,t)$ data taken on a 48^3 lattice at $\mu = 0.5$, for both $k = \frac{\pi}{3} \approx 2\mu$ and $k = \pi - \frac{\pi}{3}$ (the modulus is taken because we use a log-linear scale, and C fluctuates in sign). Whereas $C_\perp$ and $C_\parallel(k = \frac{\pi}{3})$ all show behaviour consistent with algebraic decay, the decay in the $C_\parallel(k = \pi - \frac{\pi}{3})$ channel is much faster, and resembles the exponential decay expected of an isolated pole. We have fitted it with the form

$$C_\parallel(\pi - k, t) = A\exp(-\Omega(k)t) + (-1)^t B\exp(-\omega(k)t), \tag{5}$$

in most cases employing datapoints with $t \in [10,38]$. For small k, the coefficient $B \gg A$, suggesting that the correlator is dominated by a pole in the alternating channel. In Fig. 1 we plot the resulting dispersion relation $\omega(k)$ for $\mu = 0.2,\ldots,0.6$. The behaviour $\omega \propto k$ as $k \to 0$ suggests the presence of a massless pole similar to that of a phonon. Support for this interpretation comes from recasting the meson bilinear in terms of continuum-like fields $q^{\alpha a}$, $\bar{q}^{\alpha a}$ having spinor index $\alpha = 1,\ldots,4$ and 'color' index $a = 1,2$ [6]. We obtain

$$(-1)^{x_1}(-1)^t\bar{\chi}(x)\eta_1\chi(x+\hat{1}) \sim i\bar{q}(\gamma_0 \otimes \tau_2^*)q, \tag{6}$$

demonstrating that the excitation can be viewed as an oscillation of local baryon density.

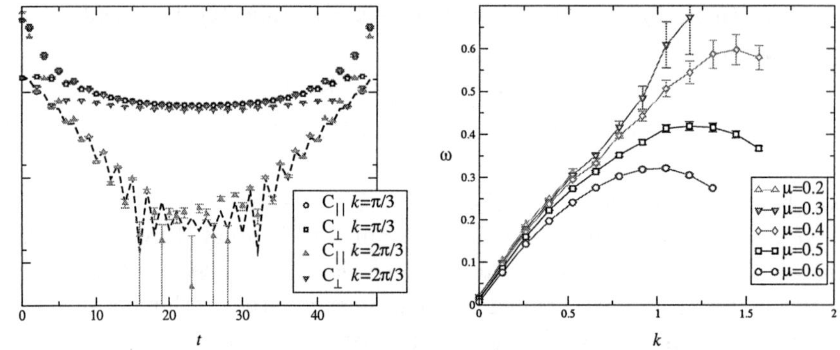

Figure 1. Left: Meson propagators $|C_\parallel|$ and $|C_\perp|$ for both $k = \frac{\pi}{3}$ and $\pi - \frac{\pi}{3}$ for $\mu = 0.5$. The dashed line shows a fit of the form (5). Right: Dispersion relation $\omega(k)$ for various μ for the channel defined by the correlator $C_\parallel(\pi - k)$. Data are from 48^3.

Although we are examining the $k \to 0$ limit, we are trying to model a Fermi surface phenomenon on the lattice; the appropriate scale with which to compare β_0 is the "bare" Fermi velocity (ie. with no discretisation correction) which for free fields with $E(k) = -\mu + \sinh^{-1}(\sin k)$ is given by

$$\beta_F^{\text{bare}} = \left.\frac{\partial E}{\partial k}\right|_{E=0} = \sqrt{\frac{1 - \sinh^2 \mu}{1 + \sinh^2 \mu}}. \tag{7}$$

The general trend in our data is that the speed ratio $s = \beta_0 / \beta_F^{\text{bare}}$ increases towards unity as μ increases (for $\mu = 0.2$ $s = 0.896(8)$ and for $\mu = 0.6$ $s = 0.959(2)$). In [3] we used Fermi liquid interaction to leading non-trivial order in $1/N_f$ to find analytical solutions to the Boltzmann equation corresponding to zero sound. The excitation is spin and isospin symmetric, and has speed ratio $s > 1$ for almost all $\mu > \mu_c$. To what extent can we be sure that our numerical results describe the same physical phenomenon? Solutions with $s < 1$ allow the possibility of emission of a phonon with momentum $\vec{q}$ and energy $\beta_0|\vec{q}|$ from a quasiparticle with momentum $\vec{k}$ and energy $\mu + \beta_F(|\vec{k}| - k_F)$. This is allowed kinematically for $s < 1$, the angle of emission ϕ satisfying

$$\cos\phi = s + \frac{|\vec{q}|}{2|\vec{k}|}(1 - s^2) > s. \tag{8}$$

All radiation is emitted within a cone of half-angle $\cos^{-1} s$ centred on the quasiparticle trajectory. This well-known phenomenon is variously known as Landau damping, Čerenkov radiation, or most appropriately in the current context, as a sonic boom.

First let us argue how our numerical results can support a state resembling a simple pole but with $s < 1$, in apparent contradiction to the above. On a spacetime lattice, the radiation process is constrained because there is a natural lower bound for the angle of emission, $\phi_{\text{min}} \sim 2\pi/L_s|\vec{q}|$. Landau damping is thus kinematically forbidden for

$$s > \cos\phi_{\text{min}} \simeq 1 - \frac{2\pi}{L_s^2|\vec{q}|^2}. \tag{9}$$

With a conservatively high $|\vec{q}| \sim \mu/2$, then on a 48^3 lattice we have no damping for $s > 0.73$ at $\mu = 0.2$, rising to $s > 0.97$ at $\mu = 0.6$. It is plausible therefore that Landau damping is suppressed in our lattice data and that the phonon is described by an isolated pole even if $s < 1$.

Simulations on volumes considerably larger than those used here, however, will be needed to disentangle the various possible systematic effects due to finite L_s, finite μ/T, and non-zero lattice spacing in order to determine the sign of $s-1$ for finite N_f. We think the most probable explanation for $s < 1$ is an effect of non-zero $T \equiv (aL_t)^{-1}$; the physical temperature of the $L_t = 48$ lattice decreases by a factor of roughly two going from $g^{-2} = 0.75$ to $g^{-2} = 0.6$, and according to our lattice data $|s-1|$ decreases by roughly the same factor. It is known empirically that the speed of sound in liquid ^{3}He *increases* as $T \to 0$ and the dominant mode of propagation changes from first sound to zero sound [7]. In principle, since our extracted value of $s \lesssim 1$ has relied on a rescaling of β_F to take account of discretisation artifacts, we also need to go to considerably finer lattices, ie with $\beta_F^{\text{bare}} \approx 1$, in order to demonstrate a clear distinction between our signal and first sound with expected propagation speed $\beta_1 \simeq 0.7$. If, however, the theoretical arguments about zero sound dominating as $T \to 0$ can be taken seriously, then for the first time we have succeeded in identifying a collective oscillation in a lattice simulation.

References

1. S.J. Hands, Nucl. Phys. A **642** (1998) 228.
2. L.D. Landau, Sov. Phys. JETP **5** (1957) 101.
3. S.J. Hands and C.G. Strouthos, `hep-lat/0406018`.
4. S.J. Hands and S.E. Morrison, Phys. Rev. D **59** (1999) 116002.
5. S.J. Hands, J.B. Kogut, C.G. Strouthos and T.N. Tran, Phys. Rev. D **68** (2003) 016005.
6. C. Burden and A.N. Burkitt, Europhys. Lett. **3** (1987) 545.
7. J.W. Negele and H. Orland, *Quantum Many-Particle Systems*, (Westview Press, Boulder, 1988) ch. 6.

IMPACT OF BARYON RESONANCES ON THE CHIRAL PHASE TRANSITION

D. ZSCHIESCHE, G. ZEEB, S. SCHRAMM, H. STÖCKER

Institut für Theoretische Physik, Robert-Mayer-Str. 8-10, D-60054 Frankfurt am Main, Germany

We study the phase diagram of a generalized chiral SU(3)-flavor model in mean-field approximation. In particular, the influence of the baryon resonances, and their couplings to the scalar and vector fields, on the characteristics of the chiral phase transition as a function of temperature and baryon-chemical potential is investigated. Present and future finite-density lattice calculations might constrain the couplings of the fields to the baryons. The results are compared to recent lattice QCD calculations and it is shown that it is non-trivial to obtain, simultaneously, stable cold nuclear matter.

1. Introduction

QCD is the accepted underlying theory of strong interactions. In the hadronic world, as well as in the region not too far above the phase transition to the deconfined and chirally restored phase, the strong coupling constant is large and thus QCD can not be treated perturbatively. From a theoretical point of view, lattice gauge calculations represent the most direct approach to investigate the QCD phase diagram of strongly interacting matter. However, lattice QCD alone does not seem to be able to completely disentangle the physics of the QCD phase transition and give an explanation of the structure and the scales of the phase diagram. Thus, effective lagrangians [1,2] are studied, representing a complementary approach to disentangle and understand the physics of the QCD phase transition (see e.g Refs. in [3]). These effective models should incorporate known features of QCD. To this end, we consider here the phase transition properties of the chiral hadronic model presented in [4]. It represents a relativistic field theoretical model of baryons and mesons built on chiral $SU(3)_L \times SU(3)_R$ symmetry and broken scale invariance. A non-linear realization (see [5,6]) of chiral symmetry is adopted. The model has been shown to successfully describe hadronic vacuum properties, nuclear matter saturation, finite nuclei and hypernuclei [4,7,8]. For a deeper understanding of the chiral phase transition in strongly interacting matter, we will systematically analyze the role of baryon resonances, which in our investigation are effectively represented by the lightest baryon decuplet.

2. The chiral model

The chiral hadronic SU(3) lagrangian in the mean field approximation has the following basic structure (for details see [4])

$$\mathcal{L} = \mathcal{L}_{\text{kin}} + \mathcal{L}_{\text{BM}} + \mathcal{L}_{\text{BV}} + \mathcal{L}_{\text{vec}} + \mathcal{L}_0 + \mathcal{L}_{\text{SB}} \ , \tag{1}$$

consisting of interaction terms between baryons respectively spin-0 (BM) and spin-1 (BV) mesons, interactions between the scalar mesons (0) and an explicit symmetry breaking term mimicing the QCD effect of non-zero current quark masses (SB). The scalar interactions $\mathcal{L}_0$ induce the spontaneous chiral symmetry breaking. The effective masses of the baryons are generated through their coupling to the scalar fields, which attain non-zero vacuum expectation values due to the self-interactions [4]. For the decuplet D we introduce an explicitly symmetry breaking mass term of the form

$$\mathcal{L} = -m_{\text{Dec}}\bar{D}D \, . \tag{2}$$

The vector couplings $g_{i\omega}$ and $g_{i\phi}$ for the octet as well as for the decuplet result from pure f-type coupling as discussed in [4],

$$\begin{aligned} g_{i\omega} &= (n_q^i - n_{\bar{q}}^i) g_{8,10}^V \\ g_{i\phi} &= -(n_s^i - n_{\bar{s}}^i)\sqrt{2} g_{8,10}^V \, , \end{aligned} \tag{3}$$

with $i = N, \Lambda, \Sigma, \Xi, \Delta, \Sigma^*, \Xi^*, \Omega$, while g_8^V and g_{10}^V denote the vector coupling of the baryon octet and decuplet, respectively. n^i represents the number of constituent q (u, d) or s (strange) quarks. The grand canonical thermodynamic potential of the system can be written as

$$\frac{\Omega}{V} = -\mathcal{L}_{\text{vec}} - \mathcal{L}_0 - \mathcal{L}_{\text{SB}} - \mathcal{V}_{\text{vac}} \mp T \sum_i \frac{\gamma_i}{(2\pi)^3} \int d^3k \left[\ln\left(1 \pm e^{-\frac{1}{T}[E_i^*(k) - \mu_i^*]}\right)\right] , \tag{4}$$

where γ_i denote the hadronic spin-isospin degeneracy factors and $E_i^*(k)$ are the single particle energies and μ_i^* the effective potentials and $\mathcal{V}_{\text{vac}}$ the vacuum energy. By extremizing Ω/V one obtains self-consistent equations for the meson fields. The sources for the scalar and vector fields mainly depend on two parameters: the number of degrees of freedom coupled to the fields and the corresponding coupling constants. For the octet, the couplings are fixed [4]. Here we will systematically study the dependence of the hadronic matter properties on the strength of the scalar and vector coupling for the decuplet. These are controlled by the explicit symmetry breaking m_{Dec} and the relative strength of the vector coupling,

$$r_{\text{v}} = \frac{g_{10}^V}{g_8^V} = \frac{g_{\Delta\omega}}{g_{N\omega}}, \tag{5}$$

respectively.

3. Results

At vanishing chemical potential the phase transition behavior depends only on the scalar coupling, i.e. on the choice of m_{Dec}. Figure 1 shows the scalar fields σ and ζ as a function of temperature T for different values of the scalar coupling. For $m_{\text{Dec}} < 260$ MeV (i.e. $g_{\Delta\sigma} \gtrsim g_{N\sigma}$), a first-order phase transition occurs, as indicated by the jump in the chiral condensates at $T_c \approx 155$ MeV. At finite chemical potential recent lattice QCD data [9,10,11] predict a

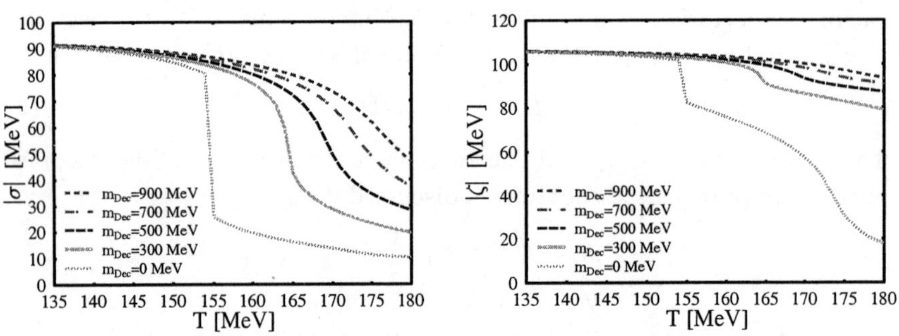

Figure 1. Non-strange condensate σ (left) and strange condensate ζ (right) as a function of temperature for vanishing chemical potential (μ_q=0). The curves correspond to different values of the explicit symmetry breaking term m_{Dec}.

critical end point at $T_c \approx 150 - 170$ MeV and $\mu_{q,c} \approx 100 - 250$ MeV. For $m_{\text{Dec}} = 260$ MeV the highest T_c is obtained for a minimal vector coupling $r_{\text{v}} = 0$. We find $T_c \approx 155$ MeV and $\mu_{q,c} \approx 70$ MeV. The resulting phase diagram is depicted in Fig. 2 (left), showing a critical point in the same region as predicted by lattice QCD. Increasing the vector coupling of the baryon resonances decreases T_c and increases $\mu_{q,c}$ (see Fig. 2 middle and right). Very similar behavior is observed when decreasing the scalar coupling and keeping the vector coupling constant. However, for all choices shown in Fig. 2 ($r_{\text{v}} = 0, 0.2, 0.4$), the chiral phase transition occurs at $\mu_q < 300$ MeV for $T = 0$, which implies a stable chirally restored phase. Hence, the current form of the model it is not able to obtain simultaneously a critical temperature in the region predicted by lattice QCD and a successful description of nuclear matter properties (density, binding energy). In Fig. 3 (left) we

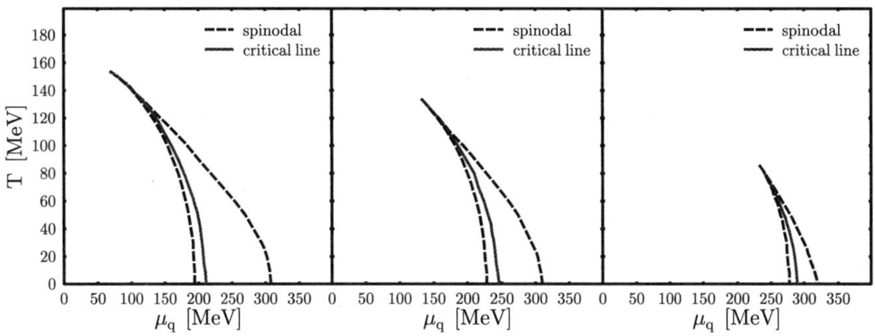

Figure 2. Phase diagram for $m_{\rm Dec} = 300$ MeV and increasing values of the decuplet vector coupling from left to right ($r_{\rm v} = 0, 0.2, 0.4$).

show the parameter regions, which, at $T = 0$, either give a first-order phase transition and an absolutely stable chirally restored phase, or a chiral phase transition but stable normal nuclear matter, or no phase transition at all (crossover), respectively. Insisting that normal nuclear matter be stable reduces the maximum critical temperature to $T_c \approx 50$ MeV, with a critical chemical potential $\mu_{q,c} \approx 290$ MeV. This phase diagram is shown in Fig. 3 (right) for $m_{\rm Dec} = 300$ MeV and $r_{\rm v} = 0.5$, the correspondingly lowest possible value for $r_{\rm v}$, which leads to stable normal nuclear matter. The resulting phase diagrams for higher values of the explicit symmetry breaking $m_{\rm Dec}$ and the corresponding smallest possible vector couplings, look nearly identical. Thus, in the present form the chiral model yields $T_c \lesssim 50$ MeV, if we insist on normal nuclear matter being stable.

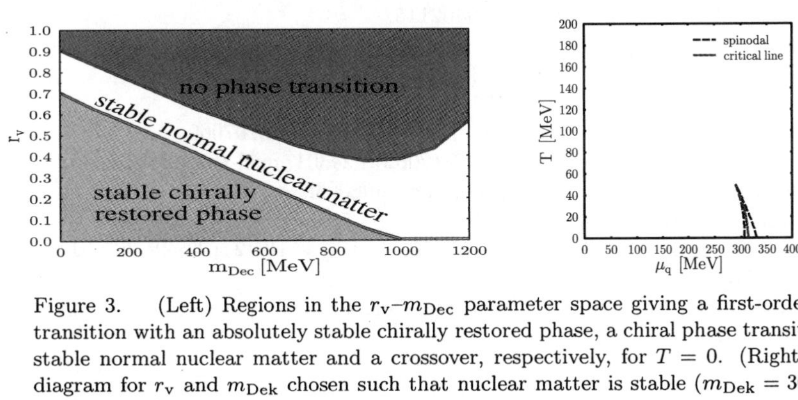

Figure 3. (Left) Regions in the $r_{\rm v}$–$m_{\rm Dec}$ parameter space giving a first-order phase transition with an absolutely stable chirally restored phase, a chiral phase transition but stable normal nuclear matter and a crossover, respectively, for $T = 0$. (Right) Phase diagram for $r_{\rm v}$ and $m_{\rm Dek}$ chosen such that nuclear matter is stable ($m_{\rm Dek} = 300$ MeV and $r_{\rm v} = 0.5$).

4. Conclusion

Using a chiral $SU(3)$ model we investigated the dependence of the QCD phase diagram on the scalar- and vector-coupling of the baryon decuplet. We found that qualitative agreement with recent lattice results can be obtained, as for example a critical point around $T_c \approx 150$ MeV and finite $\mu_{q,c}$. However, the slow decrease of the phase transition temperature with increasing chemical potential could not be reproduced. Moreover, demanding existence of a normal nuclear matter ground-state at $T = 0$ reduces T_c significantly to approximately 50 MeV. These results show that the simultaneous description of the phase diagram and nuclear matter saturation gives strong constraints on the model. Although our study neglected many contributions, especially the pions and the higher resonances, some very interesting conclusions appear. First of all, the baryon resonances may very well drive the chiral phase transition and give a structure in agreement with lattice QCD. The question how the QCD phase diagram with realistic pion masses looks like is still open. As was shown in [3], several different approaches, which do not take into account baryon resonances, yield much lower critical temperatures. This clearly suggests an important role of baryonic degrees of freedom for the characteristics of the chiral phase transition. The stronger curvature of $T_c(\mu)$ as compared to lattice results might be due to only including the baryon decuplet. It will be very interesting to see how the results obtained here may be changed by a larger or and distorted hadronic spectrum as proposed in [12].

References

1. S. Weinberg, Phys. Rev. Lett. 18 (1967) 188.
2. S.R. Coleman, J. Wess and B. Zumino, Phys. Rev. 177 (1969) 2239.
3. M.A. Stephanov, (2004), hep-ph/0402115.
4. P. Papazoglou et al., Phys. Rev. C 59 (1999) 411.
5. S. Weinberg, Phys. Rev. Lett. 166 (1968) 1568.
6. C.G. Callan et al., Phys. Rev. 177 (1969) 2247.
7. S. Schramm, Phys. Rev. C66 (2002) 064310, nucl-th/0207060.
8. C. Beckmann et al., Phys. Rev. C65 (2002) 024301, nucl-th/0106014.
9. Z. Fodor and S.D. Katz, (2004), hep-lat/0402006.
10. F. Karsch, (2004), hep-lat/0403016.
11. P. de Forcrand and O. Philipsen, Nucl. Phys. B642 (2002) 290, hep-lat/0205016.
12. F. Karsch, K. Redlich and A. Tawfik, Eur. Phys. J. C29 (2003) 549, hep-ph/0303108.

LIST OF PARTICIPANTS

Gert Aarts (Ohio State University)
Ramón F. Álvarez-Estrada (Universidad Complutense de Madrid)
Jens Andersen (Nordita)
Peter Arnold (University of Virginia)
Alejandro Arrizabalaga (University of Amsterdam)
Patrick Aurenche (LAPTH Annecy)
Christian Beckmann (J.W. Goethe University, Frankfurt/Main)
Jürgen Berges (Heidelberg University)
Szabolcs Borsányi (University of Heidelberg)
Manuel Calderón de la Barca Sánchez (Brookhaven National Laboratory)
Margaret Carrington (Brandon University)
Saumen Datta (University of Bielefeld)
Dennis D. Dietrich (Université Paris XI)
Adrian Dumitru (J.W. Goethe University)
Shinji Ejiri (University of Bielefeld)
Kari Enqvist (University of Helsinki)
Jörg Erdmann (University of Bielefeld)
Kari J. Eskola (University of Jyväskylä)
Philippe de Forcrand (ETH Zürich)
Eduardo Fraga (Universidade Federal do Rio de Janeiro)
Filipe Freire (Leiden University)
Lars Fromme (University of Bielefeld)
François Gelis (CEA/Saclay)
Andreas Gerhold (Vienna Technical University)
Pierre Giovannangeli (Centre de Physique Théorique, Marseille)
Daniel Gómez Dumm (National University of La Plata)
Angel Gómez Nicola (Universidad Complutense de Madrid)
Françoise Guerin (University of Nice)
Antti Gynther (University of Helsinki)
Yoshitaka Hatta (Brookhaven National Laboratory)
Andreas Heinen (University of Dortmund)
Ari Hietanen (University of Helsinki)
Mark Hindmarsh (University of Sussex)
Ralf Hofmann (University of Heidelberg)
Ronald Horgan (University of Cambridge)
Mei Huang (J.W. Goethe University, Frankfurt/Main)
Kay Hübner (University of Bielefeld)

Janne Ignatius (CSC, Finnish IT center for Science)
Andreas Ipp (Vienna Technical University)
Aiichi Iwazaki (Nishogakusha University)
Antal Jakovác (Hungarian Academy of Sciences & Budapest University of Technology and Economics)
Tommi Järvi (University of Helsinki)
Kimmo Kainulainen (University of Jyväskylä)
Keijo Kajantie (University of Helsinki)
Frithjof Karsch (University of Bielefeld)
Peter Kolb (Technical University Munich)
Thomas Konstandin (University of Heidelberg)
Chris Korthals Altes (CNRS/University of Marseille)
Chris Kouvaris (MIT)
Matti Krusius (Helsinki University of Technology)
Janne Kuokkanen (University of Oulu)
Edwin Laermann (University of Bielefeld)
Mikko Laine (University of Bielefeld)
Tuomas Lappi (University of Helsinki)
Jonathan Lenaghan (University of Virginia)
Daniel Litim (CERN)
Maria-Paola Lombardo (Laboratori Nazionali di Frascati)
Axel Maas (Darmstadt University of Technology)
Cristina Manuel (University of València-C.S.I.C.)
Jose M. Martinez Resco (Ohio State University)
Stefan Michalski (University of Dortmund)
Claus Montonen (University of Helsinki)
Guy Moore (McGill University)
Stanislaw Mrówczyński (Soltan Institute for Nuclear Studies)
Harri Niemi (University of Jyväskylä)
András Patkós (Eötvös University)
Bengt Petersson (University of Bielefeld)
Péter Petreczky (Brookhaven National Laboratory)
Konstantin Petrov (Brookhaven National Laboratory)
Owe Philipsen (University of Sussex)
Robert Pisarski (Brookhaven National Laboratory & Niels Bohr Institute)
Michael Plümacher (CERN)
Tomislav Prokopec (University of Heidelberg)
Arttu Rajantie (University of Cambridge)
Anton Rebhan (Vienna Technical University)

Urko Reinosa (Vienna Technical University)
Dirk H. Rischke (J.W. Goethe University, Frankfurt/Main)
Paul Romatschke (Vienna Technical University)
Kari Rummukainen (CERN & University of Oulu)
Vesa Ruuskanen (University of Jyväskylä)
Sami Räsänen (University of Jyväskylä)
Mischa Sallé (Helsinki Institute of Physics)
Christian Schmidt (University of Wuppertal)
Michael G. Schmidt (University of Heidelberg)
Andreas Schmitt (J.W. Goethe University, Frankfurt/Main)
York Schröder (MIT)
Kai Schwenzer (NC State University)
Michael Seniuch (University of Bielefeld)
Julien Serreau (University of Heidelberg)
Dénes Sexty (Eötvös University)
Jan Smit (University of Amsterdam)
Frank Steffen (DESY)
Tim Stockamp (University of Bielefeld)
Michael Strickland (Vienna Technical University)
Costas Strouthos (Frederick Institute of Technology, Limassol)
Stefan Strüber (J.W. Goethe University, Frankfurt/Main)
Zsolt Szép (Hungarian Academy of Sciences)
Anders Tranberg (University of Amsterdam)
Dionysis Triantafyllopoulos (Saclay)
Kimmo Tuominen (University of Jyväskylä)
Mikko Vepsäläinen (University of Helsinki)
Aleksi Vuorinen (University of Helsinki)
Harmen Warringa (Free University Amsterdam)
Heribert Weigert (University of Regensburg)
Steffen Weinstock (Brookhaven National Laboratory)
Urs A. Wiedemann (CERN)
Laurence Yaffe (University of Washington)
Detlef Zschiesche (J.W. Goethe University, Frankfurt/Main)

